Building Construction

Second Edition

P.C. VARGHESE
Honorary Visiting Professor
College of Engineering, Guindy, Anna University
Formerly Head, Department of Civil Engineering, IIT Madras
and
UNESCO Chief Technical Advisor, University of Moratuwa, Sri Lanka

PHI Learning Private Limited
Delhi-110092
2019

₹ 395.00

BUILDING CONSTRUCTION, Second Edition
P.C. Varghese

ISBN-978-81-203-5284-1

The export rights of this book are vested solely with the publisher.

Twelfth Printing (Second Edition) ... ... **July, 2019**

Published by Asoke K. Ghosh, PHI Learning Private Limited, Rimjhim House, 111, Patparganj Industrial Estate, Delhi-110092 and Printed by Mudrak, D-61, Sector 63, Noida, U.P.-201301.

To

Dr SARA and *Dr THOMAS*

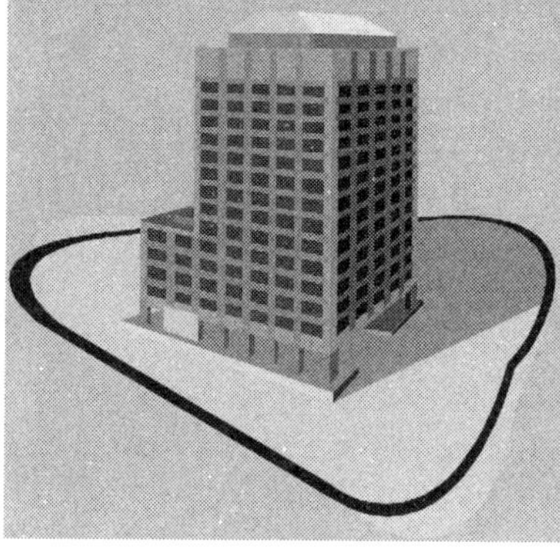

Contents

Foreword

As a teacher, researcher and educational administrator for over forty years, it is my considered opinion that the single most important prerequisite for improving and maintaining the standards of engineering education is the availability of good textbooks. It is a sad fact that textbooks in engineering and technology that are well written, appropriate and relevant to Indian conditions are a desideratum. It may be said, without any hesitation, that those who fulfill this strongly felt need are doing great service to engineering education.

Professor P.C. Varghese, who has had the benefit of engineering education in India, the USA and the UK, has long experience as a teacher, researcher, designer and consultant in foundation and structural engineering—a combination that is relatively rare. With this background, he is eminently equipped for the preparation of books that provide sound theoretical basis and insights into practical applications, a blend needed most, but often left to be desired in the books that abound on the library shelves. His earlier books—*Limit State Design of Reinforced Concrete, Advanced Reinforced Concrete Design, Foundation Engineering* and *Building Materials*—are enduring contributions to engineering education and engineering practice. They have naturally been well received by the readers.

With one book on Building Materials and other three on Building Design, Prof. Varghese now completes the circle by presenting this text on Building Construction. The book serves the felt need of three constituencies—the students, the teachers and the field practitioners. The author has enhanced the dimension of the book by adding Review Questions at the end of each chapter. A distinguished and reputed teacher that he has been, the presentation is reader-friendly, the language is lucid, and there is remarkable clarity in the discussion of principles. It is a comprehensive treatment of the subject comprising 40 chapters, dealing with every important component of building construction—from site preparation to plumbing and acoustics. In a way, it is a brief compendium of building construction.

I warmly congratulate the author, Professor P.C. Varghese, as well as the publishers, Prentice-Hall of India, on their really valuable contributions to the literature in Civil Engineering, in general, and Building Construction, in particular.

V.C. Kulandai Swamy
Former Vice-Chancellor
Anna University, Chennai, and
Indira Gandhi National Open University, New Delhi

Preface

This book on Building Construction was first published in 2007. Since then, there has been many development and very much progress in the areas of *water supply* as well as *treatment of waste water and sewage* in the buildings located far away from towns and cities with municipal supplies. In the second edition of the book, these topics have been introduced in Appendices F and G, without disturbing flow of the contents in the existing chapters.

Except in large towns and cities, most buildings in India are ordinary, classical buildings. All Civil Engineers, whatever their specialization may be, are expected to know the basics of Building Construction and Maintenance. Thus, this book has been designed in such a way that it gives the necessary basics of the building construction and maintenance.

P.C. Varghese

Preface to the First Edition

This book on *Building Construction* is intended as a textbook for the junior students in civil engineering and architecture as well as for those who are interested in the construction of their own buildings. It is a continuation of the author's book on *Building Materials,* also published by Prentice-Hall of India.

As building activities take place all around us all the time, the subject of Building Construction should ideally be taught not as a long list of information, but to act as a stimulant to the students for the observation of actual buildings already built as well as the buildings that are built around them. The students should learn more from observation and practice. This type of teaching and study will inculcate in the students the twin habits of keen observation to details and self-study which will stand them in good stead in all walks of life in later years. The book is written with these objectives.

A detailed treatment of building construction will require a much more voluminous and specialized publication than this book. Such a treatment is not attempted in this volume. This book is made simple by not dealing in detail with advanced topics such as design practice of foundations, structures, building services, etc. as they form special subjects for study in the later years of the students' academic studies.

The book has been made lecture based, each chapter dealing with only one lecture topic, to help young teachers (who are always asked to lecture on this subject in various engineering colleges in India) to prepare their lectures, if necessary, with additional materials from other references. The book has been written in accessible style for easy assimilation by junior students. Review Questions given at the end of each chapter will help them understand the fundamentals of the subject.

I earnestly hope that this book will help young engineering teachers as well as junior students and all those who are interested in the subject to understand the basic building construction practices and make them interested in civil engineering construction practices taking place around them.

P.C. Varghese

Acknowledgements

I wish to acknowledge the help I received from various individuals and institutions in authoring this book on Building Construction.

It was Dr. S.R. Sengupta, Former Director, IIT Kharagpur, who encouraged me to take interest in building construction by appointing me as Construction Engineer at IIT Kharagpur during the later part of my service in that institute. Later, while working with the Building Department, Sri Lanka, as a UN Adviser, Mr. R. Paskaralingam, Former Secretary, Ministry of Works and Housing, Sri Lanka, suggested to me the preparation of a set of "Notes on Building Construction." The format of this book is largely based on these notes. I am thankful to both these individuals in making me interested in the subject.

I also wish to acknowledge the help I received from Prof. V.C. Kulandai Swamy, Former Vice-Chancellor, Anna University, to continue my academic pursuits after retirement at Anna University. He has also associated himself with this book by writing the Foreword.

I am very thankful to the present authorities of Anna University as well as my colleagues in the Structural Engineering and Geotechnical Divisions at Anna University for the cooperation they gave me all those years.

I appreciate the assistance I received from Mrs. Rajeswari Sivaraman for word processing my handwritten manuscript and to Mrs. R. Uma for preparing all the drawings for this book.

This book is a compilation from various notes, codes, books and many other publications, too many to be mentioned individually. I am indebted to all their authors and publishers.

I thank the publishers, Prentice-Hall of India, New Delhi, and especially, the editorial and production staff, for the full cooperation I received in the publication of this book.

P.C. Varghese

Chapter 1

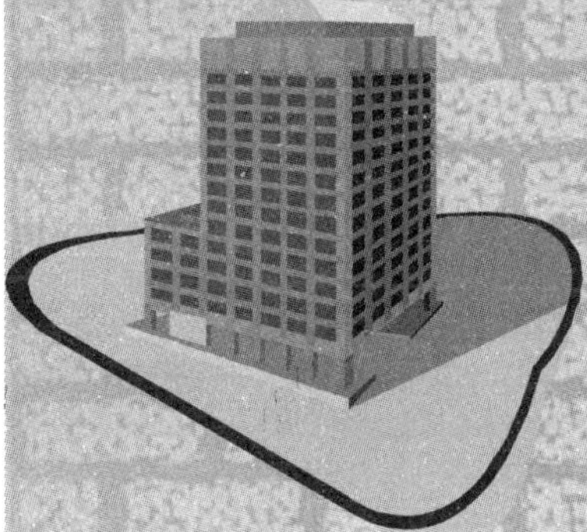

Components of a Building and Building Specifications

1.1 INTRODUCTION

All buildings have the same components such as foundation, walls, floors and roof. In addition, we have to decorate them with plaster, paint, etc. to make them aesthetically beautiful. These items of building works are called *civil works*. Water and electricity have also to be supplied to the buildings to make them habitable. These items are called *building services*. In this chapter, we will briefly examine the various components of civil works and building services.

1.2 COMPONENTS OF A BUILDING: CIVIL WORKS

The important parts of an ordinary building are shown in Figures 1.1 and 1.2. They are as follows:

1. Foundation
2. Plinth
3. Walls and columns
4. Floors
5. Lintels and *chajjas*
6. Roof
7. Doors and windows
8. Stairs and lifts
9. Finishing work (plastering and painting)
10. Building services
11. Fencing and external works

A building can be divided into substructure (foundation) and superstructure, the plinth being the dividing line between them. In building construction, we study how the civil works are carried out in the field after they have been planned by an architect and structurally designed by an engineer.

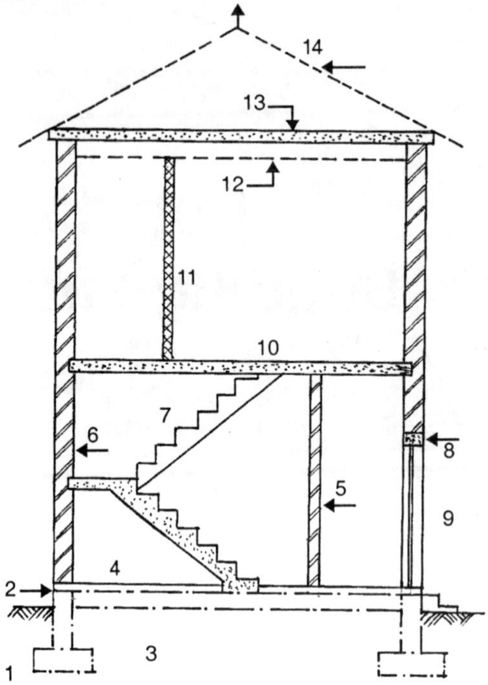

Fig. 1.1 Parts of a building: 1. Foundation, 2. Plinth, 3. Basement filling, 4. Ground floor, 5. Internal wall, 6. External wall, 7. Staircase, 8. Lintel, 9. Door, 10. Upper floor, 11. Partition wall, 12. Ceiling, 13. Flat roof, 14. Sloping roof.

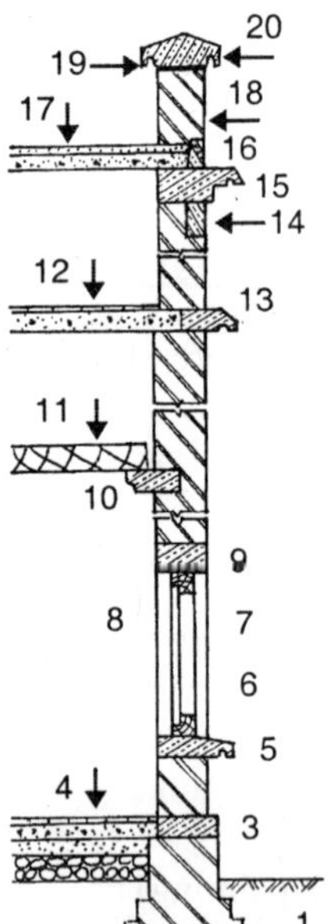

Fig. 1.2 Section through a wall: 1. Foundation, 2. Hard core (Basement filling), 3. Plinth, 4. Ground floor, 5. Sill, 6. Window, 7. Reveal (revealed vertical wall on the sides of door or window frame as inner reveal and outer reveal), 8. Jamb (vertical wall on both sides of doorway or window opening), 9. Lintel, 10. Corbel, 11. Ceiling, 12. Upper floor, 13. String course with throating, 14. Frieze (stone course below cornice), 15. Cornice, 16. Blocking course, 17. Terracing, 18. Parapet, 19. DPC under coping, 20. Coping.

We should be aware that there are many aspects that are involved in the preliminary planning and design of buildings. For example, an architect specializes in the following works:

1. Planning the orientation, layout and dimension of the rooms of the building
2. The preparation of the features to make the building attractive

Work of this nature for small buildings may be taken by non-architects also. However, the construction of a building should always be carried out under the supervision of a qualified person. We will now briefly examine the construction of the different components of buildings.

1.2.1 Construction of Foundation

Foundation is a very important part of a building. It is fully described in Chapter 4. Foundation engineering is a special subject. A foundation engineer should know how to examine the soil profile and arrive at a suitable foundation. The following are some of the different types of foundation generally used:

1. Strip foundation (shallow foundations)
2. Footing foundation (shallow foundations)
3. Raft foundation (shallow foundations)
4. Pile foundation (deep foundations)
5. Pier foundation (deep foundations)

In framed construction, we use footings as the foundation for the column and the brickwork for walls starts from grade beams connecting columns. (Grade beams on underreamed piles are also called *capping beams.*)

As the subject of foundation design is dealt with in detail in geotechnical engineering which is included in later part of the civil engineering curriculum, we will only examine the general principles of this subject. Under foundations, we will also study the anti-termite treatment for buildings.

1.2.2 Construction of Plinth

Plinth is the dividing line between the substructure and superstructure. Thus, the projecting part of the wall above the ground level to the floor level is the plinth. It is capped by a beam called *plinth beam.* The provision of a plinth beam and damp-proof course at plinth level are very important in building construction. The plinth is usually kept at least 45 cm (1.5 ft) above the general ground level of the building.

1.2.3 Construction of Walls and Columns

Walls are mostly made of masonry. It may be of brick, blockwork, stonework, etc. Hence, a study of these different types of masonry is made under this head. Construction of different types of walls such as load bearing walls, partition walls, etc. is also to be studied under this head. Buildings may also be constructed as a framed structure with columns and footings and then infilled. Most of the flats and high-rise buildings are built this way. In many places in the masonry, we use arches and lintels. Their study also forms part of masonry construction.

1.2.4 Construction of Floors

We have to study the details of construction of the ground floors constructed on the ground as well as the top floors. These top floors are nowadays usually made of reinforced concrete. A detailed study of concrete work, formwork and placing of steel reinforcement comes under this head. Different types of floor finishes for these floors are also to be studied.

1.2.5 Construction of Roof

Roof is an important part of all buildings. The most important item in housing is to have a "roof over one's head". Depending on the finances available and also the climatic conditions, we can have different kinds of roofs. Roofs can be sloped or flat. Many types of roofs and roofing materials are available nowadays. A study of these is absolutely essential for a building engineer. Another important study is how to make the roofs waterproof and heat- or weather-proof.

1.2.6 Fabrication of Doors and Windows

Openings are necessary in buildings for passages inside and outside the buildings. We also need windows for lighting and ventilation. A detailed study of doors and windows is an important part of building construction as the expenditure on this item alone can go up to 15 to 20 per cent of the total cost of civil works. Traditionally doors and windows were made of wood and hence this work is sometimes referred to as *woodwork* in buildings. Other woodworks like provision of cupboards are also important but it comes under the subject of interior decoration.

1.2.7 Stairs and Lifts

Nowadays most buildings are made more than one storey high. A knowledge of various elements of a staircase and the construction of simple staircase is essential to all those involved in building construction. Study of the layout and design of ornamental staircases is a special subject. Usually vertical transportation devices like electric lifts are to be provided in buildings having more than four floors including the ground floor. We must also be familiar with these devices.

1.2.8 Building Finishes

The final appearance of a building depends very much on its finishing. We have to deal with the following:

1. Plastering and pointing
2. Painting of walls, woodwork, grillwork, etc.

It is essential that we have a knowledge of the finishings to be used on the various materials of construction such as plaster, wood, metals, etc.

1.2.9 Building Services

Water supply drainage, sanitation, electric supply lifts, external works, construction of cupboard, etc. are considered as items outside civil works and are called *building services.*

Water supply, drainage and sanitation (building service). These works are considered separate from civil works and are also estimated separately. Design and details of these will be studied in public health engineering under building services. However, an elementary treatment of the subject is always included in basic building construction.

Electrical works (building service). Lighting and supply of electricity for various pieces of equipment used in buildings also come under building services. The supply and distribution of electricity in a building is a specialized work to be carried out by an electrical engineer. However, the basics of these works are also usually dealt with in elementary building construction.

Usually a lump sum equal to 25 to 30 per cent of civil works is estimated as the probable expense for the above two items taken together.

1.3 BUILDING SPECIFICATIONS

Building plans that are made for approval by the municipal authorities show only the arrangements and dimensions of the rooms, passage, etc. with very *brief descriptions* or *general specification* of the construction of various parts such as foundations, floors, walls, doors, windows, roof, etc. Details regarding the materials and workmanship are given separately in what is known as *detailed specifications*. The drawings and specifications together define the type of construction. Each item of work and the total work are then estimated on the basis of the drawings and detailed specifications. Thus, specifications are of two types—general or brief specifications and detailed specifications.

General specifications give only a general idea of the whole work and are useful in estimating the approximate cost of construction. They give a general description of the different parts of a building. These specifications will depend on the type or class of building to be built. Examples of general specifications for Class I, Class II and Class III types of buildings are given in Table 1.1. At the outset itself, we should make up our mind on the specifications that we would like to adopt for the construction of a given building as the cost of the building will depend on its specifications. Five-star hotels and cinema theatres should have expensive specifications, while residences of middle class people should be planned for moderate specifications they can afford. Similarly, factories for light and heavy works will be built on specifications much different from those of residences. Low-cost and temporary housing projects will have different specifications. Examples of general specification for permanent residential buildings are given in Tables 1.1 and 1.2 taking into account the seven important items of construction.

Table 1.1 General Specifications for Residential Buildings

S.No.	Item	Specifications		
		Class I building	*Class II building*	*Class III building*
1.	Foundation and plinth	First class brick in 1:6 cement mortar over 1:4:8 cement concrete	First class brick in lime mortar over lime concrete	Second class brick in lime mortar over lime concrete
2.	Damp-proof course	2.5 cm cement concrete 1:1½:3 with one kg impermo per bag of cement and painted with two coats of bitumen	2 cm of crude oil concrete 1:2:4 with standard water proofing material	2 cm thick cement mortar 1:2 mixed with standard water-proofing material
3.	Superstructure	First class brickwork in cement mortar 1:6 with RCC lintels over doors and windows finished with cement plaster	Second class brickwork in lime or cement mortar with RCC lintels over doors and windows finished with cement plaster	Second class brick in mud mortar with arches of second class bricks over doors and windows without plastering or with mud plaster
4.	Roofing	Minimum height 3.15 m (10.5 ft) RCC roof with lime concrete weather-proof course covered with tiles pointed joints [For sloping roofs, 2.55 m (8′ 6″) at eves]	Minimum clear height of 2.7 m (9 ft) RCC roof in situ or precast with lime concrete terrace and brick tiles	GI or AC sheets or thatching
5.	Flooring	Marble, ceramic or mosaic (terrazzo) floors with dado. Bathrooms with marble or ceramic tiles	Cement concrete floors with red oxide top coat or floors of brick tiles	Brick on edge floor on well-rammed earth and cement plaster
6.	Finishing	Inside walls painted with plastic emulsion paint; outside with waterproof cement paint (like unique)	Inside walls with snowcem and outside whitewashed or colour washed	Inside and outside walls whitewashed
7.	Doors and windows vanished	Teakwood doors and painted with enamel paint	Good treated country or other wood with enamel paint	Country wood painted with ordinary oil paint

Note: Usually the above *seven items* are taken to classify buildings with respect to civil works. In buildings having two to three storeys, the specifications for staircases and in buildings having more than four floors including the ground floor, the specifications for stairs and lifts are also to be provided. A framed structure with infilled walls may be also planned. For example, a set of residential flats in a city may have the *general specifications* given in Table 1.2.

Table 1.2 Examples of General Specifications for a Set of Flats

Civil works

S.No.	*Item*	*Specifications*
1.	Foundation	RCC footing foundation
2.	Damp-proof course	RC plinth beams with bitumen water proofing below ground floor
3.	Superstructure	RCC columns and beams with 9 inch external brick walls with 4½ inch partition walls, plastered with cement plaster and painted with cement paint primer
4.	Roof	RCC slab with lime brick jelly concentrate and floor tiles
5.	Flooring	Rooms with grey mosaic (terrazzo) tiles with marble chips and mosaic skirting or joint free ceramic tiles for floors; kitchen with black mosaic and glazed tile skirting or marble floor and skirting
6.	Finishing	Oil-bound distemper (OBD) or equivalent for inside walls; synthetic enamel paint for doors and grills; unique supercem paint for external surface
7.	Doors and windows	Melamine finished flush doors (exterior grade) with teakwood frames; glazed sliding shutter in aluminium frames for windows
		Services
1.	Sanitary	Baths, white washbasins, European closet, shower with provision for fixing geysers and exhaust fans; ceramic tiles for floors and walls up to 5 ft
2.	Electrical	Three phase concealed wiring on ring circuits with circuit breaker on main lines with 5 amp and 15 amp plug points
3.	Lift	Five passenger capacity lift and staircase to rooftop

Detailed specifications give details of each of the different types of work in the order in which the work is carried out at the site. "Specification (Ref. 69, 70)" published by CPWD India, "Tamil Nadu Building Practice", published by Tamil Nadu PWD in 1977 or specifications published by other state PWDs are examples of such detailed specifications. They form part of the contract documents and all works for the governmental organizations have to be carried out according to the respective specifications.

As already pointed out, the first task of the owner constructing a building is to decide on the general specifications of the work depending on the money available for the construction and the manner the building will be put to use.

1.4 CONTROLLING COST IN BUILDING CONSTRUCTION

Controlling the cost of construction of a building should be thought of from the planning stage itself. The main methods used for control are discussed further.

Planning and standards of accommodation. Sizes of rooms, height of the ceiling, number of windows and doors, sizes of rooms, etc. are important components in cost of buildings. There should be only minimum areas for corridors, lobbies, etc. for economy. Attention to them in the planning stage can decrease the total cost of the building to a large extent.

Specifications of materials. The specifications used for floors, woodwork, painting, etc. considerably affect the cost. For example, teak woodwork will be very much more expensive than other ordinary types of wood. Similarly, ordinary glazed tile or mosaic tile floors are cheaper than expensive marble or vitrified tile floors. Painting with expensive paints can also increase the cost of a building.

Structural and foundation design. Economic design of foundation and structural elements such as beams and slabs can considerably decrease the cost of a building, especially as the costs of cement and steel are nowadays very high. The design of the components of the building should always be carried out by a qualified engineer. This will not only reduce the cost but also ensure safety of the building.

Administration. The cost of supervision of work and payments will depend on the method of construction. Contractor's profit nowadays, because of the taxes such as service tax, etc., has to be more than 20 per cent, whereas cost of simple supervision by a qualified civil engineer will cost only about 7 to 10 per cent of the works. Hence, the method adopted for getting the building constructed is also important.

1.4.1 Approximate Costs of Various Components of Construction

Table 1.3 gives the *average percentage* cost of civil works for a residential building. It also gives the possible savings that can be made. Actual values will obviously depend on many factors.

Table 1.3 Percentage Cost of Civil Works

S.No.	*Work*	*Percentage*	*Possible saving on total cost (%)*
1.	Foundation	10 to 15	1 to 3
2.	Structure (wall and roof)	30 to 40	4 to 13
3.	Doors and windows	12 to 20	1 to 8
4.	Flooring	10 to 20	2 to 7
5.	Finishes, plastering, painting	5 to 20	1 to 8
6.	Waterproofing (roof, wet areas)	8 to 12	1 to 2

In addition, the cost of services as estimated as a *percentage of civil works* is taken as follows:

1. Water supply and sanitation — 12.5 per cent
2. Electrical works — 12.5 per cent
3. External works — 5.0 per cent

So, the total cost of the services is up to 30 per cent of the cost of total cost of civil works.

1.5 VARIOUS STEPS IN CONSTRUCTION OF A RESIDENTIAL BUILDING

The following can be considered as the nine steps involved in construction of a residence:

Step 1. The first step is visit to the site by the owner and architect to evolve a suitable plan for the building which can satisfy owner's requirements. Investigation of foundation condition also has to be made. We have to arrive at a brief specification of various works as given in Tables 1.1 and 1.2.

Step 2. The second step is to get the approval of the drawings by the competent authority.

Step 3. During this time, the structural drawings are made. Detailed estimation of works and adjustment of specifications to suit the budget are also done. Finalization of electrical and plumbing diagrams is also to be completed.

Step 4. Site preparation, construction of sheds, fences, etc. is the next step.

Step 5. Taking up and completing civil works is the fifth step. It should be preferably started just after the rains when ground conditions are ideal or started at least three months before the rains are due so that part of civil works can be completed before rains. (This is the major part of construction described in detail in this book. See Appendix D for the major items of work.)

Step 6. Starting and completion of electrical works

Step 7. Starting and completion of plumbing works

Step 8. Construction of interior features like shelves and other carpentry works

Step 9. Construction of boundary walls, fences, gates, etc.

Each of the steps involved in construction of a residential building is a specialized activity, but in this book, we will deal principally with the elementary aspects of the execution of civil works only, which are common to all types of buildings. Other works will only be described very briefly.

SUMMARY

Before we start a building work, we should properly plan its layout and decide on the specifications for each major items of work so that the cost of the construction can be kept within the estimated value. For economy, the type of foundation superstructure and roof that will be used for the building should be carefully selected in advance. These should not be overdesigned. Specifications for finishes like types of flooring and painting should be carefully drawn up to reduce cost. Wood work in doors, windows, ventilators etc is also a major item in cost of buildings. Nowadays teak (which was extensively used in ordinary buildings in the past) is very costly. Alternate materials like chemically treated wood or use of other materials like aluminium must be considered for reducing the total cost. In short, advanced planning of each item is necessary for cost reduction and in many parts of India, can lead to considerable economy in construction.

REVIEW QUESTIONS

1. Indicate by sketches and mark the various parts of a reinforced concrete two-storey flat roof building.
2. What are the seven items usually taken to classify building works? Give a short specification for a Class I building.
3. Explain why it is necessary to draw up a short specification of the proposed building right at the start of the project.
4. Describe in brief the methods for controlling the cost of construction of a building. What are the percentage costs of the various civil works and how do you estimate the cost of services?
5. What are the items of works under civil works and services in the construction of a residential building?
6. Define the following terms:
   (a) Reveal
   (b) Jamb
   (c) Corbel
   (d) Coping
   (e) Frieze
   (f) Sill

Chapter 2

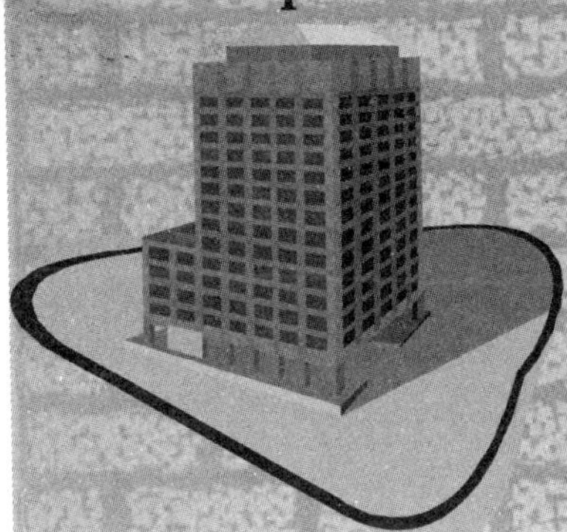

Site Preparation and Setting Out of Works

2.1 INTRODUCTION

The first work to be taken before the actual construction of a building is checking the dimensions of its boundaries as soon as the site is made available for construction. The vital boundary stones should be in their position and they should be checked with reference to the survey plan. Any difference that may be found regarding front, rear or side dimensions should be reconciled before the work is started (see Chapter 34).

2.2 SITE LAYOUT

The site layout for construction consists of the layouts of access roads, sheds, etc. They should be made as follows:

Access roads. An examination of the site drawing will determine the best layout for access roads. Wherever possible, access to the site for lorries and carts should be the shortest and capable of carrying materials either to a central place or various places of work, as may be desired.

Sheds. A study of the site drawing will indicate where weatherproof sheds must be erected for storage of materials such as cement, lime and other perishable materials. If the cement stores have to be large, they should be provided with two separate doors, one at each end—one for accepting delivery and the other for issue of materials.

2.3 SITE CLEARING

Site clearing means any one or all of the following works:

(i) Surface cleaning of grass, trees, anthills, hillocks, etc.

(ii) Cleaning of obstructions which may be above or below the ground level such as old foundations, old drainage works, old septic tanks, pit-type latrines and soak pits

(iii) Cleaning of obstructions belonging to other organizations such as drainage or water supply lines, underground electric or telephone cables

In case of the first two items, they should be suitably removed and filled up to the ground level by good earth or sand. In case of the third item, the concerned party should be notified well before the commencement of the work. If they are not to be disturbed during the construction, proper arrangements should be made for its protection.

2.4 ENCLOSING THE SITE

The building site is enclosed firstly for safety of the public. If any person falls into an excavation made for the building without an enclosure, the supervisor will be put to blame, whereas if a person falls into an excavation in an enclosed area, that person is a trespasser. Secondly, by enclosing the site, it becomes more secure from any theft. We should also ensure that the methods of storing of materials are safe. For example, if material is stored against a boundary wall, the wall should be strong enough not to fall down. If gates are provided, they should be wide enough and properly located for the lorry to come in. They should open inwards (not outwards) from the road.

2.5 WATER SUPPLY FOR CONSTRUCTION

Water is an important building material. If groundwater is available, it should be tested suitably for various uses. Cost of water comes to about *one to two per cent of the cost of civil works*. If no groundwater is available and water connection can be obtained from the municipal authorities, the pipes should be so laid that they will become part of the permanent water supply system after completion of the building. If suitable water has to be brought by lorries, temporary or permanent water storage tanks should be built for storing sufficient quantity of water for each day and also for discharge from the supply tanker. If good groundwater is available, sinking of a proper tube well (which can also be used later) at a suitable place will ensure a good supply of water during construction.

2.6 ELECTRICAL SUPPLY

Electricity supply is necessary for modern building works. The points to be observed are as follows:

1. Cables should be of good quality and suitably supported.
2. The switch board, etc. must be properly enclosed and capable of being locked.
3. The switch board should be about 1.5 m above the ground level.

2.7 PROTECTION OF EXISTING SERVICE LINES

Provision should be made to protect water, electricity, telephone and other public distribution lines which may be running through the property and be affected by foundation works. We should mark all the existing service lines so that workers can identify them and can be careful not to disturb them during the construction. (There are many instances when underground electric lines have been disturbed, causing fire to nearby installations.) Sometimes, pipe ducts or concrete coverings are provided to protect these services if they cannot be removed from the site to other alternative places.

2.8 INITIAL CHECKS ON DRAWINGS

It is very important that before commencing actual setting out of the building on the ground, the following checks should be made *on the drawing from which the layout is to be carried* out:

1. The sum of the intermediate dimensions should match with those of the overall dimensions.
2. The levels of various constructions (such as ground level with respect to road level, floor level, etc.) are clearly shown in the drawing.
3. Check whether the proposed building will actually fit into the plot with the mandatory distances specified by the municipal authority.

2.9 SETTING OUT OF BUILDINGS

Setting out of buildings consists of the following two operations:

1. The first operation is *the setting out of centre lines.* This means establishing the centres of the walls in case of a building with load-bearing walls, or the centre of columns in case of a framed building.
2. The second operation is *the setting out of trenches* or establishing the excavation lines for proceeding with the excavation.

We will separately describe the procedures used for the buildings with walls and that for the framed buildings.

2.9.1 Setting Out Centre Lines for Bearing Walls

The step-by-step procedure *for setting out the centre lines of walls* can be stated as follows (as shown in Fig. 2.1):

Step 1. Establish a *benchmark* from which all levels for the various parts of the building can be established and which will not be disturbed during the building operations. This can be done by driving down a 50 mm × 50 mm angle 2 m long or a steel rod of suitable diameter and 2 m length in a previously dug hole so as to project about 10 cm from the ground level and then concreting the base to a suitable depth below the ground level to form a pedestal around it.

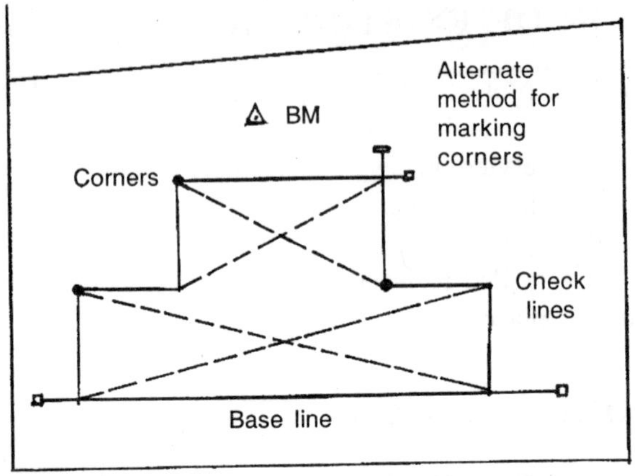

Fig. 2.1 Setting out centre lines and checking out corners.

Step 2. The second step is to mark a *baseline* from which all dimensions can be measured. The centre line of the longest outer wall of the building is usually taken as the baseline. This is marked with respect to the boundary.

Step 3. The third step is to start from the baseline and mark the *corner points* of the centre line of walls of a building by means of 50 mm × 50 mm wooden posts driven firmly to the ground projecting 25 to 50 mm above the ground. A nail or saw cut is placed on the peg to indicate the exact centre point. Setting dimensions are measured with steel tapes and ranging roads between corner posts. It is important that the 90 degree angles at corners are measured by using a builder's square or by using the 3 : 4 : 5 principle or a theodolite. Check whether all the dimensions of the diagonals tally.

Step 4. Using the corner points, transfer the centre line to the ground with dry lime by stretching lines between the pegs.

2.9.2 Setting Out of Trenches for Excavation of Bearing Walls

Having set up the centre line of corners and checked the dimensions of the building on the ground, we proceed to set out the lines for trenches using the centre line already established.

The aim of *setting out trenches* is to mark the direction and width of excavations to be carried out and also to mark the width of the wall to be built. This is carried out by using pegs or by profile boards. These are masonry pillars or timber boards fixed to the ground some distance away from the exaction on which the excavation and wall boundaries can be marked as shown in Fig. 2.2. These are set up at least 2 m clear of the excavation as shown in Fig. 2.2. *The profile boards may be masonry or timber.*

The level of the top of the profile boards should be related to the side datum level and fixed at a convenient height if boning rod (a traveller) is used to control the depth. The centre line, wall width and trench width are marked on the profile board. The trench width is marked on the ground by lime powder after stretching strings between the profile boards.

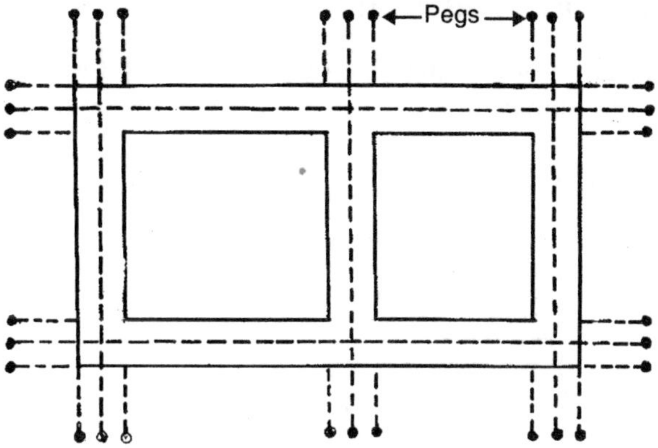

(a) Use of pegs

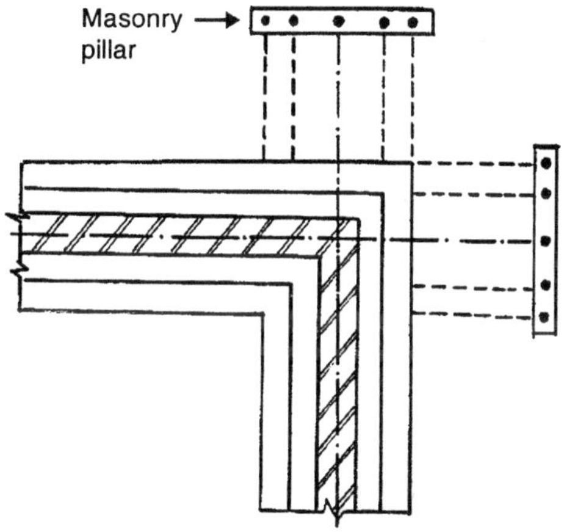

(b) Use of masonry pillars

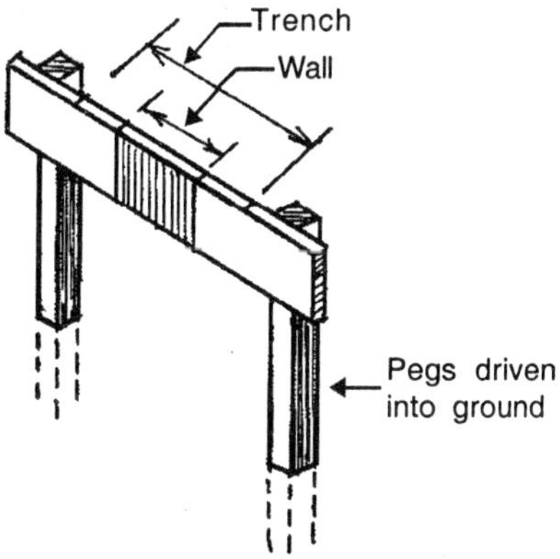

(c) Use of profile boards

Fig. 2.2 Setting out of trenches.

2.9.3 Setting Out a Framed Building

For setting out the foundation of a framed building also, we first establish the benchmark and set out the centre line of columns. This is usually carried out by a theodolite as the column centre lines are usually marked *on a grid* as shown in Fig. 2.3—one axis is marked as 1, 2, 3, 4, etc. and the other as *A*, *B*, *C*, *D*, etc.

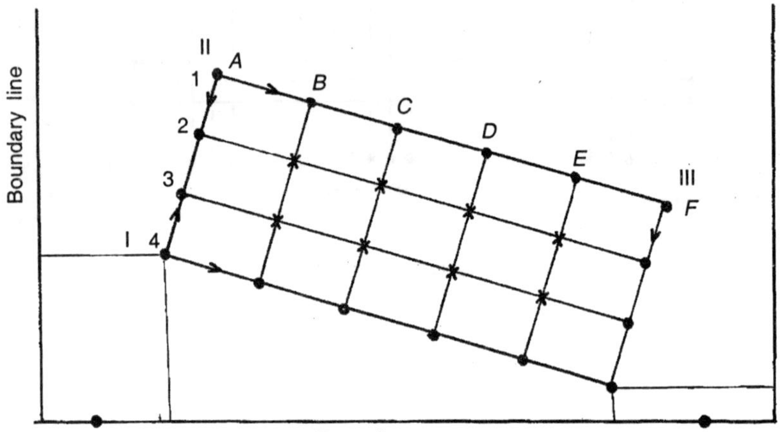

Fig. 2.3 Setting out footings of framed buildings by theodolite (I to III positions of theodolite).

In this case, *we first fix all the peripheral points* as shown in Fig. 2.3. We first fix point 4 with respect to the boundaries of the plot. Then we station a theodolite at point 4 and fix *F*4, *E*4 to *B*4. Turn 90° and fix *A*3 to *A*1. Secondly, fix theodolite at *A*1 and fix *B*1 to *F*1. Thirdly, station theodolite at *F* and fix *F*2 to *F*4 (check *F*4). The internal points can then be fixed easily by theodolite or with stretched lines. Once the grid has been set out, profile boards can be fixed clear of excavation work to carry the excavation of the footings.

2.10 METHODS TO DETERMINE DEPTH OF EXCAVATION

For construction of foundations and sewer drains, the depth of the base of the excavation is usually set out by means of sight rails and *boning rod* (also called *traveller*) as shown in Fig. 2.4. In foundation, construction of the base is to be levelled and for drain construction, it is to be laid to the required gradient. This operation is fully dealt with in surveying.

Another method that can be used for levelling of foundation is the use of *a water level.* A plastic tube is filled with water and is used as a water level. First the required depth of excavation is excavated in one place. In all the other sites, the level is determined by means of the water level with reference to this point. One more method is the use of the traditional levelling staff.

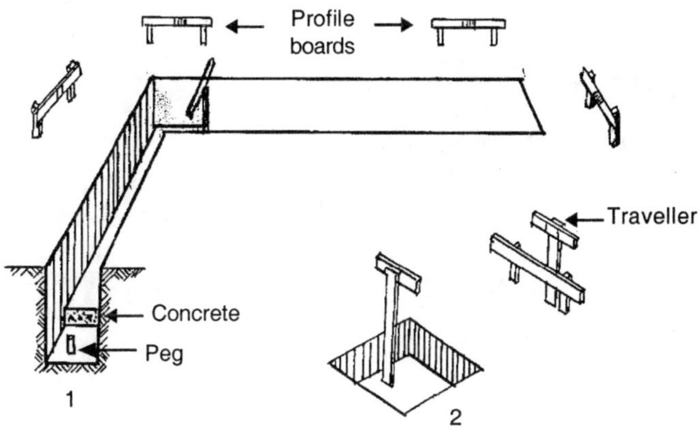

Fig. 2.4 Determination of depth of excavation by boning rod or traveller and profile boards: 1. Strip foundation, 2. Column footing.

SUMMARY

The procedure of site preparation and setting out of works is an important item of work as the final dimensions of the building and the sizes of various rooms depend on this operation. It should be carried out with precision by an experienced person.

REVIEW QUESTIONS

1. What is meant by setting out of buildings?
2. What are the preliminary works to be carried out before we set out a building?
3. Describe briefly the work of setting out of the plan of a building and the lines of excavation.
4. Describe briefly two methods you can adopt to establish the levels of excavation of the trench for the foundation of a building with load-bearing walls.
5. Write short notes on:
   (a) Profile board
   (b) Boning rod
   (c) Benchmark at a building site
   (d) Methods of checking corners and diagonals used to set out plan of a building
   (e) Site preparation for construction

Chapter 3

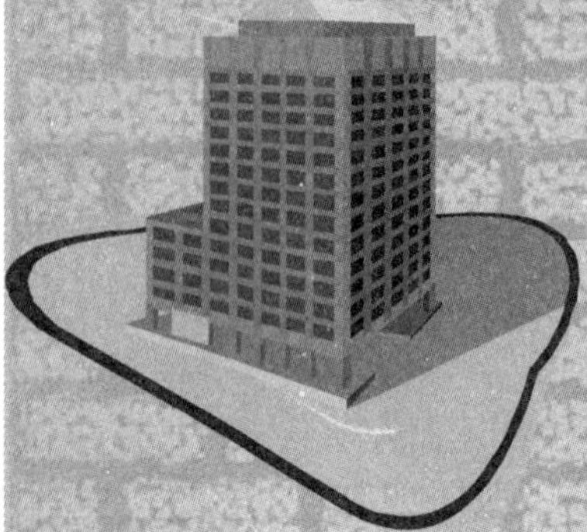

Earthwork and Anti-termite Treatment

3.1 INTRODUCTION

After setting out the trenches, we proceed with the excavation for the foundation. The items to be considered in this work are the following:

1. Excavation of foundation
2. Strutting of excavation
3. Filling in trenches and underfloors
4. Anti-termite treatment

3.2 CLASSIFICATION OF SOIL FOR EXCAVATION

For payment to contractors for excavation, the earthworks have been classified into the following categories:

1. Soft/loose soil
2. Hard/dense soil
3. Ordinary rock not requiring blasting
4. Hard rock where blasting is allowed
5. Hard rock where blasting is not allowed

If excavation work is given on contract, there should be mutual understanding of these definitions between owner and contractor as payments for excavation of the different categories differ very widely. In many cases, thorough wetting of the soil can make the excavation work easy.

3.3 EXCAVATION FOR FOUNDATION TRENCHES

In South India (by tradition), building work in the form of excavation starts from NE corner of the proposed construction. (NE is given great importance in Indian *Vastu Sastra*). For ordinary buildings with continuous (wall) footing foundation, excavation is made in trenches. For a building having a basement, the whole area in the plan of the building has to be excavated to the desired depth. For column footings, excavation is made only around the column. *The perimeter walls for framed buildings start on grade beams which may be placed much above the foundation level of the column footing.* This aspect is very important in saving cost of foundation, especially in buildings where the foundations of the columns are very deep. The minimum depth of a foundation to be adopted is given as the depth not affected by climatic changes (as in black cotton soil) and also in walls, a depth in which rodents will not borrow through and enter the building. The minimum depth of 40 to 60 cm may be adopted for temporary buildings but the minimum value usually adopted is 90 cm to 1 m for permanent buildings. It should be noted that in newly developing areas, the general ground level of buildings should be kept well above (not less than 45 cm) the *proposed road level* of the locality. Otherwise, with raising of the road level, which is bound to happen, the surface drainage of the plot will become a problem in subsequent years. Hence in newly developing sites, the decision regarding plinth level should be made on the basis of both safety and economy *with reference to the future final ground level.*

If the excavation has to be made deeper than the minimum required level to meet good soil condition or if the filling of the site around the building has to be very large, we can save brickwork in foundation if fill up part of the depth of the foundation with soil of good bearing capacity like compacted coarse sand or hard material (such as brick jelly concrete or lean concrete) instead of the expensive brickwork being started from a great depth.

3.4 PROCEDURE FOR FINAL EXCAVATION FOR FOUNDATION

Foundation of buildings usually starts with sandfilling and base concrete on top of the sandfill. Generally foundations are first dug to the approximate depth from the ground level and it is then levelled with sandfilling, to take care of variations of level in the excavation. We proceed as follows. Pegs are first driven in less than 3 metre intervals in the excavation already made to the approximate depth required so that the levels of the top of these pegs are at the level of the *top of the foundation concrete.* This can be accomplished by a level and a levelling staff, or by a water level or a boning rod. When using the boning rod, the end of the rod is set on the peg and the peg is driven until the cross head is on a level with the previously set profiles which is easily seen by sighting across them. Digging is then continued till the bottom of the trench reaches the required depth. The pegs are left to serve as a guide for sandfilling and levelling the base concrete. The levels of the top of the pegs can also be checked with the help of a 3 m straight edge and a spirit level. In order to reduce any inaccuracy to a minimum, the straight edge and level should be reversed each time a fresh level is taken.

If the excavation is shallow and the local soil will allow the sides to stand vertical, it can be carried out by trenches with vertical sides. Excavations which are deep or in loose or in weak soils should be made with *sloping sides* or provided with supports which are called *timbering*. Excavation work can be carried out by manual labour or machinery.

3.5 TIMBERING OF EXCAVATION

It is very important that when vertical trenches are excavated deeper than what the soil can sustain itself in a vertical cut—and in any case, more than 1.5 m (the average height of a person)—the sides must be given some form of temporary support, the extent of which will depend on the depth of the trench, nature of the soil, the season and period for which the work lasts. A number of accidents often occur due to lack of supports of excavations. Traditionally, this has been done in timber and hence the work is called *timbering of excavation*. Excessive timbering with too many struts between the sides must also be avoided because of the difficult working condition that can occur in such situations.

Deep and large excavations, as those required for basements, require specially-designed structures for withstanding the earth pressures. The conventional types of timbering for ordinary excavations are the following (Figures 3.1 to 3.3):

1. Vertical poling boards (open or close spaced) and horizontal waling pieces and strutting for fairly firm soils. (Wood in contact with earth is called *poling board.*)
2. Close horizontal sheeting (poling boards) with vertical waling pieces and strutting for deeper excavations in good soils. The vertical waling pieces can be anchored back to stakes driven into the ground at the back of the excavation. (Wood that support poling boards is called *waling*).
3. Close vertical sheeting introduced in stage is called *stage sheeting*. Wooden sheet piles with iron shoe and strutting with longitudinal walings for soft and loose soils is called the *runner system*. The soil is to be excavated after driving the piles.

For ordinary excavations, we use items (1) and (2). For deeper excavations, we use the expensive methods such as stage sheeting, runner system with special wooden planks with steel shoes, steel I soldier beams with horizontal sheetings and steel sheet piling, etc. In case of vertical sheeting (usually applied to deep excavations), it will be useful to extend the sheets above the ground to act also as a protection for people falling into the excavation. [These special systems of timberings are described in books on *Foundation Engineering*.]

Simple types of timbering for moderate depths are indicated in Figures 3.1 and 3.2. After completing the excavation, the base concrete is laid and the foundation brickwork is constructed as described in Chapter 4. Method of timbering of excavation for basements is shown in Fig. 3.3.

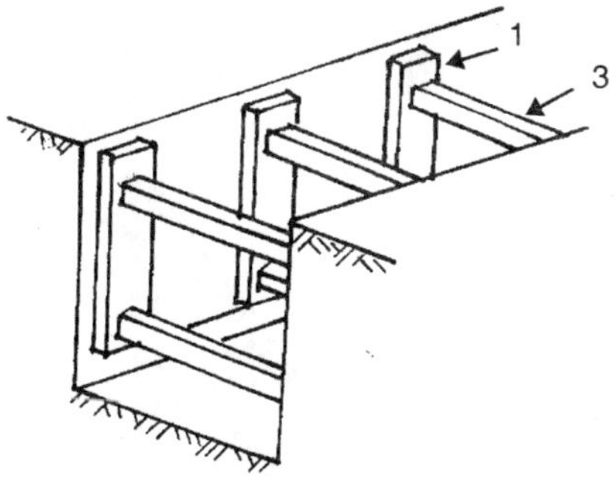

(a) Fim subsoil: open vertical timbering with poling boards and struts [Struts in each poling board can be replaced by horizontal walings as in (b).]

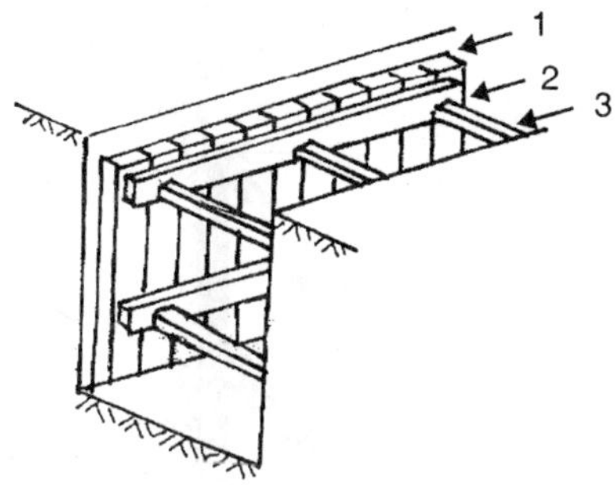

(b) Close vertical timbering

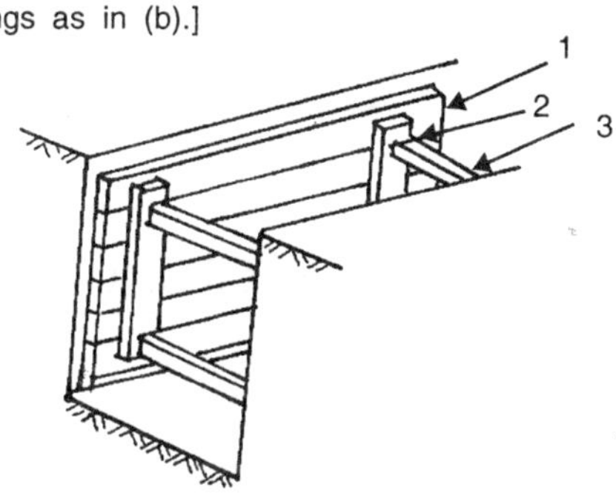

(c) Close horizontal timbering

Fig. 3.1 Timbering of excavations of moderate depth: 1. Poling boards, 2. Walings, 3. Struts [Vertical punchions as in Fig. 3.3 may also be provided for (b).]

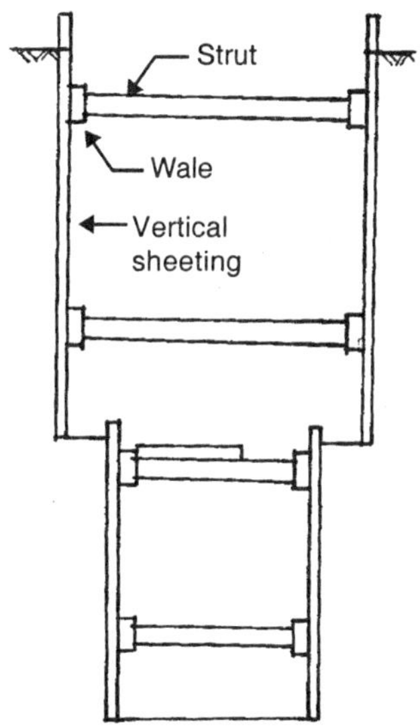

(a) By use of vertical sheeting

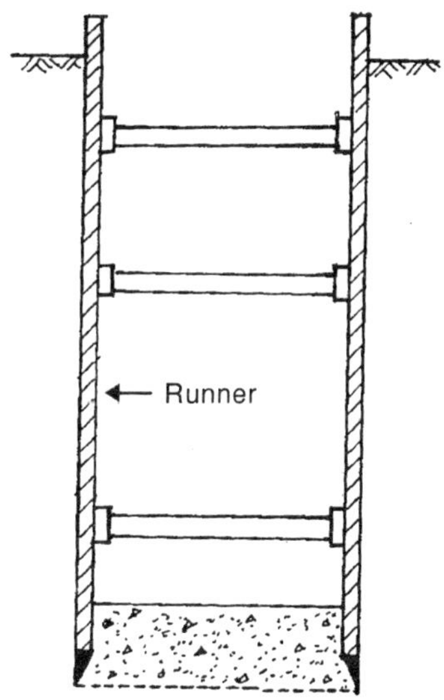

(b) By use of runner system (vertical sheets to project above GL for safety of workers)

Fig. 3.2 Timbering of deep excavations.

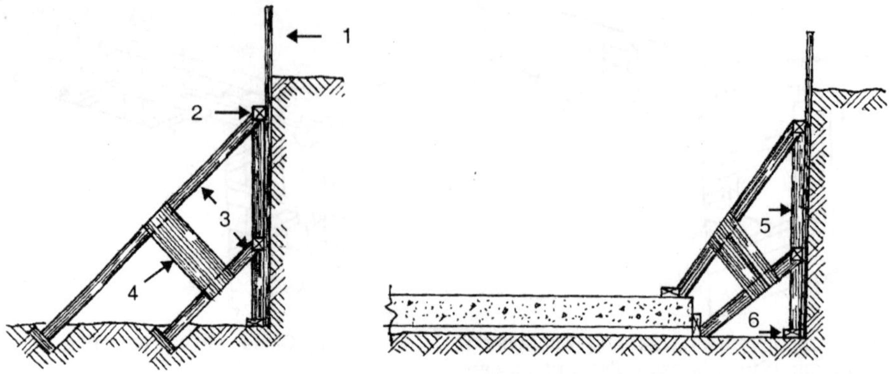

(a) First stage (b) Second stage

Fig. 3.3 Timbering of excavation for basements: 1. Vertical face timbering projecting above G.L. for safety of workers, 2. Horizontal waling, 3. Raking struts, 4. Binders to struts provided on both sides, 5. Vertical punchions at 2 m centers, 6. Sole plate.

3.6 FILLING IN FOUNDATION TRENCHES

After construction of the foundation and the walls to a height above the ground level (as will be described in Chapter 4), the vacant spaces in the trenches are to be refilled to the original surface of the ground with approved materials. The filling earth should be devoid of clods and should be well watered and rammed. They are filled in regular layers of not more than 250 mm in thickness and consolidated by the addition of water to each layer. Expansive soils and highly clayey materials should be avoided.

3.7 EARTHWORK AND SANDFILLING IN BASEMENT

Earthwork. Another item for earthwork is "filling-in of the basement or the portion below the floor level." The surface to receive the filling should be free from roots, vegetation or spoil and should be wetted first. Earthfilling is carried out to the plinth level, and it should proceed in layers. Filling should proceed along with the construction so that the fill is also consolidated by the people walking on it. It is a good practice to fill it in layers of not more than 15 cm to a height of at least 75 mm *more than the required height.* After thorough consolidation, sufficient quantity is then removed to the level required for laying the flooring. For rooms, it is made level and for *verandas,* a slope of 1 in 48 towards the edge is usually provided for drainage.

Sandfilling. Flooring concrete for ground floors should not be laid directly on the original earthwork described above without sandfilling except in cases where the foundation soil itself is sand. A sand layer with thickness of 30 cm for very clayey soils and at least 15 cm for soils other than sand should be placed above the fill. It should be compacted in layers by flooding. The purpose of sandfilling under floors is to break the upward movement of capillary water under the floors which can lead to floor sweating of the ground floor during rainy season. Sandfilling also reduces settlement of floors in expansive soils such as black cotton soils. The sand should be as coarse as available. As sand has become expensive where good earth is not available, quarry dust, which is a waste product in stone quarries, is becoming more and more popular for this purpose. However, capillary effect is present in quarry dust also. Hence greater part may be filled with a quarry dust but a coarse sandfill should be provided for the last 10 to 15 cm at least immediately below the floor.

3.8 FILLING AROUND THE BUILDING

Another major item of earthwork is earthfilling around the building. When the area around the building is to be filled to a height of more than 60 cm above the existing level of the site, the filling inside and outside the building should preferably be carried out simultaneously to reduce earth pressures on the masonry. Care should also be taken in fixing the general ground level around the building. As already pointed out in Section 3.3, it should be well above the future or existing adjacent road level at least by 45 cm. This is especially true in newly developing colonies where roads will be built after building of the houses. A number of such cases have occurred where the ground level of the property has been reduced in the course of time to much below road level because of subsequent road construction leading to problems of rainwater drainage.

3.9 DISPOSAL OR USE OF EXCAVATED EARTH

Plans must be made from the start of the project whether the excavated earth is to be used for the work or disposed of from the site. Good soil is difficult to get and if the soil at site is good, it should be planned to be stored at the site to be used later.

3.10 ANTI-TERMITE TREATMENT OF FOUNDATIONS

The subject of anti-termite treatment of buildings is usually divided into the following three parts:

1. Pre-construction measures to be taken
2. Pre-construction treatment
3. Post-construction treatment

IS 6313 Part 1 deals with the general constructional measures to be taken, Part 2 deals with pre-construction treatment and Part 3 discusses post-construction treatment (treatment of existing buildings). They give many constructional features to be provided in buildings such as for godowns, etc. to reduce termite attack which we will not deal with in this book. Here we will consider only pre-construction and post-construction treatments of ordinary buildings.

3.10.1 Termites in Buildings

Based on their habitat, termites are classified into the following two types:

1. Ground nesting or subterranean termites
2. Wood nesting or non-subterranean termites

Ground nesting or subterranean termites are more common in India and their presence in and around a building is indicated by swarms of winged reproductions flying from the soil or wood lying around the site just at the beginning of the monsoon. Wood nesting or non-subterranean termites live in the dry wood inside buildings and is recognized by their pellets of digested food or blisters on the wood surface as the wood is eaten away. In their search for food, these termites damage not only wood (cellulose) but also substances like rubber, plastic, underground cables, etc. Both varieties can be terminated by the same anti-termite chemicals.

3.10.2 Recommended Chemicals for Treatment

Even though Aldrin, DDT and many other chemicals were once prescribed in old IS codes, many of them were found to be very toxic and leave toxic residues in the soil and ground walls which give health problems over a period of time. Hence many of them have now been banned in the new revision of the BIS code. The following emulsifiable chemicals have now been recommended:

1. Chloropyrifos concentrate 1.0% by weight
2. Heptachlor concentrate 0.5% by weight

Note: Chlordane once prescribed before has now been prohibited.

Chloropyrifos in 1% concentration by weight conforming to IS8944 (1978) is a popular chemical being used. They are available with agrochemical agencies as they are used to control termites in crops such as cotton, coconut, citrus, etc. Durmet is a brand name of one of the chloropyrifos available in the market *in various concentrations*. Another brand name is Dustban TC. These can be diluted to form 1% concentration as desired in water for soil treatment and in kerosene oil for treatment of wood.

Generally 5 parts of Durmet 20 EC is diluted with 95 parts of water (or kerosene) to give an emulsion of 1%. As the solution is toxic, if it comes in contact with the skin, it should be washed out immediately with soap and water. The methods of treatment (pre-construction treatment and post-construction treatments) are briefly described below.

3.10.3 Pre-construction Treatment with Durmet

Hand operated pump is used for uniformly spraying of the chemicals at the specified rate. To facilitate the right dose, graduated containers should be used. Treatment should not be carried out when it is raining or soil is wet with subsoil water. It should start when foundation trenches and pits are ready and just before concreting of the foundation. The treated barriers for termites should not be disturbed after treatment. The treatment is shown in Fig. 3.4 and carried out in seven stages as described further.

First stage: Treatment of wall trenches and basement excavation. As the first step, all termite hills found at the site during site clearance should be sprayed with chemicals. All surfaces of pits (*sides and bottom of wall trenches and basement excavations*) should be treated to a height of 30 cm from the bottom with the solution at the rate of 5 litres per square metre of surface area.

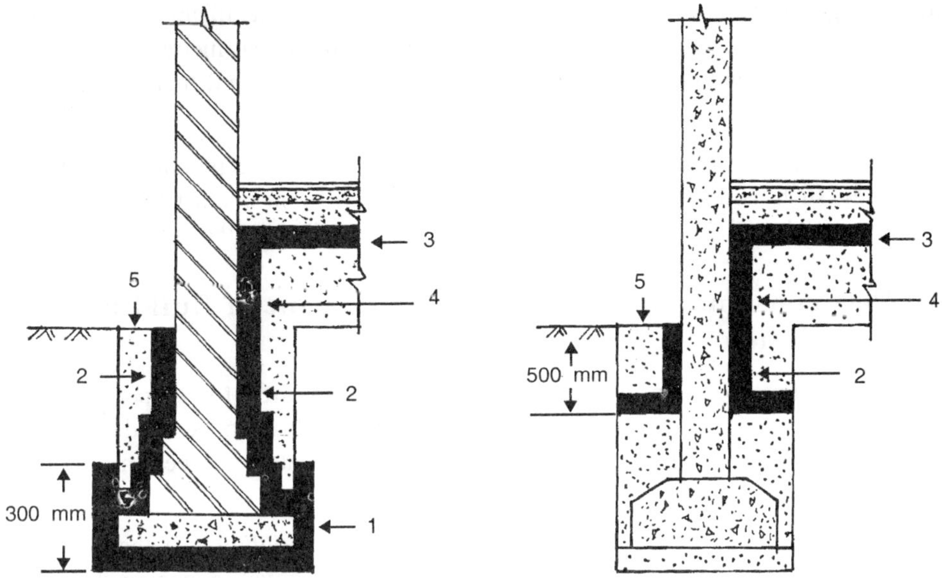

(a) Load bearing wall foundation (b) RCC column footing

Fig. 3.4 Anti-termite treatment of foundation and earthfill below ground floor.

Second stage: Treatment of refill in contact with foundation. All the refill earth in the excavation *immediately in contact with both sides of the wall footing* and all four sides of a column footing should be treated for a distance of 30 cm (as shown in Fig. 3.4) at the rate of 3 to 5 litres per linear metre vertical surface of the wall. If water is used for ramming the earth, then treatment should be done after the consolidation by rodding (making holes) in the consolidated earth at close intervals of 15 cm close to the wall or columns and applying the chemical close to the wall. It is preferable to treat all the earth adjacent to the foundation for a width of 30 cm. Similarly in framed structures, excavations for the plinth beams should also be treated.

Third stage: Treatment of soil below floors. The earthfill below the floors up to the plinth level has also to be treated after the fill has been made by putting holes 5 to 7.5 cm deep at 15 cm centres in a grid pattern and filling the holes with the solution at the rate of 5 litres per square metre of treated surface.

Fourth stage: Treatment of junction of floor and wall. Before laying the subgrade, channels of 3 cm wide and 3 cm deep are dug along the junctions of floor and wall and treated at 15 litres per square metre of wall surface by putting holes at 15 cm apart along the channel and allowing the chemical to seep through to the bottom. The soil is tamped back in position after the operation.

Fifth stage: Treatment of soil along external perimeter of building. After the building is completed, holes are made along the external perimeter at intervals of 15 cm and depth of 30 cm. These holes are filled with the chemical emulsion at the rate of 5 litres per metre length of wall.

Sixth stage: Treatment of other locations. Anti-termite treatment should be made at expansion joints after the subgrade has been laid at 2 litres per linear metre of expansion joint. Similarly, when pipes and conduits enter the building, the soil around them for a distance of 15 cm and depth of 7.5 cm should be loosened and treated.

Seventh stage: Treatment of wood surfaces. It is also a good practice to paint all fresh wood surfaces, such as door and window posts which will be in contact with masonry, with two coats of the chemical in kerosene oil before it is installed in the building.

3.10.4 Post-construction Anti-termite Treatment of Building Foundation

For post-construction operation, a pressure pump will be required for proper penetration of chemicals into the surface to be treated. Proper check should also be made regarding the adequacy of the quantity of chemicals (chemical mixed with water) pumped in. The following procedure is usually adopted:

Treatment of foundation of outside walls around the building. If there is no apron, we make trenches 50 cm deep and equal to the width of a shovel, exposing the foundation near the external walls. Holes 15 cm apart and 50 cm deep are made by an iron rod. Emulsion at the rate of 5 litres per square metre of vertical surface of substructure is to be used for each side of the floor. One-half of the quantity is pumped through the holes and the other half is poured along the trench. If there is a concrete apron, 12 mm diameter holes are dug as close to the wall as possible about 30 cm apart and the chemical is pumped into these holes at the rate of 5 litres per linear metre. Similar treatment is to be made for column and plinth beams.

Treatment of soil under floors. For this purpose, 12 mm diameter holes 30 cm apart are made deep enough to reach the soil below along the junctions of wall and floor, along cracks in the floor and along the construction joints. The emulsion with water is pumped into these holes to soak the soil or at the rate of one litre per hole. The holes are then sealed with cement mortar 1 : 2. (The floor may also be drilled and the ground below treated but this is a cumbersome process.)

Treatment of masonry at plinth level. The movement of termites through the walls can be stopped by drilling holes in the masonry at plinth level at an angle of 45° on both sides of the wall (and especially where wood such as door post is present) at 30 cm centres. The emulsion is pumped till the masonry is soaked or at a maximum of one litre per hole. The holes are then sealed with 1 : 2 cement mortar.

3.10.5 Post-construction Treatment of Woodwork

In post-construction, we may have to deal with buildings in which the woodwork such as doors or fittings (such as shelves) have been attacked by termites

It is better to treat woodwork with the chemical mixed in kerosene oil. By this process, we get better penetration of the chemical into the wood. First the masonry around the wood is treated with the chemical and secondly the wood itself is treated by drilling 6 mm diameter holes at 15 cm intervals and infusing it with the kerosene-based chemical. Two coats of the chemical can also be applied on the surface of the wood before applying the paint.

3.11 DEWATERING OF FOUNDATION

In places having high water table, excavation can be carried out only after dewatering of the foundation. The following methods are adopted:

1. By ditches and sumps for shallow excavations
2. By simple and multistage well point systems with suction pumps
3. By shallow wells of 30 cm or more diameter dug below the foundation level and water pumped up by suction pumps
4. By deep well pumps for deep foundations
5. By combination of deep wells and well points

The lesser used methods of vacuum method and electro-osmosis are also possible. Here we will deal only with the simple methods used for shallow foundations and also the use of well points.

3.11.1 Dewatering by Ditches and Sumps

This method is used mostly in sandy soils which can drain easily. Shallow pits called *sumps* are dug at locations beside and along the periphery of the excavation as shown in Fig. 3.5(a) beyond and to a level below the excavations. Water collected in these sumps is constantly pumped out by a sump pump. If there is a tendency for the bottom of the sump to rise up, it is usually weighted down by a *reverse filter* which consists of successive layers of fine to coarse material with coarse materials on the top.

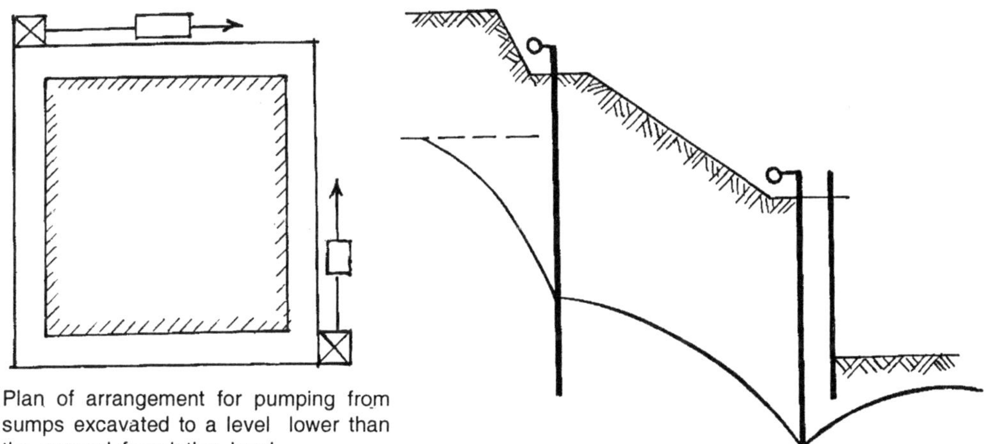

(a) Plan of arrangement for pumping from sumps excavated to a level lower than the general foundation level.

(b) Section of arrangement of multistage well point system for deep excavations.

Fig. 3.5 Dewatering of excavation.

3.11.2 Dewatering by Well Point Systems

Single stage and multistage well point systems can be used in difficult situations of groundwater. They essentially consist of perforated pipes of 5 cm to 8 cm diameter with filter points at their ends installed in the ground at 1 to 2 m spacing to depths well below the depth of the groundwater. These pipes are connected to a common leader connected to a suction pump. The system is kept under suction. The pump will be running all the time round the clock till the work in the excavation is completed. This system pumps up the groundwater and keeps the excavation dry as shown in Fig. 3.5(b). Details of these methods will be studied under geotechnical engineering.

SUMMARY

For earthwork in foundation and anti-termite treatment, there are many details that should be carefully attended to in building construction. All excavations should be planned in advance and timbering and dewatering should be provided, if necessary, for safety of workers and the structure. Only a very brief account of these have been given in this chapter.

REVIEW QUESTIONS

1. Give a brief description with sketches on how to timber an excavation made up to 3 m in dry ground at a site (where the foundation soil is mainly sandy) for laying a pipeline.
2. Enumerate the methods used to dewater foundations. Explain the method of dewatering by ditches and sumps.
3. Give the name of a chemical used for anti-termite treatment. Describe the pre-construction anti-termite treatment of the foundation of an ordinary residential house built on load bearing walls.
4. Write short notes on:
   (a) Timbering of excavations
   (b) Dewatering of foundations
   (c) Anti-termite treatment of buildings
   (d) Waling pieces, poling pieces and struts as used for earth excavation
5. A sump (water tank below the ground level) for house water supply 2.1 m × 1.5 m in plan and 3 m deep is to be built on a sandy soil. Sketch the type of timbering you will use for the excavation.

Chapter 4

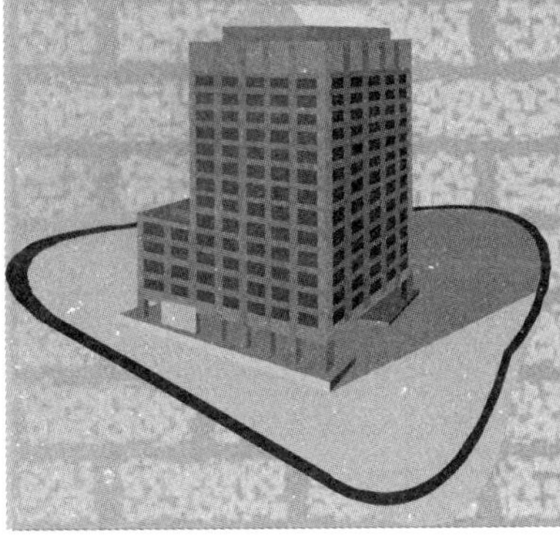

Construction of Foundation

4.1 INTRODUCTION

The most commonly used foundations for buildings are the following and they are shown in Fig. 4.1.

1. Strip foundation under walls (stepped RC wall footings)
2. Spread or pad footing foundation under columns (rectangular or sloped or combined footings)
3. Raft foundation
4. Pile foundation
5. Pier foundation (as used in bridges)

There are many other types of special foundations such as piled rafts and floating (compensated) foundations. Their layouts and designs are studied in geotechnical engineering.

The subject of foundation engineering is a specialized subject to be studied under geotechnical engineering. In this chapter, we will deal only with the simple strip foundations. A strip foundation consists of a continuous strip under the load-bearing walls. Construction of spread foundations under columns also follows the same principles as strip foundation and does not need separate treatment.

After excavation and levelling the foundation, a sandfilling of at least 15 cm (preferably up to 30 cm) is provided if the foundation soil is not sandy. In cases, where it is necessary to fill the foundation to a higher level from the excavation for starting the foundation, the filling is done by a hard core consisting of sand and gravel or sand and brickbat mix compacted well to the required level. This is sometimes called a *hard core*. This is then blinded with coarse sand. This will facilitate levelling and also act as a cushion for the foundation. The next step is to lay the levelling course of lean concrete, over which the structural foundation (mass concrete, reinforced concrete or any other type) will be built. This chapter deals with the procedure for the construction of a strip foundation under the following headings:

1. Evolution of strip foundation
2. Placing levelling course of sand layer and lean concrete
3. Setting out of brickwork in foundation
4. Foundations of partition walls

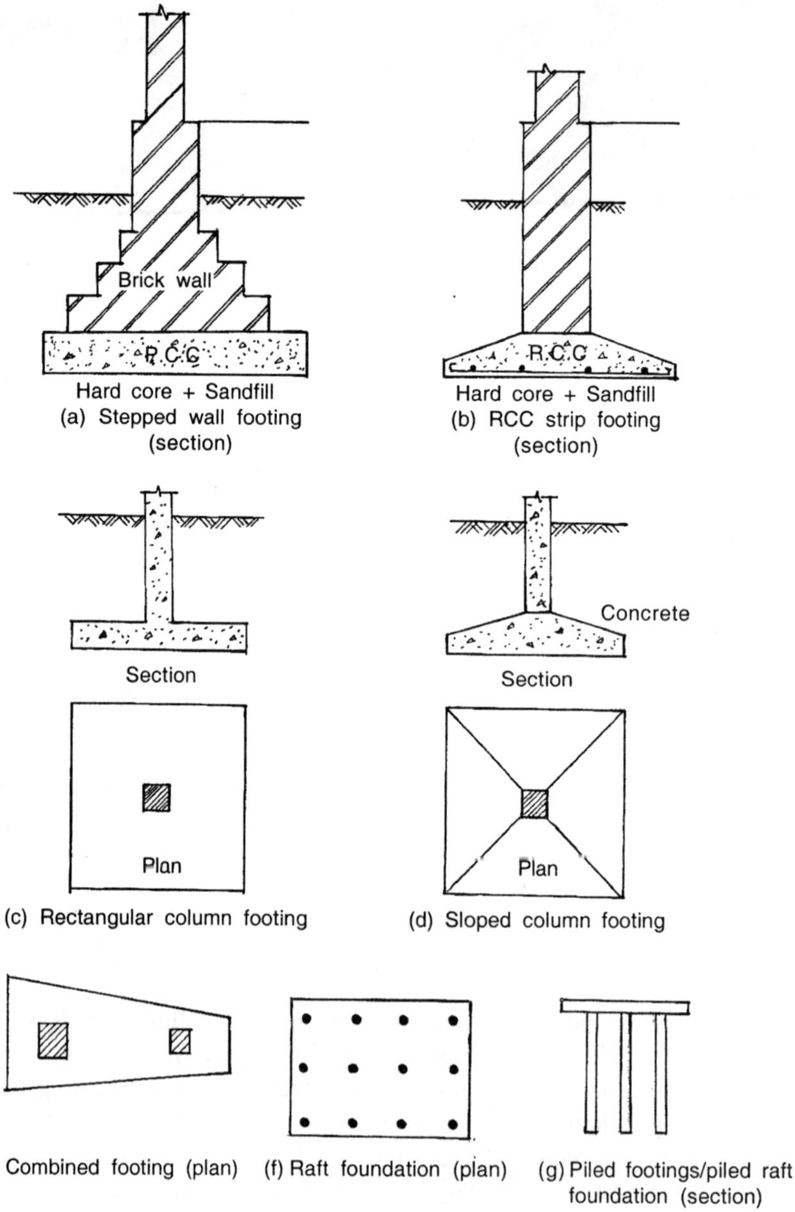

Fig. 4.1 Common types of foundations.

4.2 EVOLUTION OF STRIP FOUNDATION

Before the advent of cement, lime mortar was traditionally used in India for foundations. Brick walls were extended below the ground and rested directly on the subsoil or on a bed of broken brickbats in lime concrete. As an improvement, the masonry below the ground was built in steps, the brick offset being half a brick (4½ inches)—a quarter brick on each side. Thus a 9 inch wall was carried through 13½ inch, 18 inch and 22½ wall steps, each step being two or three courses of bricks of 3½ inch high with the mortar joint as shown in Fig. 4.1(a). They are called *stepped footings*. However, with the availability of good cement, the stepped brick foundation has generally been replaced by mass cement concrete with reduced number of steppings. Brickwork of sufficient width is placed on a bed of mass concrete 1 : 3 : 6 or 1 : 4 : 8 of sufficient depth to distribute the load safely to the foundation. We assume an angle of dispersion of 45 degrees. 1 vertical to 1 horizontal in cement concrete and 2 vertical to 1 horizontal is assumed in lime concrete. The minimum width of excavation is 1 m, which gives enough space for the workmen to work. The thickness of concrete then depends on thickness of the brickwork in foundation, but the least thickness of concrete provided is 15 cm.

Nowadays, with the advent of RCC in modern construction, reinforced concrete footings are mostly used in strip and pad footings. Such foundations do not require much thickening of the brickwork at the base, but can be carried straight from the RCC strip. However, in order to give a plinth offset at plinth level, the wall is usually given at least one offset from the plinth level downwards to the level of the reinforced concrete foundation. Thus for one brick wall, a one and one-half brick wall with one-fourth brick offset on either side is all that is wanted. This considerably saves brickwork in foundation. Thus, nowadays, we can go for a plain concrete block stepped foundation or a reinforced concrete strip foundation depending on the nature of the soil and the load on the wall, as shown in Fig. 4.1.

Care should be taken to choose the right materials for foundation construction. All specifications state that brickwork used in foundation should be first class bricks in 1 : 3 or 1 : 4 cement mortar. *In places liable to flooding, one may preferably use rubble work* if stones are easily available. Similarly in places where sulphates are present in soil or in the subsoil water, care should be taken to avoid ordinary Portland cement in the foundation. Special sulphate-resistant cement or blended cement can be used for such foundation work.

4.3 DETERMINATION OF WIDTH OF FOUNDATION

The width of a strip footing will depend on the load to be carried by the wall and the safe bearing capacity of the soil. It is given by the following formula:

$$\text{Width required} = \frac{\text{Load on unit length of wall}}{\text{Safe bearing capacity}}$$

but should not be less than 1 m, the minimum width required for excavation.

4.4 PLACING LEVELLING COURSE OF CONCRETE IN FOUNDATION

Two levelling courses are in general provided in the foundation—the sand levelling course that we have already described and the concrete levelling course. As already stated the first is sand levelling course not less than 100 mm to level up the foundation. Over this levelling course, the second course of the plain concrete is placed. Its thickness will depend on whether we are using plain or RC foundation. (See Chapter 15 for more details.)

The second levelling course can be of lime-brick jelly concrete 1 : 2 : 6 preferably with surki or sand as fine aggregates or lean cement concrete 1 : 3 : 6 or 1 : 4 : 8 of large size aggregates 50 mm. For foundations in clayey soils, we should prefer to use lime as it has the property of reducing the expansive properties of clays. For brick jelly concrete, the aggregates (the brick ballast) must be red or copper in colour and should be clean, free from dust and foreign matter. Fine aggregate can be surki, sand or cinder (free from dirt and dust). Surki is preferable as fine aggregate with fat lime and with hydraulic lime, sand is commonly used. The proportion for lime concrete usually used is 1 : 2 : 6 (lime : surki : coarse aggregate). When using a reinforced concrete strip foundation, we first place a mud mat of lean concrete over the sand layer and then place the steel with necessary cover (75 mm) over it.

4.5 SETTING OUT FOR BRICKWORK OF A LOAD-BEARING WALL

After the base concrete foundation has set, the lines giving the projections of the bottom course of the footings will be transferred from the profile board to the bottom of the trench on the concrete surface as described in Chapter 2. This is accomplished by using a plumb line suspended from the ranging line and its position marked on a mortar screed spread on the concrete, marks being made at each end of the wall being set out. In this way, first the positions of each corner are marked out and *the corners are built*. It is always advisable to check the wall line again with ranging lines strained between the *profiles at the corners* before the ground level is reached so that slight adjustments can be made good at this stage of construction below the ground level.

4.6 FOUNDATION OF PARTITION WALLS

Foundations of external and internal *load-bearing walls* should always be placed at sufficient depths so as not to be affected by climatic changes. However, *half-brick internal partition,* which does not carry any load (except very light load and its dead weight), do not need such elaborate arrangements. For ordinary buildings, the practice in non-clayey soils is to build these walls without going through the full depth of foundation on a widened concrete base forming part of the floor layout (which do not go for the full depth of the foundation), on the cohesionless filling under the floors. The following detailed guidelines are commonly used by dividing partition walls into four types (type 1 to 4) and the following foundations as shown in Fig. 4.2 can be provided on nonexpansive soils:

Type 1 walls. These are half-brick walls (11 cm) taken to ceiling heights only [Fig. 4.2(a)]. It should be of the conventional type. For these walls, the foundation should consist of one-brick walls (23 cm). They are constructed below the floor level and taken to 30 cm below the original ground level. A nominal concrete foundation of width 30 cm and depth 15 cm is provided beneath it. A sandfilling of 15 cm is also provided under the concrete.

Type 2 walls. These are half-brick walls taken only up to 1.73 m (5′ 9″) and are used generally for storage shelves, etc. [Fig. 4.2(b)] Where the ground is not made up or expansive, the wall can be made to rest directly on a concrete foundation 30 cm wide and 20 cm thick provided just below the base concrete for the floor. A sandfilling is also provided under the concrete. If the length of the wall exceeds 2.4 m, then a Z-type layout will give better rigidity to the wall.

Type 3 walls. Partition walls in framed structures can be easily built on connecting beams made just below the floor concrete level and connected to other beams such as grade beams supported on columns. Similar arrangements can be made in residences provided with RC plinth beams at plinth level by connecting beams joined to plinth beams.

Type 4 walls. Minor partition walls of low height such as those in kitchen or bathrooms without any loading can be directly built on the top of the floor, as shown in Fig. 4.2(c), if a good depth of sandfilling over the original ground has been provided below the floor beneath the wall. Alternately, a lightly reinforced shallow beam resting on sandfill at the level of the floor can be provided to support this wall. (See also Section 5.11.)

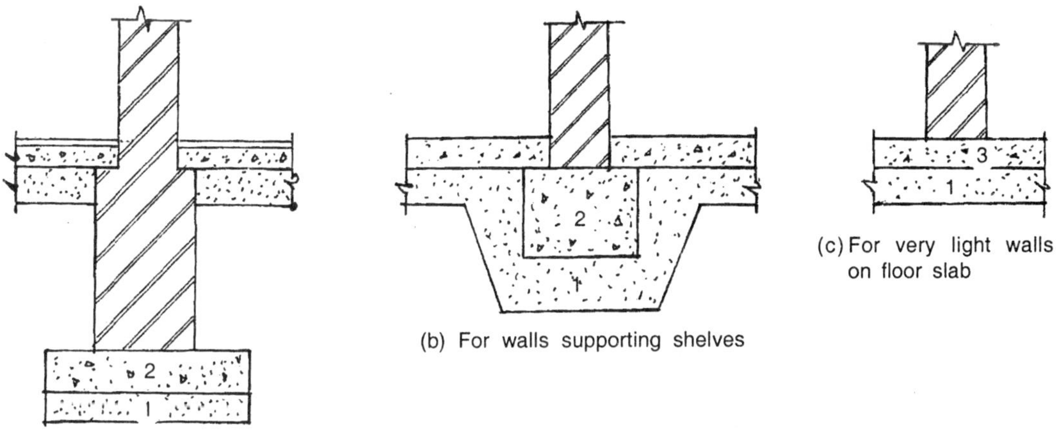

(a) For walls taken to ceiling heights

(b) For walls supporting shelves

(c) For very light walls on floor slab

Fig. 4.2 Foundations for lightly loaded half-brick walls or good soil: 1. Sandfilling, 2. Foundation concrete, 3. Floor slab.

4.7 FOUNDATIONS OF FRAMED BUILDINGS

Framed buildings are built with columns and beams forming the framework. The columns are built on reinforced concrete footings which have construction similar to the reinforced concrete strip footings and may be placed at sufficient depth below the ground level to cater for heavy loads on columns. The walls need not be built from the foundation level of the footings, but the walls are usually carried on *beams called grade beams built at the minimum depth below the ground level* as shown in Fig. 4.3. This construction considerably reduces the cost of brickwork in foundation.

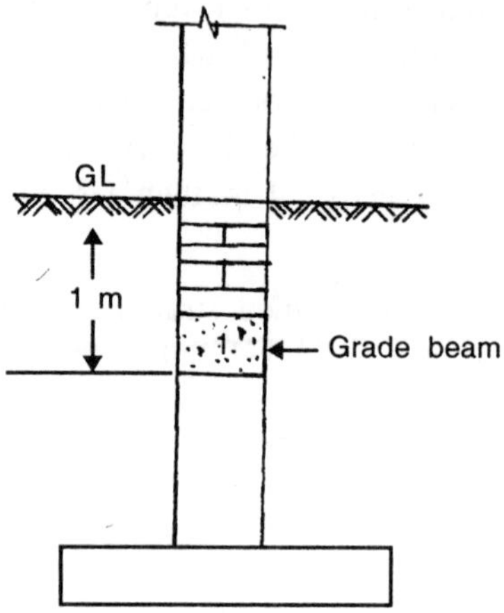

Fig. 4.3 Location of grade beams: 1. In framed buildings.

4.8 FOUNDATIONS FOR STAIRCASES

As foundations of staircases inside the building are not subjected to climatic changes, they need not be taken very deep unless the soil is very clayey. Usually in good soil conditions, the foundations are laid about 40 cm (18 inches) below the original ground level. However, if the site is to be filled up for a large height, *special care should be taken so that* the soil below the foundation of staircases is well compacted while filling is made.

4.9 PLINTH BEAMS

We have seen that the projecting part of the wall immediately above the ground up to the ground floor level is known as *plinth.* It gives an appearance of additional stability to the building and also the clearance from the ground level. The level of the top of this wall is called *plinth level.* In first class buildings, for getting crack free walls, especially in clayey soils, an RCC beam is usually provided in the main walls above the ground level and just below the ground floor level. This is called a *plinth beam.* Under normal condition, it is made 10 to 15 cm in depth and extending the full width of the upper wall. Two numbers of 8 mm (or three numbers of 6 mm) high strength steel, both on top and bottom, bound by 6 mm stirrups at 15 to 23 cm centres (depending on depth of the beam) are provided as reinforcement for the plinth beam (see also Section 8.7). In addition, DPC (damp-proof course) of bitumen coating is usually provided on top of this plinth beam. Plinth beams should always be incorporated in buildings to reduce differential settlement and are also mandatory for earthquake resistance.

In case of cheaper houses, the plinth course can be formed of bricks in any one of the following ways:

(*i*) *Brick on end.* It consists of bricks laid on end with plinth projection of about 20 mm.

(*ii*) *Brick on edge.* It consists of bricks on edge. It gives strength and rigidity.

Dampproofing of these surfaces have to be carried out as described in Chapter 9.

4.10 ECCENTRICALLY-LOADED FOUNDATIONS

In construction of compound walls, their foundations may have to be built in one's own property. Eccentric footings can be adopted for these walls only if the scil is sandy. As the pressure on the foundation at the boundary side is higher, as shown in Fig. 4.4, in this design, there will be greater settlement at the boundary. Hence, the wall has a tendency to tilt and

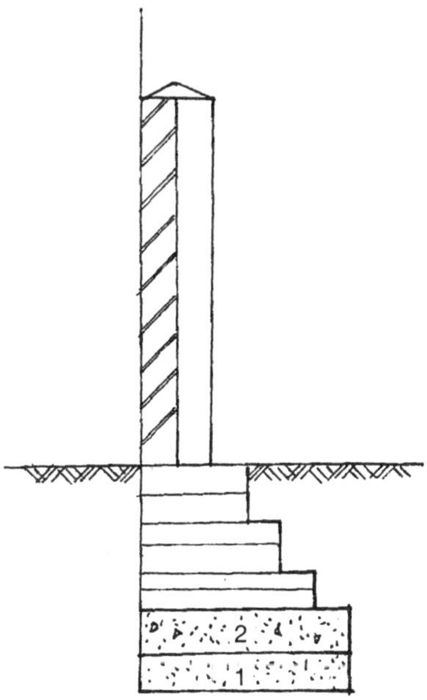

Fig. 4.4 Footing foundation for compound walls in good soil: 1. Sandfill, 2. PCC.

overturn in course of time if the foundation is clayey. Underreamed or ordinary piles with a grade beam is a better choice for such walls in clay soils. Construction of underreamed piles is shown in Fig. 4.5.

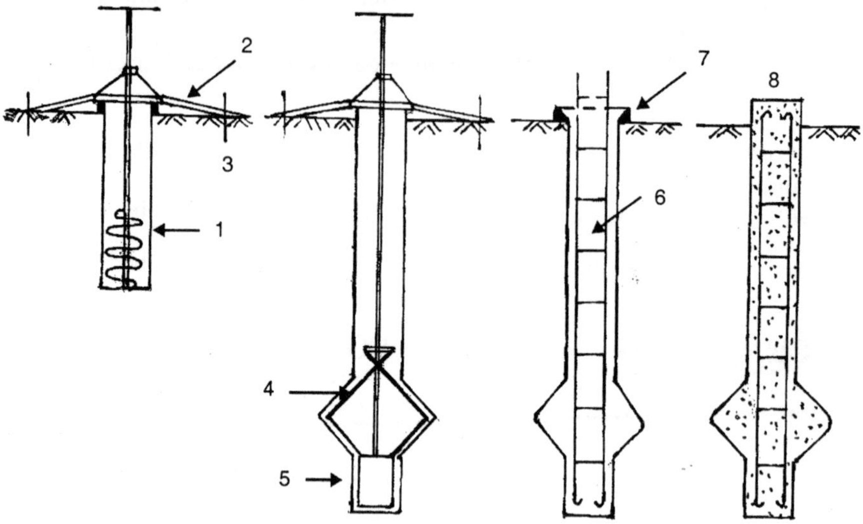

Fig. 4.5 Stages in construction of underreamed piles: 1. Spiral auger for making the hole of required diameter, 2. Boring guide for aligning the centre of pile, 3. Spikes for anchoring boring guide, 4. Underreaming tool, 5. Bucket to receive soil removed during underreaming, 6. Pile reinforcement, 7. Steel funnel laid above hole for pouring concrete in the hole, 8. Completed underreamed pile.

SUMMARY

Construction of a foundation over the excavated virgin ground starts with a sand layer and a lean concrete over the sand layer. The thickness of the lean concrete depends on whether we are using an RC foundation or a mass concrete foundation. In the latter case, the lean concrete will be thicker. In places where the brickwork can be submerged, only very good bricks or rubblework should be used in foundation.

For framed buildings with columns and beams, the columns rest on footings and the brickwork in between the columns can start from a basement beam or grade beams built between the columns at shallow depth from the ground level. The walls need not start from the deeper foundation level of the footings.

REVIEW QUESTIONS

1. Describe with sketches how you will start constructing the foundation for a load-bearing brick wall construction:
   (a) On mass concrete base
   (b) On an RCC slab foundation
2. Explain with the help of sketches the following types of foundation from ground level to foundation level:
   (a) Stepped brick footings and load-bearing wall
   (b) Reinforced concrete strip footing and load-bearing wall

(c) RC column footing for a framed building
(d) Raft foundation with columns
(e) Pile foundation on columns

3. What are plinth beams and grade beams? What are the advantages of using RC plinth beam in load-bearing wall construction? Explain the use of grade beams.
4. Sketch the details of foundations of the columns and walls of a framed building for a residential flat three storey high.

Chapter 5

Brick Masonry

5.1 INTRODUCTION

In this chapter, we will examine construction of walls with clay bricks. Clay bricks are still available in many places in India only in the traditional type 9″ × 4½″ × 3″ (230 mm × 14 mm × 76 mm) nominal in size (with a 3/8″ or 10 mm maximum mortar thickness). The metric bricks are 200 mm × 100 mm × 100 mm nominal in size. As brickwork is an important part of building construction, we will consider it in greater detail than other topics.

5.2 BRICK LAYING TOOLS (LAYING, SHAPING AND CUTTING TOOLS)

Brick trowel. Brick trowel is usually about 30 cm long with a steel blade and wooden handle [see Fig. 5.1(a)]. The tang of the blade sometimes passes completely through the handle and the end is completed with a washer, the end of the tang riveting over. Although this is a strong method of connecting the blade and the handle, it may sometimes damage soft bricks when the handle is used for tapping the bricks into place. The blade has one straight edge usually on the left when looking towards the point and one curved edge. The curved edge is the cutting side of the trowel. The common or soft bricks are usually cut by this tool.

Bolster. Bolster is a chisel having a blade slightly wider than an ordinary building brick [see Fig. 5.1(b)]. This makes it easy to cut the bricks without too many blows. It may also be used for cutting chases or channels in bricks or walls to receive pipes or electrical conduit.

Brick hammer. The cutting of bricks to provide suitable shapes for the bonding is done with the club hammer and bolster [see Fig. 5.1(c)]. A club hammer may vary from 1 to 1.5 kg in weight.

Spirit level. This is the most important tool since with its help the brickwork can be *levelled.* It consists of a hardwood stock with anything from two to six bubbles formed in it [see Fig. 5.1(d)]. These bubbles (as they are termed) are small sealed glass tubes, slightly curved and containing a quantity of alcohol insufficient to fill the tube. This leaves a bubble of air. When the tube is held horizontally, the bubble rises to the centre of the curved portion. There are usually two red lines on either side of the centre of the tube, and when the tube is correctly set in the wooden stock, the bubble will come to rest between these lines when the level is perfectly upright, or perfectly horizontal depending upon which way round the level bubble has been placed. White lead paste is a good medium for setting the bubble tubes in the stock, and if care is taken, it is possible to set these bubbles at complete level. It is important that it should be correct, otherwise all brickwork done with it will be out of true level.

Plumbing is carried out to ensure that the brickwork is completely *vertical.* Even though this verticality may be obtained with the spirit level, it is more accurately carried out by using the plumb line method, though it is a little more cumbersome and difficult to use.

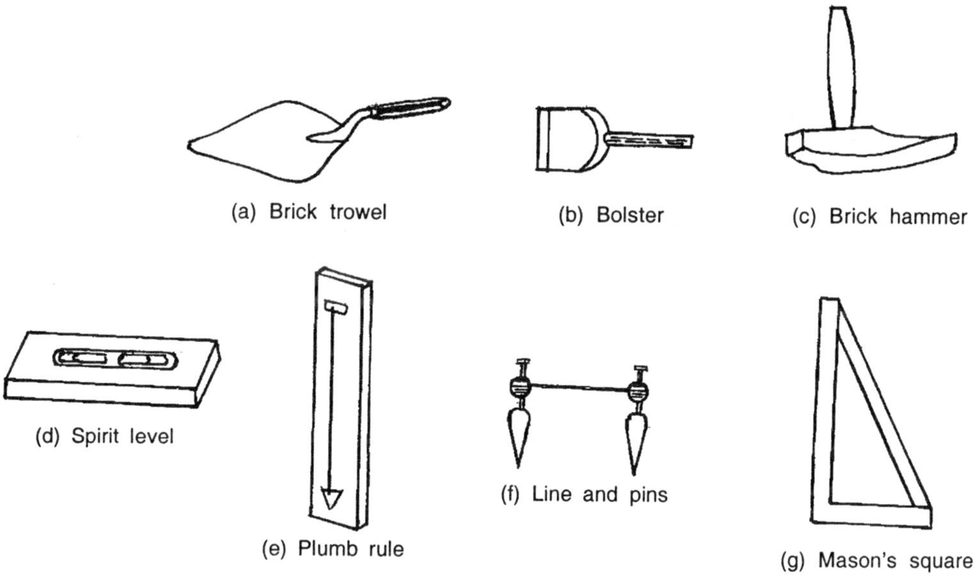

(a) Brick trowel (b) Bolster (c) Brick hammer (d) Spirit level (e) Plumb rule (f) Line and pins (g) Mason's square

Fig. 5.1 Commonly used bricklayer's tools.

Plumb rule. Plumb rule consists basically of a piece of milled, 25 mm thick straight grained wood ruler about 1.35 m long and 100 mm broad, with a piece of string which is fixed in the wood at one end in a saw cut and with a metal plumb bob at the other end of the string [see Fig. 5.1(e)]. The wooden ruler has perfectly straight and parallel edges and has a line marked exactly down the centre. When the plumb bob is suspended and is allowed to hang freely, it is perfectly vertical. Thus, when bringing the marked centre line on the ruler in line with the line of the string on the plumb bob, both edges of the ruler are also vertical. It is by means of these plumb rules that the verticality of brickwork is tested.

Line and pins. The two pins used for brickwork have flattened and blunt point ends for inserting into the joints of brickwork as work proceeds and to take the strong bricklayer's line which is coiled onto it [see Fig. 5.1(f)]. Both pins should contain part of the string line so that alterations in the length or position of the line can be made at either end.

Mason's square. This is sometimes referred to as a *building square.* It is used to set out a right angled corner and must be carefully employed when setting out the first course of bricks as all subsequent courses are plumbed from this course. A square is rather like a large triangle having two legs of approximately 4 ft length and a brace [see Fig. 5.1(g)].

End frames. These are less commonly used and are L-shaped frames made of hardwood, preferably teak. An end frame has a small base 32.5 cm × 22.5 cm and an upright 125 cm in height. It has a number of saw cuts as the upright to take the bricklayer's line. As the name implies, it is used at the two ends or corners of *brick walls to hold the line instead of pins* and is very effective. It can be changed quickly from one course to another. The end frames can be fixed to the wall at different heights by MS flats. It is usual to build the corners or ends of walls first, making sure that each course of bricks starts at the correct height. Brickwork at a corner must be perfectly plumb in both directions. Thus, stretching the bricklayer's line from the end frames between ends or corners, course by course, ensures straightness of the wall between points and also limits the amount of plumbing necessary in the main body of the wall. Sometimes on an extra length of the wall, the line tends to sag. To avoid this temporary small-sized column of bricks is roughly built up in one or more convenient positions to enable the line to be held at the correct height and prevent the sagging.

5.3 TERMS GENERALLY USED IN BRICKWORK

The terms generally used in brickwork with mortar are as follows (height in 3 in):

Header. A brick laid with its 4½ in × 3 in (with end of brick) parallel to the face of the wall.

Stretcher. A brick laid with its 9 in × 3 in (with side of brick) parallel to the face of the wall.

Bat. Any portion of a brick cut or broken *across* its length. For example, *half bat* will be 4½ in × 4½ in × 3 in size. *Three quarters* are bricks in which one-fourth of the length is cut off.

Closer. The portion of a brick cut along the lengths in such a way as one long face remains intact. When it is cut into two equal halves, it is called a *queen closer.* A brick cut at the corner along the mid-points of the adjacent sides is a *king closer*, which is at times used in junctions of walls.

Bed. The bottom surface of a brick which rests upon the mortar spread to receive it.

Frog. The indentation on one or both of the 4½ × 9 in surfaces of the brick.

Arrises. The edges of the brick where its surfaces intersect.

Course. A complete layer of bricks laid on the same bed.

Perpends. These are the short vertical joints in the face of the wall that fall vertically over one another in the alternate courses. Instead of perpends, a practical term frequently used is *cross joints*.

Quoins. The stones used at the corners are quoins. The word *quoin* is also used to mean the corner (see Fig. 5.2e).

Junctions. The meeting place of a longitudinal wall and a cross wall is called a *junction*.

Plinth course. The horizontal course of stone or brick provided at the base of a wall at floor level above the ground level is called *plinth course*. The top level of this course is called *plinth level*. It is practically the bottom floor level in ordinary buildings.

Sill. It is the portion immediately below the window opening. (Refer to Fig. 1.2 for this and the following items.)

Lintel. It is the horizontal member placed above the doors, windows, etc. to carry the load above the opening.

Jambs. These are the vertical sides of an opening for doors and windows.

Reveals. The exposed vertical faces on the sides of openings of doors and window frames after they have fitted in position.

Copings. These are stones, concrete, brick, etc. placed on top of parapet walls to prevent seepage of water into the walls.

Corbels. A projecting part in stone or brickwork. An ornamental corbel placed below the roof and projecting out is called a *cornice*.

Throating. The grooves provided at end of corbels, lintels, etc. for discharging rainwater clear of walls is called *throating*.

Freeze. Coarse of stones placed immediately below cornice to improve appearance is called *freeze*.

(The size of modular bricks is $190 \times 90 \times 90$ mm^3 and that of old type of bricks (non-moduler bricks) is $229 \times 110 \times 70$ mm^3)

5.4 MORTARS TO BE USED

We have studied mortars in detail in the subject Building Materials. Table 5.1 gives the recommended mortars for different brick strengths. As a rule, the strength of a mortar need not be more than that of the bricks and it can be made of coarser sand. Plasters should be more plastic and made of finer sand. Sand for mortar should pass through a standard sieve at 8 meshes to an inch, i.e. 64 meshes to a square inch and sand for plaster should pass 12 meshes to an inch, even though Mortar Development Support System (MDSS) has specified that it should pass through 18 meshes to an inch.

Table 5.1 Recommended Mortars for Brickwork

Brick strength (N/mm²)	*Mortar mix* *Cement or cement lime*	*Mortar strength* (N/mm²)
Below 5	1 : 6 or 1 : 2 : 9	3
5 – 15	1 : 5 or 1 : 1 : 6	5
15 – 25	1 : 4 or 1 : ½ : 4½	7.5
> 25	1 : 3 or 1 : ¼ : 3	10

Notes (i) Cement lime (combination mortar) is more plastic than cement mortar.
(ii) Some authorities recommend mortars of very low strength for low-strength bricks as follows (for low cost housings):

Cement and sand mortar:	1 : 8
Lime and sand mortar:	1 : 3
Cement, lime and sand mortar:	1 : 4 : 14
Lime, surki and sand mortar:	1 : 2 : 6
Cement, lime, surki and sand:	1 : 2 : 4 : 20

However, we should always use a workable mix as mortar for brickwork. Lime gives workability to mortar.

5.5 BONDING OF BRICKS

The art of *bonding brickwork* consists of the orderly arrangement of the bricks in such a way that continuous or *through joints at right angles* to the face of the wall are eliminated and longitudinal through joints along the wall are also reduced to a minimum. Bonding helps in the distribution of loads as shown in Fig. 5.2. Bonding is carried by use of closures (in the header courses) or three quarters in the stretcher courses as shown in Fig. 5.2.

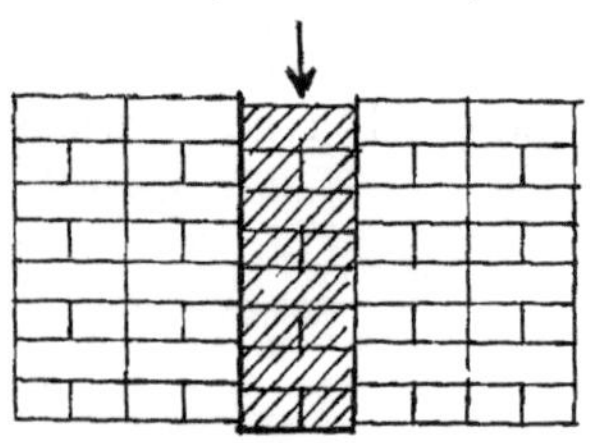

(a) Lack of distribution of load with no bonding

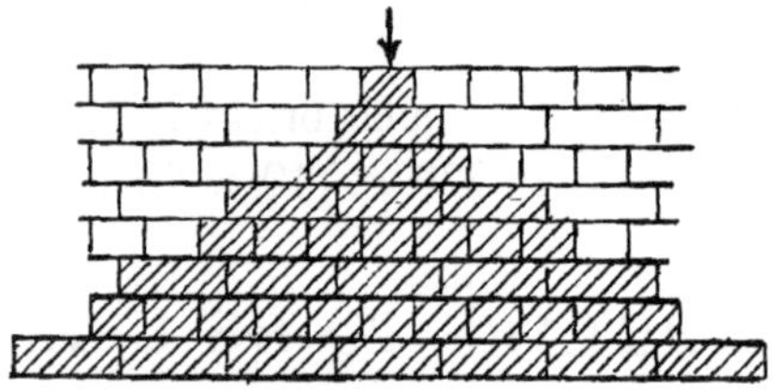

(b) Good distribution of load with bonding

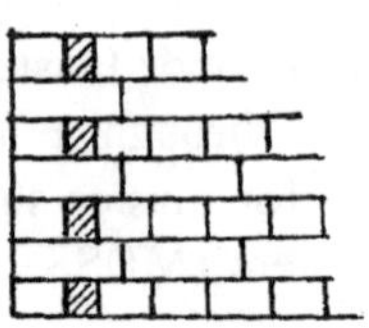

(c) Bonding by using a queenclosure (bricks cut lengthwise into two halves)

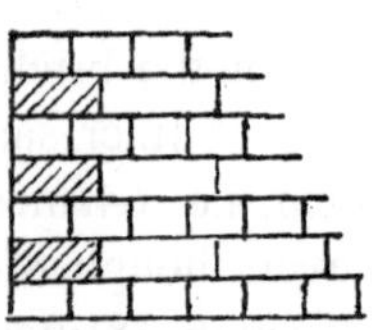

(d) Bonding by using a three quarter bat (bricks cut across its width at three-fourths of the length)

Fig. 5.2 Bonding of brickwork.

As bricks of different sizes are used in practice in various places, the width of brickwork is measured by bricks and not by actual measurement. Thus using 9″ × 4½″ × 3″ sizes as headers a half brick wall, usually taken as 4½ inches in thickness will be obtained. The thickness of one brickwall is taken as 9 inches. The thickness of mortar joints is usually taken as not more than 6 mm (1/4 inch) for very good bricks and 10 mm (3/8 inch) for ordinary bricks.

Even though the object of bond is primarily to give strength to masonry, *it also creates artistic effects when the brickwork is left unplastered.* The bond mostly used in practice is the English bond described further. Other bonds may also be specified. Usually one type of bond is carried throughout the wall. At junction of walls also, the arrangement of bricks in alternate courses shall be such as to *ensure through bonding*. Bonding avoids through joints, which may create lines of weakness and result in fractures in the wall. Bonding is of special importance when building walls in lime mortar, which has very little tensile strength. *However, with cement mortar, which in many cases will be stronger than the bricks themselves, the structural value of the bonding is of comparatively lesser importance.*

5.5.1 Types of Bonding

Unless some special purpose is to be served, brickwork should be bonded by one of the traditional bonds. A brief description of the principal types is given below:

English bond. It is the most commonly used bond for all wall thicknesses. On elevation, this consists of *alternate courses* of all headers (length of brick) and all stretchers (sides of bricks)(see Fig. 5.3). It is the most straightforward of all bonds to lay and *give greater strength than any other* as it results in fewer through joints and entails the use of a

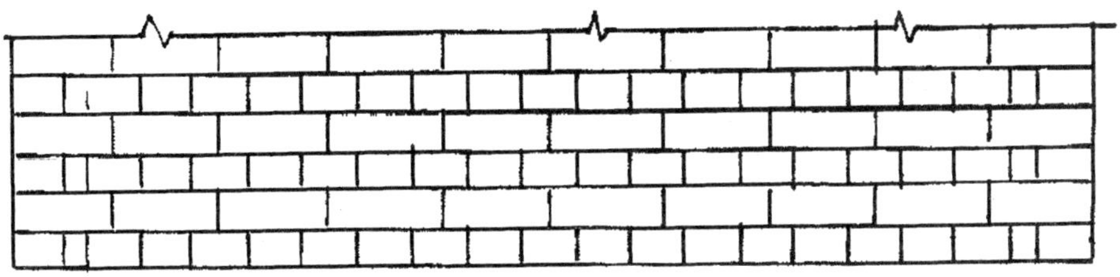

Fig. 5.3 Brickwork in English bond (elevation) – alternate header and stretcher courses.

minimum number of brick bats. The disadvantages are that it is sometimes uninteresting and it requires more facing bricks than other bonds. When it is used for a one brick thick wall we can make only one side fair face because of the inherent variation of sizes of bricks.

Flemish or double Flemish bond. On elevation, it consists of alternate headers and stretchers in every course (see Fig. 5.4). It is perhaps not quite as strong as English bond. However, this bond is preferred where special bricks are used for facing work on the grounds of *greater economy and more interesting appearance*. It is economical because it requires fewer facing bricks. In an unplastered wall interesting elevation can be created by only

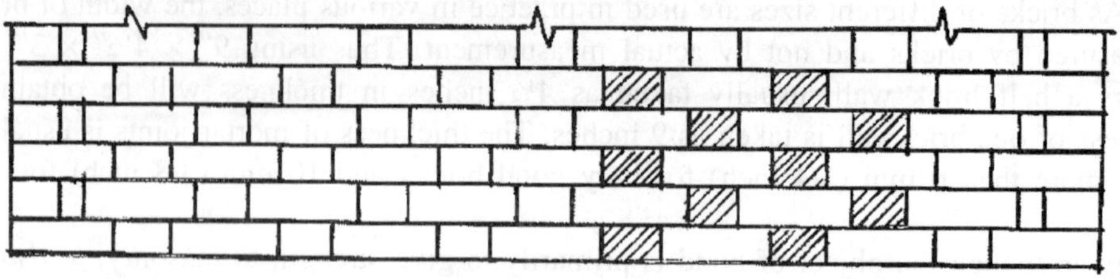

Fig. 5.4 Brickwork Flemish bond (elevation) – alternate stretchers and headers in the same course.

plastering the headers as shown in Fig. 5.5. This type of exposed brickwork is very popular in Kerala. Double Flemish bond has same appearance from inside and outside.

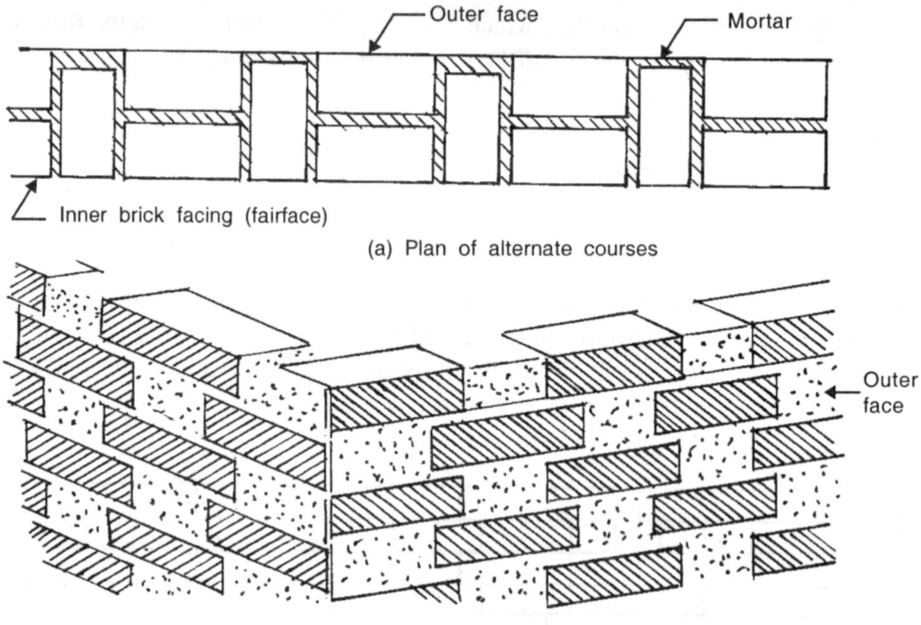

(a) Plan of alternate courses

(b) View of brickwork

Fig. 5.5 Flemish bond with unplastered brickwork 'fairface' on inside with a pleasing pattern of bricks and mortar on the outside.

Single Flemish bond. Single Flemish bond is sometimes applied to a combination of two bonds using Flemish on one face and English for the backing, when building walls are thicker than one brick. It is supposed to combine the merits of both. However, it is rarely used in practice.

Rat-trap bond. This bond is shown in Fig. 5.6. It is said to have some of the advantages of cavity wall described in Section 5.9. It uses brick on edge courses. Hollow pockets or voids inside the wall reduce the weight of the wall. As we use brick on edge, there is also some saving in the number of bricks to be used.

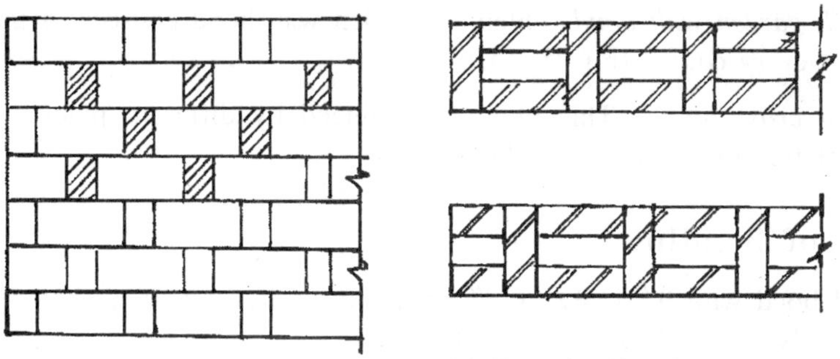

(a) Elevation (b) Plan of odd and even courses

Fig. 5.6 Rat-trap bond with bricks built on edge.

Stretcher bond. A bond consisting solely of stretchers, except that headers or 3/4 bricks are used or inserted where necessary for breaking bond or to make up lengths between openings, etc. It is used for half-brick (4½ inch) partitions and for the half-brick leaves of cavity walls.

5.5.2 Special Bonds

There are many other special bondings which are discussed briefly below. They are used for ornamentation.

Garden wall bonds. This name is applied to the types of bond which at one time was frequently used for building one full brick thick garden walls. Some of them are as follows:

(a) *English garden wall bond.* English Garden wall bond consists of *three to five courses* of stretchers to one course of headers [Fig. 5.7(a)].

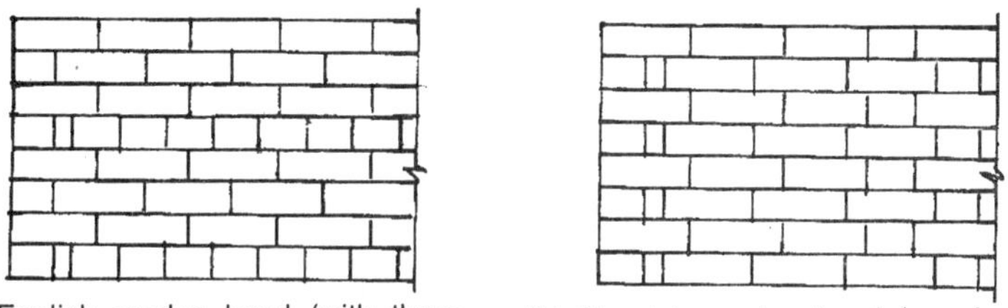

(a) English garden bond (with three-stretcher courses to one-header course)

(b) Flemish garden bond (one header to three stretchers in all the courses

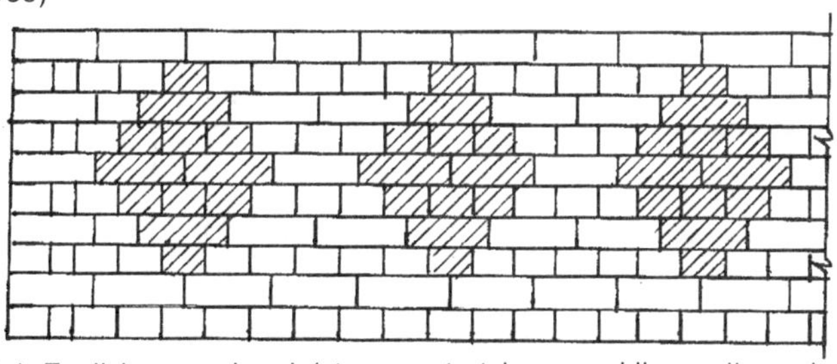

(c) English cross bond (stagger stretchers enabling patterns to be formed with coloured bricks or by painting bricks)

Fig. 5.7 Special bonds in brickwork.

(b) *Flemish garden wall bond.* Flemish garden wall bond consists of three stretchers to one header *in every course* [Fig. 5.7(b)].

(c) *English cross bond.* This bond is produced by different placing of bricks to produce different patterns as shown in Fig. 5.7(c).

5.5.3 Rules of Bonding

To ensure good bond with 9″ × 4½″ × 3″ bricks, the following rules should be observed. (These rules are also applied to metric bricks.)

(i) The bricks should be uniform in size, and the proportion of length to breadth be such that the length becomes twice the width plus one joint. Good bond is impossible otherwise, as the lap would not be uniform.

(ii) The minimum amount by which the bricks in one course overlap the bricks in the course below should be 2¼ inch along the length of the wall and 4½ inch across the thickness of the wall.

(iii) The vertical joints in the alternate courses should fall in a plumb (vertical) line from the top of the wall to its base, whether on the face or in the interior of the wall.

(iv) Bats should be used as little as possible and where used, should be evenly distributed throughout the whole of the work.

(v) The bricks in the interior thickness of the very thick walls should be laid with their length across the wall (i.e. headerwise).

The details of junctions and corners of the English bond are shown in Figures 5.8 and 5.9.

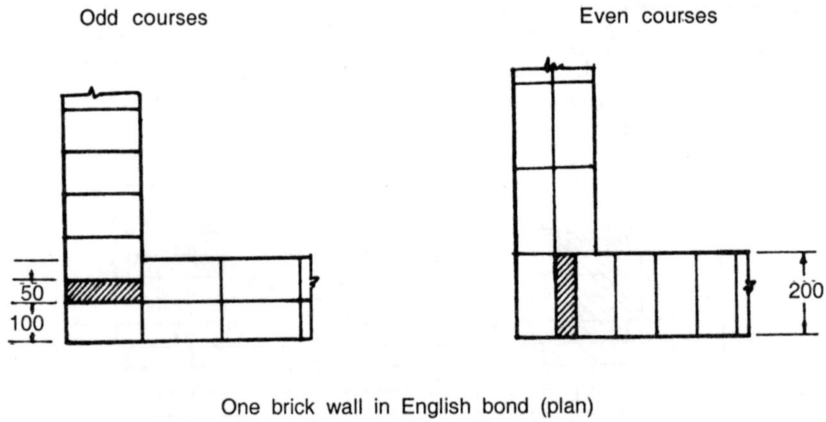

One brick wall in English bond (plan)
(Dimensions are for metric bricks).

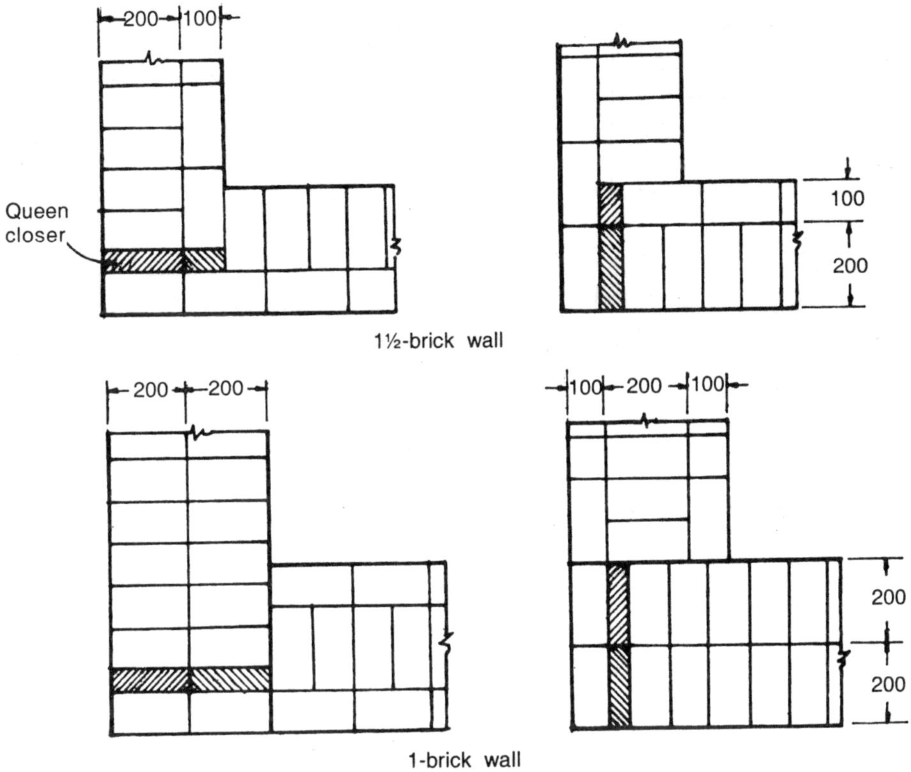

Fig. 5.8 Details of construction of corners of walls in English bond.

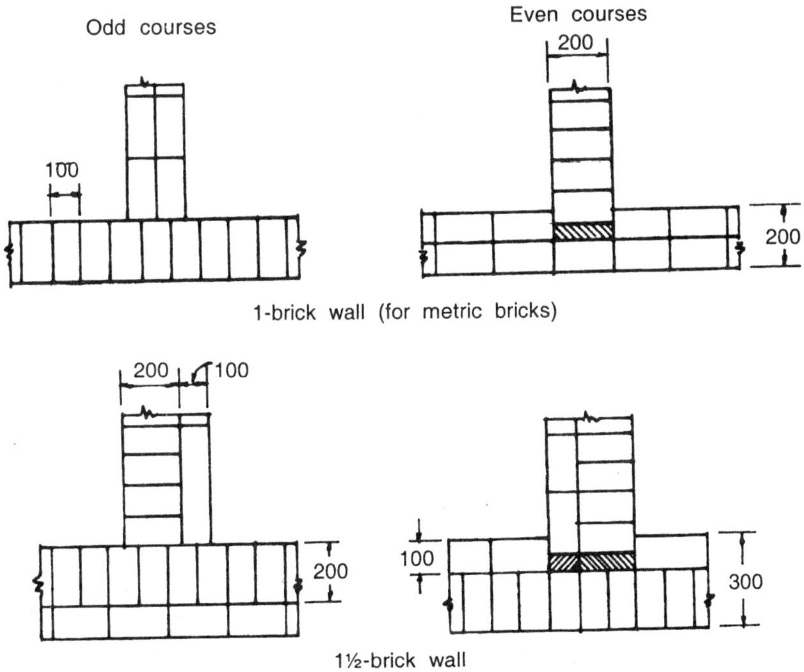

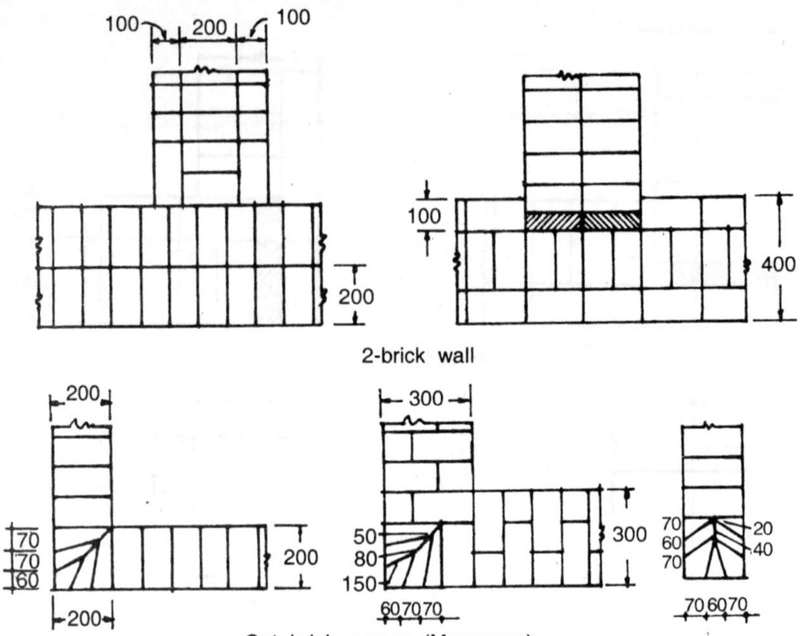

Fig. 5.9 Details of cross wall and corner construction in English bond.

5.6 METHOD OF LAYING OF BRICKS

Tests to find quality of bricks have been studied in the book on *Building Materials*. Good bricks should not disintegrate when placed in water. This is very important as when these bricks are used with cement mortar, we *should thoroughly soak them before laying*. Only when fat lime or clay mortar is used or when one is forced to use bricks that are not well burnt, this soaking rule has to be relaxed. The time of soaking can be found from a field test. In all cases, bricks should not be handled in baskets or in any other mode which will destroy the sharpness of their edges.

5.6.1 Brick Laying Operation

The operation of brick laying in ordinary general work is as follows. A layer of mortar is spread to cover the full width of the wall for a suitable length of the lower course. The end brick is then laid on it. Each brick is then properly bedded usually with frog up. (Bricks with 10 mm deep frog are to be laid with frogs up and bricks with 20 mm deep frog are usually laid with frog down to save the quantity of mortar.) For this purpose, the inside faces of each brick are buttered with mortar before the next brick is laid and pressed against it by tapping it with the handle of the trowel or wooden mallet. On completion of a course, the *vertical joints are fully filled with mortar from the top*. Finally, we press the side bulging mortar in firmly to be level with the face of the wall if it is to be left unplastered.

In the common practice adopted by many masons, a row of bricks is first placed on a thin layer of bed mortar leaving the cross joints empty. Then a heap of mortar is thrown over the top of the bricks and the cross joints are filled with mortar placed on the top with a trowel or straight edge while adding a copious amount of mortar. Strictly this is not a good practice. (However, when using cement mortar of high strength, the loss of strength due to violation of these rules may be only marginal.)

The walls are raised truly plumb. All courses are laid truly horizontal and all vertical joints truly vertical. Vertical joints in alternate courses should come directly one over the other. The thickness of the brick course is to be kept uniform. For this purpose, wooden straight edge with graduation giving thickness of each brick course including joint can be used for guidance.

For a thick wall, the above operation is repeated along both faces of the wall and the interior filling bricks for the thick wall are laid in a similar manner. The open joints inside are filled with mortar as each course is completed. This operation of filling open joints is termed *flushing-up*. It is wrongly omitted for several courses on some badly-executed jobs and only done afterwards (not after every course) in an inefficient manner. It is not a good practice and should be avoided as it is essential that every course should be flushed up to the level if good work is required.

5.6.2 Grouting

When brickwork is set with a fine joint, it is usual to fill the interior joints with a thin liquid mortar. This process is called *grouting*.

5.6.3 Larrying

In heavy engineering works or buildings, where the walls are very thick, the facing bricks are laid in the usual manner. For laying the inner bricks, mortar is shovelled into the interior of the course, spread out and water is added at the same time to thin out the mortar. The filling bricks are then squeezed into position. The mortar rising and filling the vertical joints completely and forming an exceedingly strong and solid wall. This is known as *larrying*.

5.6.4 Building Rat Trap Bond

For a rat trap bond work (as in construction of cavity walls), if mortar is placed carelessly on the brick, some of it will fall into the cavities and will be wasted. To avoid this, a piece of wood (about 1 metre long and 75 × 20 mm in section, as used for cavity walls) can be held in the hole over the middle of the wall to cover the cavities while applying the mortar.

5.6.5 Other Points to be Observed

(i) At the end of the day's work, the top of the brickwork shall be thoroughly cleaned of all mortar and the frogs, if laid up, are kept exposed to their full depth so as to provide proper keying for the next course. The face of the brickwork shall also be cleaned of all mortar droppings, etc.

(ii) When circumstances render it necessary to carry on a portion of a building in uneven courses, the work shall be built back (according to the bond used on the work) at an angle not steeper than 45 degrees so as to ensure a uniform and effective bonding. It should not be left toothed.

(iii) If facework is to be left unplastered, every joint should be neatly struck at the close of the day's work and before the mortar has completely set to give a good appearance. Otherwise, for faces to be plastered, finishing of the face joints should be carried out as discussed in sub-section 5.6.6. (This is very important.)

(iv) The walls should be uniformly raised all around not leaving any part one metre (three feet) lower than the other. A day's work should not be more than 1.5 m high.

(v) All iron fixtures such as hold fasts, pins, etc., to be built in the wall should be embedded in cement mortar or cement concrete in their correct positions preferably during the progress of the work itself or arrangements should be made to fix it without breaking the wall.

5.6.6 Preparation of Brick Surface Meant for Plastering

When the facework is to be later plastered or the joints alone are to be pointed, *the joints must be raked while the wall is being built.* It should be raked to a minimum depth of 12 mm by a raking tool during the progress of the work itself, when the mortar is still green. If this raking is not carried out during the erection of the wall, the plaster to be added later may detach or fall off due to lack of grip especially from smooth machine moulded bricks. If plastering or pointing is not envisaged, the joints must be struck flush and finished at the time of laying itself, as already stated.

5.7 INSPECTION OF BRICKWORK

All field engineers should know how to supervise construction of brickwork. When inspecting brickwork in construction, the following points should be closely checked:

(i) The sand should be clean, free from clayey matters or other impurities and suitable for mortar.

(ii) The proportion of the mortar ingredients (lime, cement, sand and water) should be such that it will ensure adequate strength for the type of work. It should, however, be remembered that the mortar need not be stronger or denser than the bricks or stone bonded in it and the two should have a similar texture.

(iii) The mortar should be thoroughly mixed, as indicated by uniformity of the mix throughout the mass.

(iv) If cement is used, its condition should be sound. On no account, cement mortar be used after the initial set has commenced. The mixed mortar should be used within an hour or two of mixing. Even then, the mortar should be kept agitated frequently with the addition of a little water, if necessary, so that the setting does not take place.

(v) The bricks should be sound, whole, laid in regular pattern or bond with minimum of bats. Bricks with the usual 10 mm frogs should be laid FROG UP, where strength is required in the wall. All vertical joints should be flushed up with mortar.

(vi) When using cement mortar, the bricks should be well wetted before laying to prevent absorption of moisture from the mortar.

(vii) All brickwork should be made according to the rules specified in Section 5.6.

5.8 PROTECTION AND CURING OF BRICKS

Brickwork should be protected from rain by suitable covering when the mortar is green. Ideally curing of brickwork, especially in cement mortar, is done (as in concrete work) by thoroughly saturating it with water after setting of mortar (preferably after covering it with straw, hessian, gunny bags, etc.) for fourteen days. For important works, a further curing by wetting once a day is to be continued for another seven days. However, in actual practice, the curing is carried out by keeping the work moist for a *minimum period of seven days only.* Brickwork in lime mortar also requires gradual drying out but does not require as careful curing as brickwork in cement mortar. (Lime mortar acquires strength by carbonation.)

5.9 SCAFFOLDING FOR BRICKWORK

Scaffolding for brickwork is described in Section 11.4. Ideally double scaffolding with two sets of verticals is to be used for exposed and important works. For ordinary works, single scaffolding with one end of horizontals resting on the wall being built is permitted. Where the scaffolding pole rests on the wall, only *one header brick* shall be left out so that the bond of the wall can easily be made complete and perfect after the scaffolding is removed. Such holes shall not be allowed in places such as in pillars or columns (less than one metre in width) carrying heavy loads.

5.10 CONSTRUCTION OF CAVITY WALLS

Construction of cavity walls is shown in Fig. 5.10. The use of cavity walls in India is not very common. They are common in cold countries. However, one should know how they are constructed, and why. The normal 275 mm (11″) cavity wall which is suitable for buildings not exceeding two storeys in height, consists of two half brick (4½ inch walls with 2 inch cavity). They are useful in two ways. Firstly, the cavity prevents the dampness from the outer leaf percolating into the inner leaf. Secondly, they provide good insulation from heat and sound. However, when used in tropical countries, sufficient precautions must be taken to see that they do not become breeding places for lizards and insects. The inner leaf is found to take greater portion of the imposed load transmitted by floor and roof. Hence, the two leaves of the wall are bonded together with ties (wall ties) usually placed 900 mm apart vertically and 450 mm horizontally in every 6th course staggered. This gives more than two ties per square metre.

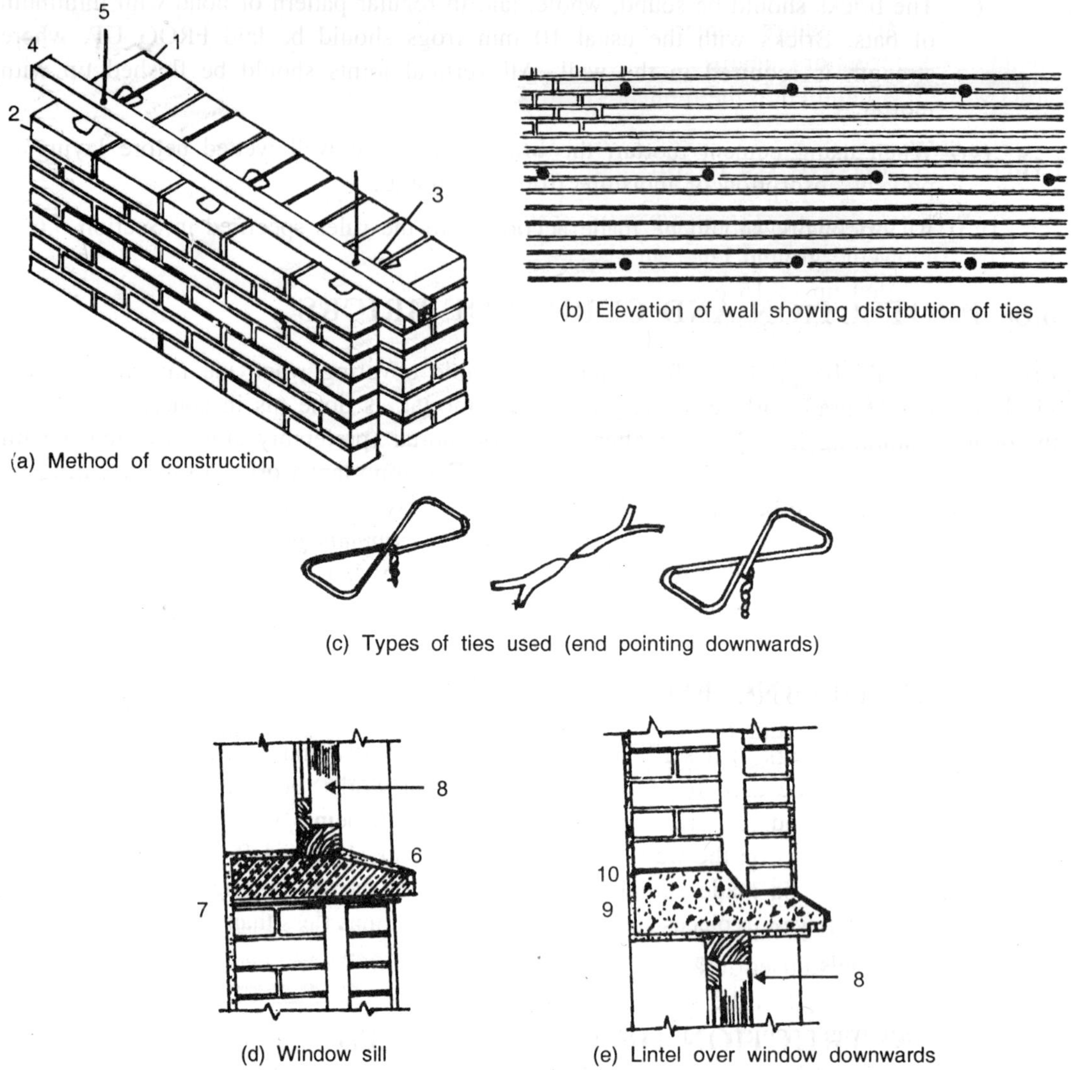

Fig. 5.10 Construction of cavity walls: 1. Inner wall, 2. Outer wall, 3. Wall ties, 4. Wooden batten, 5. Batten lifting device, 6. sill, 7. DPC under sill, 8. Window, 9. Lintel, 10. DPC over lintel.

These wall ties are made from mild-steel wires of 3 to 4 mm diameter or MS bars and fabricated to shapes as shown in Figure 5.10. They are dipped in hot tar and sanded or made from galvanized steel to prevent rusting. In very important works, copper may be used. Wire ties are placed with their twisted end down to allow water, that may seep inside, to drip down in the cavity.

The bond to be used for both the leaves, when thickness is half brick, is the *stretcher bond*. Where the inner leaves are made thicker for carrying heavy loads, English bond can be used for that part.

Cavity walls require considerable care in design and supervision, otherwise their efficiency as insulators against driving rain, heat and cold will be seriously impaired. Particular care is required at the top and bottom of the walls and around openings in the wall.

5.10.1 Special Points to be Observed in Cavity Wall Construction

(i) The cavity should extend to 15 cm below the damp-proof course level.

(ii) Below the ground level, the walls are built solid, or preferably the cavity should be filled up to 15 cm above the damp-proof course with fine concrete.

(iii) Under no circumstances, should the DPC be laid to span one leaf only. It should cover both leaves of the wall.

(iv) The upper part of the wall where it ends should also be built solid for two or three courses below the wall plate or roof line, to stiffen the head of the wall and distribute the load over both leaves.

(v) The wall ties must be kept free from mortar droppings by means of a timber batten suspended in the cavity and raised as the work proceeds during its construction. Some bricks may be temporarily left out at the ground floor level to form openings to permit the bottom of the cavity to be cleared of mortar droppings at the end of each day's work.

(vi) In exposed positions, it is desirable to leave a few vertical joints in the outer leaf open, at the bottom of the cavity to permit water to drain away.

(vii) A certain amount of ventilation to the cavity is desirable to prevent stagnation of air and excessive humidity. It can be provided by vents, say 150 × 75 mm (6″ × 3″) at intervals, near the base and top of the wall, by leaving a few joints open.

(viii) The cavity walls should not be built solid at the jambs [the sides of door and window openings (see Fig. 1.2)] unless a vertical "damp-proof course" is inserted to prevent water driving to the inner face. Thus solid jambs without DPC are only permissible in fairly sheltered sites and where the wall surface is roughcast.

(ix) A lead, galvanised iron or other suitable material made to form a trough or gutter, may be placed in the cavity above all openings for exposed doors and window to collect water which may drive through the outer leaf.

(x) The cavity wall should not be built solid below window sills also and a damp-proof course is desirable at this point also.

5.10.2 Masonry Diaphragm Walls

Very tall walls for single storey buildings such as warehouses or sport centres have to be stiffened with beams if the walls are thin. Otherwise, because of the high slenderness ratio, they can crack or fail. They are usually constructed with close pilasters (forming T sections) or using masonry diaphragm walls as shown in Fig. 5.11. They form box or I sections made of thinner walls. These walls give the required rigidity for the large height. These diaphragm walls can also be used as retaining walls and boundary walls. The voids may be filled with mass concrete or stones to give extra strength and stability. When used as boundary walls, shrubs and plants can also be grown in them for good appearance.

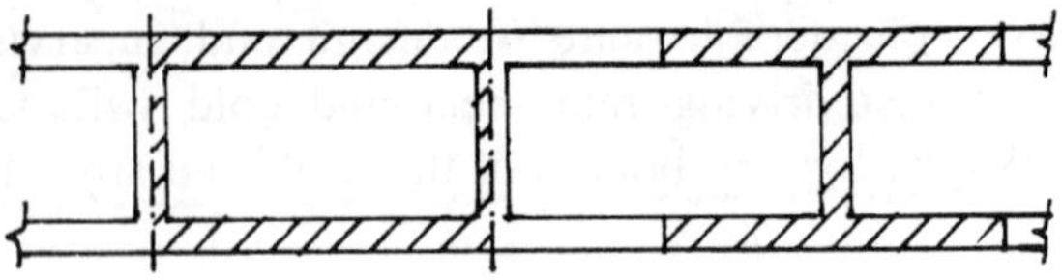

Fig. 5.11 Masonry diaphragm walls.

5.11 CONSTRUCTION OF HALF-BRICK MASONRY WALLS

Half-brick walls *tend to crack* unless care is taken in its construction. Brickwork in half-brick masonry is to be carried out with stretchers in 1 : 5 mortar. In special cases, if it is considered necessary to reinforce the same (as in long lengths of walls) with two 6 mm MS bars provided at every third or fourth course in 20 mm rich cement mortar 1 : 3. Half the mortar joint (10 mm) is first laid and the other half (10 mm) is laid after the reinforcement is placed so that it is fully embedded in the mortar. (Any contact of the reinforcement with the brick will hasten corrosion of the steel and damage of the wall.) These walls are generally plastered with cement plastering 1 : 5 with thickness of 12 mm. This plastering also adds to the strength. If these walls are planned to be stopped at door height (as usually done in tiled roof with ceilings), the walls should not be stopped exactly at top of the door level and abruptly left at that level. One layer of bricks should be laid over the door frame and throughout the length of the wall and then only the partition wall should be finished. A top layer of steel in cement mortar is also advisable.

5.12 HONEYCOMBED BRICKWORK

This is a simple brickwork with openings. This type of frickwork is usually provided in compound walls, partitions, etc. The thickness of honeycombed brick is half brick or one brick as specified. The openings must be equal and alternate in every course, the bearing width of each side being minimum 20 mm. The band used is stretchers throughout in half-brick thick work and heading throughout in one-brick thick work. The bricks at joints should be fully embedded in mortar and the jointed edges stuck flush and finished smooth as the work proceeds. The top course is a continuous brickwork. This is shown in Fig. 5.12.

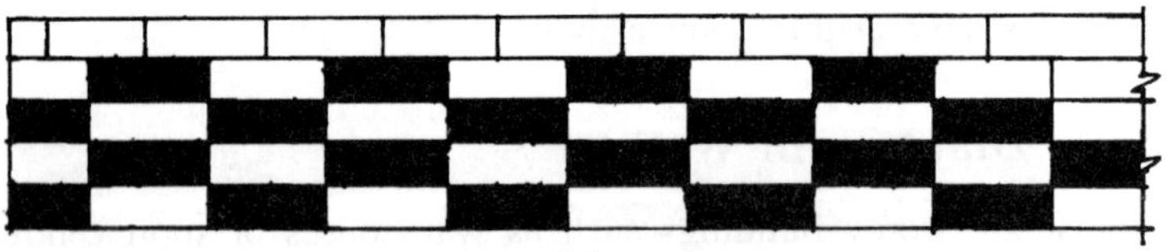

(a)

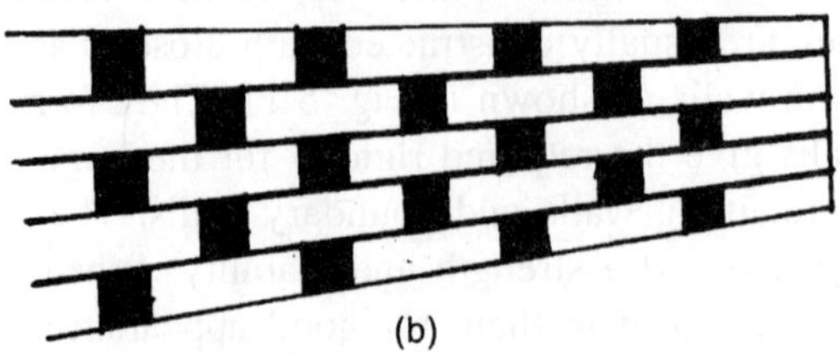

(b)

Fig. 5.12 Honeycombed brickworks.

5.13 BRICK ON EDGE COPINGS

(The top course of a masonry wall is called *coping*.) Ending the top of brickwork with brick on edge is considered to give good bearing strengths. Places such as the top course of all plinth, top of wall below the reinforced concrete parapets, steps, at window sills, etc. are usually constructed by brick on edge with extra fine vertical joints not exceeding 3 mm thickness. Bricks at the corners in such works are to be properly radiated and keyed in position.

5.14 TREATMENT OF WALLS AT THE END OF STEEL BEAMS AND TRUSSES

The ends of steel beams and roof trusses should be provided with a recess having 15 cm of side space for free circulation of air and preferably be provided with zinc sheet side protection. They are also usually supported on a concrete bed block to distribute the concentrated load. Steel or wood should not be encased in brickwork.

5.15 CORBELLING

Corbels are architectural features built into and projecting from walls. These are made by corbelling where *one-fourth of the brick projections* is built in every course for ordinary work and only *one-eighth of the brick projections* is made where greater strength is required.

5.16 BRICK JALIWORK

If only ventilation and light are needed for a room such as in a low-cost housing, we can provide a brick jali instead of expensive windows. Various designs are possible. Jaliwork is carried out by leaving open the vertical joints in definite patterns as shown in Fig. 5.13.

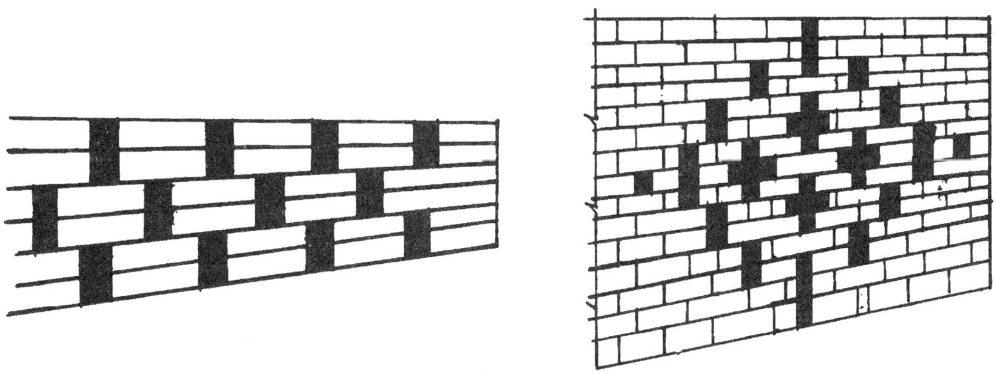

Fig. 5.13 Jaliwork in brick masonry for ventilation used in low-cost housing instead of windows.

5.17 BRICKWORK IN MUD MORTAR

Bricks when used with mud mortar require little or no wetting with water prior to use. The mud mortar should be prepared as described in the book on *Building Materials* (see also Section 13.6). The joint thickness is generally kept 10 mm. When using such brickwork, the top 30 cm of all walls of the building and 30 cm of walls around doors and windows and other openings are generally constructed in lime or cement mortar. With enough eves provided for protection of the wall from rains, they fare well. This is the type of construction generally used for temporary sheds like watchman's shed or cement godown shed at a building site. With brickwork in mud, the bricks can also be reused when the shed is pulled down after its use to be built at another site. Brickwork in mud is also used for temporary works such as centring of small arches (see Section 8.4).

5.18 HEIGHT OF WALLS IN BRICK MASONRY

The slenderness ratio of masonry walls is expressed as the ratio of effective length as well as effective height to thickness of masonry (see Appendix A). For brick masonry walls, the value of slenderness ratio should not exceed 20. In most cases, it should be kept below 14.

5.19 BRICK PIERS

Brick masonry piers are provided in two ways namely as isolated (as used for supports for beams in verandas or support for gates) and attached piers (or pilasters) where the piers are attached with walls as used to support roof trusses in a factory building. They can be built in English or Flemish bond. As the joints should break bond we use closures and bats as shown in Figures 5.14 and 5.15 (see bat and closer in Section 5.3).

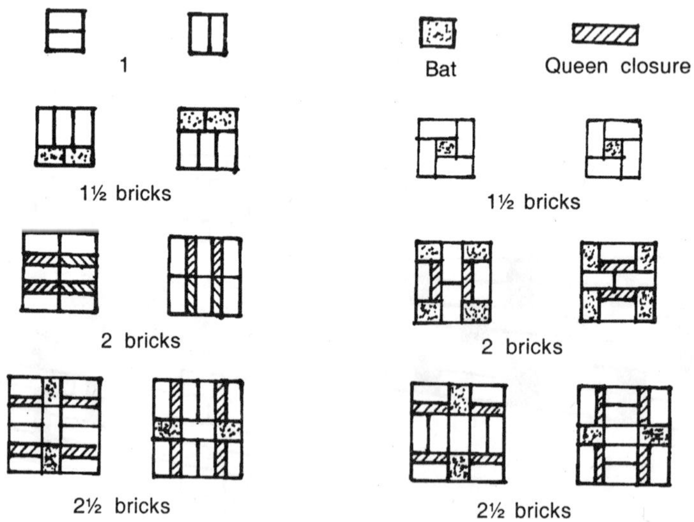

(a) English bond (b) Flemish bond

Fig. 5.14 Construction of brick piers shown in alternate courses.

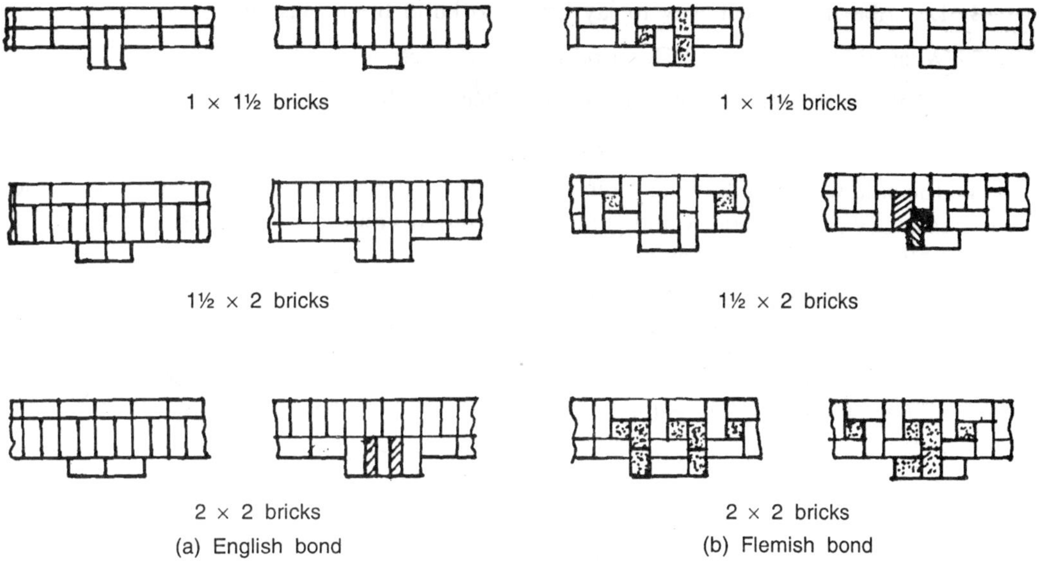

Fig. 5.15 Construction of pilasters (piers attached to walls) shown in alternate courses.

5.20 CLASSIFICATION OF BRICKS

Bricks are classified on the basis of their minimum compressive strength. Thus, a class designation 75 means, its minimum compressive strength is 75 kg/cm^2. These are generally used in building construction in India. Also, water absorption should not be mere than 20%.

SUMMARY

Brick masonry work is an important item in building construction. Construction supervisors should know very clearly the important points to look for during inspection of the work.

REVIEW QUESTIONS

1. Specify the cement mortar you will recommend for brickwork using bricks of strength 5 N/mm^2 for the following:
   (a) Brickwork in foundation
   (b) Brickwork in superstructure
   (c) Cavity walls
2. Sketch and describe briefly the following giving the merits of each:
   (a) English bond
   (b) Flemish bond
3. Describe rat-trap bond. What are its advantages?

4. Describe the correct way to lay bricks to build brick masonry.
5. What are cavity walls? Describe their construction.
6. What are masonry diaphragm walls? What are their uses?
7. Write a short note on brick jaliwork.
8. The side wall of a drawing room of a residence is 6 m in height. What types of brick masonry construction would you recommend? If you use a one-brick (235 mm thick) wall, what modifications will you make?
9. A compound wall 50 m in length is to be built along the length of a house property. What type of arrangements of wall would you recommend?
10. What is meant by "raking of joints" in brickwork? Describe in what situations, raking of joints is obligatory and state how it is to be carried out.
11. Give the methods and duration of curing brick masonry in (a) cement mortar and (b) lime mortar.

Chapter 6

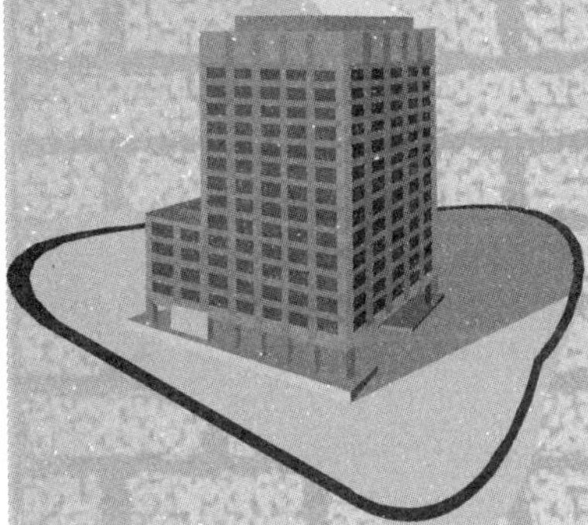

Block Masonry

6.1 GENERAL

With increasing cost of burning of clay bricks, blocks made out of soil cement, cement mortar, concrete, aerated concrete, precast stone concrete (large stone chips embedded in concrete) are becoming more and more popular. These blocks can be solid or hollow. However, the proper use of these materials, especially concrete hollow blocks, is still not understood by most users. Strict technical supervision is required in construction of buildings with blocks, as otherwise it will lead to heavy cracking of the walls. Concrete block as a material of construction is dealt with in the book on *Building Materials.* We will now consider its use in masonry. IS 2185 (Parts 1 and 2) gives the construction details and the standard sizes recommended for blocks to be used in India. The standard sizes prescribed for blocks are length of 400, 500 or 600 mm, height of 100 or 200 mm and width of 50, 75, 100, 150, 200, 250 or 300 mm. We use the larger widths for the main walls of the building and the smaller widths for compound walls and partition walls. The commonly made sizes are 400 mm × 200 mm × 200 mm and 400 mm × 200 mm × 100 mm for main and partition walls respectively. These are nominal dimensions including 10 mm mortar joints so that the actual sizes can be 10 mm less. A variation of ±5 mm is allowed as tolerance.

The main types of cement blocks available are the following. They may be solid or hollow. (Refer books on building materials for thickness of walls, etc.)

1. Concrete blocks
2. Soil cement blocks
3. Light-weight aerated (cellular) concrete block
4. Fly ash cement block

Notes:
1. In addition to these, we have also hollow blocks made out of ceramics which are generally not used for walls, but for roofs.
2. Blocks made of aerated concrete (which are usually made solid) are called *cellular blocks*.

We will restrict our study to the more commonly used concrete blocks in this chapter. Similar rules are applied far other blocks also. Concrete blocks are made of concrete not mortar. They come in many types, namely solid, hollow, or core type. They are called "*hollow*" when the percentage of void is more than 25 per cent (clay bricks in which holes are made for less than 25 per cent of its face area to increase the burning efficiency of the fuel are called *perforated bricks*). Hollow blocks are those with one or more formed holes or cavities which run through the block. Blocks with one end of these holes closed (so that it can be laid uppermost to provide a continuous surface to spread the mortar for the next course) are called *core blocks* or *closed cavity blocks*. Stabilized soil blocks and cement mortar blocks are generally used for low cost housing and non-load-bearing walls. They have very little strength, only of the order of 1.5 N/mm^2. Hollow concrete block construction without additional steel reinforcements performed so poorly and led to so many fatalities during the cyclones in Sydney (Australia) that its use without additional steel reinforcements is now prohibited in Australia in regions prone to cyclones. Only reinforced blockwork (for strength) with core filling (for stability) can withstand such high cyclonic winds. This will also be true for regions of high earthquake intensity. Similarly the cracking performance of soil cement block walls in the tropics with concrete roofs on top of the walls due to movement of top slab has been very poor. It is better to roof over these soil cement block walls, usually used for low-cost construction, with AC sheet, tiles, etc., which are isolated from the walls and produce no thermal movements on the walls.

Concrete hollow blocks as well as solid soil cement and aerated concrete blocks are frequently used in place of brickwork. Concrete hollow blocks have the advantages that they can be made strong by introducing steel reinforcement or core filling. Reinforced blockwork is quick in construction, and its performance under all conditions such as cyclones and earthquake loads has been good. Aerated blocks are light and perform very well as filler blocks for multistorey framed buildings. They can also be used to carry light loads.

To comply with 9 inch brick sizes, hollow blocks of normal size — 450 mm × 225 mm × 225 mm (18" × 9" × 9")— can also be produced. To build a masonry wall of one square metre, only 10 such blocks are required as against 115 clay bricks. In addition to the above work, it requires only 13 litres of cement mortar as against over 45 litres for brickwork. It can also be built by an average brick layer four times faster. However, experience shows that the selection of blockwork should be more for its architectural effect than for its lower cost. In actual practice, plastered blockwork tends to be as expensive as brickwork but can be faster in execution. However, the most popular size of block for buildings is the 400 mm × 200 mm × 200 mm (nominal size) blocks, which can replace 8 standard metric bricks 200 mm × 100 mm × 100 mm (nominal size). For partition walls, we use blocks of the same length and height as used for load-bearing walls but only 90 mm or 100 mm thick.

6.2 CONSIDERATIONS FOR USE OF HOLLOW CONCRETE BLOCKS

We have already seen that we choose hollow block construction more for architectural appearance and quickness of construction than for low cost. The second factor that needs consideration nowadays is the non-availability of good quality bricks and the large improvement that has taken place in the quality of concrete blocks which are being produced in India. However, we should remember that if blockwork is to be efficient for construction, the dimensions of rooms, walls, etc. should fit the block size modules so that there will be a minimum number of blocks cut and wasted. We should also note that full advantage blockwork will not be realized unless the complete range of block shapes necessary for the construction is available. In addition to regular blocks, blocks for bond beams for lintels, and blocks for pilasters (rectangular columns in walls) should be available. In construction of buildings for concealed electric wiring or for fixing of washbasins, etc., special arrangements should be made. Otherwise for easiness of construction brickwork must be prescribed in such places.

Masonry construction with solid and cellular blocks is very similar to brick masonry. Hollow concrete block walls are constructed in the *following three different ways*:

1. Simple masonry (similar tc brick) with the hollows untreated—the blocks are laid one over the other breaking bonds in strecher courses.
2. Infilled hollow block masonry with plain concrete infilling.
3. Reinforced hollow block masonry with steel reinforcement and concrete infilling for higher strength (used for lintels, retaining walls and walls for multistoreyed buildings).

The two important *structural features to remember* when using concrete blocks are as follows:

(i) There is the inherent weakness in these blocks for shrinkage and moisture movement. (Hence blocks should be well cured and absorption should be less than 100%.)

(ii) Blocks being larger in size and more brittle than bricks, the cracks due to settlements in blockwork will be larger in widths than in brickwork. Hence allowable differential settlements should be small.

Closed cavity blocks, in which the holes are closed at one end, ensure better load distribution, better insulation properties in cold places and minimum wastage of mortar in blockwork. With open cavities in hot humid regions, the air current set up in open holes can be considered as an advantage.

6.3 LAYING OF BLOCKS

Blocks are laid as stretchers breaking bond as shown in Fig. 6.1 (p. 64). As already stated, the most important thing is that concrete blocks should dry out thoroughly before use. It requires at least 28 days under normal conditions of normal curing. Solid and closed cavity blocks with one end closed are laid in walls with full mortar bed as in brickwork. If the walls are constructed of hollow blocks with through holes, economy in mortar can be made by "shell bedding" where the mortar is laid only on a 5 mm (2 in) strip around the outer edge of the hollow block. *Blocks should not be wetted before use* (as we specify for brickwork in cement mortar) as the concrete can take up water and then shrink. Excess mortar is gently removed to give a good appearance. The bottom layer of blocks on the foundation is always laid on a full bed of mortar.

The third important point to remember in hollow block construction is that the *mortar used for blockwork should not be too strong*. With strong mortars, when the wall shrinks, it tends to crack up with a few large cracks. With moderately strong mortars, the movement results only in a number of small cracks distributed on the whole length of the wall which will not be noticeable. If cement mortar is to be used, a mix 1 : 5 with plasticizer is found to be sufficient. 1 : 1 : 6 or 1 : 2 : 9 cement lime plaster is preferable to pure cement plaster. The ideal mortar should have the same density, strength and drying shrinkage as the block.

Pointing is usually done as in exposed brickwork. The longer the pointing tool, the smoother and better the joint results. The fresh joint is left for an hour to set, and then it is worked on for pointing.

6.4 EXCLUSION OF RAINS

The three general classifications of exposure of a wall to rain are referred to as *exposed but sheltered*, *moderate* and *severe*. Unrendered walls of hollow blocks of dense or light weight aggregate concrete not less than 203 mm thick, with only shell-bedding as mortar joints has been found to be resistant up to moderate exposure to rain. It is not fully resistant to "severe" rain conditions. Walls with even solid blocks unplastered were not resistant to severe rain penetration. Cavity brick walls commonly used in cold climates (Section 5.10) are found to perform better in this respect. Cement lime combination mortar bedding has higher resistance to rain penetration than mortar with cement mortar only.

In tropical countries, if solid blockwork is to be completely waterproof, two or more coats of external plastering with 1 : 6 cement or 1 : 2 : 9 cement lime plaster is generally recommended. For plastering, the joints of blockwork should be raked during its construction in the same way as we treat brickwork, to provide a proper key to the plaster.

Bureau of Indian Standards special publication SP-20, *Explanatory Handbook on Masonry Code*, deals with the design of all types of masonry including concrete block masonry. ACI Standard 531–58, *Building Code Requirement for Concrete Block Masonry Structures,* along with its commentary, also gives good guidance for design and construction of plain and reinforced concrete block masonry.

6.5 HOLLOW CONCRETE BLOCKS WITH CONCRETE INFILLING

Hollow concrete blocks (especially those with two through holes) form continuous vertical holes when *placed with proper bonding* one over the other. They are ideal for concrete filling, as these holes can be completely or partly filled with insitu concrete as the blockwork proceeds. Mixes with large size aggregates can be used for this purpose. This will add greater strength to the walls. Tests show that filling of hollow blocks with plain concrete will give basic strength approximately halfway between these for unfilled blocks and solid blocks with the same quality of concrete.

6.6 REINFORCED CONCRETE HOLLOW BLOCK MASONRY

The vertical holes in conventional block construction can accommodate steel reinforcement and can thus be made of reinforced concrete construction when infilled. Similarly *special blocks can* accommodate horizontal reinforcements so that horizontal beams such as lintels and horizontal strengthened reinforced concrete bands can be built up with these blocks (see bond beams described below). Such construction is known as *reinforced concrete hollow block masonry construction.* ACI standard 531–581 and its commentary give valuable information for design and construction of such works.

6.7 SPECIAL FEATURES OF CONCRETE BLOCK MASONRY

In addition to using fully-cured blocks and lean mortar joints, the fourth most important feature to be incorporated in blockwork, for buildings as different from brickwork, is the need to provide the three features, namely *control joints, band beams* and *joint reinforcements.* Unless these are provided, block masonry is bound to result in random cracking due to shrinkage and moisture movements. These are briefly described further.

Control joints. Concrete has higher coefficient of expansion than brickwork. Control joints, are continuous joints, usually vertical, built into concrete block masonry walls for controlling thermal movements. They should be located at change of sections in the form of vertical mortar joints to minimize cutting of units. They can be shrinkage joints and expansion joints. The shrinkage joints are continuous vertical joints approximately 18 mm (3/4 inch). These joints are later *raked and caulked* (filled in). Expansion joints are located at spacing of 45 to 60 m in suitable locations. These are filled with *joint fillers* which are compressible materials that close the gap but allow movement.

Bond beams. Bond beams are beams such as lintels that are to be provided on top of openings or top of walls without top loads such as compound walls. It is usually a reinforced masonry course, which is constructed out of specially channel-shaped units which are filled with concrete and reinforced [Fig. 6.1(b)]. It usually serves as a structural element and as a means of crack control on top of openings. For lintels, the reinforcement usually consists of at least two 12 mm bars placed at the top and bottom. For crack control, they are made discontinuous at control joints. Where continuity is required for structural reasons, a dummy joint should be given to control the location of the probable crack.

Joint reinforcement. Horizontal joint reinforcements in the form of minimum two No. 9 gauge wires at intervals can be used as in reinforced brick work for crack control. They are expensive and are not generally provided. However, when used as load-bearing walls with RCC slabs resting on the walls, the top three courses at least should be provided with joint reinforcement as shown in Fig. 6.1(a).

6.8 COMPOUND WALLS IN BLOCKWORK

Unplastered concrete blockwork which is cheaper than plastered brickwork is used nowadays extensively for constructing compound walls as shown in Fig. 6.1(c). There are many variations in the construction of these walls. One of the popular methods of construction is to use under-reamed piles at convenient intervals of 3 to 4 m with grade beam on top of the

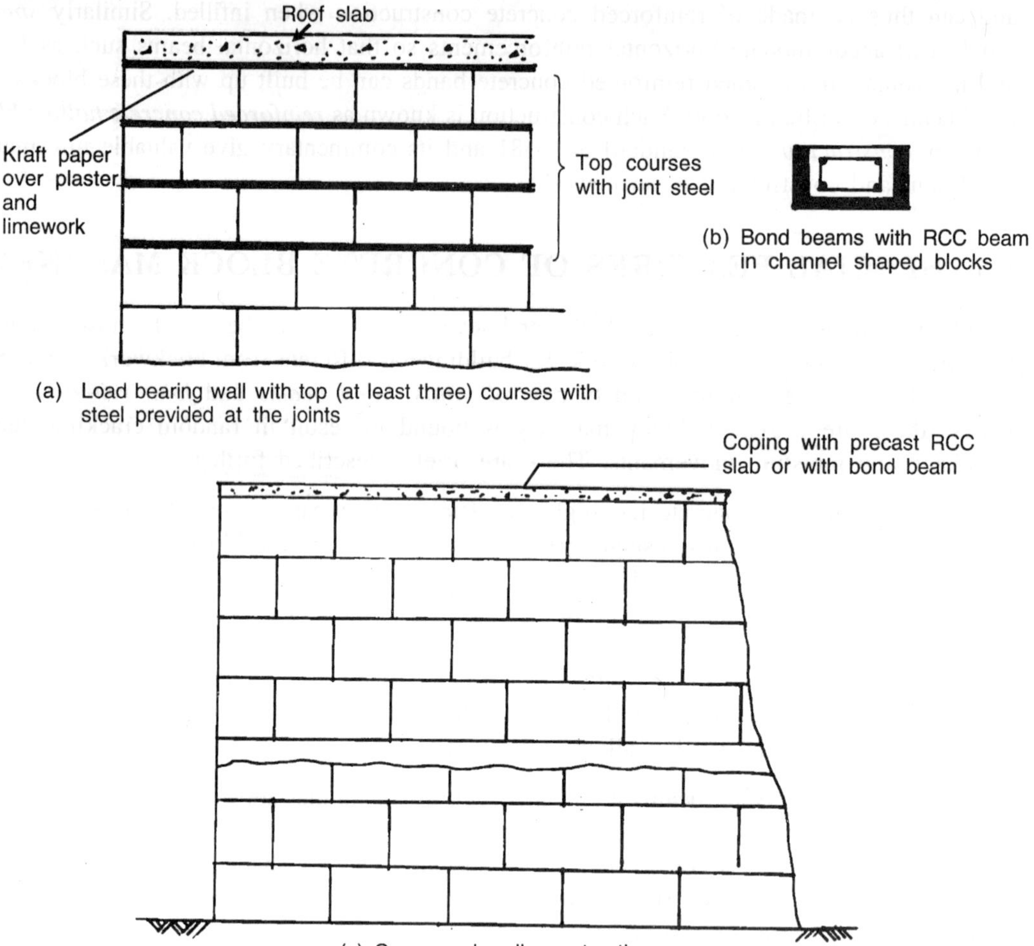

(a) Load bearing wall with top (at least three) courses with steel previded at the joints

(b) Bond beams with RCC beam in channel shaped blocks

(c) Compound wall construction

Fig. 6.1 Concrete block work.

piles and the blockwork built on top of the grade beams. Vertical reinforcements can be embedded in the hollow of the block at regular intervals from the under-reamed piles, thus strengthening the wall along its length giving lateral resistance to wing loads or earth pressure. The top of these walls should be finished with a coping as shown in top of Fig. 28.2. The coping as shown in Fig. 6.1 can be a slab or a bond beam (see Section 39.5.1 page 409).

SUMMARY

Concrete blocks are commonly used in place of clay bricks resulting in reduction in cement mortar and also increasing the speed of construction. As manufacturing clay bricks require good soil and high energy, it is also more environment friendly. However, there are many special points to be taken care of in good blockwork masonry as compared to brick masonry. They have been briefly described in this chapter.

REVIEW QUESTIONS

1. What are the different types of cement blocks used in block masonry? State where each of them can be used. How is concrete block superior to other blocks?
2. State the advantages and disadvantages of concrete block construction over clay brick masonry construction. What are the types and sizes of the popularly used concrete blocks?
3. What type of bonding is generally used in hollow block construction? What are the precautions to be taken in concrete block construction? State the special features to be included in block masonry construction.
4. What are the three types of construction possible with hollow concrete blocks? Why is it important that the mortar used in simple block construction should not be much stronger than the concrete in the block? Would you recommend block masonry where moderate foundation settlement can take place? Give reason for your answer.
5. Write short notes on:
   (a) Classification of exposure conditions of walls to rains
   (b) Control joints
   (c) Bond beams
   (d) Compound wall construction with hollow blocks
   (e) Shell bedding

Chapter 7

Stone Masonry

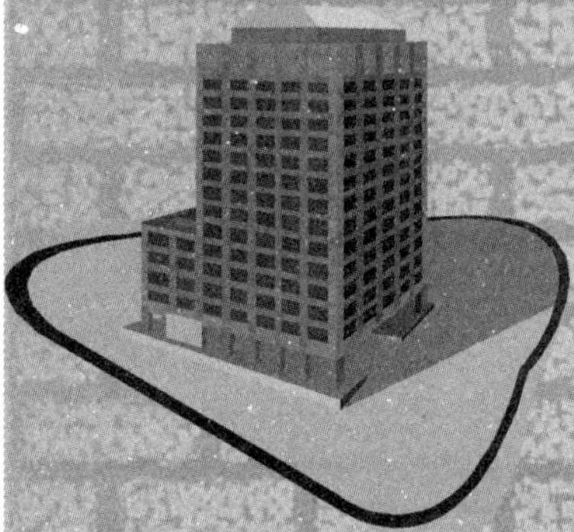

7.1 INTRODUCTION

Stone masonry has been in use in many places from ancient times. Where stones are easily available, random rubble work will be cheaper than plastered brickwork. Where the groundwater level is high and bricks are liable to deteriorate, random rubber work is preferred even today for foundation work.

The common types of stones used in India and their minimum crushing strengths are given in Table 7.1 below.

Table 7.1 Common types of stones used in India and their minimum crushing strengths

Type of stone	*Min. crushing strength* (N/mm^2)
Granite	100
Marble	50
Basalt	40
Sandstone	30
Limestone	20
Laterite	3

(*Note*: Strength of normally used cement concrete in construction is only 20 N/mm^2.)

The sizes of stones used in stone masonry work depend on the type of masonry. The stones should not be larger than what can be handled and placed by one person. For ashlar work, it is specified that the length should not exceed three times its height. The breadth should not be greater than three-fourths of the thickness of the wall or less than 15 cm. The height can be up to 30 cm. Generally, in random rubble work, larger stones are used on the faces and smaller ones at the hearting. We should be able to identify the various types of stonework as the *labour and payment* depend on the final appearance of the work. Each of them is specified for different situations. For example, whereas ordinary random rubble work can be specified in foundation, backside compound walls or garden walls, ashlar work will be more fitting in monumental buildings or even as main front walls of buildings and roadside compound walls. In this chapter, we will briefly study the construction of different types of stone masonry used in buildings.

7.2 CLASSIFICATION OF STONE MASONRY

There are many classifications of stone masonry. We will follow the one used in CPWD Specifications 77. According to this classification, stone masonry can be broadly divided into two groups:

1. Rubble masonry (three subdivisions)
2. Ashlar masonry also known as cut stone masonry (two main subdivisions and each divided into two divisions)

Rubble masonry is made of stone as obtained from the quarry with very little or no dressing. Ashlar is laid with carefully-dressed stones and will look like brickwork in stone. Though ashlar work, if properly constructed, is very strong and beautiful, it is very expensive. There are many divisions of rubble masonry, but the more commonly known ones are the following three according to the CPWD Specification 77:

(i) Random rubble masonry

(ii) Coursed rubble masonry of the first sort

(iii) Coursed rubble masonry of the second sort

Ashlar masonry is divided into the following types:

(i) Plain ashlar masonry (rough-tooled and fine-tooled)

(ii) Chamfered (also called punched) ashlar masonry (rough-tooled and fine-tooled)

Note: Chamfered means bevelled at edges.

7.2.1 Bond Stones in Rubble Work

One of the important features of rubble masonry work is the use of *bond stones*. These are long selected stones placed from front to back of thin walls or from outside to the interior of thick walls as shown in Fig. 7.1. They hold together the masonry walls transversely. If single stones of sufficient length are available, a pair of stones each penetrating three-fourths of the thickness of the wall with a minimum overlap of 150 mm, or a set of stones each overlapping 150 mm can be used as bond stones. If separate bond stones of ample lengths are not available or the available stones are porous to develop damp penetration, *concrete bond stones* made from 1: 3 : 6 concrete mix and of the specified length can be used as bond stones. In many specifications, the use of a certain number of bond stones is mandatory. The recommended specification is that it should not be spaced at more than 1200 mm horizontally and 600 mm vertically.

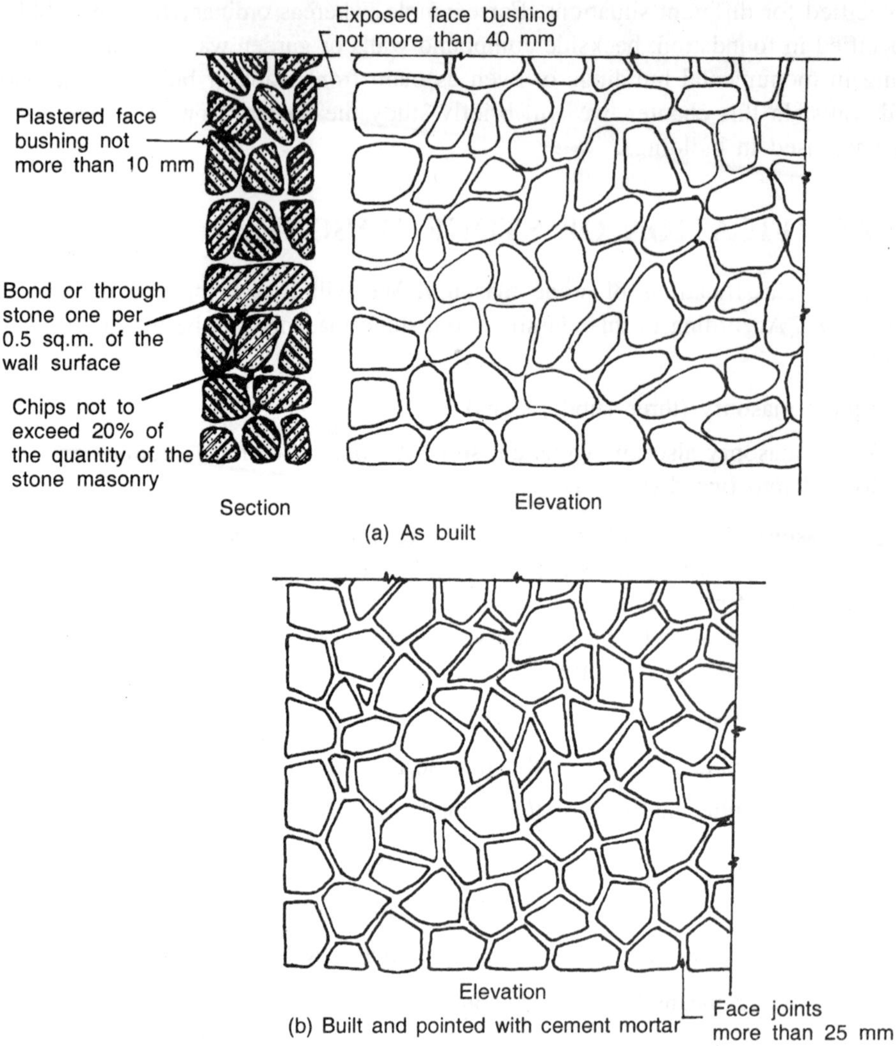

Fig. 7.1 Random rubble masonry.

7.3 RANDOM RUBBLE MASONRY

Random rubble masonry is the first variety of stone masonry. Stones are arranged at random. The minimum thickness of random rubble work that can be constructed with great care is 225 mm (9″) and with ease 300 mm (1 ft). The offsets to be provided in rubble masonry construction have to be 75 mm (3 inches) *on either side*. Work for footings has to be adjusted to these sizes. Random rubble is the roughest sort of stone masonry. The stones used are quarried in such a manner and of such sizes so that they can be lifted and placed by hand. It is only hammer dressed on the face, the side and the beds. The brushing in the face should not project more than 40 mm on an exposed face and 10 mm on a face to be plastered. Its appearance will be as shown in Fig. 7.1. The mortar for this masonry is 1 : 5 up to 1 : 8. The mortar for plaster should be 1 : 3 or 1 : 4. The stones are laid on their natural bed on a full even bed of mortar. Every stone is carefully fitted to the adjacent stone so as to form neat and close joints. Stones should be wetted before being placed.

Stonework should be brought to level at window sills and roof level *with concrete* made of one part of mortar not leaner than that used in the masonry and two parts of graded stone of 20 mm nominal size. (This practice should be adopted in compound walls also.)

Face stones, stones placed on the faces of wall, should extend and bond well into the backing. They should not be less than 125 mm in height and should be hammer dressed. These should break joints as much as possible. The hearting or interior filling of the wall should consist of rubble stones which may be of any shape but not less than 125 mm in size. These are laid carefully and hammered down with a wooden mallet into the position and solidly bedded in mortar. Chips and spalls of stones are used, if needed, to avoid thick mortar beds. The hearting should be nearly level with the facing. Where the masonry of one part has to be delayed, the work is sloped at an angle not more than 45 degrees. Toothing of stonework for joints is not recommended. Sufficient number of bond stones which are long enough to extend the full thickness of the wall as specified in Fig. 7.1 should be used in random rubble work.

Joints should be fully packed with mortar and chips. Face joints should not be more than 20 mm thick. The face can be left as built, pointed or plastered. If pointing or plastering is to be done, the joints should be raked to a minimum depth of 20 mm by a raking tool (as in brickwork), when the mortar is still green.

Single scaffolding with one set of vertical legs is allowed for construction of these walls of adequate thickness. Such holes are later filled with a proper sized stone or with cement concrete 1 : 3 : 6 with 20 mm aggregates. The walls are to be cured for a minimum period of seven days, the fresh work being protected from rain and sun. Their appearance (plain and pointed masonry) will be as shown in Fig. 7.1.

7.4 COURSED RUBBLE MASONRY OF THE FIRST SORT

This is the second type of rubble masonry and is built in courses (not random). In this type of work, the stones on both faces *are hammer dressed on all the beds and joints* so as to give them approximately rectangular block shape. This type of rubble masonry should be squared on all joints and beds, the bed joints shall *be rough chisel dressed for at least 80 mm back from the face.* Stones are set in regular courses. The height of the course should not be less than 150 mm. Work on interior face is to be the same as on the exterior face. Similarly the side joints are dressed for at least 40 mm so that no portion of the dressed surface is more than 6 mm from a straight edge placed on it. The hammer-dressed stones should have a rough tooling for a minimum width of 25 mm along the four exposed edges of the face of the stone. The bushing on the face should not project more than 40 mm on an exposed face and 10 mm on a face to be plastered. The percentage of stone chips used should not exceed 10% of the total stones used for the masonry.

In general, the work is carried out using the same mortar as in random rubble work. Its appearance when finished will be as shown in Fig. 7.2(a). *The face joints should not be more than 10 mm thick.*

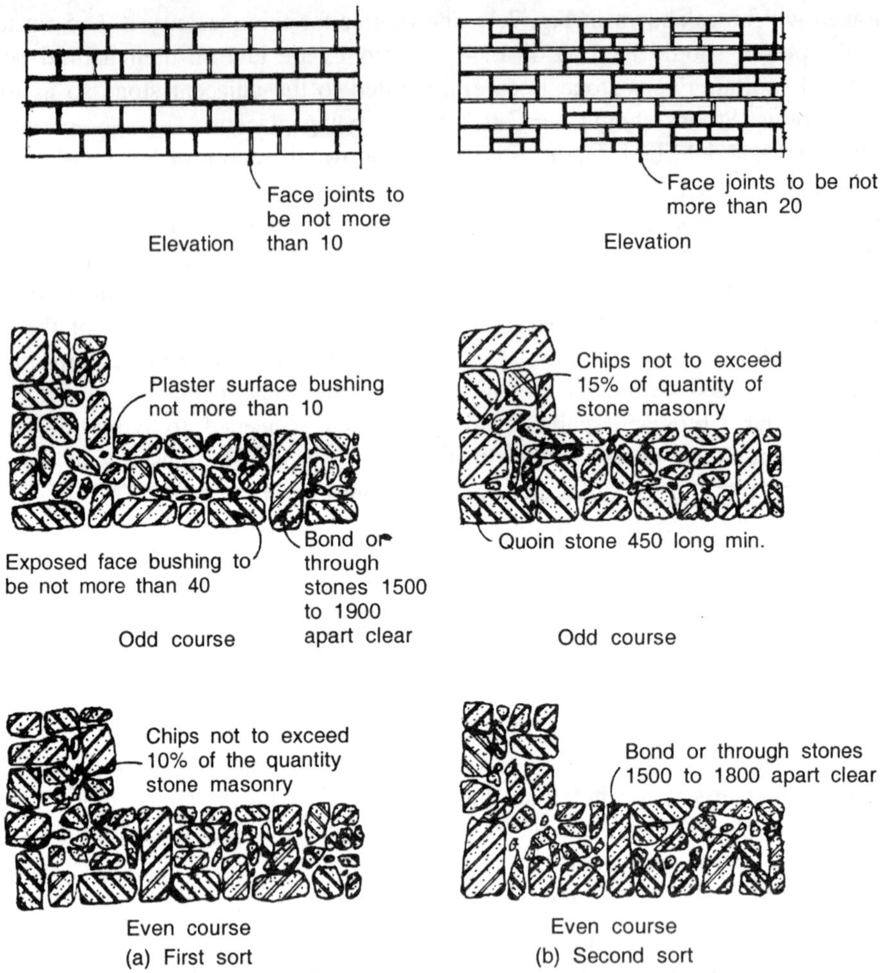

Fig. 7.2 Coursed rubble masonry.

7.5 COURSED RUBBLE MASONRY OF THE SECOND SORT

This is the third type of rubble masonry and is also built in courses. It is inferior to the coursed rubble masonry of the first sort but will look similar. No portion of the dressed surface should be more than 10 mm from a straight edge placed on it. *In this type of work, the face joints should not exceed 20 mm in thickness* (10 mm in the first sort). *Thus, the joints are allowed to be thicker than those in the first sort.* Its appearance is shown in Fig. 7.2(b). This type of work is more expensive than random rubble work but cheaper than course rubble of the first sort. The percentage of chips used is not to exceed 15% of quantity of stones in the masonry.

7.6 ASHLAR STONE MASONRY

We have seen that the difference between rubble and ashlar masonry is that in ashlar masonry, every stone must be cut to the required size and shape to give truly vertical and horizontal joints. No point on the faces should vary more than about 1 mm when tested with a 60 cm straight edge. Horizontal lines should not vary more than 3 mm and vertical lines more than 6 mm. This dressing makes ashlar masonry costlier than rubble masonry. *In short the completed work will look like "brickwork" in stone.* The two types of ashlar masonry are: plain and chamfered. They can be either rough tooled or fine tooled. Thus we have

1. Plain ashlar masonry (rough tooled and fine tooled)
2. Chamfered or punched ashlar masonry (rough and fine tooled)

Both are similar in all respects except that the chamfered ashlar will have joints beveled at a 45° a ngle to a depth of 25 mm rough tooled ashlar masonry will have a chisel dressing 2.5 mm wide all round the edges and shall be rough tooled between the drafts so that the dressed surface will not be more than 3 mm from a straight edge placed over it. Its appearance will be as shown in Fig. 7.3. In the fine tooled work the face will be fine tooled.

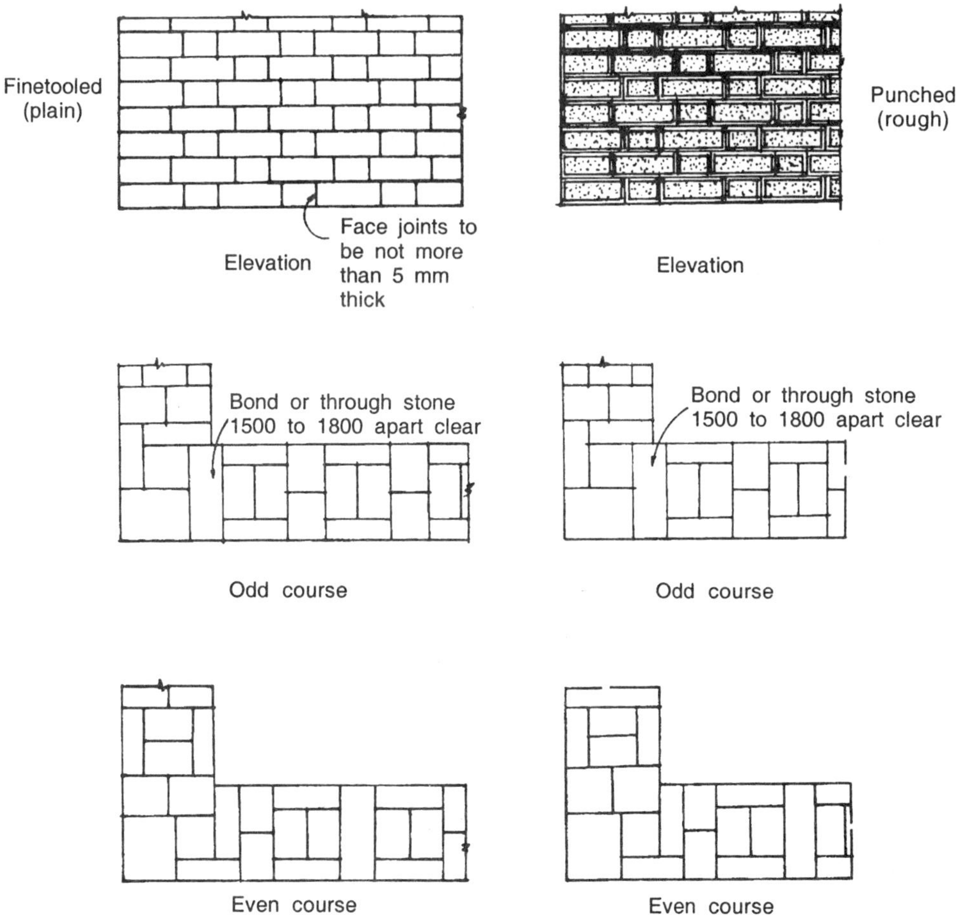

(a) Plain (fine tooled) (b) Punched (rough tooled)

Fig. 7.3 Ashlar stone masonry. (All dimensions in mm)

In constructing ashlar work, the stones should be covered with gunny bags before ropes are passed over it for lifting so that the stone edges are not damaged.

It should be noted that in the most dressed stonework, the stones in each course are laid as alternate headers and stretchers as shown in Fig. 7.3. The headers on alternate courses should be approximately in the middle of the stretchers above and below.

7.7 ASHLAR FACING WITH BACKING OF BRICKWORK (COMPOSITE MASONRY)

This type of work is executed in many places for its beauty. In these works, the face stone of ashlar in each course is laid as *alternate courses of headers and stretchers* unless stated otherwise. Face stones and bond stones must break joints in the face for at least half the height of the standard course, and bond stones should be carefully maintained throughout. Work on all the connected walls should be carried out at one level throughout, but where breaks are unavoidable, the joint must be made in good long steps so as to prevent development of cracks. This type of work is shown in Fig. 7.4.

(a) Brick backed ashlar (b) Brick backed coursed rubble work (c) Random rubble with facing brickwork

Fig. 7.4 Composite masonry.

7.8 STONE VENEER WORKS FOR FACING OF WALLS

Facing work not exceeding 8 cm in thickness is treated as *veneering*. Veneer work with marble, polished granite or other stones gives a distinctive look and is used for prestigious buildings. The stones are *secured to the backing masonry by mortar and also mechanically by means of noncorrosible metal cramps* (metal bar with bent ends) made of stainless steel, copper or gunmetal, usually not more than 60 centimeters apart. Where cramps are designed only to hold the slabs in position, the facing should be provided in addition with a continuous support to carry the weight of the stones. These supports are usually in the form of projections from or recessions into concrete slabs or beams at the desired levels. These should not be spaced more than 3.5 m apart vertically and are also provided over top of openings. They should also be provided where the thickness of facing stone changes. Alternately, the cramps should be designed to transfer all the load of the facing to the back wall itself. The mortar used for fixing the stones should be preferably combination mortar—1 part cement, 1 part lime putty combined to 1 part fine sand. The cramps are to be set in 1 : 2 (cement: fine sand) mortar. To give stability to the work, pins or cramps are also used to tie the horizontally adjacent stones together.

In the case the stones are to be fixed to brickwork construction, the backing should be constructed as far as possible along with the face work. However, in the case of RCC and stonework, the stonelining can be secured to the backing after the concrete has been cured. While laying of the stones if any hollow sounds are detected by tapping, the facing stones should be taken out from such places and re-laid.

It is advised to use modern special joint sealing compounds when the work is on the outside and is exposed to heavy rain and winds. The joints on the faces should be thin and uniform, not more than 1.5 mm. Where the joint filler compound is used, the joints should be raked out at least 25 mm after the mortar in the joints has set sufficiently and the filler or compound is applied. The work should be properly protected and cured during its execution. (CPWD Specification gives more details of this work.)

Nowadays, strong chemical adhesives that can be used to stick the stonework to brick or concrete walls are available in the market. These adhesives are used for fixing stones such as granite, marble, etc. to not too high inside walls of rooms, conference walls, etc.

7.9 MARBLE WORK

Marble is classified in many ways. According to CPWD Specification 77, it is classified as follows, depending *on* different colours of marbles and also the places of origin.

1. Plain white marble:
   - (i) Makrana first quality (without streaks)
   - (ii) Makrana second quality (with light shades or spots)
   - (iii) Abu white (with blue or grey shades)
   - (iv) Abu panther marble (with irregular blue or black spots)
2. White veined marble:
   - (i) Makrana Doongri Adanga
   - (ii) Makrana Chak Doongri
   - (iii) Abu veined
3. Black plain marble
4. Black zebra marble:
   - (i) Bhainslana
   - (ii) Kishangarh
   - (iii) Abu black
   - (iv) Narnaul
   - (v) Makrana Dhobi Doongri
5. Green marble:
   - (i) Baroda

(ii) Abu
(iii) Falna
(iv) Bundi

6. Pink plain marble (Makrana and Bar)
7. Pink Adanga marble:
   (i) Makrana
   (ii) Bar
   (iii) Bundi Adanga
8. Grey marble:
   (i) Kumari
   (ii) Bundi
9. Brown marble:
   (i) Bar
   (ii) Narnaul

Limestones of Cuddapah and other places are also sometimes referred to as *marbles in the trade,* but they should be distinguished from the real marbles.

7.9.1 Table Rubbed and Polished Plain Marble Work in Steps

Plain marble works in steps, joints, columns, kitchen platforms and other plain works are to be laid in the following way. Every stone should be cut to the required size and shape. It must be chisel dressed on all beds and joints to give truly vertical, horizontal, radial or circular joints as required. All exposed faces must be fine tooled and level so that a straight edge laid along the face of the stone is in contact with every point on it. *The surface should be rubbed smooth and polished* before they are laid. Similarly joints to a depth of 6 mm from the face should be fine tooled so that all visible angles and edges shall be true, square and free from any chippings. Beyond the depth of 6 mm from the face, the joint should be dressed with a slight splay so that it (the joint) is in the shape of an inverted V as shown in Fig. 7.5. The surface coming in contact with the backing need not be chisel dressed. The marble stones shall be wetted before laying. They shall be floated on mortar and bedded properly in position with wooden mallets without the use of chips or under pinning of any sort. Laying of marble stone in floors with thin joints is described in Section 16.3.

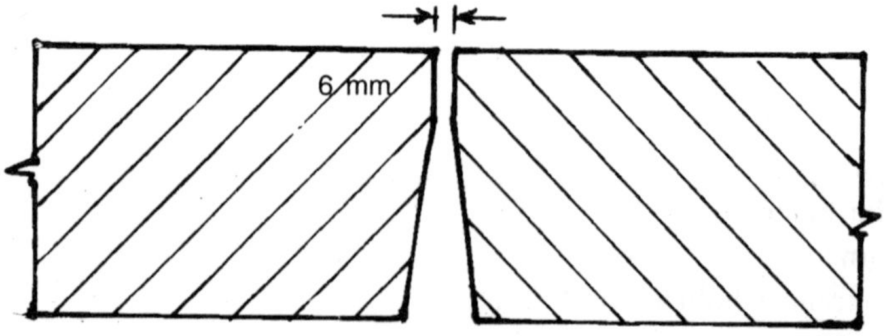

Fig. 7.5 Inverted V-shaped joint for marble work.

7.10 LATERITE MASONRY

Laterite is a residual rock formation found in hot humid regions such as Kerala, West Bengal and Sri Lanka. The IS standard blocks are 390 mm × 190 mm × 190 mm and 590 mm × 290 mm × 290 mm. In Kerala, they are quarried in blocks 30 cm or 45 cm long (1 ft stone and 1½ ft stone). The stones after curing are soft when quarried but become hard on exposure to the atmosphere due to oxidation of the iron compounds. The blocks after curing for use should not have a water absorption more than 14% after immersion in water for 24 hours.

Laterite stones are extensively used in Kerala as building blocks in the same way as bricks. Care should always be taken to protect them from rain, as exposure to water can disintegrate the material, even if it has been plastered. In all laterite walls of buildings, the eves of the roof should be extended beyond the walls so that rainwater does not fall on these walls. As the strength of laterite is low, it is advisable to use lime *surki* mortar rather than cement mortar for laterite work. If cement is used, only a lean mix is necessary. Laterite walls should be plastered only after it has thoroughly hardened and dried by exposure to air.

7.11 TOOLS FOR STONE MASONRY

Stone masonry has been developed into a fine art in India and different specialized tools have been developed for this type of work. These tools depend on the region where the work is carried out and the workmen carrying out the work.

SUMMARY

Stonework in places where stones are easily available can be cheaper than brickwork for many situations. Good stones behave well in foundations. Stone masonry with good stones can be given many facings or appearances. The six traditional types of stone masonry are random rubble, coursed rubble of the first sort, coursed rubble of the second sort, plain ashlar rough tooled, plain ashlar fine tooled and chamfered (or punched) ashlar. We should be able to distinguish between them. Stone is also used as veneer work for facing of buildings.

REVIEW QUESTIONS

1. Give the different classifications of stone masonry. Giving sketches, explain the differences between the various types of these stone masonry.
2. Give instances where you will use (a) random rubble work and (b) ashlar work.
3. How do you lay stone veneer work for a façade of a tall building?
4. What are the precautions to be taken in laterite masonry work?
5. What are bond stones and for what type of stone masonry it is used? If stones of enough strength are not available for bond stones, what alternative would you adopt?
6. Describe the various types of composite masonry used in construction.

Chapter 8

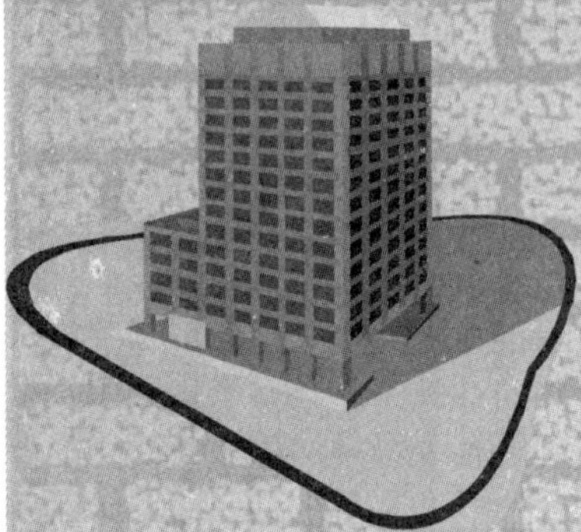

Arches and Lintels

8.1 INTRODUCTION

The Romans perfected the technique of construction of masonry arches and many of their bridges and aqueducts were carried over arches. Some of them exist even today. Hence it is an age old construction method handed down from ancient times.

Arches and lintels are used in modern buildings to bridge over the openings for doors and windows as well as wall openings provided for separation of rooms such as drawing room from dining room. Arches are also used extensively in bridge construction. They should be designed to support the load acting above the openings. Brick and stone arches are built with individual units whereas reinforced concrete and steel arches behave as one continuous unit. Arch action in brick and stone arches depends on the friction between surfaces of the constituent blocks (called voussoirs), which in turn depends on the compression between the voussoirs. Lintels, on the other hand, act mainly as beams. Concrete and steel arches are in compression and also bending. Arches, in general, transfer the thrust to their supports as inclined loads, whereas lintels transfer their loads as a vertical load only. In building construction, reinforced concrete is more commonly used for lintels. However, much savings can be made in low-cost buildings if brick arches in compression are used. (Another development for low-cost housing is the use of thin RC lintels described in section 8.6.2.) Arches can be made aesthetically beautiful and are also used for ornamentation purposes in many parts of the buildings. In this chapter we will examine the important points to be considered in building arches with bricks and also construction of reinforced concrete lintels. We will concentrate more on the construction of arches than on their action and design which are dealt with in structural engineering.

8.2 TERMINOLOGY IN ARCHWORK

The important technical terms used in archwork are the following. They are illustrated in Fig. 8.1.

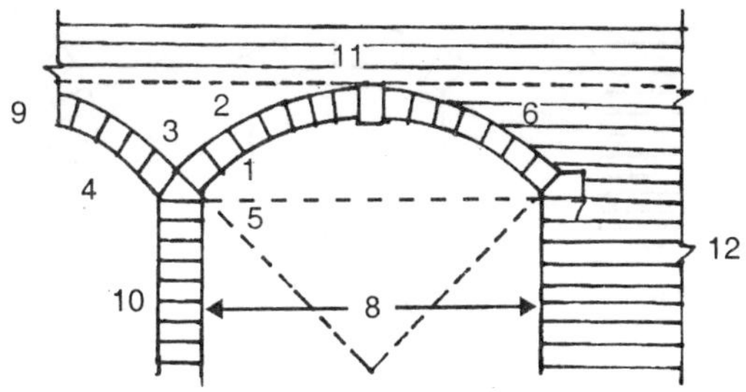

Fig. 8.1 Elements of a segmental arch: 1. Introdos, 2. Extrados, 3. Spandril, 4. Voussoir, 5. Spring line, 6. Haunch, 7. Skewback, 8. Span, 9. Depth or height of arch, 10. Pier, 11. Crown and key stone, 12. Abutment.

1. *Introdos* is the inner curve of the arch. It is also known as the *soffit.*
2. *Extrodos* is the outer curve of the arch.
3. *Spandril* is the space between the extrodos and the horizontal line through the highest part of the extrodos called the *crown.*
4. *Voussoirs* are the wedge-shaped individual units that are formed of arch.
5. *Spring line* is the imaginary line joining the starting points or *springing points* of the ends of the arch.
6. *Haunch* is the lower part of the arch between crown and skewback.
7. *Skewback* is the inclined or splayed surface on the abutment, from which the arch springs.
8. *Span* is the clear distance between the supports.
9. *Depth* or *height* is the distance between the extrados and introdos.
10. *Rise* is the clear vertical height between the springing line and the introdos of the arch.
11. *Crown and keystone*: The highest point of the arch is called the *crown.*
12. *Breadth of soffit (thickness)* is the horizontal distance between the front and back faces of the arch.

8.3 CLASSIFICATION OF ARCHES

Arches are classified according to their geometry or shape. We select the shape according to the architecture of the building. They can be made of bricks, stones, blocks, etc. There are many types of arches, of which the following are the most important.

Corbel arch. This arch is constructed by corbelling without centring [Fig. 8.2(a)].

Segmental arch. This arch is in the form of a segment of a circle [Fig. 8.2(b)].

Semicircular. This arch is in the form of a semicircle [Fig. 8.2(c)]. Other types of arches can be elliptical, parabolic and so on.

Pointed arch or Gothic arch. This arch is built around triangles and is formed of segments meeting at the apex [Fig. 8.2(d)].

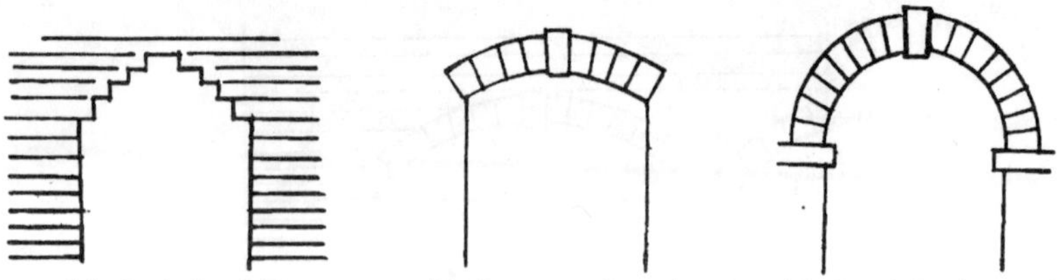

(a) Corbel arch (b) Segmental arch (c) Semicircular arch

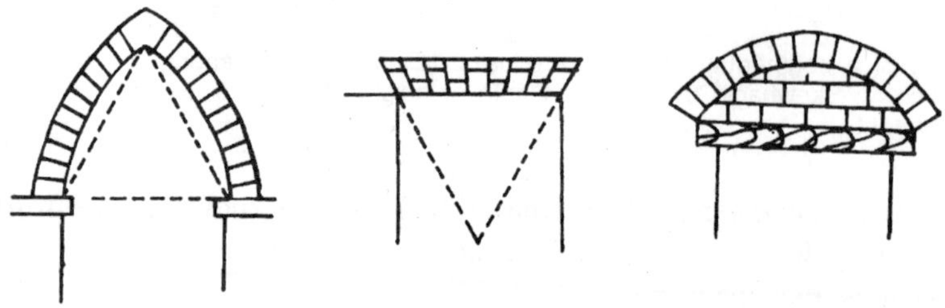

(d) Pointed arch (e) Flat arch (f) Relieving arch

Fig. 8.2 Different shapes of arches.

Flat arch. These arches are used for light loads and spans up to 1.5 m. The skew backs are set at 60° with horizontal, thus forming an equilateral triangle with the introdos forming the base. The extrados is straight, and the introdos is given a slight rise or camber of 10 to 15 mm per metre span to allow for small settlement. They are used in low-cost buildings instead of lintels [Fig. 8.2(e)].

Relieving arches. These are constructed to relieve the load from a wooden lintel or a flat arch in an opening [Fig. 8.2(f)].

(The usual types of brick arches used in buildings are shown in Figures 8.2 and 8.3. The popular types of stone arches are shown in Fig. 8.5.)

(a) Rough brick arches with ordinary bricks and thick joints

(b) Arches with wedge shaped bricks and uniform mortar joints

Fig. 8.3 Brick arches.

8.4 CONSTRUCTION OF BRICK ARCHES

Construction of brick and stone arches is an art and experienced special masons should be used for construction of masonry arches. *Centring* is the temporary structure or staging over which the arches are built.

8.4.1 Operations Involved in Construction of Arches

The three operations involved in the construction of arches are as follows:

- Installation of centring
- Laying the arch
- Removal of centring

These are briefly described further.

Installation of centring. Arches are generally built over a centring made of the shape of the needed arch. This centreing is built over a staging. For minor works, centring can be made of brick in mud masonry built over a staging in the required shape and plastered on the top with mud or lean cement plaster. For small spans and large numbers, the staging is usually made of wood as shown in Fig. 8.4(a). It can be also made of steel for very large arches. The centring made from wood piece is known as a *turning piece*. They are supported by props and

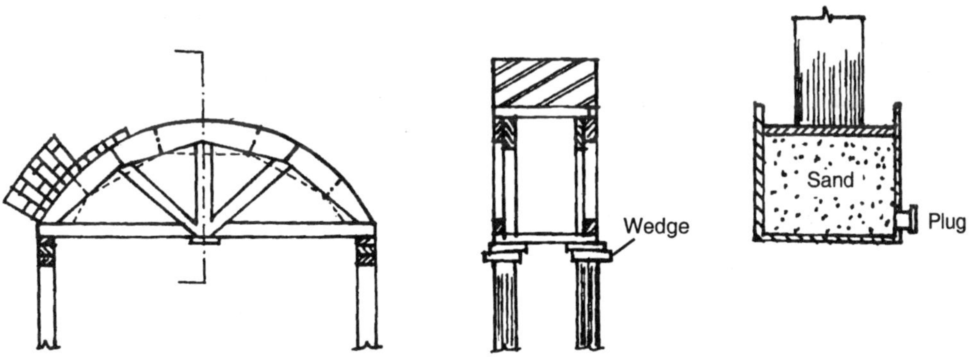

(a) Elevation (b) Decentring by using wedges (c) Decentring by sandbox method

Fig. 8.4 Decentring of arches.

a pair of folding wedges (double wedges) is provided at the top of thc props under the centring to tighten or loosen the centring.

Laying of arch. The skewbacks at the springings are first prepared and placed very carefully in the correct angles. The voussoirs are then laid and the work proceeds from both sides of the crown. The bricks can be placed as showing either "stretcher face" or "header face". Semicircular arches are often built with the bricks showing header face. There can be more than one ring for the arches. The joints should be filled with mortar not less than 5 mm and not more than 15 mm thick. Finally the key-stone is inserted to lock the whole arch in position. It is important that the masonry above the skewback at the haunches should be loaded by filling up the spandrels up to the level of the crown of the arch or at least two-thirds of the height of the arch to counteract the thrust. This should be done on both sides. Before the mortar hardens, the centring is slightly slackened by means of easing the wedges under the centring on top of the props by only 2 to 3 mm to permit the voussoirs to sit on their beds properly. This initial release, as soon as the arch is completed, is very important for the proper seating of the arch to enable friction to act between the blocks. The archwork should be kept wet during the work so that no portion fully hardens and sets before the whole arch is completed.

Removal of centring. The ancient sandbox method of release of formwork invented by the Egyptians enables the centring to be removed without shock. The free flowing fine sand is allowed to flow out slowly as shown in Fig. 8.4(c). Other methods such as use of wedges can also be adopted Fig. 8.4(b). It is important that the centring should be designed to be capable of being lowered with ease.

8.4.2 Specification for Brickwork in Arches

Bricks for archwork. They shall be specially selected and shall be free from defects of any sort. The brickwork shall be carried out from both ends simultaneously and keyed in the centre with the keystone. The bricks shall be buttered with mortar and well pressed into their positions so as to squeeze out a part of the mortar and leave the joint thin and compact. The thickness of the joint shall neither be less than 5 mm nor more than 15 mm, and the voussoir joint shall be normal to the curve at these points. The archwork shall be quickly and evenly completed and kept moist so that no portion of the arch sets before the whole arch is completed and initial release made.

The centring. Centring should be strong enough to take the load without any appreciable deflection. Arrangement shall be made for slackening the formwork without vibration being transmitted to the arch. In case of a series of arch being constructed, the centres should be slackened simultaneously.

Time of slackening. The time after which the initial slackening (not the final removal) for centring to be carried out shall be carefully decided. If lime mortar is used, it is slackened 24 hours after completion of the arch. The time should be decided so that the mortar in the last joint should not be too soft to squeeze out, and at the same time, all the other mortar joints should be still moist to allow the arch to compress itself and bring all the joints to the full bearing.

Final release of arch. In case of arches which transmit *inclined loads* such as the segmental arch, great care shall be taken to see that the skewbacks are secure and they are given a week's time to set before the archwork is fully released. In case of semi-circular, elliptical or other types of arch, where the springing is from a horizontal surface, the final release can be much earlier as the loads on supports are vertical. As already stated, in all cases, the adjacent wall shall be built upto two-thirds of the height of the arch before slackening the centre.

8.5 STONE ARCHES

In monumental buildings with stone facings masonry, the provision of stone arches adds to its beauty. Depending on the workmanship, stone arches can be of two types (like stone masonry), namely

(i) Rubble arches
(ii) Ashlar arches

Rubble arches are made from hammer-dressed stones and are useful only to small spans—up to one metre. They are also not very pleasing to look at. *Ashlar arches* are made from stones which are cut to proper shape of the voussoirs and are fully dressed. The voussoirs are made for the full thickness of the arch. The shapes are cut to fit the shape of the arch. They are very pleasing in their appearance. Nowadays it is used only for expensive classic buildings. It is also a specialized stonework which needs experienced stone cutters and supervisors for the construction. These are shown in Fig. 8.5.

(a) Circular stone arch

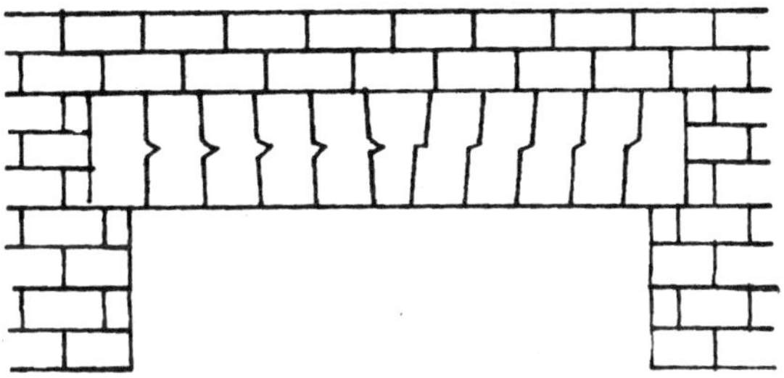

(b) Tooled and rebated flat stone arch

Fig. 8.5 Ashlar arches.

8.6 LINTELS

A lintel is a horizontal member used to span openings in walls as shown in Fig. 8.6. Lintels can be classified according to the material it is made of as follows:

1. Timber lintels
2. Stone lintels
3. Brick lintels (or flat-brick arch)
4. Steel lintels
5. Reinforced concrete lintels

The most commonly used lintel is the reinforced concrete lintel.

Structural action of lintels. Lintels depend on beam action (bending and shear) for their strength. In old buildings (such as temples), stone lintels were commonly used for doorways and openings. Nowadays stone lintels are used only for small openings and in low-cost houses. As stones are weak in tension, they are not efficient as beams. As reinforced concrete lintels are the most commonly used type, we will study its construction in more detail. In buildings, they are usually combined with sunshades and also RC bands or belts that are mostly provided in walls.

8.6.1 Layout of Continuous Lintel Bands

Instead of lintels over openings only, *continuous lintel bands* are usually provided in buildings. The following practice is generally recommended in buildings when continuous belts are cast at top of window and door levels. Such lintels are necessary as in *expansive soils* or *for earthquake resistant design*. The minimum thickness of concrete in the stretches without openings should be a 1 : 2 : 4 concrete layer of 75 mm (3 inches) thickness with one layer of reinforcement as shown in Table 8.1. Cross reinforcement of 8 mm at 30 cm centres is also provided. In special situations a beam similar to the plinth beam is also adopted in continuous lintel. Along the outside of the building over windows and outside doors, the lintels are generally combined with cantilevering sunshades. In such cases it is preferable to have a minimum thickness of concrete as 150 mm (6″) for the lintel. The recommended steel in the concrete bands for earthquake resistance is given in Table 8.1.

In ordinary buildings built in earthquake regions, these continuous lintels above the door and window level is mandatory. In difficult situations, where they may run at two different levels, concrete should be cast vertically to tie it up, if necessary, with the roof or floor slab.

Table 8.1 Recommended HSDB bars in RC lintel bands

(Ref. Table 6 IS 4326 (1993), Earthquake Resistant Design and Construction of Buildings, Code of Practice)

Span of band between cross walls (m)	*Building category*			
	B	C	D	E
5 or less	2 of 8 mm	2 of 8 mm	2 of 8 mm	2 of 10 mm
6	2 of 8 mm	2 of 8 mm	2 of 10 mm	2 of 12 mm
7	2 of 8 mm	2 of 10 mm	2 of 12 mm	2 of 10 mm
8	2 of 10 mm	2 of 12 mm	2 of 10 mm	2 of 12 mm

Notes: 1. Category of building corresponds to the earthquake intensity region. (Category E is of high intensity.)
2. Use of 6 mm rods in exposed situations is not recommended.

8.6.2 Lintels between Openings

Lintels between openings can be provided as precast lintels or cast in place lintels. There are two ways of designing cast in place lintels—conventional lintels and thin lintels. They are shown in Fig. 8.6. A bearing of one tenth of the span and a minimum of one full brick length should be provided for the lintels.

The main considerations in construction of these lintels are discussed further.

Conventional lintels. Conventional lintels are designed according to regular beam theory with a load from a triangle at 60° to the horizontal above the lintel, when the walls on both sides and above the lintel are more than half the effective span of lintel in height so that arching can take place [see Fig. 8.6(a)]. Otherwise we assume a rectangular distribution.

Thin lintels. According to the thin lintel theory, the member is designed to act as a composite beam. The brickwork above the concrete lintel is assumed to take the compression and major part of the shear. The reinforced concrete lintel acts only as a tension member in

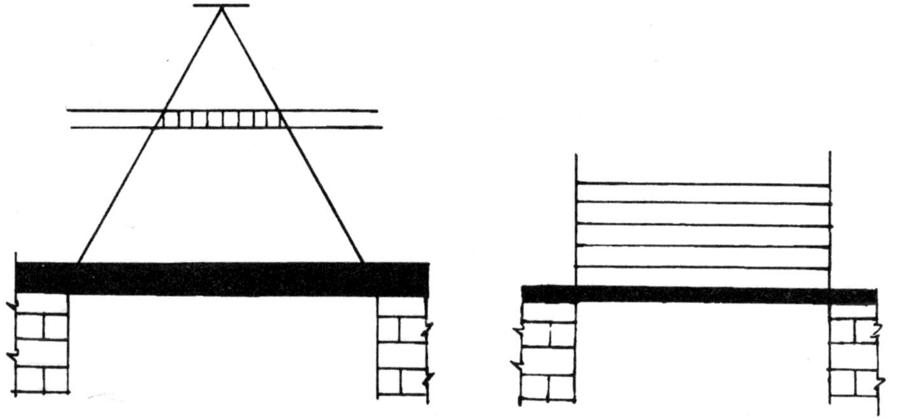

(a) Conventional lintels and assumed loading (b) Thin lintels with the necessary brickwork on top for thin lintel action

Fig. 8.6 Reinforced concrete lintels.

the composite brick and concrete lintel beam [Fig. 8.6(b)]. It is important to remember that this theory assumes that a brickwork of at least five courses will be built in 1 : 6 cement mortar, out of bricks of strength at least 10 N/mm^2 before the bottom support of the lintel is removed. The lintel is to be supported (for 30 cm length) at its middle till the brickwork is completed and cured for at least seven days. This design has been found applicable up to a span of 1.83 m (6 ft), beyond which the shear stress in the masonry becomes too high to be safe.

Details of thin lintels. As already stated, for the thin lintel designs for spans up to 1.8 m, we use a concrete slab one-brick course high (3 inches or 75 mm) acting as a tension member. The 3 nos. 10 mm rods with 6 mm distributors at 30 cm centres are placed at the centre of the slab. The reinforcement cage as a beam need not be made for thin lintels. This can be either prefabricated with roughened top face or cast in place. It should be supported in its middle till a brickwork of 45 cm (1.5 ft) height or 5 courses are built over it and set properly.

Details of conventional lintels. The conventional design results in beams of the following dimensions. For spans up to 1.5 metres, we use a thickness of 15 cm (6 inches or 2 courses of brickwork) with 3 nos. of 8 mm bars (middle bar bent) as bottom bars. Also 8 mm or 6 mm stirrups at 30 cm spacing are provided with two hanger bars of 8 mm given at the top. It has been found from experience that 6 mm bars in thin sections, especially in exposed situations, corrode fast and it is better to use a minimum size of 8 mm in thin sections.

8.6.3 Lintels with Sunshades

The bearing for individual lintels should be at least one brick (23 cm) on each side. Similarly, the sunshades are also to be extended for this length of 23 cm on either side of the external opening. General considerations in planning of sunshades are as follows. The sunshades should project for a length not less than that required to protect the window or door leaf, that may open out, from rains. As a shading device, it is taken as a minimum of 60 cm for the main windows of residence and up to one metre in the front side for the entrance doors. Side drops are sometimes provided on the north side for the sunshades of windows (but never for doors) for the top one-third of its height as a shading device. Similarly facia for sunshades can be provided generally in the front side and vertical drops can be provided on western side windows, if necessary, as a shading device. *Thus the whole layout of sunshades will depend on the shading required from the sun as well as the protection of the window leaves opening out from rains*. Ventilators and openings in bathrooms or kitchen usually do not require sunshades but can be given an outside border.

Details of sunshades. The sunshades are made 75 mm (3 inches) at the cantilever end and tapered upwards to the depth of the lintel (Fig. 8.7). On an average, the projection will be 60 to 90 cm, but it should be sufficient to safely cover the opened window shutters from rain (see Section 23.1). The main rods of the cantilever of the sunshade can be 8 mm rods at 10 cm centres with 8 mm distributors at 15 cm centres. Enough cover (at least 20 mm) should be provided for the reinforcements as it is fully exposed at the top. When we use continuous lintels or belts with the constant thickness at lintel level, then the criterion that lintel thickness should correspond to the height of the brick coursework need not be obeyed. In such cases, as we start brickwork all along the wall from this level, no mismatch in levelling of brick courses will occur (see also Fig. 10.4.)

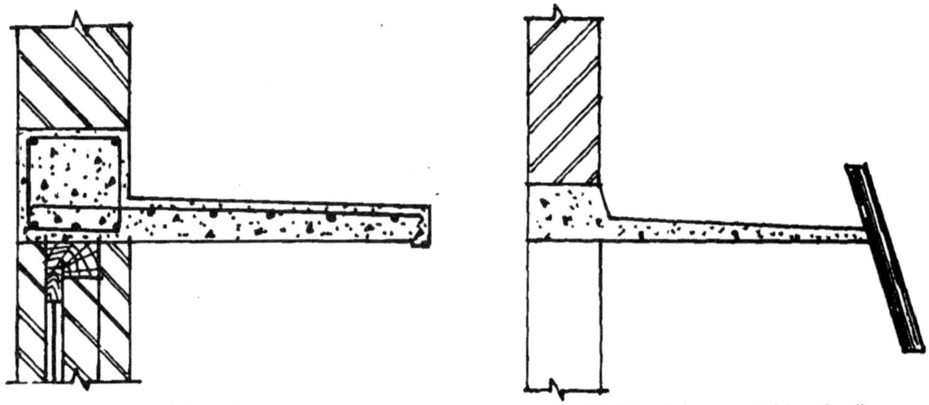

(a) Normal type (b) Used for westside shading

Fig. 8.7 Reinforced concrete lintels with sunshade (*chajja*).

8.6.4 Boot Lintels or Lintels for Cavity Walls

RC boot lintels as shown in Fig. 5.9 (Chapter 5) are provided for cavity walls. It is also necessary to provide a DPC covering over both the leaves of the wall to prevent moisture seeping down the cavity.

8.7 PLINTH BEAMS AND GRADE BEAMS

Plinth is the portion of the building between the ground and surface of bottom floor. We will briefly examine the plinth beams and grade beams also in this chapter.

Plinth beams. Plinth beams [Fig. 8.8(a)] are provided at plinth level (on top of the plinth). In load-bearing masonry walls they even out settlements of buildings. They also serve to provide the DPC course. In buildings in seismic areas, they are prescribed as a continuous band at plinth level in addition to other similar beams at lintel and roof levels. The minimum recommended depth of plinth beams is 10 cm (4 inches) to 15 cm (6 inches) and it should be for the full width of the wall above the plinth beam. The minimum reinforcement recommended is 3 numbers of 6 mm MS bars (or preferably 2 numbers 8 mm Fe 415 bars) at top and bottom as longitudinal bars with 6 mm stirrups at 23 cm (9 inch) spacing. (See also Section 4.9.)

Grade beams. Grade beams [Fig. 8.8(b)] are provided in between isolated foundations. Thus we connect underreamed piles or columns on isolated footings with grade beams and construct the walls over grade beams. These beams should be stronger than plinth beams. The minimum recommended depth of a grade beam should be 150 mm and a minimum of 3 numbers of 8 mm Fe 415 bars should be provided at the top and bottom.

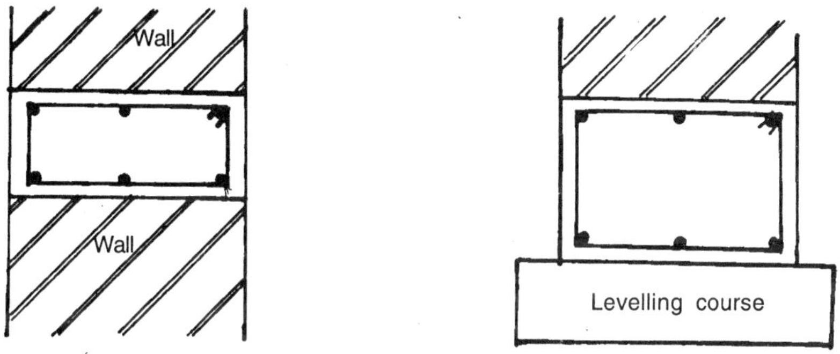

(a) Plinth beam

(b) Grade beam on top of underreamed piles or between columns

Fig. 8.8 Beams under walls.

SUMMARY

Use of conventional brick arches of moderate or small span over doors and windows reduces the cost in low-cost housings. Stone arches are used only in prestigious buildings for appearance. RCC lintels are extensively used in modern building construction. Plinth beams are constructed in masonry construction at plinth level and grade beams between columns below the ground level. According to recent IS Code design requirements for earthquake, continuous reinforced concrete bands have to be provided at various locations (plinth, lintel and roof levels) for earthquake-resistant design in all low rise, non-engineered buildings built in most parts of India.

REVIEW QUESTIONS

1. (a) What is the difference between the actions of a brick arch and an RCC arch?
(b) What are lintels? Sketch a lintel combined sunshade for an external doorway.

2. (a) Sketch roughly five common types of brick arches used in building construction. Of these, which one would you use over a window one metre wide for a low-cost house?
   (b) In what situations, would you use brick arches in a building?
3. Describe how brick arches are constructed. In what situations, would you use a concrete arch instead of a brick arch in building works?
4. What are the two types of lintels that can be used in buildings? What is meant by continuous lintels, and in which situations would you use them?
5. Write short notes on:
   (a) Spandril of an arch
   (b) Flat arch
   (c) Methods used for the release of centring of arches
   (d) Continuous lintels in buildings
   (e) Design of sunshades in buildings
6. What are plinth beams and grade beams? Give a short account of their uses in building construction.

Chapter 9

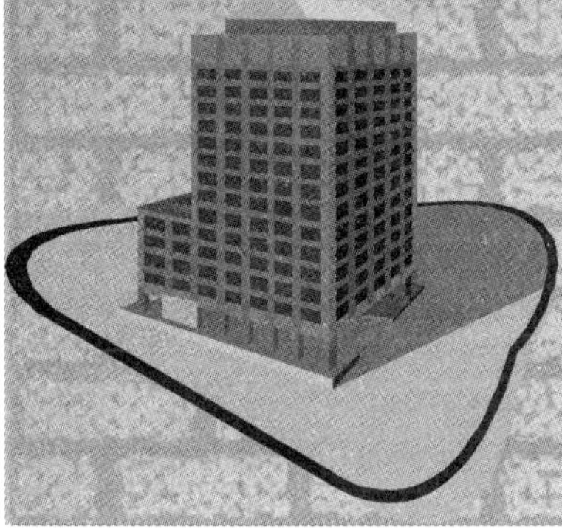

DPC and Waterproofing of Basements

9.1 INTRODUCTION

The construction for prevention of migration of moisture inside buildings by capillary action is generally called *dampproofing* and that for prevention of flow of water inside buildings due to hydraulic pressure (from rains or groundwater) is called *waterproofing*. Dampproofing and waterproofing are needed in buildings in many situations such as:

1. Prevention of entry of dampness and moisture from ground upwards through *wall and floor* or from *roof downwards through wall* by capillary action come under dampproofing.
2. Prevention of entry of water from the *outside through walls* in basements is waterproofing.
3. Prevention of flow of water or moisture from bathrooms or wet areas of upper floors downwards is waterproofing.
4. Prevention of flow of rainwater in flat reinforced concrete roofs is waterproofing.

Dampproofing has to be done in walls as well as in ground floor. DPC in walls is described in Section 9.2 and dampproofing of ground floor is described in Section 9.3. Waterproofing of basements is described in Section 9.4. Waterproofing of roofs and wet areas is dealt with in Chapter 27.

9.2 DAMP-PROOF COURSE (DPC) IN WALLS

Damp-proof courses are built in masonry walls to prevent dampness penetrating through the wall either from the foundation upwards or from the roof downwards or from other exposed surfaces inwards in the building. Thus DPC has to be provided on top of the plinth above the ground level (GL) in all walls—in some cases, at the intersection of the roof with walls, below chimney stacks at roof level, above lintels over cavity walls, etc. The lower DPC is built at the plinth level, which should never be less than 15 cm above the ground level as soils may be deposited along the wall after construction and then water can seep from the soil above the DPC. (Generally the DPC is kept at 45 cm from the GL at the plinth level.) Special care should also be taken to prevent entry of moisture from construction of flower beds, etc. along walls to ensure that water pour for plants does not rise through the walls from them.

In non-clayey soils, DPC is provided only on a plinth beam under the walls for prevention of moisture moving upwards [see Fig. 9.1(a)]. In clayey soils, DPC may also have to be provided under the walls and under the floors [see Fig. 9.1(b)].

The wall above the roof can also be a source of dampness. If parapets are provided by continuing the wall above roofs with the roof slab not extending the whole thickness of the wall and the coping is of pervious nature, a flexible damp-proof course must always be provided under the coping or above the finished top level of the roof. This DPC is carried through the full thickness of the wall above the roof level. [DPC is provided in parapet walls for prevention of misture moving downwards] [see Fig. 9.1(c)].

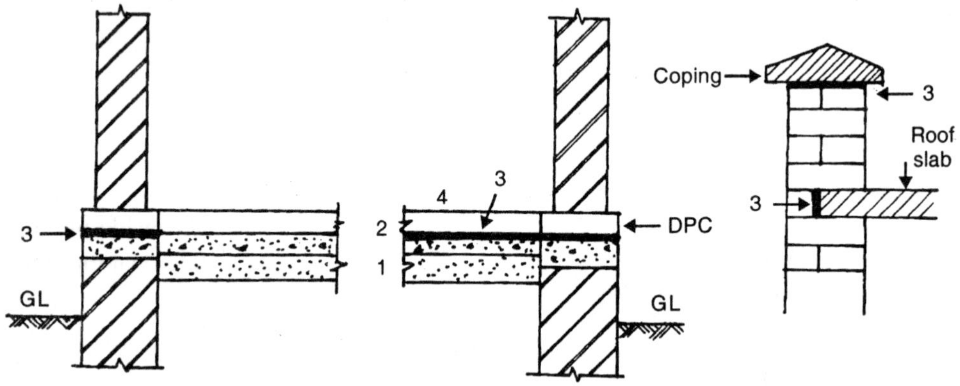

(a) DPC in non-clayey soils (b) DPC in clayey soils (c) DPC in parapet walls

Fig. 9.1 Provision of damp-proof courses: 1. Sand filling, 2. Lean concrete, 3. DPC (shown in dark), 4. Floor.

DPC consists of a continuous layer of any of the impermeable material discussed in section 9.2.1. Even though in the past metals such as lead, copper, etc. in sheet forms were used over plinth, nowadays bitumen as paint or membrane laid on rigid cement concrete or on dense waterproof mortar is used. These are much cheaper than metals. They are also easy to apply and give good results. For buildings of minor importance, a damp-proof course may consist only of half-inch layer of 1 : 3 portland cement mortar with waterproofing compounds laid at the junction of the wall and the underface of the floor when there is a sandfilling under the floor. Otherwise it is provided just above the floor level. The DPC in cavity walls is to be installed in both leaves of the wall as dealt with in Chapter 5.

9.2.1 Detailed Specifications for DPC in Walls

Any one of the following types of damp-proof courses depending on the nature of the ground and importance of the building can be used. Some of them are cheap applicable to low-cost houses and others are expensive to be used in important permanent buildings.

1. Two courses of dense machine-made bricks set in 1 : 3 cement mortar. The vertical joint between bricks is left unfilled.

2. A layer of well-burnt ordinary bricks soaked in hot pitch or bitumen.
3. One layer of non-porous stone about 2 inches thick laid for the full width of the wall over a bed of 1 : 3 cement mortar.
4. Two layers of non-porous stone slabs laid in 1 : 3 cement mortar to break joints (i.e. the two layers should break joints).
5. A 20 mm thick 1 : 2 cement plaster over the full width of plinth walls and two coats of hot bitumen applied over the mortar at the rate of 1.5 kg per square metre for each coat after the plaster has dried.
6. 25 to 50 mm of 1 : 1½ : 3 or 1 : 2 : 4 cement concrete (without reinforcements) with two coats of bitumen after the concrete has been cured and is dry.
7. A reinforced concrete plinth beam of dense concrete with two coats of hot bitumen applied at the rate of 1.5 kg per square metre for each coat.

Among the above specifications, the item 5 is popular with cheap construction and items 6 and 7 are popular with more important works. In all construction of buildings, the specification to be adopted for DPC should be indicated in the drawings. We will examine the item 6 in more detail. Plinth beams have already been dealt with in Sections 4.9 and 8.7.

9.2.2 Description of Damp-proof Course with 1:1½:3 Concrete and Bitumen

Materials for this work (which is provided when no plinth beams are planned) will consist of concrete of 1 : 1½ : 3 (25 mm) proportion with 2 per cent integral waterproofing compound (at the rate of 1 kg per bag). The dampproof is applied at the plinth levels. The level of the surface of plinth is first checked longitudinally and transversely. At the top of the walls, the brick course should be laid with frogs down or on edge. The brickwork should be thoroughly wetted before concreting over it. Side shuttering with planks of 2.5 cm thickness must be fixed properly and firmly on both sides of the wall and the concreting is to be carried out with all the necessary precautions usual in concreting. This concrete damp-proof course should be laid continuously without joints. If unavoidable, joints can be provided below sills of doors or openings. These joints should be sloping and it should be treated with cement grout before a new concrete work is started.

Two hours after laying the surface of concrete, it should be made rough so that it will form a key with the wall above. After a day, the shuttering is removed and the edges are immediately made up smooth. The concrete is to be cured by keeping it wet for at least 7 days.

After curing of the concrete, generally, two coats of asphalt paint is applied—the first coat at 1.5 kg/sq m and the second coat at 1.0 kg/sq m. The concrete should be dry when it is to be coated. The painted surface is blinded with coarse sand immediately after painting and the surface is lightly tamped. The second coat is applied after the first coat has dried. It is also blended with sand immediately with coarse sand and is lightly tamped.

9.2.3 Other Damp-proof Courses

With 20 mm 1 : 2 cement mortar. In cheap construction, instead of concrete, a rich cement mortar can be used. The damp-proof course of cement mortar is to be laid as above using 1 : 2 cement and coarse sand mortar with standard waterproofing compound at the rate of 1 kg per cement bag. This mortar layer is then coated with asphalt.

With plinth beam. The damp-proof course *on plinth beam* consists of simple treatment of top of the reinforced concrete plinth beam described in Section 8.7 with two coats of bitumen as described above. Care should be taken to have the top of the plinth beam sufficiently above the ground level.

9.3 DAMPPROOFING OF FLOORS

We have also to see that water does not seep up through ground floors laid on the ground by capillary action. Unless the floor is properly constructed, sweating of ground floor can occur due to capillary water rising from the soil below the floor. (The capillary rise of water in clayey soils can be as much as 10 to 15 metres.) Dampproofing is specially important in floors on clayey soil. In normal cases, it can be prevented by the provision of a coarse sand layer under the floor above the foundation soil. However, in extreme cases, we may add a porous layer of lean cement concrete or brick jelly concrete layer above the sand layer and below the floor concrete and give bituminous membrane waterproofing applied on it as shown in Fig. 9.1(b). Alternately, a continuous plastic sheet (PVC) can be laid on this lean concrete layer to prevent penetration of moisture.

9.4 WATERPROOFING OF BASEMENTS

When basements (which are below the ground level) are built in areas where the water level is high (or where water can rise high during rains), they must be waterproofed. Concrete, even with additives, when placed in the field with interruptions of works (resulting in joints) cannot be fully made watertight. Hence *waterproofing of basements* should be carried out wherever there is a possibility of the groundwater rising above the basement level (Fig. 9.2). There are,

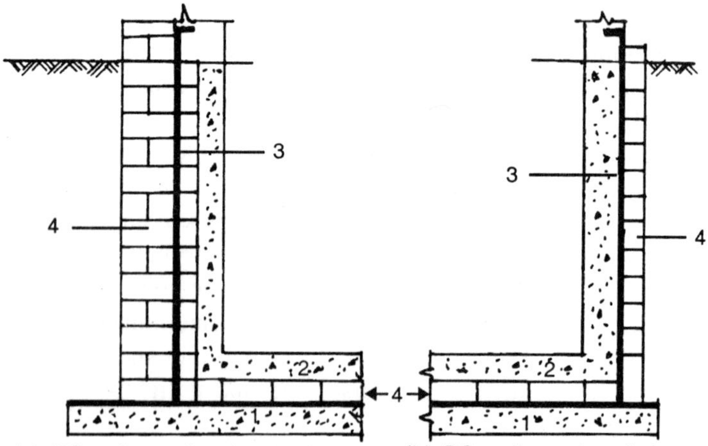

(a) RC wall cast last in places with limited space in excavation

(b) RC walls cast and then covered with mastis and brickwork

Fig. 9.2 Tanking (water proofing) of basements: 1. Lean cement concrete, 2. Floor slab, 3. Dampproofing, (in black colour), 4. Brickwork.

in general, three ways by which such waterproofing can be accomplished and any one method, depending on local situation, can be adopted for construction. These are described below.

Use of waterstops at joints along with dense concrete construction. In this method, the basements are made of waterproof concrete and all the joints (construction and expansion) are provided with waterstops. Plastic PVC waterstops are durable and can take movements of 12 mm or more according to their designs. Rubber waterstops can withstand much larger movement. Copper waterstops are useful if the movements are small. Provision of waterstops at joints and use of dense rich mix of concrete with admixture carefully concreted between these waterstops can give reasonably dry basement in usual situations where the water pressures are not high.

Tanking using asphaltic coating (membrane waterproofing). If one wants a basement absolutely free from water (even when basement is under groundwater level), then *membrane waterproofing should be resorted to*. This construction is also called *tanking*. In this method, an asphaltic membrane is given right around the basement *at the positive side,* i.e. at the outer side from which water tries to flow in.

In this method, *blown asphalt,* melted down at a high temperature, is used for the asphalt coating. Alternately, various types of ready-to-use ready-made asphaltic membranes can be applied. For asphalt lining to be fully effective to resist the penetration of groundwater, it should be in sandwiched position between concrete or between brickwork and concrete as described below. Under these conditions, it can withstand water pressures up to 10 metre high. The details of construction is the same as for the following (mastic) method.

Tanking using trowelled on asphalt mastic (membrane waterproofing). The British practice for membrane waterproofing is to use successive coats of asphalt *mastic* or *bitumen mastic* 6 to 10 mm thick by trowelling or mopping on the surface on the positive side. Usually three coats making up to 30 mm is used on horizontal surfaces and three coats making up to 20 mm is used on vertical surfaces. (Bitumen mastic is a mixture of bitumen, mineral fillers such as limestone dust or cement and crushed rock to give a semi-fluid consistency when heated to 180°C.)

9.4.1 Details of Tanking

The details of construction (tanking) are as follows:

Construction of horizontal layer (floor). The horizontal asphalt layer is spread over a levelled course of mass concrete and this in turn is sandwiched with 50 mm thick protective coat of cement screed, concrete or a layer of brickwork over which the structural floor is laid. This horizontal layer should extend at least 15 cm beyond the line of the vertical layer so that there will be leaking at the corner.

Construction of vertical face (walls). For the vertical surface, two types of construction, depending on availability of space around the building, are used. These are as follows:

1. If there is enough space, the vertical concrete structural retaining walls can be first cast, the asphalt is applied on this concrete and then protected by a half-brick thick protecting skin around the vertical walls before earth is filled in.

2. However, if this space is not available, then the brick wall is first built and covered with cement plaster. Then the asphaltic layer is placed on the cement plaster to be followed by a mortar coat and then the structural wall concrete is placed.

In all cases, the vertical asphalt lining should extend up to at least 15 cm beyond the ground level of the building.

9.4.2 Basic Principles of Waterproofing of Basements

The following are the basic principles to be followed in waterproofing of underground basements and water tanks:

1. The concrete used in the construction should be with integral waterproofing materials. The water-cement ratio should not be greater than 0.4 and the concrete should be well compacted.
2. Tanking or external waterproofing must be carried as described above.
3. The concreting operation should be continuous. If construction joints are planned PVC waterstops should be incorporated in the joints. These joints should be regrouted after construction.
4. Corrosion-resistant treatment should be given on the inside floor and sides. In extreme cases, a 35 mm concrete screed as cover is applied on the wall before plastering. A thick plastering (up to 25 mm for larger tanks) with 1 : 3 cement mortar with latex is also recommended.
5. The bottom slab should be designed for the necessary uplift pressure. Otherwise lowering the ground water table by providing perforated drainagepipe system at the foot of wall should be considered. Alternately pressure releasing systems should be provided.

9.5 DAMPPROOFING AND WATERPROOFING OF OTHER WET AREAS

There are many other areas such as bathrooms and roofs in a building that require special treatment in waterproofing. Details of these are given in Chapter 27.

SUMMARY

Simple methods can be used to dampproof buildings. Waterproofing of basements is a major work and should be carefully carried out to be effective. Many other areas such as roofs, bathrooms and other wet areas also need special waterproofing treatments.

REVIEW QUESTIONS

1. Distinguish between dampproofing and waterproofing.

2. Enumerate three popular specifications for construction of DPC in walls and under floors. Which of these will you adopt for:

   (a) A low-cost housing project
   (b) A middle-class residential house built in clayey soil

3. How do you dampproof the floor of a residence built in a clayey area? Give a sketch to illustrate your answer.

4. Describe briefly how basements in places where the groundwater level is high can be made waterproof.

Chapter 10

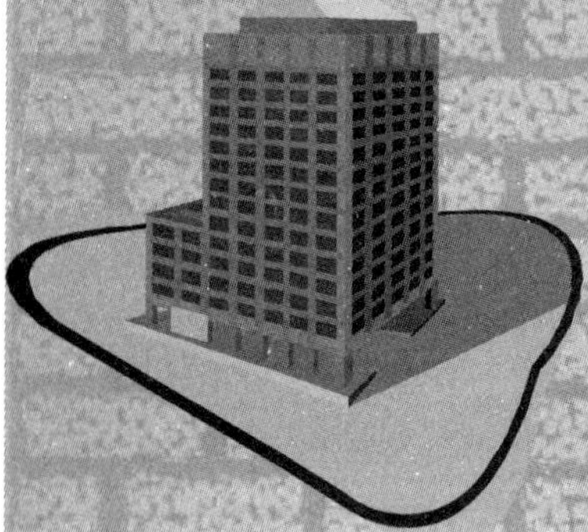

Concrete Work

10.1 GENERAL

Concrete is one of the very important and widely used materials in modern building construction. In most places in India, the material is prepared at the site from factory made cement and locally available sand and coarse aggregate by construction labour. In small works, it is hand mixed and in large works, machine mixing is used. In large cities in India, ready-mixed (factory mixed) concrete is available nowadays as RMC (ready-mixed concrete). A good understanding of making, casting and curing of concrete is absolutely necessary for successfully using concrete in construction. We have already studied in the book on *Building Materials* that the grade of concrete is specified by its 28 day's cube strength. Thus M20 (20 mm) concrete means the concrete whose cube strength in 28 day's standard curing is 20 N/mm^2 and the maximum size of coarse aggregate used is 20 mm. The minimum amount of cement in all RC structural work should be 300 kg/m^3. Concrete used on works is specified according to IS 456 (2000) Clause 9.3 by one of the following three methods:

1. As nominal mix concrete for given strength
2. As designed mix concrete
3. Concrete mixed by traditional method by volume

Method 1: As nominal mix concrete. It is used for concrete of M20 and below. The proportion of nominal mix should be as prescribed by BIS which is shown in Table 10.1.

Table 10.1 Proportion of nominal mix
(IS 456 – 2000 Table 9)

Grade of concrete	*Mass of total aggregates per bag of cement* (kg)	*Fine to coarse aggregate ratio*	*Water in litres per bag*
M 20	250	1 : 1½ (for 10 mm)	30
M 15	330	1 : 2 (for 20 mm)	32
M 10	480	1 : 2½ (for 40 mm)	34
M 7.5	625	Generally 1 : 1.5 to 2.5	45
M 5.0	800	Generally 1 : 1.5 to 2.5	60

Method 2: As designed mix concrete. For the concrete required for high quality work, where field staff for quality control is available, we use designed mix. (Mix design is described in the book on *Building Materials.* It specifies the following:

(i) Grade of concrete
(ii) Maximum size of concrete
(iii) Minimum cement content in kg of cement per cubic metre of concrete
(iv) Maximum water-cement ratio
(v) Type of control assumed on the work

An example of designating such a mix will be as follows: "M20 (20 mm) concrete, minimum cement content 300 kg/m^3—Maximum w/c ratio 0.5—good control". This means that a concrete of 28 day's strength of 20 N/mm^2 with 20 mm maximum size of aggregate and not less than 300 kg/m^3 cement. The control at the site will be "good".

Method 3: By traditional method. In this traditional method, the mix is designated by volume batching according to Table 10.2. This practice is still followed at many sites in rural India. A mix 1 : 2 : 4 will mean a concrete made with one volume of cement, two volumes of fine aggregate (sand) and four volumes of coarse aggregate (stones).

Table 10.2 Assumed strength of various volume batched mixes of concrete with 20 mm aggregate

Mix	*Grade*	*Quantities per 50 kg of cement* (0.035 m^3) *yield* (cm)		
		Fine agg. (m^2)	*Coarse agg.* (m^3)	*Water* (lit.)
1 : 1 : 2	30	0.035	0.070	21
1 : 1½ : 3	25	0.053	0.105	23
1 : 2 : 4	20	0.070	0.140	25
1 : 3 : 6	10	0.105	0.210	32
1 : 4 : 8	–	0.140	0.280	32

Notes: 1. We use gauge boxes for measurement of fine and coarse aggregates (see Section 10.5).
2. Approximate yield is taken as two-thirds of the total volume of the mix. Yield for 1 : 2 : 4 mix = 2/3(0.035)(1 + 2 + 4) = 0.16 m^3 per bag of cement. (One bag of cement is taken as 35 litres in volume.)

10.2 THE CONCRETE CHAIN

The process of making concrete may be regarded as a chain of eleven links, as shown in Chart 10.1, the chain being as strong as its weakest link. We should not overemphasize or ignore any one of the links.

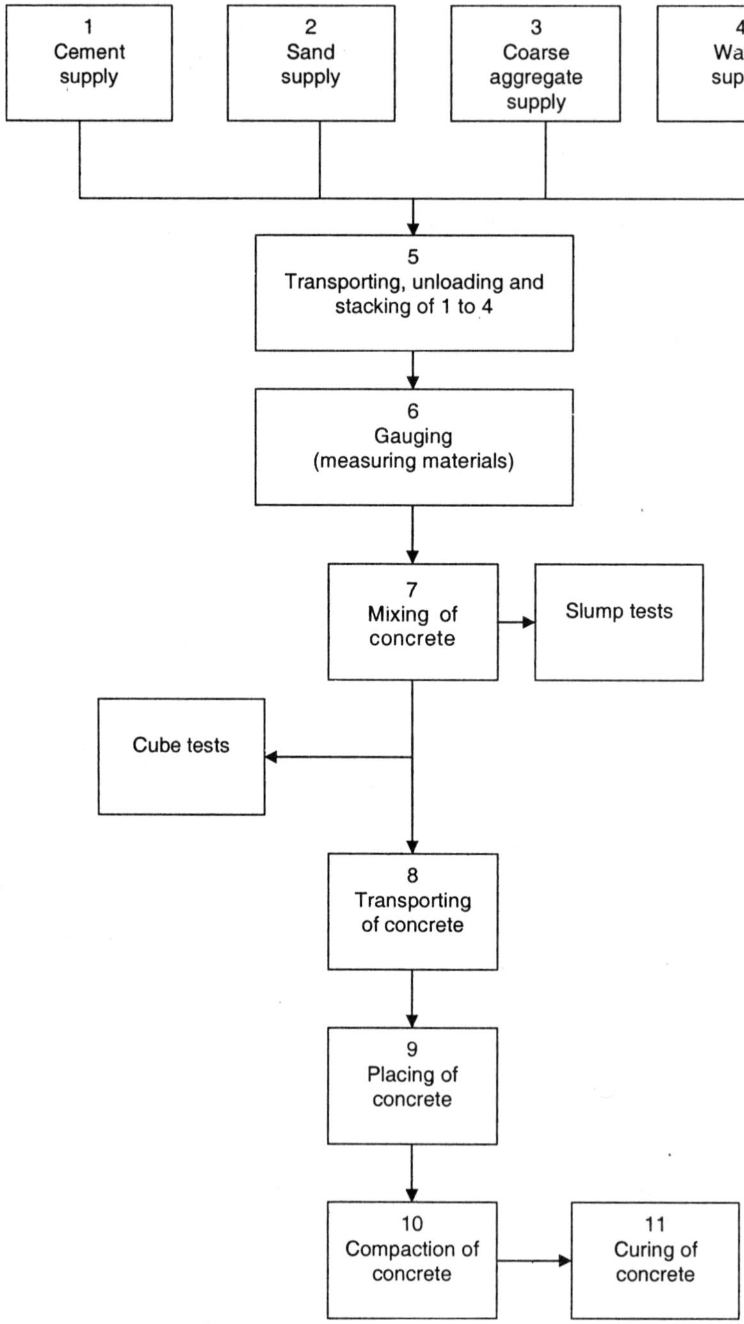

Chart 10.1 The process of making concrete.

The most important factor that constitutes the strength of concrete is its *water-cement ratio* and the two important properties that govern the ease of placing concrete are the *workability* and *cohesiveness*. Workability increases with water content and cohesiveness increases with the amount of fine particles (especially cement) in the mix, but increase of fine particles reduces workability for a given water content. For good concrete, care should be taken during all the three stages, namely making of concrete, concreting and curing of concrete, when concrete is hardening. Some of the aspects regarding the eleven links are discussed below.

10.3 CEMENT SUPPLY

The most important ingredient of concrete is cement. There are different types and brands of cement. The appropriate type should be used for the job to give the desired result. It should be properly stored. These are dealt with in the book on *Building Materials*. Nowadays in India, we get only blended cement (Portland cement mixed with pozzolana or ground slag) in the market. Cement is specified as grade 33, 43 and 53 which denotes its 28 day mortar strength. Special cements (sulphate resistant, quick setting, etc.) are also available.

10.4 AGGREGATE AND WATER SUPPLY

Coarse (stone) and fine aggregates (sand) as well as water also need careful inspection for quality. The aggregates should be stored properly. These are dealt with in the book on *Building Materials.*

10.5 GAUGING

The term *gauging* refers to *measuring the quantities of coarse and fine aggregates and cement* for making the concrete. In large works, aggregates and cements are weigh batched using batching plants. In smaller works, portable concrete mixers give good results. In very small works and in remote areas in India, concrete is hand mixed. For small works, the aggregates are usually measured by volume. It is ideal if mixing is done in units of bags assuming one bag weighing 50 kg and occupies 35 litres or 0.035 m^3. Alternately, cement can also be volume batched.

Measuring coarse and fine aggregates. When concrete is batched by volume, there is always a danger of variation between one batch and another. We should never allow the aggregate to be measured by the shovel. A shovelful, as a measure, is bound to be inaccurate. Whenever possible, have *gauge boxes,* made up especially for each job, to hold exact quantity required for the batch (or half or one-third of that quantity, if the whole would be too heavy to handle). Make sure the box is always completely filled, and the top struck off level with a levelling rod (Fig. 10.1).

Making the gauge box. In most building sites, gauging boxes are used only for measuring sand and aggregates. Cement is measured by bags. The inner dimensions of the

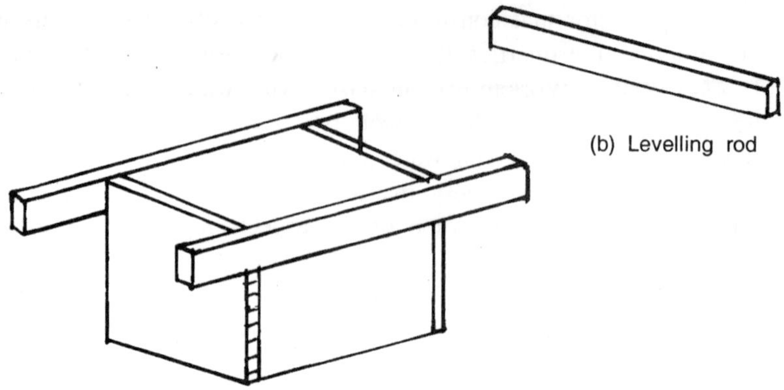

Fig. 10.1 Gauge box with handle and levelling rod.

boxes for aggregates are usually made 400 × 350 mm and 250 mm in height (which will be 35 litres equal to the volume of one bag of cement). If cement is also added in boxes, the height of the special box for cement is made 290 mm in height to allow for the bulking of cement. The boxes should be made of 30 mm (1¼ inch) thick prepared timber, which gives a good and strong job. Joints should be tongued and grooved. This prevents any dirt getting in if joints are open through shrinkage or rough handling. The faces of the joints should all be painted with red lead in linseed oil before the box is assembled. It is advisable to flash the top edge of the box with sheet metal to keep a clean level edge for striking off. It will also be of help if the box has a metal lip on the side which will rest on the mixer hopper when the contents are being tipped into the mixer.

Bulking of sand. A mix specified by volume is generally based on the weight of one bag of cement given as 1 bag (50 kg or 0.35 cubic metres or 35 litres as loose volume) of cement to so many cubic metres of sand to so many cubic metres of coarse aggregate as given in Table 8.2. These figures are for dry sand. But the damp sand, that is usual on most jobs, has a good deal of *bulking,* i.e. larger in volume than the same amount of sand when it is dry or fully submerged. The space taken up by a given weight of damp sand can, in fact, be as much as one-third greater than that taken by the same weight of sand, either dry or saturated. If this 'bulking' is not taken into account when batching by volume, the mix will be seriously under-sanded and will contain more cement per cubic metre than was designed. *So if we are using damp sand, we should increase the quantity of sand added. In ordinary situations, a quarter more than as much as is specified will account for the bulking of sand.* In important works, one will have to test the sand as described in the book on *Building Materials* to find the precise bulkage to adjust the mix.

Measuring water. The most important job of water in concrete is to make the cement set hard. Its first use is to chemically combine with the cement. The less the water, the stronger will be the concrete if it can be well compacted. The second use of water is for easiness of placement to make it easier to be placed at the site. The more the water content, the easier it is to place the concrete.

The first chemical process needs only very little of the water in the mix. Most of the water that is added is simply to make the concrete workable enough to be placed properly. All this extra water will evaporate out as the concrete dries, and as it dries, each drop will leave tiny holes in their places. Millions of drops, so small they can't be seen (something like the drops you can't see in mist or cloud of steam) all leaving millions of tiny holes and turning the concrete into a sort of solidified sponge. The more holes there are, the weaker the concrete will be, as obviously it will not stand up to crushing as a solid nonporous mass would. Hence the aim is to strike a balance and one must use just enough water to make the mix workable for the method of placing the concrete.

After deciding how much water to put into each batch (this amount has been carefully worked out by the engineer), we must stick to it. Our aim is to turn out concrete that is of the same quality right through the job. If one batch is dry and the next wet, we will have one batch strong and the next weak, and the strength of the structure will be that of the weakest concrete placed.

We have also to make corrections for water in the sand. Sand that is only just *damp may* have a moisture content 2 to 3 per cent but *very wet sand* may have as much as 10 to 12 per cent. That makes a big difference to the total water in the mix and to keep the concrete the same, we will have to add either more or less water at the mixer. We will be able to make an adjustment within the given limits, but if even after the adjustments, the concrete mix is coming through too wet or too dry, we should report it at once to the superior.

On large works, a gauge is incorporated on the mixer to measure the quantity of water dispensed to each batch. In such cases, we must make sure before work is started that the gauge delivers the right quantities. When the mixer gets jolted about or is standing on uneven grounds, the gauge can easily get out of order. It is important to see that the delivery pipe for water is not blocked at any time of the operation of the mixer. To check this, we should see first that the mixer is level. Then fill the tank and draw off a quantity of water we can measure, and see if the gauge shows that amount less in the tank. If not, set the gauge right before the commencement of the work, or take the difference into account to allow for the differences. If the mixer has no gauge, a container like a bucket that holds a definite measure, say two or three gallons, may be used. Measure the amount of water carefully the first time, and if it works out correct, stick to that measurement.

10.6 MIXING OF CONCRETE

Even though in India, on important large jobs, concrete is machine mixed, on small works and in remote places, it is hand mixed. Hence all civil engineers should know how to hand mix concrete. The following procedures of hand mixing (manual mixing) give good results.

10.6.1 Hand Mixing of Concrete

Hand mixing must be done on a clean well-built water tight platform (of iron sheeting, timber board or brick paved floor) large enough to provide space for the size of mix contemplated. If concrete is mixed directly on timber, tongue and grooved boarding shall be used to prevent leakage of mortar. The platform should have a ledge (projection) all around, to prevent loss of materials, especially the liquid sand cement phase.

The mixing platform should be as near as possible to the work and the maximum time for transporting the mix should not exceed 60 min from the time water has been added to the mixture.

The sand and broken stone are measured by volume in gauge boxes. Cement is preferably to be measured by weight or "mixing by bags" may be adopted. The methods of hand mixing of concrete are as follows:

Method 1. The preferred method of hand mixing is as follows:

(i) Measured quantity of sand is spread evenly.

(ii) The cement is dumped on the sand and distributed evenly.

(iii) The sand and cement are mixed intimately with spade turning the mixture over and over again until it is even coloured throughout and free from streaks.

(iv) The sand cement mixture is then spread out and measured quantity of coarse aggregate is spread on its top. Alternately, the coarse aggregate is spread out and the cement-sand mixture is spread on its top.

(v) The above layers are mixed together at least three times by shovelling and turning over by twist till the coarse aggregate gets evenly distributed in the sand-cement mixture. *It is then heaped into a pile.*

(vi) A depression is made in the middle of the mixed pile to receive the water.

(vii) Three quarters of the total quantity of water required is added and simultaneously the material is turned in towards the centre with shovels. The remaining water is added with a water can fitted with a rose head, while slowly turning the whole mixture over and over again until a uniform colour and consistency is obtained throughout the pile.

(viii) The mixing platform must be washed and cleaned at the end of day's work.

Method 2. Alternate method. The following method is also used in practice. Coarse aggregate is first spread on the platform; sand is then spread uniformly over it. Cement by weight or in units of bags is then evenly spread on the top. The materials are then mixed together by shovelling over to an adjacent part of the ledge and then back to the original position. The shovel being given a turn so as to spread out the material and mix up the ingredients. After thorough mixing, the mixture is heaped and water is added as specified above.

Alternately, in either method, after mixing of the concrete small portions are taken from the heap and then again mixed and shoveled from the platform in suitable containers to be placed on the work.

10.6.2 Machine Mixing

Good concrete can be obtained more easily when concrete is mixed in a machine. To get concrete that does not vary with the batch, it is to be mixed for the same length of time. This time is fixed by trial and error and is usually specified. In general, the minimum mixing time for small drum mixers is 2 to 3 minutes after all the materials, including the water, are in the drum. For larger drum mixers, it should be a bit longer. Pan mixers require less time (see Chapter 33). A batch which is mixed for less time than required will definitely be weaker than the one mixed for the full length of time.

A mixer that will hold materials for at least one bag of cement should be used on the works. *We should have approximately 250 litres of dry mix capacity for mixing one bag of cement.* The coarse aggregate is added first, then the sand and finally the cement. The dry materials are mixed at least four times and then water is added gradually while the drum is rotating. The following points should be noted:

The first batch. The first batch of concrete will probably be harsh and stony because some of the cement and sand would have stuck to the sides and blades of the mixer. Such a mixture must not be used for structural work. There are three things one can do about it. We can prevent that harshness in the first batch by putting more cement in the first mix with more water to keep up the water-cement ratio to make up for the amount that will stick to the mixer blades. Otherwise we can put in less (say half as much) coarse aggregates. Or, again, we can use the harsh first batch in a trench or some place where the quality of the concrete is not so important.

Measuring the consistency of concrete. In the field, the consistency and workability can be felt by making a ball of the mixed concrete and examining it. Workability is measured in the field by slump test studied under "Building Materials".

Upkeep of mixer. Good job needs good tools, and hence the mixer must always be properly maintained. Empty the water tank on the mixer every night, and fill it up afresh in the morning. During "lunch-time break" and also after completion of the day's work, it is a good practice to put a shovelful of stones and water in the drum and make the mixer running for about ten minutes. This will clean it out and prevent it from getting clogged. Clean the drum out thoroughly when work is stopped for the night. It will help one to keep the machine free of caked cement if it is rubbed over with an oily rag for cleaning it. Make sure the blades are not bent or broken and always keep the working parts of the machine well greased.

10.7 TRANSPORTING CONCRETE

Concrete is transported by pans, wheel barrows, dumpers, lorries, etc.
The following factors should be considered in transportation of concrete:

10.7.1 Important Factors

Drying out of concrete. Drying out can happen in very hot weather, if we are not careful. In hot sunshine or a strong wind, if the position of placing is at a large distance away, or if for any reason, we have to leave the concrete for a time transport, the concrete in a barrow, skip or lorry on its way from the mixer.

Segregation. Wet concrete consists of particles varying from cement powder to coarse aggregate size. Unless it is properly handled, the larger particles tend to separate from the smaller. This is called *segregation of concrete.* Segregation is a danger with too wet a mix.

Jolting the concrete on its way from the mixer to the construction site makes the larger stones sink to the bottom and the water and fine material separate off at the top. Segregated mix gives concrete that is patchy and badly mixed. Such concrete will be weak in some places and strong in other.

Should segregation occur by chance during transportation of the concrete, it should be remixed before placing (However concrete which has commenced setting should not be used). As pouring or tipping concrete into place from too great heights also produce segregation, it is usual to specify that ordinary concrete should not be placed from height not greater than 1 to 1.2 metres.

Consolidation. Consolidation during transportation is caused by jolting combined with a long stay in the container. It is more likely to happen with large batches of concrete. Dry concrete transported by lorry has been known to compact so hard that it had to be hacked out with a pick. It may be that a slight adjustment of the mix can in many cases put matters right.

Care in transporting the first batch of concrete. When using large containers for transport, one should be aware that the first batch of concrete when work is started or restarted has a risk of its own. The cement and fine materials may stick to the sides of containers leaving the concrete that is to be placed harsh and stony. To prevent this, we should wash out every container with water before starting the day's work. It is also desirable to pass cement grout through all containers before the concrete is placed in them.

Always clean out all containers thoroughly at the end of the day's work, or whenever a long break is likely, hose out barrows, lorries and dumpers, bucket and skips before the old concrete has time to dry on them.

10.7.2 Methods of Transportation of Concrete

The following are the usual methods of transportation of concrete:

By pan. In most small and medium-sized buildings in India, concrete is transported by labourers standing in a row and passing the mixed concrete from the mixer to the spot for placing concrete. This requires a large number of labourers and the work is also slow. The output depends on a number of factors including distance of the work from the mixer. On an average, 8 labourers are counted for 100 cft concrete on ground floor— 2 to 3 more labourers added for each upper floors.

By wheelbarrows. When concrete is transported by wheelbarrows in order to prevent segregation in wet mixes or compaction in dry mixer, it is important to avoid jolting the concrete. The wheelbarrows should run smooth, and scaffold boards should be securely clipped. The wheels of the barrows themselves should be kept at the correct pressure. Concrete should be covered in hot or windy weather to prevent its drying out and also in heavy rains from getting wet.

By dumpers and lorries. It is when dumpers or lorries are used that one should have to watch out more for segregation or compaction. The weight of the concrete itself is enough to compact the concrete, if it is in the container for a considerable time and jolting must be guarded against as much as possible to prevent segregation.

By hoists. Before a hoist is used, a check should be made to see that the guide raise are vertical and securely held against the scaffolding, that the base is fixed firmly, and the winch is in line with the bottom pulley. The whole apparatus should be examined periodically to see that all fixings are secure and that it is not pulling out of line.

By concrete buckets or crane skips. To avoid segregation, it is necessary to release the concrete as close as possible to the point of placing. Dropping the concrete from height not only encourages segregation, but may damage the formwork. Unless the concrete is specially designed for large drops, concrete should not be dropped more than 1 metre in height. Some crane skips have opening bottoms; others have to turn over to discharge the concrete. If this latter type of crane skips is being used, make sure there is enough space for the bucket to turn over without striking against the reinforcement on the formwork and displacing them.

By chutes. If a chute (particularly a long chute) is being used, the concrete mix must be of the right consistency to flow easily, without segregation. If the mix is properly designed, the concrete can be chuted direct to the place of pouring. One way of avoiding segregation is to discharge the concrete from the chute into a storage hopper and transport it from the hopper to the point of placing, in wheelbarrows or other conveyors. After any break in concreting, the chute should be inspected all along its length to be sure that it has not become blocked.

By concrete pumps. With the advent of ready-mixed concrete (RMC) in cities, concrete is usually placed by concrete pumps. It is a very convenient and quick method of placing concrete [see also Chapter 33 (Section 33.3)]. This method should always be encouraged as concrete can be placed most effectively and quickly with very few labourers at the site by this method.

10.8 PLACING OF CONCRETE

Placing of concrete refers to those factors that ensure proper *distribution of concrete* in the structure. First of all, one should remember that initial set of portland cement takes place in about 30 to 60 minutes after it is mixed with water. Hence concrete must be placed in position as soon after mixing as possible and not more than half an hour after mixing. This 30 minute period can be divided roughly as 10 minutes to be placed in the formwork and 20 minutes for working the concrete into position (for compacting it).

In RMC, concrete mixed in a central plant is transported in rotating drums mounted on lorries to the site. The constant agitation of the mix in the rotating drums *and addition of retarding agents* delays the initial set. As long as mixed concrete is kept agitated, it will not set for considerable time. (The same principle is used in cement plastering. The concrete mortar is reworked frequently during its use to prevent it from setting.) The important points to be noted in placing concrete are the following:

1. The inside of formworks should be inspected and approved before placing concrete.
2. Concrete should be placed as soon as possible.
3. There should be no segregation of concrete when it is placed.

4. It should be thoroughly compacted in position in layers. Each layer should be bonded to the next by vibrating through both.
5. The concrete should not be dropped from a large height but placed from the minimum height possible. The maximum free fall should not be more than about 1.5 m.
6. When concreting columns are more than 3.6 m in height, concrete should not be poured from the top of the column. Arrange to pour the concrete for parts of the height through intermediate openings. Pouring of column up to 3 m *above the bottom of the beam* or girder can be a continuous operation.

10.9 COMPACTING OF CONCRETE

10.9.1 General

Concrete of 125 mm (5 inch) or more slump is best placed by poking or rodding, but concrete with less than 40 mm slump can be placed satisfactory only by vibration. Vibrations give proper compaction to a concrete of medium to low slump. Heavily-reinforced sections cannot be compacted thoroughly by hand without a great deal of hard work. Because of the lesser need for water in vibrated concrete, mechanical vibration will also save as much as 15 per cent cement to produce the same strength of concrete. (However, under no condition, the cement content should be reduced to less than 300 kg/m^3, the minimum specified by code for durability.) With the advent of concrete chemicals, it is now possible to make self-compacting concrete which will fill the formwork such as water without compaction. Cast-in-situ concrete piles are always made of self-compacting concrete with high slump.

10.9.2 Compacting by Rodding

Concrete to be compacted by rodding should be free flowing and placed in thin layers. The surface of each layer should be worked over immediately before the succeeding layer is deposited, in order to break up any scum or film. Concrete should be thoroughly compacted and worked into place around the reinforcement and in the corners of formworks.

10.9.3 Compacting Concrete by Vibration

Vibration is the ideal and the only satisfactory way of compacting good concrete of moderate slump, but if it is used on a mix that is too wet (high slump), it can actually do harm. In a wet mix, the excess mortar will be squeezed out from between the lower stones. This will lead to the rise of the mortar to the top, making a surface layer of weak concrete that is short of coarse aggregate and more liable to shrinkage than the rest. So before using the vibrator, we must make sure that the mix is dry enough to be used successfully.

As a general rule, the vibrator should not be used to move concrete into place. Such procedures can cause the stones to separate from the mortar. When placing concrete that is to be vibrated, we should see that its surface is kept levelled as far as possible over the whole lift, so that the concrete will not need to be moved from one place to another by the vibrator.

Formwork for placement by vibration. All formwork to be used with vibration needs to be extra stout and well made as the strain on it is much greater than with hand compaction. Timber forms will need extra battens and bracings, and wedges should be nailed in. Special formworks should be made up with screws rather than with nails. It is surprising that how easily a six-inch nail can be shaken out by vibration. If steel forms are being used, the panels must be sufficiently braced to prevent bulging. Extra bracings and wall ties may be needed for this purpose. Also when assembling the formwork, we should be especially careful to see that joints are close fitting so that under vibration, cement and water will not leak out through any gaps, however small. Otherwise we will be left with honeycombed concrete.

10.9.4 Checking Depth of Concrete in Slabs

It is very necessary that the depth of concrete placed in slabs should be exact. Sufficient cover for both top and bottom steel should be provided. The depth of placement of concrete in slabs can be controlled by using an 8-mm reinforcement bar bent as a depth gauge as shown in Fig. 10.2. It is pushed into the concrete at various locations to control the depth.

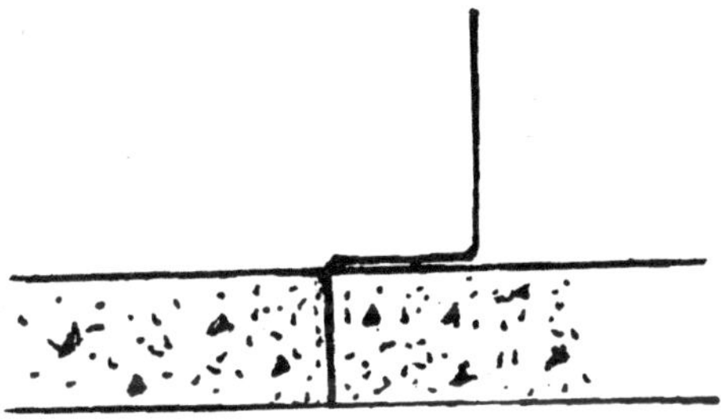

Fig. 10.2 Depth gauge made from bent reinforcement rod to check depth of concrete.

10.10 JOINTS IN RC CONSTRUCTION

Joints provided in reinforced concrete construction can be classified as follows:

1. Construction joints
2. Expansion joints
3. Contraction or control joints (dummy joint)
4. Sliding or slip joints

Construction joints. These joints are provided at places where placing of concrete has to be stopped for some reason during construction. In general, their locations are different from the location of the expansion and contraction joints which are incorporated in the design of the structure.

Expansion joints. These joints permit expansion and contraction. They are provided to allow for the movement of the structure and hence they come under movement joint.

Contraction or control joints. These joints allow only contraction. They generally consist of a simple butt joint without any bond. They are shrinkage joints to allow shrinkage. Otherwise they will lead to excessive cracking. A dummy joint is a partial joint used to allow contraction.

Sliding joints. These joints are usually formed by introducing smooth layer of plaster between the two surfaces. (Thus roof slabs are cast on brick walls after plastering them with a smooth surface and applying a coat of whitewash.)

A brief account of the desirable positions of construction and expansion joints is given below.

10.10.1 Construction Joints

Construction joints in beams. All beams (as well as slabs) shall be preferably filled to the top surface in one continuous operation from the beam bottom to top of the floor. Work should be so planned that, as far as possible, there will be no stoppage of work in the beams in T-beam constructions. If absolutely necessary, stoppage can be planned and construction joints can be introduced as follows:

In beams and girders, some codes advocate the work to be stopped at points of minimum shear. This is at the centre of beam with a vertical plane at right angles to the direction of the main reinforcements. However, to avoid the possibility of separation due to lack of care in joining, most engineers prefer to have the construction joints in beams at the centre of supports (columns) or within the middle third of the span. It should never be near the support (i.e the region where the shear is large), unless extra shear reinforcement is also provided to take the entire shear without assistance from concrete.

Where there is a cross beam intercepting the girder, the construction joint to the girder is offset a distance of at least twice the width of the cross beam. The joint is finished against a properly fixed stop board (Fig. 10.3), slotted to allow the passage of reinforcement. Great care should be taken in joining the new construction with the old, as otherwise separation of the work may be shown up later at the place where the work was stopped.

Construction joints in slabs. In *one-way slabs,* the natural stopping place is along the longitudinal centre of the supporting beam or wall, making the planes perpendicular. Special timber with holes for the reinforcement to pass through will be needed for this purpose. In the case of *two-way slabs,* concreting may be stopped somewhere within the middle third region in either direction.

Slabs continuous over supports such as walls (continuous slabs) can be stopped over the middle of the supports if care is taken to have enough continuously bars connecting the two sides of the slab together. Slabs have to be stopped for expansion joints which should be provided for every 15 to 18 m length of slabs. Roof slabs expand more than the floor slabs.

In cases of roof slabs resting on outer walls and where the outer face of the building is plastered, the RCC slabs should be as a rule be carried through to the outside of the walls and a small parapet built on it. Where the outer faces are pointed, the RCC slab may have to be stopped short of the face by half a brick with a small provision for waterproofing, and the space so left should be built up with bricks to suit the pointing. These places can be a source of leakage and special care should be taken to treat this place. The joint should be treated with asphaltic composition (see Figures 9.1, 27.1 and 27.2.)

Construction joints in a floor made of main beams, subsidiary beams and slabs, the construction joints should be planned as shown in Fig. 10.3.

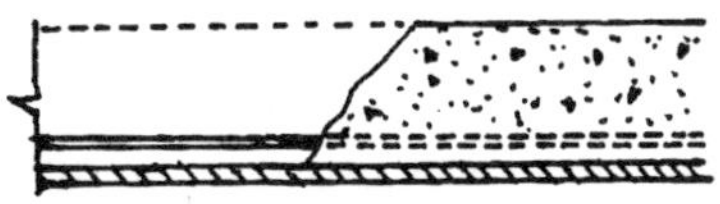

(a) Prohibited type of joint

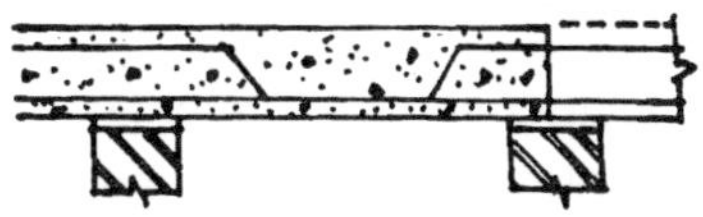

(b) Recommended joints in beams or slabs over supports

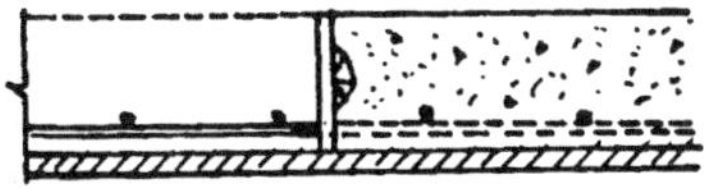

(c) Allowed type of joints in slab with provision of stop board

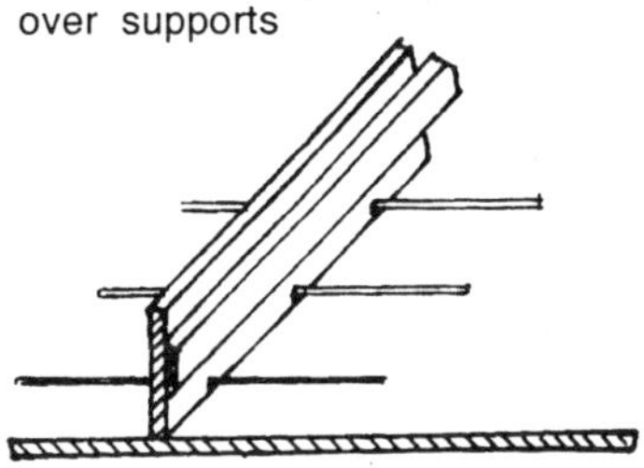

(d) Details of stop board

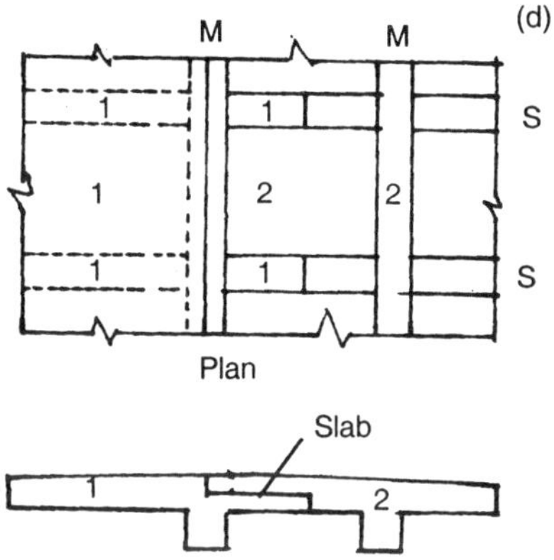

(e) Recommended joints in RCC slab, scondary beam and main beam construction (S = secondary beam, M = main beam): 1. First placing, 2. Subsequent placing.

Fig. 10.3 Joints in construction of reinforced concrete.

Construction joints in columns. Nowadays with good formwork and specially designed cohesive mixes (rich mixes), it is possible to cast one-storey high columns (3 m) in one lift. Columns are first stopped at a horizontal level below beams to allow concrete to settle before beams are constructed. Otherwise the column settlement and shrinkage may cause separation between the two parts. The portion of the column with the beam is concreted later with the beams. The old practice for continuing the column was to lap the bars just above the slab. For this purpose a kicker (i.e. a bed for placing the starter steel in the column above the floor level) is also provided above the level of the slab when concreting the slab. It should be at least 70 to 75 mm high above the floor level and carefully constructed. The formwork should be designed to facilitate the formation of the kicker. The optimum time for treatment to form the kicker is usually two to four hours after placing of the concrete. The kicker enables the column steel to be easily placed and also prevents water from accumulating at column base during cleaning up operation of the column formwork before concreting. However, *according to the new code for ductile detailing to resist earthquake forces, the lapping of bars in columns should be at mid-height of columns only.*

10.10.2 Expansion Joints in Building Construction

Expansion joints consist of gaps(10 to 40 mm wide) depending on the type of structure expected temperature movement, etc. provided at various places of the building. Movement can take place due to variation of temperature, moisture, etc. Temperature effects will also depend on the time of construction, whether it is summer or winter.

Thus in masonry walls in buildings where cross walls restrict the main walls, the joints need not be closer than 25 to 40 m apart. However, in long compound walls exposed to the sun, the joints should be closer. In reinforced concrete construction, IS 456 (200) clause 27 stipulates that expansion joints to be provided in changes of direction and structures more than 45 m in length are to be designed by providing one or more expansion joints. It is also customary to provide for expansion of slabs bearing on walls by providing slideable bearings (see Figures 21.5, 27.3 and 27.4). The recommended expansion joints in continuous lintels are shown in Fig. 10.4.

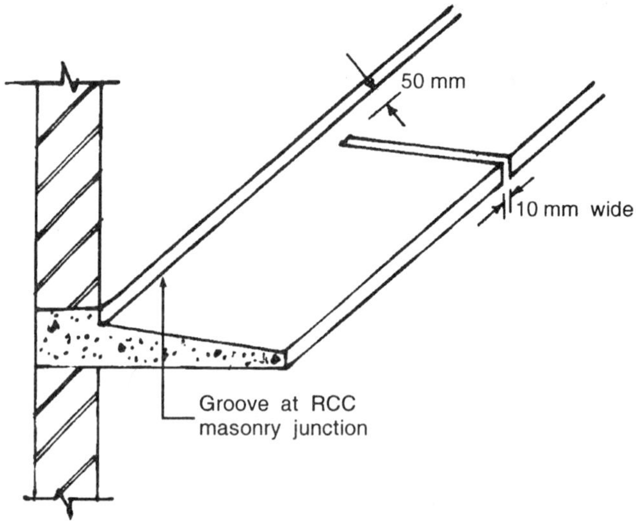

Fig. 10.4 Expansion joints in long sunshades for prevention of cracks due to temperature effects.

10.11 CURING OF CONCRETE

10.11.1 Key Factors

After placing concrete in position, it should remain protected for sufficient time to become strong and durable. Where durability is critical, keeping the concrete damp is very important. If much of water is allowed to evaporate, from the fresh concrete the skin of the concrete will be of poorer quality than the "flesh". Producing a good skin is the key to make a durable and impermeable concrete especially for the roof and this can be accomplished easily by proper surface curing. This aspect is very important in casting roof slabs.

10.11.2 Methods of Curing

Curing is different from removal of shuttering, which is called *striking of formwork*. In most cases, curing is continued even after the shuttering is released. In ordinary curing, the exposed surfaces are kept damp for at least 14 days and, if possible, for 28 days after placing of the concrete. The concrete should also be protected from hot sun and drying winds for this period. There are many methods of curing but the most common and effective method is the water curing.

The common methods of curing are the following:

1. Water curing by ponding, say 50 mm depth over surfaces such as slabs.
2. Water curing by spraying of water on surfaces covered with hessin straw or other absorbent materials. Horizontal surfaces such as slabs can be covered with 50 mm of sand and this sand is kept wet by sprinkling water over it at regular intervals.
3. Curing by application of curing compound
4. Steam curing

Curing of horizontal surfaces. The ideal form of curing horizontal surfaces *such as roof slabs* is to allow water to stand on concrete to a depth of 50 to 75 mm as soon as the concrete surface is hard enough to be watered without injury. This is done for not less than a week and then the concrete covered with wet straw, wet sacking or any other non-conducting material for three weeks. This considerably helps develop the skin of the concrete and prevent leakage of roof slabs.

Curing of vertical surfaces. Vertical surfaces of concrete generally has formwork and these can generally be kept in position for several days after the concrete is placed as it will serve the purpose of preventing rapid loss of water and thus assist the curing of concrete. In very dry weather, it may be necessary to spray the formwork with water to assist the curing of concrete. Vertical surfaces of columns, etc. can be covered with straw and water cured.

Curing by curing compound. Another method is to spray a chemical curing compound which will form a membrane which prevents moisture to escape. It is not generally used for ordinary building works in developing countries as it is costly.

Steam curing. For factory-made products such as railway sleepers, a large amount of money has to be spent on the moulds (formwork) for the products. For quick release of moulds within 24 hours, we resort to steam curing at atmospheric pressure. The necessary strength to demould the product can be obtained by steam curing for 10 to 14 hours. Another superior method as used for aerated light-weight concrete production is curing by high pressure steam which we call *autoclaving* (steam under pressure). These are not used at building construction sites.

10.12 STRIKING OF FORMWORK

Care should be taken before concreting to see that all formwork is strong enough to carry the weight of the wet concrete to be placed on it, without bulging or deflection. When the cement is wet, the shuttering should be tight enough to prevent the loss of cement grout. All struts supporting the formwork should be braced and wedged, preferably at the bottom, so that they can be removed, as necessary, without vibration. Struts for some of the immediate lower floor of buildings may have to remain in position even after the job on that floor is completed, as the lower floors will invariably be used for the storage of building materials. If bricks are stacked on floors, it should be well stacked and should be over beams and near their supports. In order to reduce the total amount of supports and shuttering, that are needed on the project and thus save cost of construction, it is a good practice to plan the construction of shuttering in such a way that the shuttering materials that are allowed to be removed earlier (such as those on sides of beams) can be removed with ease and used in other places.

IS 456 (2000) Clause 11.3.1 stipulates that forms should not be stuck until the concrete has reached a strength at least equal to twice the stress to which the concrete may be subjected to at the time of removal of formwork. In normal conditions where ordinary Portland cement is used, forms can be removed as shown in Table 10.3. Whereas bottom struts of beams should be removed only after 14 to 21 days, props of slabs can be removed in 7 to 14 days, sides of columns and beams can be removed in less number of days.

Table 10.3 Removal of formwork after concreting

Description of work	*No. of days*
Vertical faces of structural members such as columns, walls, beams	16 to 24 hours
Soffits (lower surface) of slabs (props left under)	3 days
Soffits of beams and (props left under)	7 days
Removal of props under slabs:	
(a) Spanning up to 4.5 m	7 days
(b) Spanning over 4.5 m	14 days
Removal of props under beams:	
(a) Spanning up to 6 m	14 days
(b) Spanning over 6 m	21 days

10.13 FINISHING OF CONCRETE

10.13.1 General Repair

Immediately on removal of forms, the concrete work must be examined and the defects, if any, should be made good. Work that does not satisfy strength should be rejected. Minor surface defects can be rectified immediately after removal of formwork as follows. (Care should be taken to adjust the colour of the surface, if required, by using cement of same colour.)

(i) Bulges and ridges due to formwork, etc. can be removed by careful chipping or by a grinding stone. Honeycombed and other defects should be chipped out with edges being cut as straight as possible and perpendicular to the surface or preferably slightly undercut to it to provide a key at the edge of the patch.

(ii) Shallow patches are first treated with a coat of thin grout composed of one part of cement and one part of sand and then filled with mortar of the concrete work. The mortar is placed in layers of not more than 10 mm thick and each layer given as scratch finish to secure bond with the succeeding layer. The last layer is laid to match the surrounding concrete by floating, rubbing or tooling while the mortar is still plastic.

(iii) Holes left by bolts are filled with the same mortar as of the concrete work with just enough water and are packed as tightly as possible. The surfaces must be cured as soon as possible after the patch is finished to prevent early drying and shrinkage.

10.13.2 Surface Preparation for Plastering

In most buildings in India, RC work like underside of slabs (ceiling) is plastered to give a neat appearance and for convenience in painting. Reinforced concrete should be so carried out that the thickness of plastering required should not be more than 10 mm. It is very important to remember that the surfaces to be plastered should be properly roughened *immediately after removal of shuttering by roughening by hacking at close spacings.* Concrete chemicals are also available which can be painted on the concrete to assist bonding of plaster. The use of these chemicals, though costly, is more convenient than hacking. Proper bonding of concrete and plaster is very important under floor and roof slabs (of ceilings). There have been many field cases where hacking of concrete has not been effective as it was taken up after a long time after removal of shuttering and by that time, the concrete had hardened too hard to make indentations.

Where brickwork is to be built over concrete (as in water tanks), it is the usual practice to place one *brick over the concrete while the concrete is still wet.* This ensures the brick to stick strongly to the concrete and ensures a watertight joint between concrete and brickwork. This detail is very important in construction of sumps and overhead water tanks for buildings.

10.13.3 Surface Preparation for Floor Finishes

The surface of floors *on which screed or other finish is to be laid later must be* roughened with brushes or a stiff broom at the end of concreting itself while the concrete is still green, without disturbing the bottom concrete.

10.13.4 Special Curing and Preparation of Roof Slabs

It is advisable to allow the smooth skin to be developed on the top surface of the roof slab to increase its water tightness. This can be ensured by proper curing. The top surface should also be finished even and smooth with a *wooden trowel before the concrete begins to set.* Steel towel tends to create a smooth cement rich surface which tends to crack on drying. This roughness will also assist to lay the weathering course above.

10.14 BONDING OF NEW CONCRETE TO OLD

The main points to be noted in bonding of new to old concrete are the following : If bonding of old and new concrete is not satisfactory, with differences in shrinkage of the old and new materials, the joint separates. Faulty joints result in weakening of the whole structure and making it possible for water to percolate through these joints. Oil, grease, dirt, clay, scum, etc. must not be allowed to come into contact with the surface of the older concrete against which the new is to be placed, as these substances prevent good adhesion. As a rule, concreting should be carried out as continuously as possible but, as interruptions are inevitable, the following points should be observed.

Case (i). Where the concrete against which the new concrete is to be placed is *not more than four hours* old. At the end, where work is stopped as layer a "laitance" film is usually formed, this being a chalky mass of non-cohesive material, loose in texture. It is actually very fine cement in the presence of an excess of water and it results in poor adhesion of new to old work. It is, therefore, essential that this shall be removed before placing the new concrete against the old. One of the methods is to fill the forms slightly more than required and the excess is removed before the next concreting. This results in the removal of the poor concrete. This should be done not later than 4 hours after it has been placed in position.

An alternate method for joining to fresh concrete surface, when the concrete has set but not hardened (is not more than 4 hours old), is by spraying the surface with a fine spray of water and brushing with a stiff brush to remove the outer mortar and expose the larger aggregates without disturbing them. The new concrete is then added, the mix being sufficiently plastic to enable it flow sluggishly in position when tamped. If the mix is too dry, the concrete at the bottom of each layer will be porous and if it is too wet, there may be segregation of the aggregates and the formation of laitance, and the work will shrink excessively on setting and hardening.

Case (ii). Where the old concrete has been in position *for more than four hours but less than three days*. In this case, the laitance should be removed as before; the old work brushed with a steel wire brush and thoroughly washed with clean water to remove loose particles, dirt, dust, etc. Cement mortar of composition similar to that in the concrete mix, excluding the large aggregate, should now be applied in a plastic consistency and thickness of about 12 mm (1/2 inch). The new concrete is then placed upon this mortar grout and well punned towards the joint. Some prefer to use a neat cement slurry at the rate of 2.75 kg of cement per sq.m. instead of the above mortar.

Case (iii). If the concrete is too hard and where the above treatment is not practicable, sand blasting or a needle gun must be used to remove the surface skin and laitance. Hacking of the hardened surface should be avoided as much as possible.

Notes: 1. In case of rapid hardening cement, the same methods can be adopted, but, owing to the rapidity with which hardening takes place, chipping is necessary for bonding to concrete over 24 hours old. On no account, must aluminous cement be bonded to ordinary cement concrete which has only partially hardened or vice versa, since an interaction may occur which will greatly reduce the strength. If, however, the old work has hardened thoroughly, there is no objection to bond it with a different type of cement.

2. Nowadays there are latex-based concrete chemicals that will assist bonding of new concrete to old concrete as used in repair of concrete. [refer Chapter 39 section 39.8.2].

3. Another method of designing two stage works (such as concreting a fixture to a slab after the slab has been cast and cured) is to concrete the initial part using the lost formwork described in section 11.3.4.

10.15 UNDERWATER CONCRETING

The old practice of underwater concreting was to fill cement bags with dry or semi-dry mixture of cement and aggregates and deposit these bags in a specified order (as in masonry work) by a diver. This method is not satisfactory as some cement is washed away and there will be voids in between the bags. It is also not suitable for the situations such as concreting of piles. Two other methods of concreting underwater are described below.

10.15.1 Tremie Method

The tremie method is the method used for concreting piles. The word *tremie* in French means hopper. This method uses a hopper and a tremie pipe which is a pipe of about 20 cm (8 inches) in diameter. The tremie pipe is made of easily fixable and easily removable watertight sections of steel pipes. It extends to the bottom of the hole in the water. Initially the pipe is plugged at its end with an easily removable plug. The tremie pipe is placed over the place to be concreted (say the bottom of a hole for a pile). The pipe and hopper are filled with concrete of high slump (15 to 20 cm). The pipe is then slowly lifted up a little when the plug detaches and the concrete flows out. Further concreting is carried out with a part tremie pipe removed but with the bottom of the tremie pipe always inside the concrete by at least 1 m.

This type of concreting works best when the concreting has to be done in a bentonite solution as in bored piles. The concrete is placed by displacement and the concrete that is always in contact with the water or bentonite is finally spilled over and wasted. This method of underwater concreting has been reported to give excellent results in pile work.

10.15.2 Preplaced Aggregate (Colcrete)

Another method of concreting underwater is to use the preplaced or *prepacked concreting*. This method has been used for construction of large bridge piers and in places where large amount of concreting has to be done without any stoppage underwater. In this method, the aggregates are first placed with grout pipes in a formwork. Cement sand grout is then pumped with the voids of the aggregate being filled by grouting techniques.

SUMMARY

Mixing, transporting, placing, compacting, curing and removal of formworks are important operations in concreting. These should be carried out with sufficient care to get the best results. Similarly enough attention should be given to stopping of concrete and also joining old concrete surfaces with new concrete.

REVIEW QUESTIONS

1. How do you specify the concrete mix to be used for works according to IS 456–2000?
2. Indicate the old practice of specifying concrete mix. What is the mix required for M20 concrete according to this method? How will you calculate roughly the yield?
3. What are the sizes of measuring boxes in the field when concrete is proportional by volume? What is the volume of one bag of cement?
4. Describe the procedure of:
   (a) Hand mixing concrete
   (b) Machine mixing concrete
5. Where do you stop concreting in:
   (a) A slab
   (b) A beam
   (c) A column
6. What is meant by curing of concrete? Give the various methods of curing.
7. When do you remove shutting in:
   (a) Slabs
   (b) Beams
   (c) Columns
8. How do you join new concrete to old concrete?
9. How do you prepare the underside surface of a concrete slab (ceiling) for plastering?
10. Describe the tremie and pre-packed methods of underwater concreting.
11. How do you concrete a column of a building 3 m high? How do you detail the stoppage of concrete of the column beam junction?
12. What are the types of joints introduced in concrete construction? Draw the arrangement of a continuous sunshade with provision for expansion due to temperature effects.

Chapter 11

Temporary Works: Formwork and Scaffolding

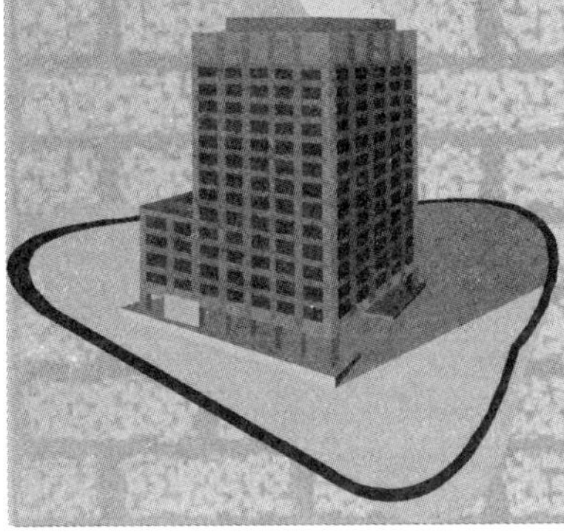

11.1 GENERAL

When undertaking works such as concreting, brickwork, plastering, erection of prefabricated members, it is necessary to install some works to carry loads temporarily or to give access for workmen to the works. These are called *temporary works*. Such temporary works are indispensable for success of any project and form a good part of the cost of the structure so that every civil engineer should have some knowledge of the types of temporary works that are used in practice. Centring, formwork, scaffolding and shoring are the principal types of temporary works. In this chapter, we will briefly examine their use in building construction. Appropriate temporary works are needed for the safety of the workers.

11.1.1 Centring, Shuttering (Formwork), Scaffolding and Shoring

There are many types of temporary works. Temporary work used for construction of arches is called *centring*. (This term is also used loosely for supports of formwork for concreting.) Centring for arches is described in Chapter 8. The temporary works used as a mould in which fresh concrete is poured for it to harden is called *formwork* or *shuttering*. Temporary works erected for construction of masonry works, plastering, painting, etc. is called *scaffolding*. Temporary supports used to prop up buildings for repair are called *shores* and such works are called *shoring*. In this chapter, we will briefly deal only with the temporary works for concreting (formwork) and scaffoldings.

11.2 TYPES OF FORMWORKS FOR CONCRETING

Formworks can be grouped into three types, namely elemental, standard and special.

Elementary type of formwork. In this type of formwork, each element of the temporary work is not more than about 30 kg, requiring not more than two men to lift each unit. These components are joined together to form the temporary work. Thus wooden planks, steel panels, etc. (for formwork), wooden beams, light open web steel trusses, etc. (for horizontal members), ballies (bamboos or jungle wood) or adjustable steel posts (for vertical members) can all be joined together to form the temporary work.

Standard formworks. These are standard elements up to 70 kg requiring up to three persons or cranes to lift them. They have been standardized for different types of works.

Special formwork. Formworks such as slip forms, moving forms, special formwork for water tanks, which require jacks, cranes, etc. to move them around, come under this class.

11.2.1 Lateral Pressures on Formwork

The magnitude of lateral pressure on formwork from the fluid concrete depends on many factors such as its consistency, rate of concreting, degree of vibration, etc. One of the commonly adopted pressure variation with depth is as shown in Fig. 11.1. It has a linearly varying hydraulic pressure from zero at the top to 0.4 W newtons per sq m at 0.4 m below the top, where W is the weight of concrete in N/m^3 (Fig. 11.1). Another practice is to take pressure as liquid pressure due to concrete poured in 3/4 to 1 hour only as by then, the concrete below that level would have set and became solid.

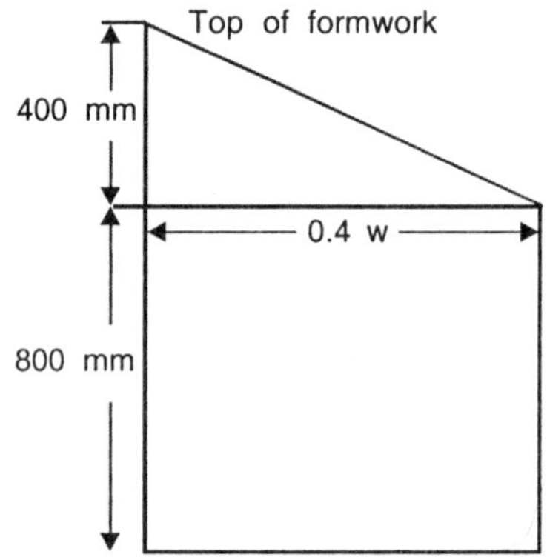

Fig. 11.1 Lateral pressure on formwork due to placing of concrete (w = unit weight of fresh concrete).

11.3 TEMPORARY WORKS FOR CONCRETING

It is estimated that 20 to 25 per cent of the cost of construction in buildings and much more in bridges goes for temporary works. Formwork is usually paid in square metre basis. The components of the temporary work for concreting are as follows:

1. *Formwork* or shuttering
2. *Props* or scaffolding (also called centring) to support the formwork (also loosely called centring)
3. *Access scaffolding* which gives a safe means of access and working place including space to support the necessary equipment for placing and compacting the concrete.

A number of fatal accidents have happened due to faulty formwork, faulty method of supporting them and wrong order of concreting the work so that one should always be watchful to properly design temporary works and make them vertically as well as laterally strong. Lateral strength of vertical supports, especially when concreting large areas, should be given special consideration in design of formwork. There are a number of cases of failure where lateral movement of formwork while concreting is not envisaged.

Even though in large cities and towns in India, where work is in plenty, the use of steel formwork is becoming more and more popular, a large part of the work away from towns is still done in traditional timber formwork.

11.3.1 Temporary Works Using Timber

As we have already seen, temporary works for concreting include the formwork or shuttering as well as the scaffolding or centring. One should always remember that these works should be constructed in such a way that removal of formwork can be done easily as planned. The materials and sizes of timber usually used for the different parts of the temporary work are given in Tables 11.1 and 11.2 respectively. A discussion of normal works of height less than 3.5 m and span less than 4.5 m is given further.

Table 11.1 Materials for temporary works

S. No.	*Material*	*Specifications*	*Application*	*No. of reuse*
1.	Timber	Soft wood and hard wood planks, as well as Ballies (for props)	All uses	1–10
2.	Plywood	Shuttering plywood 18–20 mm without wood backing, 4–10 mm plywood with wood backing	Form facing (liner)	10–100
3.	Steel	3 to 4 mm plates and Class C tubes of 40 to 48 mm diameter	All uses	100–600

Table 11.2 Timber sizes for formwork

S. No.	*Use*	*Sizes in mm*
1.	Thickness of planks for beam bottom, sheeting, wall boarding and sides of beams and columns	25 to 40
2.	Joists, ledges (cross section)	50 × 75 to 50 × 150
3.	Raw poles, Ballies (diameter)	Diameter not less than 100 at midlength and 80 at thin end.
4.	Posts (cross section)	75 × 100 to 100 × 100
5.	Column yokes (cross section)	50 × 75 to 75 × 100
6.	Diagonal braces	150 × 32 to 150 × 150

Formwork. Nowadays for formwork, special types of plywood (with water-resistant bonding agents) or steel are extensively used. Plywood panels are formed with 100 × 25 mm dressed timber backings nailed with short and thin nails up to 150 to 250 mm centres. When using thin plywoods (4 to 10 mm thick), they are given a solid backing of timber nailed along the four edges and intermediate outside surfaces. The edges of plywood sheets being joined are tacked to the same backing board to ensure a smooth joint. When using intermediate thick plywood (10 and 16 mm) also, they are nailed to a skeleton backing of dressed timber. Thicker plywood (20 mm) can be nailed direct to cross pieces at a maximum spacing of 450 mm. For spacing larger than 450 mm skeleton backing will be required.

Centring. Round poles, bamboos and cashurina poles are extensively used for propping and centring. Bamboo and similar poles should not be used for heights or lengths exceeding 3 m unless they are properly braced. In no case, one should use round wooden poles of diameter less than 100 mm at mid-height and 80 mm at the thin end. The maximum permissible spacing should be 1.2 metres centre to centre. The pole should rest on square wooden *sole plates or base plates* of 40 mm thickness and minimum area of 0.1 sq m. (1ft × 1ft). *Double wedges* should be provided between the sole plate and the wooden props to facilitate tightening and easing of shuttering without jarring the concrete. Brick masonry pillars of adequate section are also used instead of props. In such cases, wooden sole plates must be provided at the top of pillars and double wedges inserted between the sole plate and bottom shuttering.

In building construction, the base concreting of the ground floor is recommended to be completed before the upper floor or roof is concreted. This will provide a firm base for the centring to be erected for the construction of the floor or roof above.

In constructions where lateral stability is to be ensured by presence of already cast columns and peripheral beams, only light bracing between the vertical props are required. However, in situations where there is no proper lateral restraint as in large panels of floor to be cast along with beams (or slabs built in quick succession with beams), special care should be given to the lateral stability of the system. Inclined shoring from outside together with strict watch over the programming of pouring concrete (so that as little of eccentric loading as possible is produced) should be exercised. There are case histories of accidents where this possibility of lateral instability was forgotten.

As already remarked, when constructing multistoreyed buildings, the normal props for first floor slabs should be provided, if possible, on the already cast and cured ground floor base concrete below. Usually for the slabs above the first floor (top floor slabs), support in line with the props is given to the supporting floor also by the floor immediately below the supporting floor. Planks should be provided at the top ends of these props so as to give an even distribution of load. Usually the floors to be used for support shall not be less than 14 days old after complete curing. In case of balconies and cantilever beams placed one above the other, the propping should be such that the load is distributed *to at least two completed floors* below.

Before concreting, all the props and wedges should be checked. During concreting, at least one carpenter or fitter must be readily available at the site to keep a constant watch over the supports so that the props and wedges do not get loose during the concreting.

On removal of shuttering, the loads are transferred to the concreted structure. Removal of forms in structures such as domes and shells should be so programmed that the order of loadings on the finished structure is safe. Symmetrical removal gives a better distribution of these loads. Figures 11.2(a) and 11.2(b) show temporary works used for a concreting column and for a beam and slab construction respectively.

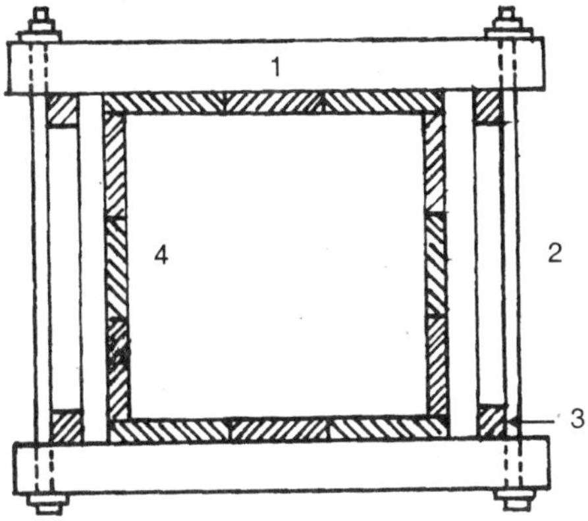

(a) Plan of formwork for column

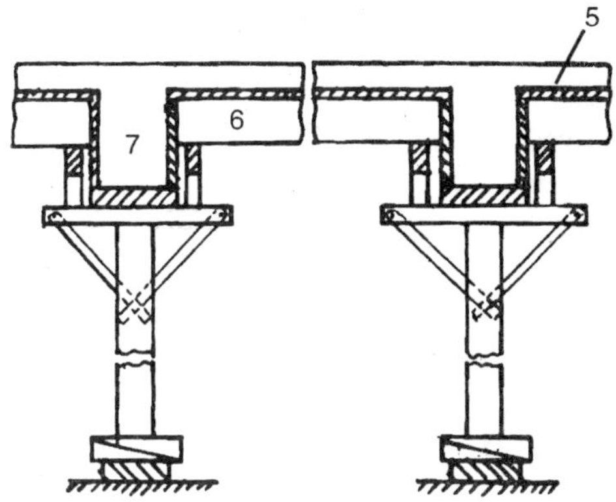

(b) Section of formwork for a floor with slab cross beams and mainbeams.

Fig. 11.2 Types of formwork RC construction: 1. Yoke, 2. Bolt, 3. Wedges, 4. Planks for sides of column, 5. Slab, 6. Cross beam, 7. Main beam.

11.3.2 Steel Proprietary Systems for Formworks

There are many proprietary systems of *steel formwork* generally consisting of panels made up of steel sheet and light steel angle framing, in sizes that can be easily handled. Special telescopic units for beams which can be easily removed, props with adjustable ends designed to facilitate easy erection, etc. are available in the market. In general, they are so designed that they can be formed into horizontal and vertical frameworks which is very stable against lateral movements. These systems must always be erected according to the manufacturer's details.

11.3.3 Surface Treatment of Shuttering

The surface of wooden shuttering that comes into contact with concrete should be well treated before concreting. In all cases, adhesion of concrete must be prevented. The surfaces should also facilitate plastering. They can be done by any of the following ways:

1. Coating with a soap solution prepared by dissolving yellow soap in water to get the consistency of paint.
2. Coating with raw linseed oil or other "form oil" of approved quality. (Unapproved oils should not be used.)
3. Coating with special form coating compositions available in the market.
4. Provided with a material such as polythene sheet.

These formulations should be removed from underside of slabs (ceilings, etc.) immediately after removal of formwork and hacking of the underside of slab carried out soon.

Otherwise the bonding of plaster to the ceiling will be very poor and special treatments may have to be carried out later to ensure good bonding.

11.3.4 Lost Shuttering

Lost shuttering as different from temporary shuttering is left with the cast concrete and not removed and reused after concreting. It is a valuable device when holes are to be left in members such as slabs which later may or may not have to be filled up with special devices. In such cases, steel mesh of close spacing, which can retain the aggregate, lines the holes. The mesh retains wet concrete and offers a rough key surface for the subsequent pours without the need for scabbing or other surface preparation. The technique in a modified form is used in countries where labour is costly in removing and re-oiling temporary formwork.

11.3.5 Provision of Camber

Suitable camber in horizontal members should be provided *to offset effect* of deflection of member after removal of term work. The usual cambers are 1/250 (4 mm per metre) for beams and slabs. We provide 1/50 of the projected length for cantilevers at the free end. Camber has to be intentionally provided for drainage of roofs.

11.3.6 Standard Types of Formwork for Concreting

Standard types of formwork are used for structures such as:

1. Columns of any shape (rectangular, circular, etc.)
2. Beams and slabs
3. Stairs
4. Walls

We will not go into further details of the subject as it is a specialized topic.

11.3.7 Slip and Moving Forms

Slip forms and moving forms are special types of formwork.

Slip forms. For construction of structures such as tall cylindrical chimneys or rectangular/circular towers for water tanks, it is very important that there should be no weak construction joints, which may separate during high winds and cyclones. Stage-by-stage construction of heights not more than about 2 m using ordinary formwork, with carefully treated construction joints is the conventional system in non cyclonic areas. Modern continuous concreting without joints by using slip forms are nowadays insisted on in important works in cyclonic areas.

Slip forms are continuously moving formwork (slipping along the poured concrete). The entire assembly of formwork is moved by jacks, on jack rods located within the cast concrete. The movement is such a slow rate that the exposed concrete below the formwork attains the required strength to withstand all the loads that come on the concrete at the time it comes out of the formwork. The concrete should also be so designed that it attains the required strength between the time it is compacted in the formwork and the time it emerges out of the formwork. The usual speed of movement is the order of 15 to 25 cm (6 to 10 inch) per hour. The frictional force between green concrete and specially treated steel forms is taken as about 0.4 t/m^2.

These slip forms can be straight slip forms, tapered slip forms or forms for special applications such as for blocks of flats, bridge piers, etc. The special advantages of these constructions are that there will be no weak construction joints and the concreting can be well controlled as the rate of pour need not be very high. Such work gives a very accurate, economic and aesthetically pleasing work. A conical water tank in which the cylindrical portion was cast by slip form is shown in Fig. 11.3. It is important to note that the work of concreting should be arranged to take place all the 24 hours of the day start to finish of the structure. Continuous jacking arrangements are also needed.

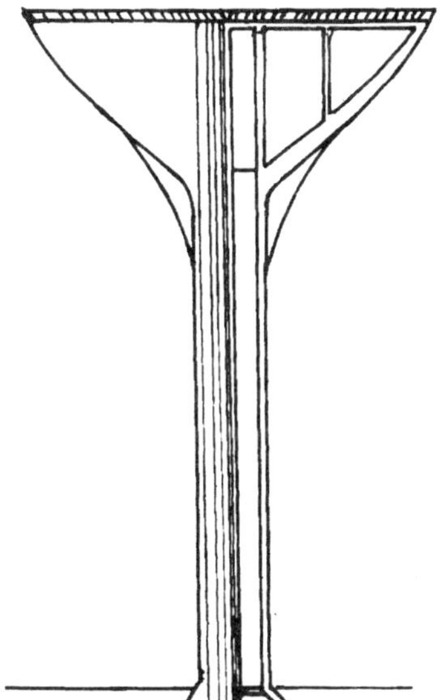

Fig. 11.3 Water tank tower built by using slip forms.

Moving forms. Moving formworks are those which can be moved from one place to other. When constructing the floor of a framed building over already constructed columns in a multibay building, the formwork can be constructed for a limited number of bays. After setting and curing of the poured concrete, the formwork can be lowered vertically down and then moved horizontally to other locations to cast other bays. Such formworks are called *moving formworks*. When casting chimney or water towers vertically, forms which are moved vertically by jacks after each separate stages of concreting, setting and curing of concrete to higher locations are called *jump forms*.

11.4 SCAFFOLDING FOR CIVIL WORKS

While the term *formwork* is generally used for all temporary works for holding concrete, etc., the term *scaffolding* is generally used for the temporary structure used to support the concrete formwork or more popularly for supporting the workmen during the construction (brickwork, plastering and painting or renewal or repainting of structures, etc.). It comes under safety regulations on construction sites and standards have been prescribed for its layout and use. For example, when a person is liable to fall more than 2 m, the platforms of the scaffolds are to be provided with a guard rail to a height of 1 m and also a *toe board* at least 200 mm above the platform. Scaffoldings are built around the periphery of the building and in multistorey buildings, it has to be assembled to very large heights. As about 35 to 40 per cent of all the accidents that take place in building construction sites is due to faulty scaffolding and consequent fall of persons from heights, strict supervision should be taken in putting up of scaffoldings according to standard practices. The types of commonly used scaffoldings are shown in Figures 11.4 and 11.5.

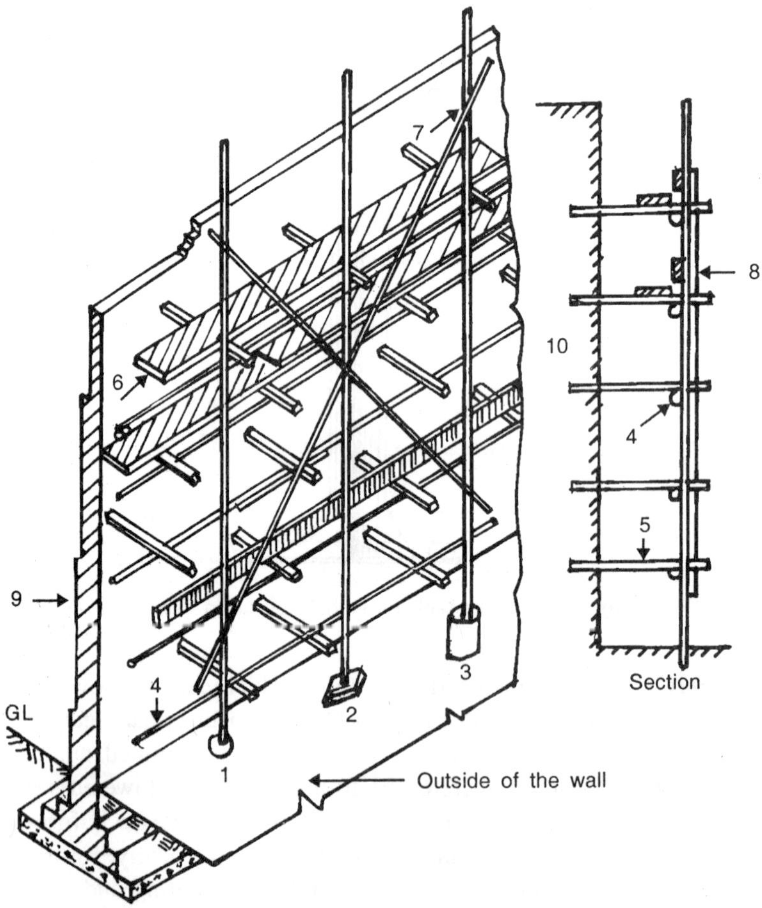

Fig. 11.4 Single or bricklayer's scaffolding: 1. Standard buried in the ground, 2. Standard on base plate, 3. Standard placed in drum with soil, 4. Ledgers, 5. Putlogs, 6. Planks, 7. Braces, 8. Guard board, 9–10. Wall.

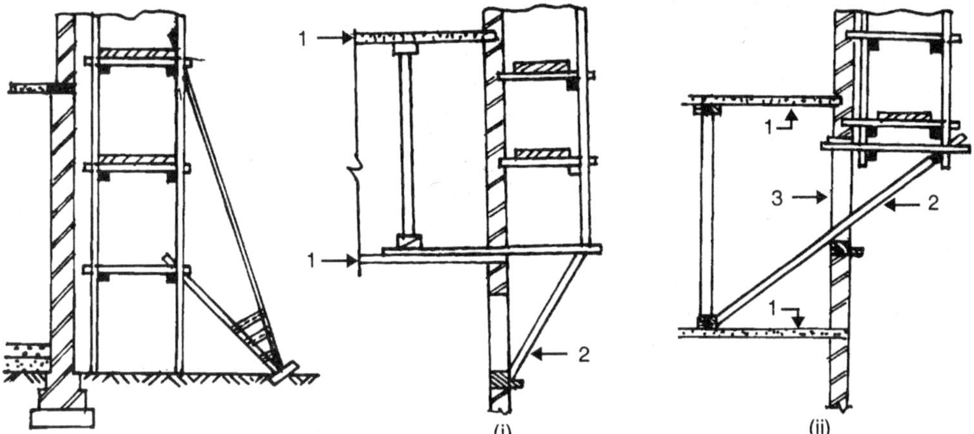

(a) Double or mason's scaffolding with raking shore

(b) Cantilever or needle scaffolding: 1. Floor slabs, 2. Strut, 3. Window

Fig. 11.5 Other types of scaffoldings used in building construction.

Scaffolding contains the following parts (see Fig. 11.4):

1. *Standards:* The vertical posts (also called uprights)
2. *Ledgers:* The horizontal members parallel to the wall
3. *Braces:* The bracing system
4. *Putlogs:* Horizontal members normal to the wall
5. *Transoms:* Putlogs whose both ends are supported on ledgers in double scaffoldings
6. *Bridle:* Member used to bridge openings (see the description of single scaffolding below)
7. *Boarding:* Planks on which workmen stand
8. *Guard rail:* Rail provided at about 1m level to guard the men working on the boarding
9. *Toe board:* Boards placed parallel to boarding near the wall to give protection to workers
10. *Base plate* or *sole plate*. Plates on the ground supporting standards

Some of the commonly used types of scaffoldings are described below:

Single scaffolding or brick layers scaffolding. This type of scaffolding is mostly used for brickwork and consists of an outer row of verticals (called standards) to which longitudinal members are tied at different levels of working (see Fig. 11.4). The cross members (putlogs) are tied to the standards at the outer end and rest on the walls being built inside. The platform is carried on the putlogs. If the putlog level coincides with an opening in the wall, it cannot be placed on the wall. Hence it should rest on a cross piece (called *bridle tube*) tied at the wall end of the adjacent putlogs. Cross bracing in the vertical planes between the verticals are also to be introduced for lateral stability.

Where the verticals posts cannot be placed on base plates or in holes made in the ground, it should be suitably placed and braced for lateral stability. A popular method in such cases is to place the vertical post in a steel barrel of 60 cm in height and of suitable diameter filled with compacted earth as shown in Fig. 11.4.

Double scaffolding or Mason's scaffolding. This is a supporting system used by masons for plastering. It has pairs of (inner and outer) verticals. The putlogs instead of being supported on the wall are supported on an inner system of verticals and longitudinals. To give additional lateral stability to the narrow and tall framework it is tied to the building at intervals when used for multistorey building. See Fig. 11.5(a).

Cantilever or needle scaffolding. For construction of the upper part of a tall residential building, cantilever scaffolding eliminates unwanted scaffolding at lower levels keeping the space free for vehicles, etc. It is similar to the double scaffold, except that it is supported at the bottom by a cantilever prop. See Figure 11.5(b).

Birdscage scaffolding. It is used for internal work and consists of a simple cage supported on four verticals from which the workers can work. It is *movable* and can be easily moved from place to place.

Ladder or trestle scaffolding. This is used for light work and is usually *portable*. It consists of two ladders with the top ends joined together by planks so that a person can work on this platform.

Suspended scaffolding. In tall buildings for working at high levels, such as painting of tall buildings, it is more convenient and economical to suspend the working platform from above the ground than to support it from the ground. This is especially so when the height is more than 30 m. Suspended scaffolds are of three types—fixed, operated by pulleys and operated by winches. (Provision for stability should be provided for these when using them in places subjected to heavy wind forces).

(i) *Fixed type suspended scaffolds.* These are the scaffolds attached to a truss or the roof truss above the site of work by means of ropes, chains, tubes, etc.

(ii) *Suspended scaffolds operated by pulleys, etc.* These are like the platforms of the window cleaners and painters of buildings.

(iii) *Suspended scaffolds operated by winches.* These are heavy platforms hung by wires from temporary outriggers and operated by external cranes.

11.5 IMPORTANT POINTS IN SCAFFOLDING

The following are some of the important rules to be remembered in scaffolding work:

1. The uprights (standards) should not be spaced more than 1.8 m (6 ft) for heavy work such as masonry and not more than 3 m (10 ft) for light work such as painting.
2. The putlogs in single scaffolding *should bear well* on the wall opening. Short members nailed onto the wall should not be used as putlogs.

3. The platforms should be wide enough. It should not be less than 425 mm (17 inches) in width if the height is more than 1.8 m. If materials are stored on it, it should not be less than 850 mm (34 inches) in width. Each board should be 50 mm (2 inches) thick and 200 mm (8 inches) wide for the putlog spacing of 1.5 m or more. It can be 38 mm (1½ inches) thick for 0.9 to 1.5 m and 25 mm (1 inch) for less than 1.5 m putlog spacings.
4. The plank should not be defective. To prevent sudden failure of the planks, the grains in the wood used should not be more than 10 degree with the length of the plank. *There should be no overhanging (or cantilevering) of these planks.*
5. All scaffolding should be properly stutted so that they do not fall away from the wall laterally. Similarly, it should be stable longitudinally also.
6. Workmen should not work under the scaffoldings.
7. Lorries should not be allowed to come near scaffoldings.
8. Safe ladders should be provided to climb on to the scaffolding.
9. If the height is more than 2 m, guard rails should be provided for the scaffolding.

11.6 USE OF STEEL TUBE SCAFFOLDING

For ordinary low-rise buildings, it is common to use bamboo or casuarinas tied togelter by coir ropes for scaffolding as such works tend to be cheap. But it should be remembered that these poles loose their strength after exposure to the sun for about six to nine months and may have to be replaced. Moreover, coir rope joints should be examined periodically for their tightness.

For large heights, for use for long periods, steel tube scaffoldings or specially made wooden systems with special fixing devices are preferable. Steel scaffoldings which are strong and can be easily joined are the best for use in multistorey buildings. It will also prove to be wise as accidents are minimal in such cases.

11.7 SELF-CENTRING CONCRETE FLOORS

These consist of roof and floor elements constructed with precast RCC elements and hoisted to roof or floor level. They need no supports except the bearing at their ends on the walls or beams. As they require no centring, they are called *self-centring floors*. Some of them are shown in Fig. 21.4 in Chapter 24.

11.8 SHORING

Temporary supports used to support unsafe buildings (or used to support buildings while they are repaired) are called *shores*. (The term *shoring* is used for the work of timbering of excavations also.) They are classified into three classes according to *their positions in space* as follows:

1. Raking (or inclined) shores
2. Horizontal (or flying) shores
3. Vertical (or dead) shores

They are extensively used during repair and underpinning operations.

11.9 UNDERPINNING

The term *underpinning* is used to the repair, strengthening or renewal of the foundation of an existing building. During underpinning, the building is properly shored. There are several methods of underpinning of foundations of which the three popular methods are as follows.

1. *Pit method.* It involves strengthening by excavation and building below the existing foundation.
2. *Pile method.* It involves supporting the building on newly built piles of various types without excavations.
3. *Pier method.* This is a combination of pit and pile methods where we excavate below the foundation and build a pier for the new foundation.

SUMMARY

Temporary work is a major part of a building's normal construction. *Formwork (shuttering), centring and scaffolding* are temporary works built for carrying out the construction of buildings. Such temporary structures when used to support temporarily an unsafe existing building for its repair are called *shores*. The term *underpinning* is used for the work of supporting a building to strengthen or renew its foundation. The cost of these temporary works is a good percentage of the total cost of the concerned construction. Hence considerable planning should be done in their design and use. To bring down its cost, we should plan for many reuses of the equipments in building construction. With more reuse of equipments, the cost comes down. The present trend in big cities for carrying out these works is to entrust the work to a subcontractor who owns these equipments and specializes in these types of work.

REVIEW QUESTIONS

1. Explain briefly the following terms:

(a) Centring
(b) Formwork
(c) Scaffolding
(d) Temporary works
(e) Shores
(f) Underpinning

2. Describe the components of temporary works used for concreting a beam and slab floor for an interior span. What are the principal considerations in setting up the temporary works for a slab to be cast:
   (a) For the first floor
   (b) For a balcony slab for the third floor of a building for residential flats
3. Sketch and name the components of a double scaffolding. What are its uses? What type of scaffolding would you use for construction of a brick masonry wall?
4. Write short notes on:
   (a) Cantilever scaffolding
   (b) Temporary works for concreting
   (c) Temporary works for brick masonry construction
   (d) Slip forms, moving forms and jump forms

Chapter 12

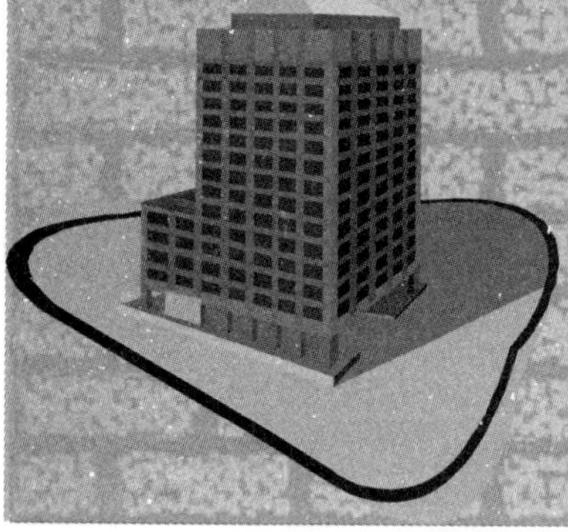

Bending and Placing of Reinforcement in RCC Works

12.1 INTRODUCTION

In many places in India, steel rods are bent manually on the bending table, one by one and placed in the formwork and tied together. There is also a trend for large works, such as in construction of bridges, to prefabricate the steel work as much as possible in a specialized place and with the help of lifting machinery, to place them in the formwork. Detailing of structures to be constructed by these preformed cages has to be done in a special way so that enough continuity bars are provided at the junctions of the reinforcements to tie them together. Generally the payment for placing reinforcement is calculated on the basis of weight (tons) placed. In this chapter, we will deal with the work of placing steel in RCC works.

12.2 ORGANIZATION OF WORK

To get the right amount of cover, the work of cutting and bending the reinforcements to exact dimensions is as important as correct placing of the steel. Also after the bars have been fixed, it is equally important to see that they do not move, either before or during concreting due to persons walking over them or by concrete dumped and vibrated between them. The whole work of bending and fixing is made easier if the site is well organized from the start. Conditions on different sites vary a good deal, but the following are a number of general points that always help keep the work going smoothly.

1. Get as much room as possible for storing the bars and as a working area for cutting and bending bars.
2. Store the bars properly off the ground and if they are to be in stock for long periods, provide some covering from rain. Slight rusting cannot usually be avoided but bars that are very rusty should be freed from loose rust by brushing with a wire brush. Otherwise the bond between steel and concrete will be poor and the strength

of the member may be seriously reduced. Torsteel cold worked bars corrode faster than MS or TMT bars.

3. Stack the bars so that the lengths and sizes required can be found easily. Bars which have been bent should be re-bundled and marked with their original bar list member on a paper-covered metal label. In this way, the fixer can find the correct group of bars without having to move several tons of the wrong type.
4. Arrange the stacks of bars, the bending tables and the cutting machine in the most convenient positions. For example, turning a long bar end for end is inconvenient and can be avoided by using two bending benches or machines, one at each end of the bench.
5. Do not allow bars to be thrown onto a stack from a distance. This will probably give them some extra bends which were not specified.
6. On some jobs in large cities, the bars may be delivered already cut and bent. They should all be carefully checked and any discrepancies with the bar schedule should be reported immediately.
7. Before the reinforcement is fixed, the bars should be examined to see that they are free from paint, loose scale, cement mortar, etc.
8. Nowadays bars are marked in drawings in standard notations such as 55Y 12-01-250 B. The letter 'Y' indicating high yielding steel (and symbol R, mild steel rounds) and B indicating bottom (and T indicating top, etc.) The above notation means 55 Nos. high yield 12 mm deformed bars with bar mark 1 at 250 mm centres placed at the bottom of members.

12.3 BENDING OF REINFORCEMENT

The important points in bending of bars are discussed further.

Bends and hooks. Bends and hooks are, in many cases, provided at the ends of the bars (except at laps) for anchorage. *When smooth MS bars* are used, all bars have to be *bent into hooks* at their ends. With high bond bars, it is generally enough to provide simple right angled bends or no bends at the ends instead of hooks. However, all beam stirrups and column ties should always be provided with *standard hooks.*

The usual practice is to use bent bars for beam bars but straight bars are used for slab reinforcements. The bending schedule normally shows the finished bar as a single line. Dimensions given should follow the convention of measuring from "outside to outside" (except in the case of links and binders where the important dimensions are the internal ones), which fix the position of the main bars. Figures for hooks and bends are usually omitted, since they should comply with the standard specification. Where special bends and hooks are required, as in the case of stirrups for ductile detailing for earthquake effects, the designer will include their details on the schedule. The specifications for bends and hooks for main bars and stirrups are as shown in Fig. 12.1.

Specifications state that for high yield bars, the internal diameter of a *bend or hook* should be four times the diameter of the bar. The straight portion beyond the bend or hook

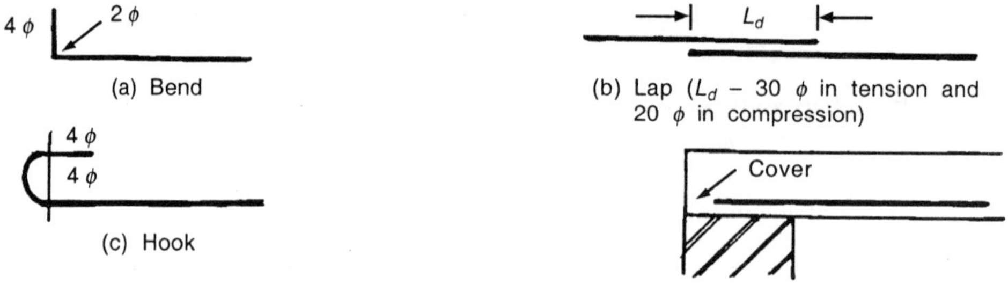

(a) Bend

(b) Lap (L_d – 30 ϕ in tension and 20 ϕ in compression)

(c) Hook

(d) Anchorage with cover

Fig. 12.1 Bends, Laps, Hooks and Anchorage in reinforced concrete construction.

shall have a length of not less than four times the diameter of the bar for all the main bars except the stirrups. For normal stirrups this extension beyond should be 4 ϕ for 180° hooks, 8 ϕ for 90° bends and 6 ϕ for 130° bends where ϕ is the diameter of the bar. For earthquake resistant design the stirrups are to be bent at 135° and extended by 10 ϕ instead of 6 ϕ for normal stirrups. *Hooks* are provided only at the non-continuous ends of steel reinforcements. At lap points no hooks are to be provided. Similarly no hooks are to be provided in column bars.

Cutting of bars. A steel tape must be used when measuring the length of a bar to be cut or bent as it reduces the possibility of error, especially on long bars. It is also convenient to have a "marking out bench" marked off in feet or metres, as the case may be for cutting the bars. Either hand or power cutters may be used on site for cutting the bars.

The additional lengths to be provided for cutting of steel lengths for hooks, cranking (bending up) and lapping of bars can be taken as the following:

1. For each standard look (4 ϕ) : 9 ϕ
2. For each 30° crank of depth d : $0.3d$
3. For each 45° crank of depth d : $0.42d$
4. Lapping: 30 ϕ to 47 ϕ (in compression and tension)

Marking bars for bending. Cut bars which are to be bent are usually delivered with an extra one inch or so in length. This extra length should be taken up in end hooks or other details and not allowed to affect the internal dimensions of the bar. For this reason, it is best to mark off the length from the centre of a bar towards each end, if the bar is to be hooked or bent at both ends. Bars hooked at one end only should be marked off from the straight end towards the hooked end, so that the extra length may be taken up in the hook.

The dimensions of the bar will have to be adjusted when its "creep" during bending is known. For this purpose, we mark out the first bar according to the given dimensions and also check it after bending. Then base the dimensions of all subsequent bars on this first bar, after making alterations, where necessary, on the first bar. This is particularly important where a bar has a number of bends, each of which is liable to "creep". The actual dimensions used will depend on the type of bending machine and the method of positioning the bars in relation to the mandrel of the machine.

Some hints on bending. The following precautions should be taken while bending bars:

1. Use the correct stops and pegs to bend the bar (see Section 33.5).
2. Make sure the stops and pegs are stout and rigidly fixed.
3. Place the bar behind the peg, not against it. This helps prevent "creep" in the bar during bending.
4. Support the bar while it is being bent, otherwise all the bends may not be at the same place.
5. After checking the first finished bar and making all necessary adjustments, occasional checks should also be made on the subsequently bent bars to ensure that no movement of stops and pegs has occurred.

Whether one is using a simple hand bender or a power-operated machine, it should be protected against the weather and should receive regular maintenance. Any worn-out part should immediately be replaced.

12.4 FIXING THE REINFORCEMENT

12.4.1 Organization of the Work

Reinforcement often gets roughly handled during concreting, and secure fixing is essential to ensure that the bars remain in the correct positions. The reinforcements are tied together by means of annealed steel binding wires, No. 16 or 18 SWG (0.9 or 1.6 mm diameter) in the standard manner of tying of reinforcements (Section 12.4.3). Fixing the reinforcement, erecting the formwork and concreting should be organized together to maintain a steady flow of work. Reinforcement once fixed should not be left in the open air for very long time as the bars get rusty and the ties go slack. On the other hand, fixing of reinforcement should not be also rushed simply to permit concreting to be continued, as this usually results in the bar being badly tied and often tied out of place.

Nominal cover is defined as the depth of concrete cover to all steel reinforcements *including links and stirrups*. The cover to be provided should depend on the type of the member (slab, beam, etc.) and depend also on the exposure conditions of the members (details will be studied under RC design). The nominal covers to be provided according to IS 456–2000 for mild exposure are as shown in Table 12.1. Increased cover is to be provided for more severe environmental conditions such as nearness to sea coast.

Table 12.1 Nominal cover for concrete members IS 456–2000

	Member	*Cover for mild exposure condition*
1.	Slabs	20 mm
2.	Beams	25 mm
3.	Column	40 mm (but with 12 mm main steel, we may adopt 25 mm)
4.	Foundations	50 mm

12.4.2 Placing the Reinforcement with Proper Cover

In shallow beams and similar members the formwork can be completed before the reinforcement is placed. In deep sections such as walls, on the other hand, one side of the formwork may be assembled first, then the steel fixed and the remaining side of the formwork erected last. In columns, it may be more convenient to fix the reinforcement entirely before the formwork is erected. The bars are kept in position by one of the following methods:

Method 1. In case of beam and slab construction and in columns, precast mortar pads cover blocks of cement mortar (with coarse sand) about 40 mm × 40 mm in size and thickness equal to specified cover can be placed between the bars and the shuttering [Fig. 12.2(a)]. *Mortar pads* are prepared from 1 : 2 or 1 : 3 mortar cast in moulds to exact cover height required. Usually a wire tie of 16 to 18 AWG is cast into it for tying purposes. Otherwise commercially available patent supports can be used for maintaining the cover. Always check that they are of the right type.

Method 2. In case of cantilevers and doubly reinforced beams or slabs, the vertical distance between layers of steel is maintained by support bars or *chair spacers* as shown in Fig. 12.2 at a minimum spacing of one metre between chairs to avoid sagging.

Method 3. Reinforcements for two faces of walls are kept in position by 10 mm spreader bars tied to them at about 1.8 m spacings in both directions.

Method 4. One may also use fabricated steel chairs as shown in Fig. 12.2(b) or other patent systems of supporting or spacing of reinforcements that are available in the market.

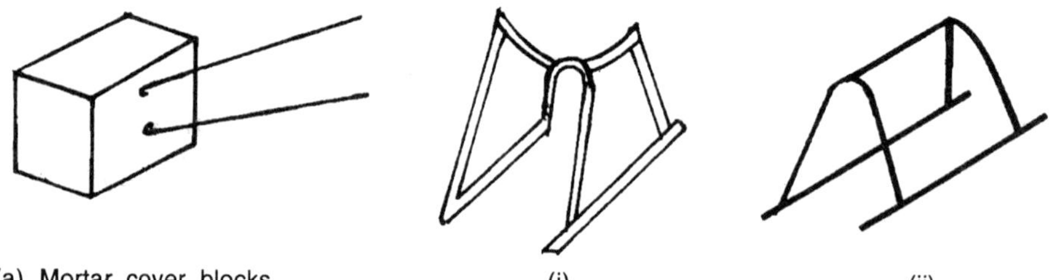

(a) Mortar cover blocks (i) (ii)
(b) Fabricated steel chairs

Fig. 12.2 Methods of providing cover to steel in reinforced concerete construction. (*Note.* Devices in plastics for cover are also available.)

When using mortar pads, we should remember that it is important to place them between formwork and reinforcement before concreting in horizontal floor slabs and also for vertical members. We should never allow the reinforcement of a slab to be laid on the formwork and temporarily raise it to give the necessary cover during concreting.

12.4.3 Tying of Reinforcement Bars

There are various types of ties used when securing reinforcing bars. It is better to use two twisted strands of annealed 0.9 to 1.6 mm or 18.9 gauge steel wire than one thick wire. The commonly used ties are snap or single tie, wrap and snap tie, saddle tie, wrap and saddle tie and figure eight tie. Some of them are shown in Fig. 12.3. Tack welding is also permitted.

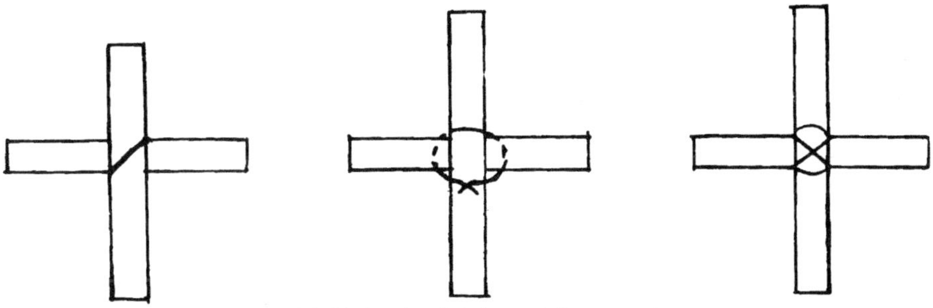

(a) Methods of tying reinforcements

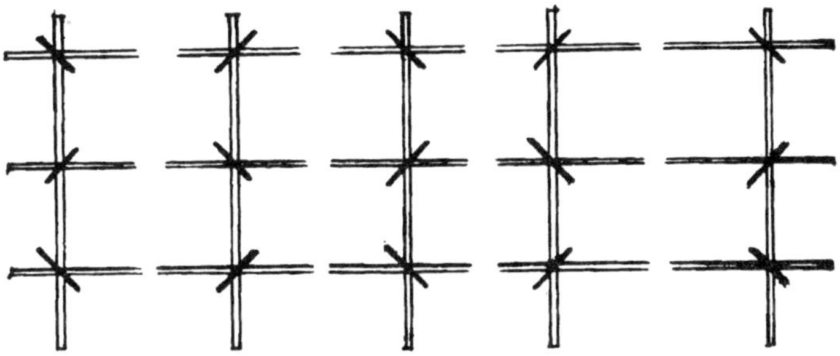

(b) Variation of direction of tying of alternate ties

Fig. 12.3 Tying of reinforcements.

It is not necessary to tie bars at every intersection of the slab reinforcement, even though same specification requires every intersection to be tied together. Ties at every 4th or 5th intersection are enough. Ends of finished ties should be bent away from face of the concrete into the concrete side and thus kept clear of the cover for concrete.

12.4.4 Junctions

Junctions where reinforcement bars of beams intersect or pass over each other, should be securely wired together by tying wires to give the reinforcement good stability. When tying these junctions, cut ends of the tying wire should be kept away from the formwork.

12.4.5 Splicing or Lapping

Splice of reinforcement means joining them by overlapping. In placing steel for continuity of bars, we will have to splice them by lapping of bars as shown in Fig. 12.1. These laps help the stress to be continued. Laps are usually provided without hooks for a length not less than 30 to 47 times the diameter of the rod under bending tension (twice this value in direct tension) and 24 to 30 times the diameter of the rods under compression. Lapping *should be staggered* and not more than one-third of the rods should be lapped at the same section. They should not also be located at points of maximum stress. Splices for bars in columns were *usually* provided immediately above the junctions of columns with the beams or the slabs. In earthquake–prone regions, column laps should be somewhere at midheight. However, in earthquake-prone regions, laps are to be provided in mid-regions of columns and staggered. In other members, it should be at places of minimum stress possible.

12.5 ENCASING ROLLED STEEL SECTIONS

It becomes necessary in some situations to introduce rolled steel sections (channels or I beams) and encase them in concrete. Such sections are used as columns, beams or in grillage. The section used should be unpainted and should be wire brushed to remove all scales/rust. Ungalvanized wire mesh or light gauge expanded metal having meshes large enough to pass aggregates of the largest size used (aggregates 12.5 mm size preferably to be used) is wrapped round the sections to be encased. The wrapping should be at the centre of the concrete with sufficient cover. In case of columns, wrapping should be carried out in stages of height not more than 1.5 m with the successive heights of wrappings tied together. The concrete cover should preferably be not less than 50 mm.

12.6 FIXING OF PRECAST CONCRETE JALIWORK

Jaliwork is usually made with 1 : 2 : 4 concrete reinforced with one 6 mm MS steel rod at the centre. The jali is set in position true to plumb (before plastering the jambs, sills and soffits of the opening) and properly grouted with 1 : 3 mortar and checked for levels. Finally the jambs, sills and soffits are plastered, embedding the jali uniformly on all sides.

SUMMARY

The placement of reinforcement for RCC work is always shown in the field drawing by standard notation. The steel is to be marked and bent to exact dimensions. Ends of MS bars have to be provided with hooks. Theoretically, ribbed bars need not be provided with bends because of the high bond strength. In practice, when using high bond bars, nowadays, slab reinforcements are detailed with straight bars only, but bent bar detailing can be used for beams. Hooks are always prescribed for beam stirrups and column ties and their bends at the ends should conform to standards. The bars bent to schedule must be placed on the works and tied together so that the necessary cover is maintained. Specifications have been laid for all these details and they should be strictly followed in construction.

REVIEW QUESTIONS

1. (a) What are bends, hooks, anchorage and laps?
(b) Why is it necessary to provide hooks at the ends in MS bars, whereas in HYD bars, only bends are specified for normal reinforcement placement?
(c) Give the dimensions of hooks and bends specified for Fe 415 bars for (a) main steel and (b) for stirrups in (i) normal design and (ii) earthquake-resistant design.

2. (a) What is meant by nominal cover to concrete? What are the IS values for slabs, beams, columns and footings for mild exposure?
(b) What are the methods used to maintain cover in reinforced concrete construction?

3. (a) What is meant by lapping of bars?
(b) What is the specification for minimum lengths of lapping of reinforcement bars in tension and compression?

4. Write short notes on:
(a) Steel chairs in placement of reinforcement
(b) Tying of reinforcement bars
(c) Methods of providing cover for concrete
(d) End anchorage of stirrups
(e) Lapping of bars

Chapter 13

Plastering and Pointing

13.1 INTRODUCTION

Plastering is the work used for covering rough surfaces of building components such as walls and ceiling with a coat of plaster to form a smooth and durable surface. It is carried out for *beauty*, *durability* and to *resist rain penetration*. The term *plastering* is used for internal work and the term *rendering* is used for work done outside. The term *plastering* may be used for both inside and outside works. The term *pointing* describes a type of finishing of the external mortar joints in masonry. Walls are pointed or plastered depending on the nature of the walls and also as to whether it is an outside or inside wall. Plastering of exterior walls makes it impermeable to rain, and plastering of inside walls and ceiling gives a smooth surface to apply paint or other finishes to produce an aesthetic surface. Pointing is usually done only on outside walls for appearances.

On a very uneven surface like rough stone masonry requiring thick plaster, one thick coat of plaster tends to sag due to the weight of the thick "still plastic" plaster so that an irregular surface results. Hence, thick plastering is done in two or three coats. In ordinary house construction, nowadays, for interior work, we use one coat of cement plaster upto 20 mm thick. For superior finish and also external cement plastering, we use two coats. The first coat is called the *base coat* or *the rendering coat* and the last as *finishing coat*. In a three-coat plastering used in lime plastering, we have the first coat called the rendering coat, the second coat called *floating coat* and the last the *finishing coat*. For external walls exposed to rain also, it is better to have plastering in two coats. Accordingly, the following are the usual specifications followed in practice.

In plastering of internal work meant for painting, smooth finish is most important. External plaster work is usually finished rough as the smooth surface shows up defects more easily than a rough surface. Also weathering is more uniform as a rough surface. Roughness is also found to help prevent penetration of rainwater in the walls. Hence, external plastering is generally made rough. As already stated in Section 5.4, the sand used for plastering should be finer than that used for making mortar for joining bricks. Coarse sand plaster will not stick to ceilings and walls as easily as plaster with finer sands. For superior class works, *the plastered surface is treated* with a layer of special putties (available in the market) or plaster of Paris before painting.

13.1.1 Specifications for Plastering

The following are the usual specifications for plastering on brickwork and blockwork (see also Sections 13.3 and 13.4):

Lime plastering specifications. General Specifications for lime plastering are as follows. (The word *thick* is usually omitted.)

(i) 12 mm or 15 mm single coat lime plaster
(ii) 18 mm or 20 mm lime plaster (two coats work)
(iii) Lime plastering 25 mm (three coats)
(iv) Lime punning on any of the above plastering

(Lime plaster must be made from lime putty kept standing under water for at least 72 hours before use so that the lime is completely slaked. We should not make lime plaster from fresh lime.) Preparation of lime plaster is covered in the book on *Building Materials.*

Cement plastering specifications. The following are the general specifications of cement plastering. (Preparation of cement plaster is covered in the book on *Building Materials.*)

(i) Cement plastering single coat (12, 15 or 20 mm)
(ii) Cement plaster (12, 15 or 20 mm) with a floating coat of neat cement
(iii) 18 mm cement plaster (2 coats work)
(iv) 6 mm cement ceiling plaster
(v) 6 mm cement plaster for slab bearing
(vi) 18 mm external rendering with cement plaster (2 coats to avoid water penetration)

Plaster specifications usually specify the number of coats such as single coat, two coats or three coats. Single coat plaster is given to low to medium cost buildings. Exterior surface in important buildings under "severe" exposure to rain, two coats will be sufficient. For buildings which require very smooth internal surface for painting, etc. as in offices, three coats of plaster will give excellent results. Generally, 12 mm plaster work is given on the rough side of single or on half-brick walls. External walls where rainwater seepage is expected, a plaster thickness of 18 mm (two coat work) or more is necessary. When plastering stonework also, the thickness of the plaster tends to be larger than on brickwork to cover all the irregularities.

It is important to remember that for external work, we can use coarser sand but it should conform to grading IV (90 to 100% passing 4.75 mm) and for internal work and the finishing coat work, the sand should be finer and conforming to grade V (85 to 100% passing 0.6 mm).

Plaster should be plastic and adhere to the masonry work. Hence it should be richer than mortar. However, it should not have more cement than required. *Otherwise it will crack due to shrinkage. It is better to use a slow strength developing cement such as C-33 than a fast strength developing cement such as C-53 for plastering.* That latter tend to crack more than the former. The commonly used mixes for plastering various surfaces are given below. (In general, cement mortar 1 : 6 for brickwork and plaster 1 : 5 is enough for most works. We must remember that higher and stronger cement tends to produce more shrinkage cracks.)

(a) Lime plaster: The mix varies from: 1 : 3 to 1 : 4
(b) Cement plaster: The mix varies from: 1 : 4 to 1 : 6
(c) Cement plaster (internal) for brickwork: 1 : 5 to 1 : 6
 1. Cement plaster (external) for brickwork: 1 : 4 to 1 : 5
 2. Cement plaster for partition (1/2 brick walls): 1 : 4
(d) Combination plaster (using fat lime): 1 : 1 : 8 to 1 : 2 : 8
(e) Ceiling plaster first coat 1 : 3 and subsequent coat: 1 : 4 to 1 : 5

In this chapter, we will examine the procedure for plastering and the various specifications for plastering.

13.2 PLASTERING METHOD

For plaster work, double scaffolding having two sets of vertical supports with horizontal pieces and scaffolding planks will be necessary. The holes left in masonry work for single scaffolding must be filled up and made good before plastering (see Chapter 11).

Preparation of surface. *All the joints of the brickwork, etc. should have been raked out properly during brickwork construction itself.* Dust, loose mortar, efflorescence, etc. must be removed by brushing and scraping. The surface must be thoroughly washed with water, cleaned and kept wet before the work is commenced.

Procedure for plastering. Ceiling plaster should be completed before commencement of wall plaster. Plastering is started from top (say left hand corner) and worked down towards the floor. To ensure even thickness and level surface, guide points are first set on the brickwork. For this purpose, guide plaster about 15 × 15 cm in size is first applied at not more than 2 metre intervals horizontally and vertically to serve as gauges. They are levelled in the plane of the finished plaster surface. After this guide plaster has set, the plaster mix is then laid on the wall between the gauges with trowel slightly more than the specified thickness. This is beaten with a wooden straight edge thoroughly filling the joints and levelled by reaching across the gauges with small upward and sideway movements. Finally the surface is finished off true with a trowel or float. A steel trowel will give a smooth surface and a wooden float sandy granular texture. Excessive trowelling should be avoided.

All corners, arises (edges formed by two planes), angles, junctions, etc. should be carefully finished with proper templates and tools. The surface levels are tested with a straight edge not less than 2.5 m long and with plumb bobs. The horizontal lines and surfaces are tested with a mason level.

One of the prime requirements of good plastering is good bond between the already-laid masonry work and the fresh plaster. (For plastering of brickwork, raking to a depth of 10 to 15 mm should have been already carried out during its construction when the mortar was still green). In brickwork, the importance of suction in obtaining bond should be remembered. Brickwork should be well watered before plastering to ensure uniform suction. Areas which dry out rapidly should be re-watered. If waterproofing material is used in the cement plaster used for the first coat, the surface should be well watered for subsequent coats also.

Plastering concrete surfaces. Plastering should bond with the concrete face. A spatter dash consisting of a thick slurry of 1 part of cement and 2 parts of fairly coarse sand can be dashed on the wetted surface to give a *good key* for very dense concrete backgrounds. In concrete and other materials hacking, as in ceiling plastering (see Section 13.4), should be resorted to before plastering. Alternately, a mixture of 1 part of cement and 1½ parts of fine clean sand with bonding agent (polyvinyl acetate suspension or latex) made into a slurry can be brushed into the surface and the surface to be plastered is made into a rough surface with a suitable brush or banister rod to form a close textured key which when allowed to dry gives a good bond. (There are also a number of cement chemicals in the market to improve adhesion of plaster to concrete.)

13.3 SPECIFICATIONS FOR PLASTERING WITH LIME PLASTER

Lime plastering can be in single, two or three coats. Plastering with lime plaster is used only in places where lime is cheap and the surface is to be whitewashed or colour washed. Where painting is to be done, it is better to use cement plaster as the alkali in lime affects most of the paints (see also Chapter 26 on Painting). The following specifications as given in most "PWD Specifications" are followed in India. As already stated, lime plaster should never be fresh but must be made from lime putty obtained by thorough slaking of lime as described in the book on *Building Materials*. Available lime should be tested for its suitability unlike the cement. Good lime in bags is difficult to get in the Indian market.

Single coat of 12 mm lime plaster. The 12 mm thick plaster is used for the *regular side of brickwork*. For plaster work, the lime plaster is applied to slightly more than the required thickness and it must be well pressed into the joints. The surface is then rubbed with a wooden float or trowel sprinkling as much water as is necessary. During the process of rubbing, an addition of lime putty (made from lime and water to the consistency of thick cream) is also applied to give an even surface.

The final thickness of the plaster is measured exclusive of the thickness of the key. The average thickness should not be less than 12 mm and nowhere less than 9 mm. Curing is to be started 24 hours after the finishing, the plastering kept wet for at least seven days and preferably for 21 days. The aim of curing lime plaster is not for curing as cement plaster but is to prevent fast drying out which will produce hair cracks in the plastered surface.

Wherever work is to be stopped for the day and recommenced, it is recommended to scrap clean the edge of the old work and the surface wetted with lime putty before plaster is applied to adjacent work. Any cracks or portions, that sound hollow when tapped or portions found soft or defective, should be cut out in rectangular shape and redone.

Single coat of 15 mm lime plaster. The work is similar to the above, except that the average thickness is 15 mm. It is used for plastering of *irregular side of brickwork* and the plastering is carried out in one coat.

Lime plaster (two coats work) (18 or 20 mm thick). The plaster is applied in two coats—12 mm undercoat and 6 or 8 mm finishing coat. The resultant average final thickness should not be less than 18 mm. For a 18 mm plastering, the 12 mm undercoat is applied first and finished rough with diagonal furrows 2 mm deep, 15 cm apart made on it. The finishing coat is applied a day or two after the undercoat has set but has not completely dried out. The finishing coat is applied slightly more than 6 mm thick and finished at an average thickness of 6 mm. The lime plaster for the finishing coat is to consist of at least 1 lime to 1.5 fine sand or other suitable proportion and ground on a roller until the mortar is reduced to a fine paste. In some specifications, two coat plaster is to be 20 mm thick.

Lime plastering three coats (25 mm thick). This work is carried out in three coats with rendering, floating and finishing coats, the first coat being 12 mm thick and thickness of each of the next coats being less than the preceding one. The total thickness of plaster is 25 mm.

Lime punning on finishing coat. Lime punning is coating with a rich lime mix to give a smooth finish. It is applied over any of the above plaster specifications. The mortar for lime punning consists of 1 lime putty and 1 very fine sand. Lime punning is carried out on an undercoat of lime plaster. It is applied when the undercoat is still green. The mortar for punning is applied in a uniform layer a little more than 3 mm between gauged pads and finished smooth with trowels. The thickness of finished punning should not be less than 3 mm.

Lime plaster need not be cured like cement plaster but it should not be allowed to dry out only slowly and not rapidly.

13.4 SPECIFICATIONS FOR PLASTERING WITH CEMENT PLASTER

Nowadays more plastering work is carried out with cement plaster than with lime plaster. Stonework and brickwork in *external walls are usually plastered with cement to resist rain penetration*. The following are the most commonly used specifications in practice in India. The thinner plaster is used for the regular face of the brickwork and the thicker one for the irregular face.

Cement plastering (12, 15, 20 mm thick in one coat). The thickness can be 12, 15 or 20 mm as specified. The plaster work is carried out as described in Section 13.2. The mixed mortar should be used within half an hour of adding water to the dry mortar. The curing is started as soon as the plaster has hardened sufficiently as not to be damaged by watering. It is kept wet for a minimum period of seven days. Sometimes thread lines or patterns are planned on the plaster. If such thread lines are planned, they should be carried out before the cement takes initial set. As already stated this is the specification used for most middle class housing.

Cement plaster (12, 15, 20 mm thick in one coat with a floating coat of neat cement. The cement plaster in this case can be 12, 15 or 20 mm thick as needed for the work. However, after the plaster has been brought to a true surface with the wooden straight edge, it is treated with a paste of neat cement (at the rate of 1 kg of cement per sq metre surface) and finished smooth so that the whole surface is covered with neat cement coating.

Smooth finishing is completed by using a steel trowel immediately and in no case later than half an hour of adding water to the plaster mix. The floating coat of neat cement is also called cement punning. The work is to be cured as in the case of other cement plastering. This gives a smoother finish than the former for painting.

18 mm cement plaster (two coats for external work). The 18 mm plaster is applied in two coats, the first coat of 12 mm as under coat and then 6 mm as finishing coat. The first coat is brought to a true surface with a wooden straight edge and the surface is left rough and furrowed 2 mm deep both ways, to form a key to the finishing coat. The finishing coat is applied when the undercoat has sufficiently set, but not dried, but in any case within 48 hours of laying of undercoat.

6 mm ceiling plaster. Ceiling plaster will be of richer mix. In order to reduce the effect of vibration set up on the slab after plastering, ceiling plaster work below a slab should preferably be commenced only after the work above the slab is completed. For example, for a roof slab, the roof work on the top should be completed before its ceiling plastering. For floor slabs, the props above the floor should be removed. Terrazzo polishing, above the ceiling being plastered, should also be programmed either after the plaster is fully cured or preferably before the plastering. However, cleaning, hacking, etc. of the underside of slab should be carried out as soon as shuttering is removed before the concrete becomes dry.

As already described in Section 13.2, to ensure even thickness, plaster pads of 15 × 15 cm region at 1.5 m intervals should be first plastered as gauge points using a plumb bob. Then the plaster is applied between these spaces, to a thickness slightly more than that required and then brought to true and even surface by working a wooden straight edge. Finally, the surface is finished with a wooden float or steel trowel as required. The average thickness of the work should be 6 mm and in no place, less than 5 mm. The work is to be cured for seven days. As already pointed out, we use a richer plaster, 1 : 3 or 1 : 4, for this work. We either hack the surface or apply special adhesion chemical for proper adhesion of plaster to concrete.

6 mm cement plaster for slab bearing. This item is very important when detailing bearing of RC roof and floor slabs. This plaster is applied over walls on which RCC roof or floor slab rests. For floor slabs, the usual specification is 6 mm cement plaster with fine sand (1 : 3) with a floating coat of neat cement. Further, a thick *coat of lime wash is also applied.* On roof slabs exposed to the sun and rains, a bedding of craft paper is also given to reduce restraint by the walls due to expansion.

External rendering with cement plaster. The types of finish normally used for external rendering are:

(a) Smooth finish
(b) Textured finish
(c) Rough cast (wet dash or harling) finish
(d) Dry dash (pebble dash finish)
(e) Machine finished
(f) Proprietary finishes

The sand for the external rendering is very important. For the undercoat, fine sand (Zone III) is suitable. However, for the final coat, the suitability will depend on the desired finish. For some textured finishes, it may be desirable to remove the coarse particles by screening through a 3.35 mm (1/8 in) mesh or 2.36 mm (No. 7) sieve while for others such as a torn texture, a larger portion of coarser than 5 mm (3/16 in) are needed. The best quantity of cement paste is obtained when the least possible water is used but enough water should be added to get a workable mix. Waterproofing agents may be added to cement to make it more watertight in external surfaces. Coarse aggregates are used for rough cast and pebble dash finishes. The size and texture will depend on the required finish.

The thickness of undercoat need not be more than 15 mm. For two or three coat work, successive coatings are made—each thinner than the former. The final coat thickness will depend on the type of finish required. In case of smooth finish, it should not exceed 10 mm and the commonly used thickness of the final coat is 5 to 6 mm.

Smooth ordinary finishes. They are either steel trowelled or wood floated. Steel trowel gives a smoother surface than that got by wooden trowel. However, smooth renderings tend to show cracking more clearly. For crack reduction, a wood float finish is usually preferred to metal trowel finish cured well as in case of other cement renderings.

Textured finishes. These are very popular with some builders. They possess advantage over smooth renderings in regard to crazing and the attractive appearance that can be achieved with very little cost. In textured finish, the top 1.5 mm of the final coat is textured with special tools after it is all allowed to stiffen for a few hours. Special tools can be made for texturing. An old hacksaw blade or wooden float faced with expanded metal sponge, hairbrush, steel combs, etc. can produce pleasing effects. If scraping to remove the surface skin of the mortar and expose the larger particles of aggregates is carried out, it is called *scraped finish*. The time to texture or scrape is important. Drying and hardening should be just sufficient for these operations. The coarseness of the scraped finish will depend on the grading of the aggregates.

Rough cast or wet dash finish. These are obtained by throwing onto the undercoat a wet mix of cement and aggregate. The method of finish varies with local labour. The plaster may be a two-coat or three-coat work. The first and the second coat in a three-coat work consist of conventional plaster (such as 1 : 5). In order to make the base more plastic, about 10% of fine hydrated lime by volume of cement is also to be incorporated to the undercoat. The mix for the final rough cast should consist of 1 : 1/2 : 3 plaster with 1½ parts of shringle or crushed stone made into a cement slurry. (The maximum size of coarse aggregate may vary from 6.3 mm to 14 mm). Sufficient water is to be added to obtain a wet plastic mix which can be thrown (dashed) onto the finished coat of rendering while it is still wet from a bucket with a laying-on trowel or hand scoop. The dash coat will be 6 mm and with the base coat the total plaster may be 25 mm. A wide and even spread should be obtained. If coloured cement is used the finish will be very attractive.

Dry dash or wet dash finish. Usually two coats of plaster are first made. The first coat consists of 1 part of cement (with approved waterproofer) and 3 to 4 parts of sand. The second coat (meant to anchor the pebbles) should consist of 1 part cement, 1 part lime and 5 parts of sand (or 1 : 1 of cement plaster) to a thickness of about 10 mm. Lime helps the mortar to be more plastic. A straight edge is then used to obtain a regular surface. Over this, selected and cleaned aggregates such as pea gravel graded from 14 mm (1/2 in) to 6.3 mm (1/4 in) should be selected and anchored for at least 50% of its size in the mortar by hand and anchored. These aggregates are then slightly tapped with a wood float to ensure a good bond. The surface gives a pleasing surface only if the aggregates have a good colour combination and are of varying sizes. A skilled mason is required for this type of work. (*Note:* This rendering on roadside walls discourages pasting notices and advertisements.)

Mechanically applied textured finishes. There are many machines which can throw plaster on the base coat and give appropriate finishes. They use the usual plaster material and are cheap in cost.

Proprietary finishes. A number of proprietary finishes of different colours such as "heritage", which claim that their maintenance costs are low, are available in the market. Though they were once popular, it is not so nowadays. Their prices are high and the laying should be done by experts to get a pleasing and lasting finish.

13.5 OTHER TYPES OF SPECIAL PLASTERS

The commonly used plasters have been described in Section 13.4. There are many other special types of plasters developed in India and some of them are briefly described below.

18 mm plastering with terrazzo finish. In this case, an undercoat of 1 : 3 (cement, coarse sand) plaster of 16 mm is given first. This is brought to true surface with a wooden straight edge and the surface is left rough with furrows of 2 mm deep diagonal scratches both ways not more than 15 cm apart. The surface is kept wet till the top coat of terrazzo is applied which should be within 48 hours of the laying of undercoat. The work is cured for at least seven days.

18 mm artificial red stone plaster. It consists of a 12 mm undercoat of cement plaster 1 : 3 over which a 6 mm finishing coat of plaster consisting of cement, marble dust and stone dust in the ratio 1 : 1 : 3 is mixed with red oxide at the rate of 15.4 kg per quintal of cement to obtain a shade matching red sand stone.

18 mm lime plaster of marble shade. This is an old Indian specification and consists of three coats of plaster. It is made of a 12 mm layer of cement plaster as undercoat with 6 mm special plaster middle coat and a finishing thin coat of special paste.

The undercoat is usually 1 : 4 cement coarse sand plaster. Middle coat plaster is made of 2 parts lime paste, 1 part fine marble chips and 1 part marble dust. The lime paste is prepared by mixing lime with water and allowing it to settle down overnight. When the water is decanted, the lime paste is left behind. The middle coat is finished smooth. This is cured for 7 days and allowed to dry for one month if the work is done in summer and two months in winter. On top of this, the finishing coat is applied.

The finishing paste is made by first mixing 16 m^3 of the above paste with 16 kg of curd. It is then screened through a fine muslin cloth. In this is dissolved, powdered crystallized sugar at the rate of 16 kg per m^3 of paste. The mixture should then be in the consistency of cream. This is applied with a brush over the rendered surface. The finished surface is kept wet for four days and allowed to slowly dry in about seven days. The surface is then rubbed with glass or mohur stones to get a fine polish. Finally the surface is rubbed with a fine muslin cloth and soap stone powder dusted over the plaster using coconut water, till a glazed surface is obtained.

Madras plaster (two coats) mirror finish. The lime plaster is made of the following items and the mortar is made by grinding these on a small curry stone or similar device.

(i) Egg shell lime: 12 parts
(ii) Fine white river sand: 9 parts (very finely ground)
(iii) Powdered marble: 1 part (fine powder)
(iv) Fermented kaduka and jaggery water

The first coat is applied and allowed to set for 2 to 3 days. The surface is scored with lines and kept wet till it is nearly set when the second coat is applied. The second coat consists of the same mixture. The surface of the second coat is then polished with trowel and then with hard stone with soapstones powder (placed in a muslin bag) and also dusted over it. Great care is necessary in joining different day's work so that the continuity is restored.

13.6 MUD PLASTERS

For cheap construction or temporary works at site where brickwork in mud mortar has been used, mud plaster may be used. The following specifications are popular:

18 mm mud plaster without waterproofing. Selected soil should be used for plastering. The clay content should not exceed 10 per cent and the plasticity index 6 to 7. (This can be adjusted by adding more fine sand particles). 10 per cent of the clay content should pass 2.36 mm and not less than 75 per cent through 850 microns. For making ordinary mud mortar, the soil should be mixed with water and left for a week or two, then only it is applied to walls and floated as for lime plaster.

The dry earth is reduced to fine powdered state and mixed with plenty of water in a pit. Chopped straw at about 33 kg per cubic metre of earth is added according to some practice. Plastering starts from top and worked towards the ground level. The total thickness is obtained by applying two coats—one of 12 mm and another of 6 mm. The first coat is finished with wooden float and must be allowed to set but not to become dry before the next coat is applied. The last coat is polished with a trowel. No curing is done for mud plaster but the work is allowed to dry slowly by protecting it with tarpaulin or mat.

20 mm mud plaster with waterproofing. For waterproofing, lime and bitumen are added to the soil. Hydrated lime 3 per cent by weight is initially added and then 2 per cent liquified bitumen penetration 80/100. (This is prepared by warming the 80/100 bitumen and then adding 25 per cent by weight of kerosene to molten bitumen). Water is added and stirred till a homogeneous mixture is obtained. The 20 mm thickness is obtained in one application or two applications and the final surface trowelled well. The applied plaster is allowed to dry slowly but completely in 3 or 4 days. For this purpose, the work should be covered with tarpaulins or mats for such period.

13.7 POINTING

Pointing work is generally used to improve the appearance. Pointing brickwork and stone masonry should be avoided as much as possible as the raking of joints is generally undesirable. Good mortar is removed by raking out the joints. The bricks or stones should also be durable when exposed to the environment. However, pointed work looks more pleasing than an unplastered external brickwork and is used in many places where the bricks used are of very high quality and can be exposed.

The joints of brickwork meant for pointing work should be raked during its construction out at least 10 mm deep neatly at the close of the day's work in all cases for cement mortar and two or three days later for lime mortar. Before pointing, the surface should be thoroughly washed with water, cleaned and kept wet, before pointing. The mortar may be lime mortar (1 : 1.5) or lime surki mortar (1 : 0.5 : 1) or cement mortar (1 : 3).

The mortar is pressed into the raked out joints with a pointing tool either flush raised or sunk according to the type of pointing required. The pointing is finished with proper tool suited for the type of pointing. Iron rod, face of trowel, etc. may be used to get the effect.

13.8 TYPES OF POINTING

The types of pointing commonly used are the following (according to CPWD Specification 77) and are shown in Fig. 13.1. Pointings can be coloured by using coloured mortar or given a coat of a matching coloured paint. A white colour pointing on red brick looks very attractive.

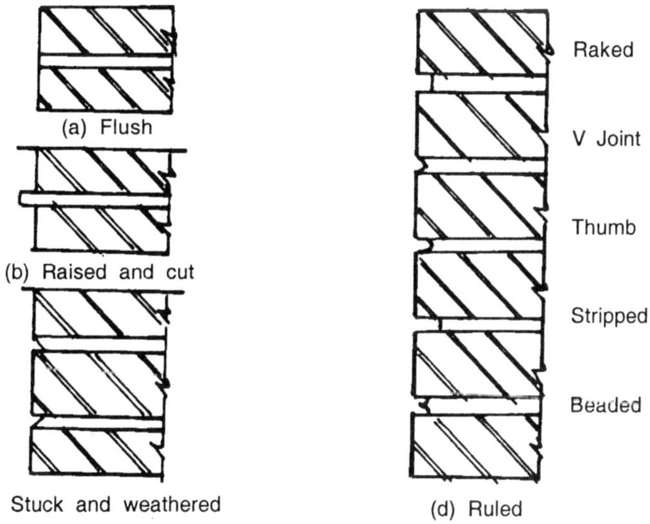

Fig. 13.1 Pointing of brickwork. (pointed face on left-hand side).

Flush pointing. This is the most extensively used pointing even though they are not very attractive [Fig. 13.1(a)]. It is more durable than the others and does not provide space for dust collection.

Protruding pointing or raised and cut pointing. A raised band of 6 mm with a width of 10 mm is provided in this type of pointing [Fig. 13.1(b)].

Slanting pointing (stuck and weathered) pointing. These pointings are finished at an angle. Mortar is pressed into the joints and then pressed back by 3 mm to form a sloping surface [Fig. 13.1(c)]. The slope should be (inwards) *from top to bottom* for stuck pointing and (outwards) *from bottom to top* for weathered pointing.

Recessed or ruled pointing. There are many types of recessed pointing. They are also called ruled pointing. For this pointing, the joints are initially formed as for flush pointing and while the mortar is still green, a forming tool of proposed shape and size is used to form the shape desired [Fig. 13.1(d)].

13.9 PLASTERING BETWEEN CONCRETE BEAM AND BRICK MASONRY

Joints between concrete surface and brickwork as happens between beams and walls and also columns and walls of a framed building require special attention. As the expansion of brickwork and concrete is different, we should either introduce a discontinuous joint where they meet while plastering them or use chicken wiremesh while plastering the joints between concrete work and brickwork.

SUMMARY

Plastering inside and outside buildings can be carried out to different specifications. Inside, the plastering has to be smoother. This smoothness of plaster depends on the fineness of sand and type of trowel used. Pointing is carried out mostly on external walls, and it should be carried out by experienced masons. Plastered surfaces are usually treated with special putty or plaster of Paris to make them smooth before painting.

REVIEW QUESTIONS

1. What considerations should be taken into account for *internal* and *external* plastering? Give the specification for these for a residence.
2. Give the usual specification for lime plaster and describe how lime plastering is to be carried out. What are rendering, floating and finishing coats?
3. Give the various specifications for cement plaster and describe how it is carried out on brick masonry including precautions to be taken in brickwork.
4. What are the effects of fineness of sand and type of trowel used on the final appearance of the plaster work on masonry walls?
5. Name five types of external rendering that can be carried out with cement.
6. What is pointing? What is the procedure to be carried out in pointing?

7. Describe the methods of plastering with mud. In what situations, would you use it?

8. What type of plastering would you select for the roadside face of a brick compound wall to discourage sticking of advertisements? Describe briefly the work.

9. What precautions will you take in the following factors to reduce cracking of plaster in a brick wall?

(a) Selection of cement

(b) Use of type of trowel

(c) Methods of curing lime plaster and cement plaster

Chapter 14

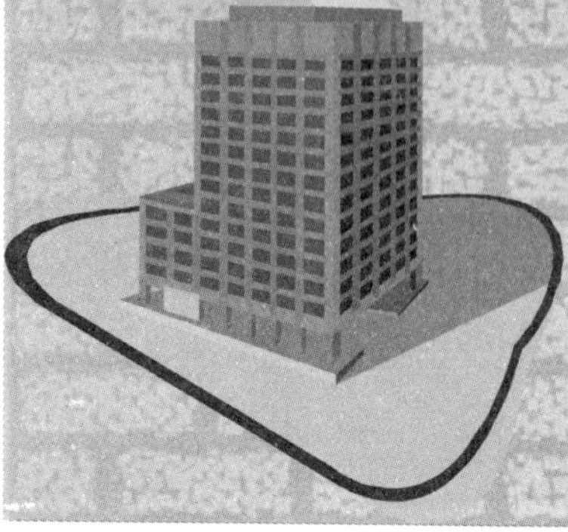

Flooring—General Considerations

14.1 INTRODUCTION

The aesthetic feelings we get when we enter a building very much depend on the finish given to the floor and the walls. Hence, much thought and planning should be given on the selection of materials for flooring and painting of walls. As we have seen in Chapter 1, the cost of flooring and painting can be very high. There are a number of instances where the expenses of the construction have gone up very much higher than the first estimate due to revision of the type of floors and finish of walls, at later stages of construction. A large number of different types of floors are available to choose from. The choice depends on owner's budget, the use to which the floor is put and its specific requirements. For example, in a residence, one may prefer to have only concrete floors with red cement plaster finish if one's finances are modest. Otherwise one may go in for mosaic floors in bedrooms, marble in drawing room, ceramic tiles in kitchen and bathrooms. In a low cost construction, we can go for a simple concrete floor which is also popularly known as *patent stone flooring*.

When considering construction of floor surfaces of a building, a clear distinction must be made between the ground floors and the basement floors, as against the upper floors (also called suspended floors). The ground and basement floors are laid on specially prepared ground whereas the upper floors are laid nowadays on structural RCC slab. As there is no circulation of air under the ground floor, unless specially provided for, there is always the possibility of rise of capillary water in the ground floors and this should be specially attended to by providing dampproofing. This chapter is only an introduction to the subject of various types of floors. The different types of floors are more fully described in the subsequent chapters.

14.2 GROUND FLOORS

Ground floors are laid as shown in Fig. 14.1(a). The components of ground floors are as follows:

1. A sub-base consisting of hardcore materials such as dry brick ballast mixed with moorum, well compacted in place and *a sand layer on its top*. The sand layer is to be placed preferably above the hardcore and it should not be less than 100 mm.
2. A base or subgrade course of 1 : 4 : 8 lean concrete not less than 100 mm thick. In addition to the sand layer, a DPC may be given over lean concrete to prevent the capillary rise of water.
3. The finished floor are prescribed when we use tiles such as mosaic or ceramic tiles. We lay them in cement slurry spread over a bed of mortar. But when we prescribe in-situ mosaic, vinyl floors, wood floor, etc., we have to provide an *underlayer* of 1 : 2 : 4 concrete with 12.5 chips as maximum aggregate size.

As a rule, the finished top level of ground floor should be at least 45 cm (1.5 ft) above the general ground level around the building so that the entrance to the building is by means of a series of steps or a ramp. Usually two steps are provided in residences from the approaching ground as shown in Fig. 14.1 so that the total steps to the house are three. Basement floors are placed below the ground level.

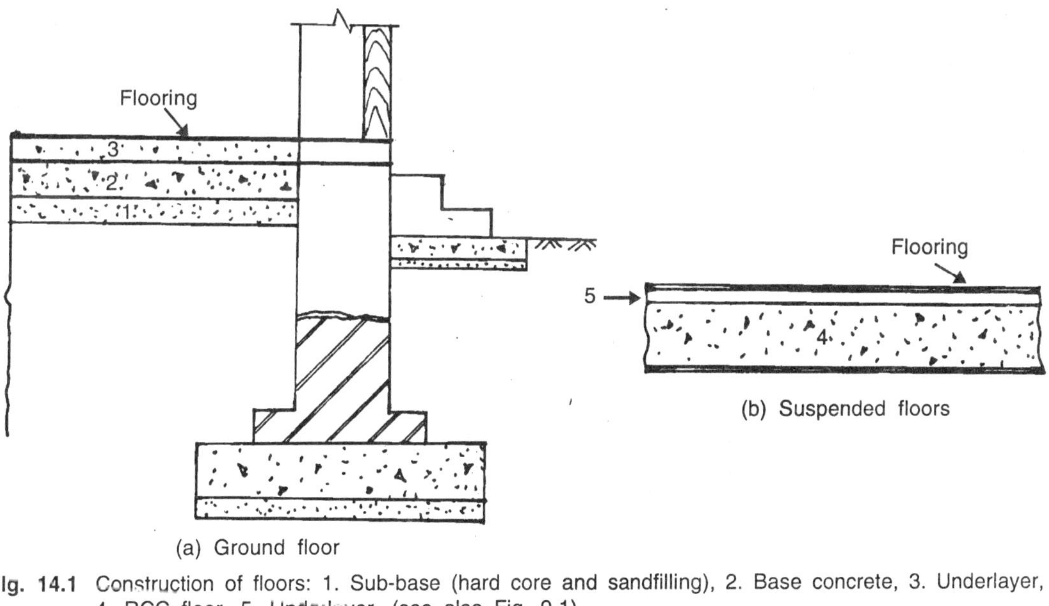

(a) Ground floor

(b) Suspended floors

Fig. 14.1 Construction of floors: 1. Sub-base (hard core and sandfilling), 2. Base concrete, 3. Underlayer, 4. RCC floor, 5. Underlayer. (see also Fig. 9.1).

14.3 DETAILS OF CONSTRUCTION OF SUB-BASE AND BASE

The ground floor has the sub-base, base concrete and the specified flooring. The following is a brief account of the construction of sub-base and base concrete.

Preparation of sub-base. In all ground floors, care should be taken to reduce capillary suction of water through the floor by providing a layer of coarse sandfilling over the consolidated ground. This consolidated ground is called a *sub-base* or *hardcore*. The sub-base to be placed should be properly selected and preferably be granular. It may be stone ballast or brick bats mixed with moorum. These layers are thoroughly compacted so that there will be no loose pockets left. Then the hardcore is covered with a layer of clear sand well consolidated to a *thickness of at least 100 mm and preferably not less than 150 mm.* (The sand layer may also be below the hardcore.)

Note: A water barrier is most important in impermeable floors such as vinyl floors when placed on the ground floor. (Vinyl floor perishes if it becomes wet.) Vinyl floors perform well on suspended floors, but unless sufficient care is taken, they may not be successful on ground or basement floors. In extreme cases, an impervious layer such as polythene or a bitumen damp-proof course may have to be specified for the successful performance of such floors at ground level. Even though, many specifications specify the polythene layer over the sand layer the expense of the membrane can be avoided, if the sandfilling and the concrete laid above it are of very high quality to cut off capillary action. However, some specifications insist on the provision of the membrane to be provided over the sub-base. Otherwise, an in-situ bitumen coating laid over a blinding layer of hardened base concrete or the sub-grade can be specified to act as a DPC (see Chapter 9).

Preparation of base concrete. Usually over the sub-base, a layer of lime surki brick jelly concrete (1 : 1 : 3½) or cement concrete (1 : 4 : 8 or 1 : 5 : 10 at least) with 38 mm aggregate is laid in one layer and beaten down by wooden rammers to the required level and grade. The average depth of this layer should not be less than 10 cm. This layer is called the *base coarse or sub-grade.* Over this base course, various types of floor surfaces are laid. However, in most cases of house constructions where a first floor or roof slab is also to be constructed, the base course is first laid and the floor is used for supporting the centring of the first floor or roof slab. The topping surface of the floor is laid during the final stages only. (In suspended floors where the floor slab is also designed as R.C. roof slabs the structural slab forms the base concrete portion).

14.3.1 Laying Floor Concretes in Strips

Cement concrete shrinks as it hardens, and the magnitude of shrinkage depends on the size and cement content. Hence large panels of floor concrete should be cast in panels to take care of shrinkage. Codes specify that the panels should be preferably of uniform size. And no dimension of the panel should exceed 2 m for indoor and 1.25 m for outdoor work such as terraces, courtyard, etc. The area of the panel should not exceed 2 sq m. The length of the panel should not exceed 1.5 times the breadth. The border panel should not exceed 45 cm in width and the joint in the floor finish should extend through the borders and the skirting. This type of panel construction of floor can be carried out in two ways as follows (also see section 15.2.1—Laying of sub-base).

Laying of concrete in alternate panels. In this procedure, the floor is divided into panels by means of steel flats or angle iron dipped in thick lime solution to prevent concrete adhesion. These formworks should have the same depth as the thickness of concrete. Concrete is laid in alternate panels in this formwork. The formwork is removed the next day and any damage to the edges repaired with 1 part cement and 2 parts coarse sand mortar and allowed to set for at least 24 hours. The alternate panels are then concreted.

Continuous laying of concrete with strips. This method is more preferable to the above methods. In this procedure strips of glass, 4 mm thick, plain asbestos cement sheet (5 mm thick) or PVC or aluminium or brass (2 mm thick) of width equal to the full depth of the concrete are fixed, to the floor to be cast, in 1 : 4 cement mortar with their tops at proper level giving the required slope. After 48 hours, the concrete is placed in all the panels without breaking into panels. This gives uniformity of colour in all the area.

From the above explanations, it is clear that if we plan the base concrete to be laid and the topping is to be placed only after a long time, we can lay the base concrete in one operation without panelling and shrinkage of lean concrete can then be made to take its own course.

14.4 SUSPENDED FLOORS

When one considers the top floors, the structural *RCC slab itself forms the base course* on which the topping is laid in a layer or two [see Fig. 14.1(b)]. In places such as bedrooms and porticos, it may also be necessary to provide a cushioning layer (described below) for noise reduction and formation of slopes. Construction on top of flat roof is specially dealt with under the chapter on Roofs. As there is no capillary rise of water in these floors, any type of floor, if properly laid, will perform well. However, wet areas such as bathroom floors should be properly treated against water leaking through these floors (see Section 27.11).

14.5 LAYING OF TOPPINGS

The preparation of the sub-base and base for ground floors is common for all types of floor finishes. Only the toppings differ. As already pointed out, in tiled floors, the tiles are laid on lime or cement mortar on base concrete on which a cement slurry is spread. But in floors such as in-situ terrazzo, resilient floors, wood floors, etc., we have to provide a under layer of 1 : 2 : 4 concrete 25 mm over the base concrete to lay them on a level surface. As already stated, the structural concrete forms the base course for the top floor and in many cases, a cushioning course is also provided for the top floors as described further.

Cushioning of layer. It may not be possible to provide the necessary slope for bathrooms, kitchen, portico, etc. in the structural concrete itself for the top floors. A cushioning layer up to 50 mm is sometimes provided with (1 part lime putty, 1 part surki and 1 part fine sand) a lime mortar layer. This is also considered as a sound absorbing layer in rooms such as bedrooms for the top floors.

14.6 CHOICE OF FLOOR FINISHES

There are a large number of floor finishes to choose from. The final choice will depend on the available finances and also the type of building under consideration. For example, in a computer room, the first choice will be for a dust free, warm resilient floor. For the entrance lobby of a high class hotel, the choice may be large size marble slabs with matching patterns or pure white colour marble rather than polished granite which may be slippery. For a large kitchen, one may prefer a vitrified ceramic floor which will not be attacked by acidic and other substances. The following are the most important requirements to be considered:

(i) Economy
(ii) Appearance and fineness of joints
(iii) Traffic to which it is subjected to
(iv) Resistance to abrasion to type of traffic
(v) Freedom from slipperiness
(vi) Smoothness
(vii) Freedom from creation of dust from floor
(viii) Low thermal conductivity (warmth)
(ix) Resistance to chemicals and acids
(x) Easiness of repair and maintenance
(xi) Durability on long time basis

When considering traffic on floors, they are generally classified into the following types:

(a) Light duty floors—floors subjected to traffic as in residences and hospitals.
(b) Medium duty floors—floors subjected to traffic as in offices, colleges and banks.
(c) Heavy duty floors as in railway stations and factories.
(d) Non-slip and special floors.

14.7 DESCRIPTION OF FLOORS

There are different types of floors. We will consider for our study only the following commonly used floors:

Chapter 15: Cement and concrete floors
Chapter 16: Stone floors
Chapter 17: Tile floors (Tiling)
Chapter 18: Resilient floors
Chapter 19: Wood floors
Chapter 20: Terrazzo floors

Only very brief accounts of these different types of works have been given in the above-mentioned chapters. For more details, references may be made to the official CPWD or PWD "Specifications" and handbooks on building construction.

14.8 USE OF ABRASIVES IN FLOORS

For slip resistances in *concrete and terrazzo floors,* abrasive materials can be incorporated (see also Section 15.8). The commonly used materials are:

(i) Silicon carbide grains (carborandum, crystaolon).

(ii) Aluminium oxide (Aloxite, Alundum) in the form of grain or aggregate (larger size) in various colours to blend with terrazzo floors. The grains are to be graded between No. 10 (1.7 mm) and No. 30 (500 μm) sieves.

It is important to note that the aluminium oxide grain or aggregate should be thoroughly soaked in water before sprinkling it on the surface or mixing it with aggregates.

In terrazzo floors, the abrasive is usually incorporated in the full terrazzo thickness in the following ratios of abrasive to crushed marble. The presence of these substances does not alter the finish of terrazzo.

(i) Light traffic: 1 : 4

(ii) Medium traffic: 1 : 3

(iii) Heavy traffic: 1 : 2

In concrete and granolithic floors, the abrasives can be introduced by two methods. The first method is the *mix method,* where it is mixed with cement and aggregates for the top layer. In the second method, known as the *sprinkler method,* the fine or coarse abrasive is sprinkled on the surface of the concrete and trowelled in when laid in-situ. The following quantities are recommended in the mix method. For light duty 3 kg/m^2 of abrasives for 20 mm thickness (for 25 mm thickness of flooring, it should be correspondingly increased). For heavy duty, 4.5 kg/m^2 for 20 mm thickness. For a heavy duty floor by sprinkle method, it will need only 1 kg/m^2. For the sprinkler method, the mixture No. 20 to 30 grain is recommended for silicon carbide and No. 10 to 30 grains for aluminium oxide. To sprinkle the abrasive, a sieve can be used. The particles are to be tamped flush with the surface by means of a trowel taking care the grain is not buried.

14.9 CONCRETE SURFACE HARDENERS

Even though a properly laid concrete floor should be dust free, the floor is sometimes treated for increasing its hardness and also to prevent dust production from the floor. Materials commonly used are silicate of soda. Magnesium silicofluoride and other patented products are also available in the market.

Silicate of soda is to be applied on a clean floor. The floor must be dry when it is applied. Old floors must be scrubbed with soap and water. New floors must be allowed to dry out for at least fourteen days after curing before the silicate of soda is applied. Two or three successive coats of the solution of the above made with water are used. The first coat will have a ratio of salt to water 1 : 4 by volume, the second 1 : 3 and the third 1 : 2, each one is applied 24 hours after the former. Unabsorbed silicate should be washed off quickly as otherwise it will cause a stain.

Magnesium silicofluoride is available in crystal form. It should be handled in plastic containers as it is acidic. To make the solution, 250 gm of the substance is dissolved per litre and the solution is applied at the rate of 1 litre per 5 sq m. Two coats are applied at 24 hour intervals. The floor should be thoroughly washed to remove all the unabsorbed liquid.

14.10 PLANNING FOR THE LAYOUT OF TILED FLOORS

For laying tiled floors (all types of tiles) in an aesthetic manner, some planning is necessary. It is always a good practice to dimension the rooms in the planning stage itself of the building in relation to the dimensions of the floor tiles to be used. We should also check whether the floors of the rooms are truly rectangular in shape. There are many ways to plan the layout of the tiles and the following gives an indication of the way it can be planned.

Case 1. For a large hall without passage, we may proceed as shown in Fig. 14.2(a). The centre lines of the length and width of the room are first marked by chalk lines. Having

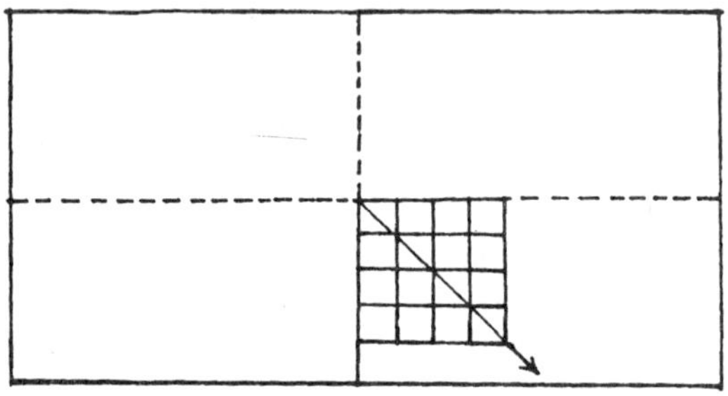

(a) Layout of tiles in a large hall with joints in line

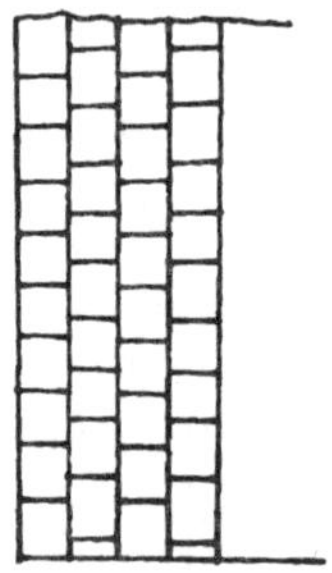

(b) Pattern of tiles laid with breaking joints usually used for roof terraces

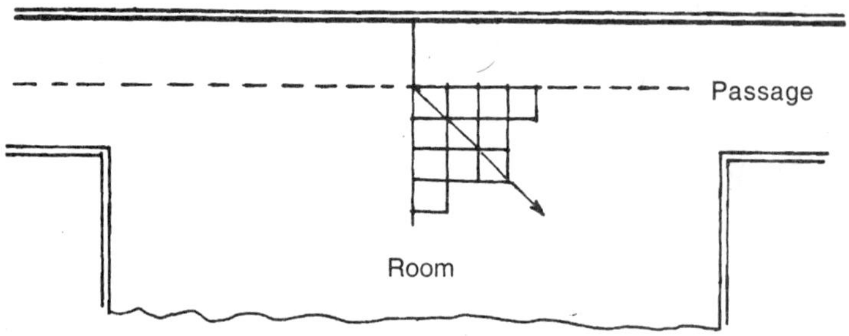

(c) Layout of tiles in a passage leading to a room

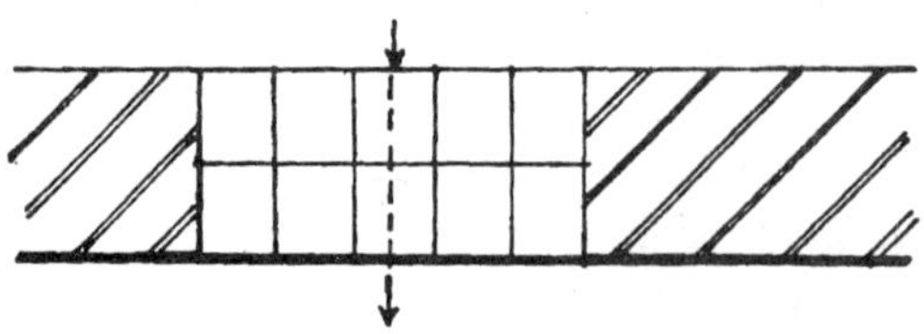

(d) Layout of tiles in an entrance doorway

Fig 14.2 Planning layout of tiled floors.

decided on the design to be laid (depending on the different types of tiles), fit the first tile with its corner at the centre and the edges along the centre lines. The other tiles are laid as shown in Fig. 14.2(a), the work proceeding diagonally towards the sides of the room. Proceed with the row till the last full tile is laid in front of the wall. The last stage of the work is to fit the part tiles round the edges. This is done by placing the tiles against the wall and cutting the overlap on the area already laid. In this design, the odd tiles will appear against the wall and at the doorways.

Case 2. Another layout for a room with a door entrance is to start with a full tile at the middle of the entrances to the room, in which case the laying should start from the entrance and work towards the sides and to the backwall [see Fig. 14.2(d)].

Case 3. For the layout for a room with passages on both sides of the entrance to the room as in an office building, we can plan the layout in such a way as to give the feeling that the passage continues into the room. For this purpose, the layout is started from the centre of the passage as shown in Fig. 14.2(c).

The pattern of tiles laid with breaking joints usually used in roof terraces is shown in Fig. 14.2(b).

In case of marble floors with different types of streaks and shades on the marble slabs, a trial placing of the slabs on the ground with the patterns is laid before fixing them on the floor. This will enable us to evolve an attractive design on the floor. Otherwise the pattern of tiles on the floor may turn out to be very unpleasant.

SUMMARY

Floors are important in building construction. Ground floors should be constructed with care. Capillary water should not rise in the ground floors during rainy seasons and more importantly, it should not subside with passage of time. Failures of ground floors due to insufficient preparation of the base and sub-base have been quite common. Similarly choice of the flooring material is also important. If possible, this should be decided at the planning stage itself to control the total cost of construction. In the upper floors, the important point to remember is the waterproofing of the wet areas so that water does not penetrate downwards (refer Chapter 39).

REVIEW QUESTIONS

1. Explain with the help of sketches the construction details of the ground floor of residential building in normal clayey sand subsoil for laying (a) ceramic tiles and (b) in-situ terrazzo floor.
2. (a) Why should you provide sandfilling under ground floors? What minimum thickness would you recommend in normal conditions?

   (b) Explain with the help of sketches the details of construction of a ground floor in mosaic tiles in a site with expansive subsoil.

3. How will you plan laying of concrete floor of a factory with AC roof in which the flooring is taken as the last item of work and the flooring work is to be carried out in one continuous operation?
4. What treatments can we give to concrete floors to better their performance as regards (a) dust prevention from concrete floors and (b) abrasiveness?
5. Describe briefly how you will plan laying of ceramic tiles in the following cases:
   (a) In a large hall
   (b) At the entrance of a residence
   (c) At the entrance of an office with passage on both sides of the entrance

Chapter 15

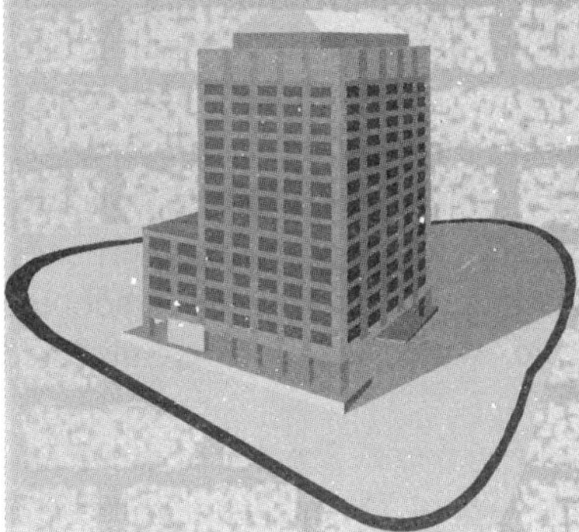

Concrete and Brick Floors

15.1 INTRODUCTION

Concrete floors are the most commonly used type of floors. They are used in factories, garages, low cost houses, etc. The ground floor consists of the *sub-base* and *base* as already described in Chapter 14 together with different types of *concrete finishes as topping* as described in section 15.2.1. In upper floors, the concrete slab acts as the base and the topping is laid over it. Clay brick floorings are used in many places such as pathways of a building and for floors in cheap housings. In this chapter, we will deal with the following types of floor finishes:

1. In-situ concrete floors with granolithic topping (also known as patent stone flooring): For light to heavy traffic situation (see Section 15.2).
2. In-situ ordinary cement concrete floors with fine finish: For ordinary low traffic situations (see Section 15.3).
3. Burnt clay brick flooring with or without concrete base (sub-grade): Section 15.4.
4. Cement concrete floors with red cement plaster finish (red oxide concrete floors): Section 15.5.
5. Cement concrete floors in courtyard and terrace. Section 15.5.

The brief description of the methods of construction of each of these floors is presented in this chapter. As a basic principle, we can take the size of the aggregate to be used for the concrete finishes depending on the thickness of the concrete to be laid. For 12 to 15 mm thick concrete layer, we use 7.5 mm maximum size of coarse aggregate. For a thickness of 15 to 25 mm thick concrete layer, we may use 12.5 mm maximum size of coarse aggregate. The usual 20 mm aggregates are used for thickness of over 25 mm.

15.2 IN-SITU CONCRETE FLOORS WITH GRANOLITHIC TOPPING

Table 15.1 gives the summary of the methods IS 2571 (1970) deals with in-situ cement concrete floors and IS 5491 (1969) with laying in-situ granolithic concrete floor topping over base concrete. The term *granolithic* simply refers to concrete with chosen aggregates. Granolithic concrete topping essentially consists of a rich concrete made *with specially selected aggregates of high quality (usually granite aggregates) suitable for use* as a wearing finish for floors. Although plain concrete floor is considered as satisfactory for many purposes, granolithic concrete is chosen because of its high abrasion resistance. Such topping is adopted for all types of floors including those in factories, garages, etc. where the floor is subjected to light to heavy loads, abrasion and impact.

Mix proportion for granolithic floor topping of 1 part of cement, 2 parts of sand and 2 parts of specially selected coarse aggregate (of maximum size 4.75 mm) by volume is written as 1 : 1 : 2 (4.75). The water-cement ratio should not exceed 0.42 by weight. *Some prefer both the fine and coarse aggregates to consist of clean hard granite chips free from dust and clay.* The fine aggregate has to have the same grading as for ordinary concrete.

In general, two methods of construction of concrete floor topping are adopted in places where the floor thickness is not less than 100 mm. The first is the monolithic construction in which the topping is laid within three hours of the base concrete where the thickness of topping needs to be small—15 to 20 mm only. The second is the partially bonded (separate or two stage) construction in which the topping is laid on the base concrete at any time after the base has hardened. In the second case, the thickness of topping has to be at least 25 mm under layer and 15 mm wearing or top layer of granolith.

As we have already seen, the construction of ground floor will be slightly different from that of top floors which are called *suspended floors*. The different practices for laying the ground and suspended floor systems are described below and shown in Tables 15.1 and 15.2.

As granolithic mix is a rich mix, there is bound to be a large amount of shrinkage of the concrete. Hence the rules regarding panelling should be followed in its construction. It is best to introduce the aluminium, plastic, asbestos cement strips and lay the floor in one operation as described in section 14.2.1. This will ensure uniformity of colour of the floor.

15.3 GENERAL PRINCIPLES OF LAYING CONCRETE FLOOR ON GROUND LEVEL

The flooring consists of the sub-base, base concrete and topping or floor finish. Each has to be laid carefully. The different types of construction are given in Table 15.1. The different operations involved in a construction are discussed further.

15.3.1 Laying of Sub-base

This has been described in Section 10.2. In case of expansive soils such as black cotton soils, lime treatment of base soil is found to give good results. Stone ballast (40 mm graded aggregate) mixed with locally available yellow or red soil or soft moorum in 1:1 proportion

Table 15.1 Various types of concrete ground floors with granular topping

Type	*Sub-base*	*Base (concrete)*	*Topping (finish) of granolithic concrete*	*Place of application*
1.	Compacted ground covered with 100 to 150 mm coarse sand	1 : 4 : 8 (40 mm) cement concrete min 100 mm	1 : 2 : 3 (4.75 mm) (minimum 20 mm thick)	Base and topping laid together, light traffic, good soil.
2.	Min. 100 mm thick hardcore on 100 mm coarse sand	As above	1 : 2 : 4 (12.5 mm) (minimum 25 mm thick)	Base and topping laid together, garage and heavy traffic, good soil.
3.	Minimum 300 mm hardcore stone ballast and moorum (1 : 1). Another 200 mm soft moorum or cinder or coarse sand	As above	1 : 2 : 3 (4.75 mm) granolithic (minimum 20 mm thick)	Base and topping laid together *for expansive soils*, moderate traffic.
4.	100–150 mm coarse sand over ground (good soil)	1 : 5 : 10 (40 mm) lime concrete not less than 100 mm thick	1 : $2\frac{1}{2}$: $3\frac{1}{4}$ (12.5 mm) (minimum 25 mm thick)	One-layer topping laid separately over hardened base for light traffic.
5.	100–150 mm coarse sandfilling over ground (good soil)	As above	1 : 2 : 6 (12.5 mm) (min 25 mm first layer) and 15 mm wearing layer of 1 : 2 : 3 (4.75 mm) second layer (Total 40 mm thick)	Two-layer topping laid over hardened base concrete for heavy traffic.

must be compacted to a layer of 300 mm (12 inches) and thoroughly saturated with water as hardcore. This surface should be further covered with another 100 to 200 mm thick layer of moorum or cinder or sand and compacted before laying the *base concrete*. Special care is necessary in consolidating the ground as otherwise the settlement of the sub-base may cause cracking of the whole floor.

In places like garages which are of less importance but where wheeled traffic comes into contact with the floor, the sub-base should consist of well compacted sand layer of 100 mm thick and an additional 100 mm thick well compacted hardcore of dry brick or stone ballast (40 mm size) blinded with moorum, disintegrated rock or coarse sand whichever is easily available (item 2 in Table 15.1 above).

15.3.2 Laying of Base Concrete

Before placing the base concrete, the sub-base is thoroughly wetted. Over the sub-base, a base layer of lean cement concrete 1 : 4 : 8 or 1 : 5 : 10 with 40 mm aggregate or lime-surki concrete 1 : 4 : 8 is laid and beaten with wooden rammers until the concrete is thoroughly consolidated. The thickness of this concrete should be a minimum of 100 mm. Any shape desired as in bathrooms, kitchen etc. is given on the base concrete itself. Plinth masonry offset is usually depressed to allow the base concrete to rest in it. For proper bonding of the topping that is to be laid at a later date the surface of the base concrete is to be made rough to provide adequate bond for the topping.

15.3.3 Laying of Floor (Topping) Finish

There are three procedures that are in practice, namely:

Case 1: Floor finish laid monolithically with base concrete

Case 2: Floor finish laid separately on hardened concrete base in one layer

Case 3: Floor finish laid separately on hardened concrete in two layers

In the first process, the base concrete and toping are laid one after the other in quick succession. In the two latter procedures, the topping is laid only after the base concrete has completely hardened. These three cases are dealt with further.

Case 1: Floor finish laid monolithically with base concrete. In this method of construction, we follow the following procedures.

Laying of base concrete. To allow for shrinkage, the base concrete of 1 : 4 : 8 is first laid in convenient panels or with strips as described in section 14.3.1. It has, however, been reported that with mechanical laying, the checker board sequence of placement, that is generally used in India, is not really effective in reducing effect of shrinkage as this shrinkage takes place rather slowly. Hence ACI recommends placing concrete in *long alternate strips* to make easy access to construction along the long strips. Sawed joints transverse to the length are made after placing all the strips to induce shrinkage to take place along preformed lines. Use of strips is recommended.

For monolith construction if the base is of *lime concrete* it is allowed to set for seven days and the topping is laid in the next three days. If the base is of *cement concrete*, the topping should commence within 48 hours of the laying of the concrete base.

Laying of topping. The concrete mix for the topping is deposited on the base concrete *in the screed strips already made for base concrete*. The mix should be stiff enough to prevent accumulation of any excess water. If any water appears on the surface it should be mopped up. The practice of allowing it to be absorbed by spreading dry cement should not be allowed.

After compaction, the surface is finished by trowelling and floating and the level tested with the straight edge and a mason's spirit. If there is any defect, it should quickly be made good. Finishing should start immediately after laying and shall spread over a period of approximately six hours in average weather conditions. The surface is to be trowelled three times to result in a uniform, hard surface, the time interval between trowelling being important. The first trowelling is done immediately after levelling to give a level surface. Excessive trowelling at this stage will bring all the cement to the surface. After a lapse of sometime (depending on setting time of cement and the temperature at site, etc.) the second trowelling is made to close all the pores and bring to the surface any excess water which should be mopped up. The third and final trowelling must be done with a steel trowel while the concrete is not too hard, but considerable pressure is required to make any impression on the surface. If to reduce shrinkage effects, concreting can be carried out either in alternate panels separated by screed strips or by using 4 mm glass, aluminium or plain asbestos strips for effective separation of panels, the base concrete and topping being laid in all panels together in one operation.

Case 2: Floor finish laid separately as one layer on hardened concrete. In this case, there are two operations—the first being the laying of the base concrete (and allowing full shrinkage) and the second, laying of the topping. These are descrifed further.

Laying base concrete. In this case, the base concrete, which is lean, can be deposited in the whole area at one stretch after wetting and ramming the sub-base properly. The surface of the base concrete shall be left rough to provide adequate bond for the topping. For this purpose, two or three hours after the concrete has been laid, the surface is brushed with a hard brush to remove any scum, etc. and swept clean and the coarse aggregates are exposed. The base concrete can be left in that state for *curing and full shrinkage to take place.*

Laying of topping. The roughened surface of the base concrete, after it has hardened, is soaked with water at least for twelve hours. The excess water is mopped up immediately before the topping is laid. The screed strips are fixed over the base concrete dividing it into suitable panels as recommended in section 14.2.1. Before placing the concrete mix for topping neat cement slurry is thoroughly brushed into the prepared surface of the base course just ahead of the topping. The topping is laid tamped thoroughly and struck off level. The surface is floated with a wooden float and finished as in Case 1 above. The surface is tested with a straight edge and mason's level. The finish can be laid in alternate panels or as separated by strips as described in section 14.3.1.

Case 3: Laying the topping itself in two layers. To obtain very smooth and dense finish, the top layer is sometimes finished in two layers. The first layer of the topping is laid but not finished smooth. It is left rough after tamping with a screed board. The top layer of wearing layer is of 1 : 2 : 3 cement concrete (or any other desired mix depending on abrasive resistance required) and of consistency stiffer than the underlayer concrete is laid immediately over the rough but green surface of underlayer, thoroughly tamped struck off level and the surface floated *with wooden float.* The top surface is then tested with a straight edge and mason's spirit level. Any defect is made good immediately. The surface is finished smooth with a steel trowel as explained in Case 1.

15.4 LAYING CONCRETE ON SUSPENDED SLABS

The construction of the concrete surface on the top floors is slightly different from these laid on ground floor. Table 15.2 gives the current practices for granolithic topping on top floors. It can be laid monolithically or as separate topping as one layer or two layers. In addition, a cushion layer of 40 mm brick lime concrete in between the structural concrete and the topping can be provided for noise reduction. The lime concrete is to consist of brick aggregate mixed with 40% mortar of lime, surki and fine sand mix in equal proportions. As the concrete is 40 mm thick, we can use 25 mm maximum size aggregate.

Table 15.2 Floor topping laid over structural slab

(*Note*: The structural slab acts as the base.)

Type	*Topping*	*Preparation of structural slab*	*Laying of topping*
1.	Monolithic topping	Topping laid with structural slab	15 mm thick baby jelly granolithic concrete 1 : 2 : 3 (4.75 mm) topping and finished monolithic with slab. Topping laid while structural slab is still green—Light traffic.
2.	Separate (one-layer topping)	Structural concrete surface brushed with coir or steel wire brush 3 to 4 hours after laying. Surface cleaned and treated with cement slurry at the time of topping	25 mm 1 : 2½ : 3½ (12.5 mm) granolithic concrete laid with strips—Moderate traffic.
3.	Separate (two-layer topping)	Structural concrete surface brushed with coir or steel wire brush 3 to 4 hours after laying. Surface cleaned and treated with cement slurry at the time of topping	25 mm concrete underlayer 1 : 3 : 6 (12.5) —Wearing layer plus 15 mm thick 1 : 2 : 3 (4.75 mm) granolithic concrete laid over green under layer and finished monolithic.
4.	Separate (two-layer topping) with 40 mm cushioning lime concrete layer over structural slab	Structural concrete surface brushed with coir or steel wire brush 3 to 4 hours after laying. Surface cleaned and treated with cement slurry at the time of topping	Cushioning layer of 40 mm lime concrete is first provided over the structural slab for noise reduction and granolithic topping is laid.

Summary. In general, the granolithic concrete can be laid *monolithic* with base concrete on ground or on structural concrete (suspended floors) in thickness of at least 15 mm. In case the granolithic is to be laid over set and hardened base concrete in ground floor or on hardened structural slab in suspended floors, it should be laid over the set layer as one layer of 25 mm or as two layers of 25 + 15 = 40 mm depending on traffic conditions.

15.5 IN-SITU ORDINARY CEMENT CONCRETE FLOORS WITH FINE FINISH

Where high resistance to wear is not needed as in low-cost housing projects, the ordinary cement concrete with available aggregates floors can be used for the topping.

Laying of topping. The topping consists of a rich mix (1 : 2 : 4) of concrete—minimum 15 mm thick. The chequer board pattern of laying concrete is sometimes recommended but the method of laying in one operation using 4 mm glass, plain asbestos, PVC or aluminium strips between panels ensures uniform colour for the slab. In moderate panels less than 2 sq m, the concrete can be laid in one operation. After depositing the concrete in panels, it is levelled and beaten with a wooden "*thappy*" till the surface is found covered with a cream of mortar. The surface is finally tested with straight edge and mason's level. Any defect found is made good immediately. Necessary slope in bathroom, kitchen and verandahs should be brought before the concrete sets.

Laying finishing. The finishing should take place immediately after cessation of beating. The surface is left for sometime till moisture disappears (or it is mopped up). Cement is mixed with water to form a thick slurry at the rate of 20 kg per sq m of area and spread over the area while the concrete is still green. The concrete slurry is then pressed and finished smooth. If there are any sunken floors, the edges of the sunk floors should be rounded with 1 : 2 mortar and finished with a floating coat of neat cement. Similarly junction of floor with wall plaster or skirting, etc. should be rounded, if so specified. Raised wooden platforms are provided for workmen to perform this work without damaging the new floorwork.

Curing. The curing must be carried out for a minimum period of ten days after the top layer has hardened. Covering with empty gunny bags, straw, etc. must be avoided as the colour is likely to be bleached with the remnants of cement matter from the bags. Curing by ponding is the most effective method.

Precaution. Flooring of lavatories and bathrooms should be laid after the fixing of water closet, squatting pans and floor traps. Traps should be plugged while laying floors and they should be opened and thoroughly cleaned immediately after the floors are cured.

15.6 BURNT CLAY BRICK FLOORING WITH OR WITHOUT CONCRETE BASE (SUBGRADE)

Depending on its use, brick floors are laid with or without base on ground floors. For heavier traffic, the bricks are laid on edge. They are called "brick on edge flooring." These can be laid in plain, diagonal herringbone bond or any other pattern.

15.6.1 Laying on Base Concrete

IS 5766 (1970) covers these works. The floor is usually laid on a concrete subgrade (base) of lean cement concrete (1 : 4 : 8) or equivalent lime jelly concrete. When using cement concrete as base, the flooring should commence within 48 hours after laying the base. Lime concrete may be allowed to set for seven days and the flooring laid in the next three days. Where the subgrade is not provided as concrete in ground floors, the earth is properly sloped, watered, rammed and consolidated.

Bricks used for this type of work should be good, uniform in colour and well burnt. The subgrade is first wetted and the bricks required for the work should be soaked in water for several hours (to reduce excessive suction) and then allowed to drain until they are surface dry. These bricks are then laid in plain, diagonal, herringbone or other suitable pattern in 12 mm thick cement mortar bed (1 : 4) and must be properly bedded flat and set by gently topping with the handle of a trowel or a wooden mallet.

The thickness of the vertical open joints should be between 8 to 10 mm. On completion of a portion of the flooring, the vertical joint *must be fully filled with cement mortar from the top*. The mortar should be 1 : 4 with 1/2 part of lime as optional to improve workability. The surface level of the flooring during laying must be frequently checked with a straight edge at least two metres long. In case of flat brick flooring, bricks should be laid with frog down. The joints should be raked out to 10–15 mm deep by a raking tool while the mortar is still green, well brushed to remove loose particles, well wetted and refilled with 1 : 3 cement mortar and flush pointed in cases where pointing or plastering is not to be done. If plastering or pointing is to be done, the joints are raked out as before to provide proper key for the plastering or pointing to be done. The face of the brick should be cleaned on the same day of the brickwork and all mortar droppings removed promptly. The cement work should be protected and cured for a minimum period of seven days. In case of masonry with fat lime mortar, curing should start only two days after laying of the masonry.

15.6.2 Brick Floors without Base with Dry Mortar

For light traffic areas such as pathways, sitouts, etc. the dry bricks can be first arranged in the desired pattern with open joints and instead of filling the open vertical joints with wet mortar, we can fill it with dry mortar (dry cement-sand mix) and wet it afterwards as described below to attain the strength. This will considerably reduce the labour.

For this purpose, we first prepare the *dry mortar mix* of 1 part of cement and 4 parts of sand. Spread the mixture over the dry surface of brick arranged on the firm ground and brush it into the joints. Tamp the mortar firmly with a piece of 10 mm plywood adding more mix, if necessary, to fill the joints. Sweep off any remaining mix on the bricks. Using an extremely fine spray, wet the paving without pools of water to be formed and without splashing the mortar out of the joints. For the next 2 to 3 hours, periodically wet the paving to keep it damp. When the mortar begins to harden, we can rake the joints to give the joint the look we want. After the mortar has set, we can wash and clean the stone of any stains.

Note: If we have a firm ground, we can also set the brick on a dry sand bed with vertical joints filled with only sand without cement. Another variation is to use a dry sand cement bed (with 1 part of cement to 6 parts of sand) instead of the pure sand bed.

Another type of dry brick flooring is made by arranging dry bricks with fine joints and filling these joints with fine sand only.

15.7 CEMENT CONCRETE FLOORS WITH RED CEMENT PLASTER FINISH (RED OXIDE CONCRETE FLOORS)

This type of floors were once very popular for inner rooms of residences. It is not fit for heavy traffic and does not fare well in direct sunlight. If it is laid well, it gives a pleasing appearance and takes good polish in course of time. It can also be polished by regular floor polish (manson polish). The work, unlike mosaic floors, can be completed in a short period also.

Preparation of underlayer. This concrete floor is laid in convenient panels using wooden strips or flat iron as shuttering or using left in strips as stated in section 14.3.1. The sub-base and base for the ground floor must be prepared as for other concrete floors. The structural slab is the base for the top floors. On the slab, a layer of cement concrete 1 : 2 : 4 (12.5 mm) is laid as described under concrete floors. This is called the *underlayer of the floor*. The underlayer is roughened with 2 mm deep diagonal lines at 75 mm spacing to form key to the topping. (Joint free flooring can also be done if proper time delay is given for shrinkage.)

Laying of top layer. The top layer of "red oxide flooring" is a 7 to 12 mm layer of special plaster *with an additional finishing layer of cement red oxide slurry*. For the *plaster*, 3.5 kg of red oxide of good quality (one should make sure that the material is red oxide and not other materials only red in colour) is added to every 50 kg of *light coloured Portland cement* (the ratio may be varied to suit the colour of cement) and the two are dry mixed well. The full quantity required for the room must be prepared in one lot in order to ensure uniform colour. (Nowadays ready-made red oxide cements are available for this purpose). A day after, laying of the underlayer of concrete, the top plaster layer is laid. For this purpose, a plaster of 1 : 3 (of the above cement and fine sand) is laid to a uniform thickness of 7 to 12 mm. It is then finished with a *cement slurry* (2 kg of above cement mixed with water for every sq m of floor). The surface is later trowelled evenly by a highly skilled mason.

Curing. Normally curing is *prescribed for a period of ten days*. Covering the finished concrete with empty cement bags is not a good practice as the red colour may get spoiled. Curing by ponding or by wet sand gives good results.

Polishing. In old days, the floor was polished by smearing with coconut oil and rubbing with a smooth husk. Powdered egg shell was also used. Nowadays mason polish, mixed with French polish (to give a glow), is used if polishing is to be done. In many cases, daily moping with water without polishing seems to enhance the beauty of the floor.

Work on skirting and dado. These are also done similar to flooring to give a uniform finish.

15.8 CEMENT CONCRETE FLOORS IN COURTYARD AND TERRACE

Floors for these places should be *rough and skidproof.* The concrete surface is finished rough with wooden floats immediately after consolidation and the moisture disappears from the top and the surface becomes stiff. Then they are given chequered finish, if needed, by pressing a piece of expanded mental of appropriate size and design onto the green concrete and leaving its impression on it.

15.9 FLOORING WITH CEMENT CONCRETE TILES

This type of floor is covered by IS 1443 (1959). The base is finished 35 to 45 mm below the final grade. These types of floors are ideally suited for quick construction of ground floors and low-cost mass housing schemes as the floors can be laid from factory-made tiles. They are laid in the same way as terrazzo tiles, described in section 20.3.1, on a suitable mortar bed.

Lime mortar is recommended in preferences to cement mortar for preparing the bed for these tiles. If cement mortar is used it need be only 1 : 6 with as little water as possible for spreading the mortar evenly. The top of the base should be rough. The mortar is spread over the well-cleaned and wetted base to a thickness not less than 10 mm and not more than 20 mm by a screeding board. *In tiling of very large areas, the level of the central portion should be kept a little (10 to 20 mm) higher than edges to avoid the optical illusion of a depression in the central portion of the floor.* After the bedding mortar has become sufficiently hard, the tiles are fixed by neat cement grout. The joints should be as close as possible (normally 1.5 mm). An area to cover 20 tiles is usually taken for fixing at each time. The day after the tiles have been fixed, the joints are cleaned with a wire brush and refilled with neat cement paste or grout of the same proportion and colour as the matrix of the tile. After fixing, the flooring is allowed to *mature for seven days* before allowing light traffic. Regular traffic should be allowed *only after fourteen days.*

The floor is also lightly polished by first thoroughly wetting it with clean water and rubbing it down especially the joints with a suitable polishing stone and plenty of water without scratching the floor. After washing it down, it is again rubbed hard with felt after the floor is moistened with water and dusted with oxalic acid powder.

15.10 CEMENT CONCRETE FLOORS WITH METALLIC HARDENER IN TOPPING ONLY

This aspect has already been described in detail in Section 14.8. For this purpose, a topping of high resistance concrete mix is provided. Metallic hardening compounds (carborundum, rust resisting steel aggregates, etc.) are used for hardening floors. One part of metallic concrete hardener to four parts of cement by weight is first mixed dry. This dry mix is mixed with small stone aggregates (6 mm nominal size) in the ratio of 1 : 2. The concrete mixture so obtained is laid as a topping coat (as thin as 12 mm thick) within 2 to 4 hours of laying the bottom layer as described in other types of concrete floors. The floor is also cured as in the cases of other concrete floors.

15.11 MAINTENANCE OF CONCRETE FLOORS

A well made concrete floor should be dust free under normal traffic. Dusting in floors can be reduced by application of a hardener, or by waxing it, or painting it with a solution of oil in spirit. Raw or boiled linseed oil (similar to that used in paint) mixed with an equal amount of turpentine has been found to be effective for such application. Proprietary products are also available for this purpose. Painting with epoxy, acrylic or chlorinated rubber base is also adopted to prevent dusting of floors (see also Section 14.8).

Wax polishing with proprietary polishes is also popular. This polish can be made at site by mixing a 25% solution of melted bees wax (or harder wax, if required) in turpentine together with pigment (iron oxide for red and chromic oxide for green) at the rate of 60% by weight of bees wax. The wax is applied to the floor. It will harden after 12 hours and then the surface should be sprinkled with French chalk and polished with dry cloth. 200 gm of bees wax will cover an area of 10 sq m.

If one wants to roughen a concrete floor, it is usually done by washing the floor with dilute hydrochloric acid. (Dilute acid is obtained by diluting a concentrated acid with five times its volume of water.) The dilute acid is applied at the rate of 0.3 litre per sq m of floor area. The acid is left for about quarter of an hour during which time the floor is brushed with a stiff brush or wire broom. Then the floor is thoroughly cleaned with water. One to three successive applications usually give the desired results. Mechanical grinding can also be used to roughen the concrete floor.

SUMMARY

Concrete floors as described in this chapter are extensively used in construction of factories, low-cost houses, etc. They are cheap in cost and perform very well under all conditions. However, good care and attention to the various details of construction are required to get their good performance.

REVIEW QUESTIONS

1. (a) Name a few places where concrete floors are ideally suited.
   (b) What are the usual specifications for concrete floors when used in the ground floor?

2. (a) What is granolithic flooring? Is there any other name by which it is known?
   (b) Describe how you will lay the granolithic flooring for the ground floor of a multistorey building where during construction, the ground floor has to be used to support the first floor concreting.
   (c) What types of difference in the sequence of construction of a floor construction would you recommend if you were to lay the granolithic floor for a factory with AC sheet roofing in comparison with the ground floor of a residential three-storey building?

3. Describe the laying of the following types of concrete floors:
   (a) Ordinary concrete with fine finish
   (b) Concrete floor with red oxide finish
   (c) Concrete floors in courtyard and terrace

4. Enumerate two different methods of laying brick floors on compacted base.

5. What are the maximum sizes of coarse aggregates you will use for a 1 : 2 : 4 concrete of the following thicknesses:
   (a) 15 mm thickness
   (b) 25 mm thickness
   (c) Over 25 mm thickness

Chapter 16

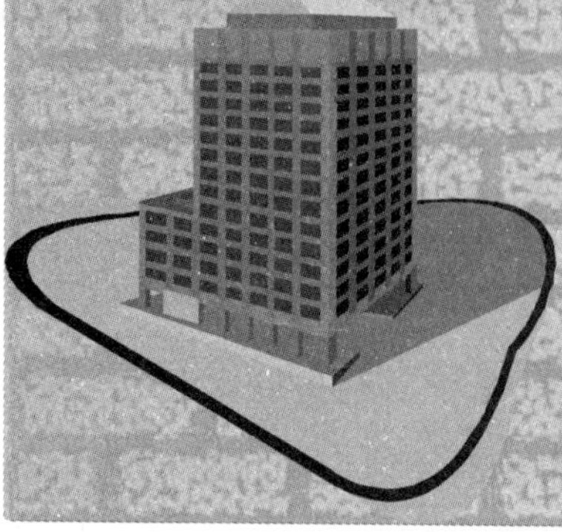

Stone Floors

16.1 INTRODUCTION

Stone floors are very much recommended for offices, educational institutions, etc. where the traffic is heavy and maintenance has to be minimal. It is also recommended in residences for its good appearance. It is also used in places such as kitchen as many of the natural stones such as granite are more resistant than cement to light acids, etc. Machine cut and polished natural stones are extensively used both as large slabs or smaller tiles. When used as large slabs, they should be much thicker than tiles. They are also available in different colours to choose from. The commonly available stones are:

1. Marble of different colours and streaks (see Section 7.9)
2. Granite stones of different colours
3. Kota stone (of thickness 20, 30 and 40 mm as specified)
4. Red or white fine-dressed sandstone (40 mm thick)
5. Shahabad stones
6. Cuddapa slabs (black in colour)

The choice of stone depends on its requirement of maintenance, etc. Marbles and sandstones are somewhat porous and get scratched if people walk on them with shoes. Marbles require repolishing once every six to seven years. Light coloured marbles are somewhat porous and do not fare well in places such as kitchens. Black marble is less porous and is found to fare better in such places. (As a rule, the darker the marble or granite, the less porous it is. So black marbles are the least porous marbles.) Granites are very durable and means of repolishing them is of very recent origin and is expensive. In this chapter, we will briefly deal only with the general principles of laying stone floors. The methods of fixing stone slabs on steps, columns, walls, etc. are given in Sections 7.8 and 7.9.

16.2 LAYING OF STONE FLOORS

For laying stone floors, the subgrade concrete in the ground floor or the RCC slab in top floors is first cleaned, wetted and mopped. The bedding of the slab is usually cement mortar 1 : 4 or lime mortar (1 part lime, 1 part surki and 1 part coarse sand) of average thickness of 20 mm but not less than 12 mm in any place. The procedure of laying stones is as follows.

Stone floors can be placed either with thin joints or with thick joints. If the joints are to be thin (3 mm), then they are laid as specified in section 16.3.2. However, if the joints can be thicker (5 mm), the slab can be placed more quickly as follows. Cement mortar (1 : 4) or lime surki mortar (1 : 1 : 1) is spread liberally under the slab after cleaning the subgrade concrete on which the slab is to be laid. The stone slabs are washed clean and laid on top. It is pressed and moved around so that all hollows underneath get filled and surplus cement works up through the joints. The top is tapped with wooden mallet and brought level with the adjoining slab. If the joints are to be pointed, they are raked up when the mortar is green to a depth of 12 mm and pointed.

16.3 LAYING OF MARBLE FLOORS WITH THIN JOINTS AND FIXING OF SKIRTING

16.3.1 General Considerations

Marble masonry work has been described in section 7.9.1. Marble is a metamorphic rock and considered an economic material for wall facing as well as for floor paving for important buildings. They are available in different qualities and costs. It is most important that the quality chosen for the floor work should fit its use. When used on walls, they can be highly polished but for floors, we need a hard and non-slippery surface. A high level of polish is slippery and the polish will be spoiled by traffic. A marble with honed and grilled surface wears well. *Marbles that contain softer veins* or planes of weakness are not suitable for floors even though they may be used on walls. *Great care should be taken for selection of marble stone for floors.* The surface should wear equally plain. *Homogeneous marbles which are naturally sounder than those with irregular patches of different colours (which are usually selected for their beauty)* are preferable for floorings.

Marbles with streaks when selected for its beauty can be put to good effect by artistic skill and matching. Thus both sides of a sawcut when sawn from the same block and placed adjacent such as an open book is called *book-matched.* In *quarter matching,* they are placed one below the other to match the grains. Marble floors may be made from large slabs or from small-sized slabs or tiles. Laying of small-sized slabs is similar to laying of tiles. However, we should be aware that this beauty due to streaks of impurities may not last as these types of stones with streaks tend to wear unequally. It has been found that in many places, they had to be replaced after about 15 to 20 years. The recommended minimum thickness of marble slab is 20 mm. Tiles can be much thinner.

16.3.2 Method of Laying Marble

When used for flooring, marble slabs are dressed and ground on all the sides and top (floor side) as bought from the supplier. A straight edge laid along the side of the stone shall be fully in contact with it. Thickness can be 20, 30, 40 mm depending on its size. The bedding for the slab is 1 : 4 cement mortar or 1 : 1 : 1 lime putty, surki coarse sand mortar as for other stone floors. The average thickness of mortar bed must be 20 mm but never less than 12 mm.

Laying of marble with thin joints is a specialized work called "marble laying". The following procedure is used when the joints are to be limited to 3 mm. The above mortar is first laid. The slab (washed clean before laying) is laid on top, pressed, tapped with a wooden mallet and brought level with adjoining slabs. The slab is then lifted and laid aside. Any hollow space in the mortar bed is corrected by adding fresh mortar. The mortar is allowed to harden a bit and cement slurry of consistency like honey is spread over the mortar at the rate of 4.4 kg of cement for one sq m of area. The edges of the slab already laid are buttered with grey or white cement with admixture to suit the colour of the floor. The slab to be paved is then slowly lowered again into position, tapped with wooden mallet till it fits closely and forms a very fine joint with the adjacent tile. Any excess of cement or mortar on the surface is immediately wiped off. The flooring is cured for a minimum period of seven days. Slabs on the floor adjacent to the walls must enter not less than 12 mm under the plaster skirting on dado, and the junction finished smooth and plane (see also section 7.9.1). Slabs which are fixed to the floor adjacent to the wall should enter not less than 12 mm under the plaster skirting on dado.

16.3.3 Fixing of Marble Slabs for Risers of Steps and Skirting

Fixing of marble in steps has been described in section 7.9.1. Marble slabs for skirting works should not be less than 18 mm thick with ±2 mm tolerance. For marble skirting work (which is usually 10 to 11 cm wide), we must ensure that 15 cm of space above the finished layer of the floor is left unplastered for installing the skirting. The marble slabs are laid lengthwise and joined in grey or coloured cement mortar 1 : 3 with the line of the slabs at least 10 mm from face of wall or riser, with an average gap of 12 mm from face of the wall. The marble slabs are first held in position at this distance by temporary MS hooks fixed into the riser or wall at suitable intervals or by bricks held against the face of the skirting tiles. The joints between adjacent slabs are filled with mortar and are left to harden. The empty rear space of the skirting is then packed after the joints are set with 1 : 3 cement mortar. The fixing hooks can be removed after the mortar in the gap has attained enough strength and the holes, if any, made good with mortar.

16.3.4 Polishing and Finishing

Terrazo work requires *grinding and polishing* whereas in marble and granite, we require only *polishing* to bring out the grains. In many cases, the stones come roughly polished when supplied and the final polishing is carried out after laying. This can be done by a machine better than being done manually. Best results are obtained by what is termed *granite polishing*, where graded grit blocks of designation 1 to 6 (the higher number indicating finer grit blocks) are used. Finally the surface is polished with tin oxide powder. Wax polishing is not carried out on marble floors especially in white marble.

16.3.5 Maintenance of Marble Floors

Marble can be maintained easily by washing with a *good detergent* and water, but ordinary soap should not be used as it is likely to leave grease patches. Dark coloured marble can be polished with good silicon wax polish. Marble is porous and light coloured marbles are more porous than dark coloured ones. Care should be taken to keep marble away from wood stains, oil and other substances. When used for kitchen floors and work tables, we should be aware that they will be damaged by citrus fruits, vinegar and other mild acids. For use in kitchens, dark colours are preferable to light colours. Black granite surfaces fare better than marble under these conditions.

16.4 LAYING OF GRANITE FLOORS

More or less the same procedure as in marble floors is used for laying polished granite floors also. In many places where the traffic is very large such as in temples and railway stations, unpolished granite stones are used as flooring and laid as described in Section 16.2. Granite floors can be polished to a very high degree by modern machines.

SUMMARY

Stone floors are extensively used in building construction. As they are expensive, great care should be taken in the selection of the type of stones to be used for flooring and in laying them.

REVIEW QUESTIONS

1. In what situations would you use stone floors? Name four types of commonly used stones for flooring. For which places would you recommend each of them for flooring?
2. Describe how a marble floor is laid with
   (a) Ordinary mortar joints
   (b) Thin mortar joints
3. How do we polish marble when laying in floors? What is the difference between finishing terrazzo floors and marble floors?
4. Describe briefly how the marble skirting for marble floor is fixed to the walls.

Chapter 17

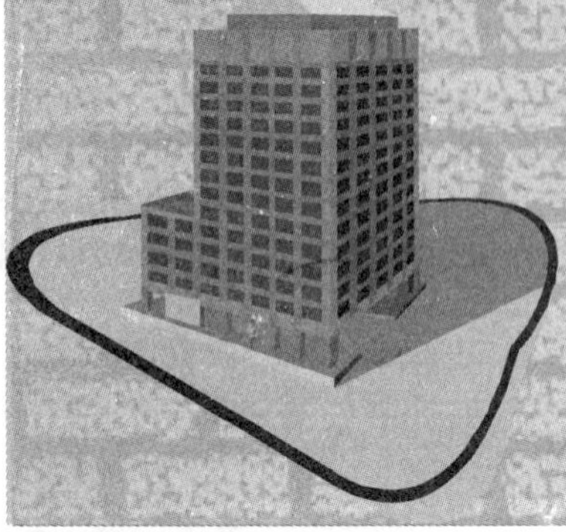

Ceramic Tile Floors and Walls

17.1 INTRODUCTION

The popular choice of flooring for residential buildings has been changing with time. In India, the red oxide concrete floors were very common long ago. They were slowly replaced by terrazzo floors to be followed by marble and other stone floors. Nowadays, large types of ceramic and vitrified tiles and wood floors are becoming more and more popular especially due to the ease with which these can be placed and also maintained.

There are various types of unglazed and glazed ceramic tiles to choose for floors and walls. Some of them are given below:

1. Earthenware tiles
2. Stoneware tiles
3. Terracotta and faience
4. Glazed tiles
5. Fully-vitrified tiles
6. Porecelain tiles

We have already dealt with the characteristics of the above-mentioned ceramic tiles in the book on *Building Materials*. Polished and unpolished (non-slippery) types of ceramic tiles are available in different colours and finishes. Formerly it was not possible to get close-jointed ceramic floors. Nowadays, special close-joint ceramic tiles are also available.

Till recently, only earthenware, stoneware and terracotta tiles were available in India. The principal differences between earthenware, stoneware and terracotta tiles lie in the basic materials. Whereas earthenware is made from selected clay, stoneware contains a large percentage of silica in the form of sand, ground flint, crushed stoneware, etc. to prevent shrinkage while being fired. Terracotta is manufactured from high grade fire clay, faience is similar to terracotta but is fired twice, once like terracotta and the second time after application of glaze. Faience has greater colour range and is generally used for enrichment of terracotta work (see the book on *Building Materials* for details).

Glazed and vitrified tiles. Recently glazed tiles have been developed. Ordinary glazed tiles were first made for walls only. They are generally made in two operations. Firstly, the bodies of the tiles are made from special white clay fired at around 1200°C. The products are called *biscuits*. Secondly, these biscuits are coated with glaze, decoration, etc. and again fired in the ovens. There are two types of glazes—earthenware glazes and coloured enamels. Coloured enamels are of two types—one with bright or glossy surface and the other with other finishes such as eggshell, vellum (parchment paper grade) or matt finish. Glazing techniques have been improved very much and nowadays, we can get glazed *tiles of different sorts fit for floors and heavy traffic also*. Glazed tiles must be distinguished from *fully-vitrified* tiles which are tiles made from a mixture of suitable clays and finely ground minerals fired at very high temperature. In fully-vitrified tiles, the whole thickness is vitrified. Being hard, they stand up to heavy traffic. *The surfaces can be polished with diamond and carbide abrasive tools.* They are said to be as hard as granite. In glazed tiles, there is only a thin glazing at the top. Hence, they are not as long lasting as vitrified tiles. (The Porcelain Enamel Institute (PEI) of UK has graded ceramic tiles into five classes based on abrasive resistance—Grade V being the most resistant against abrasion.

17.2 CHOOSING CERAMIC TILES

As already pointed out, *earthenware glazed tiles* are produced from selected clay but the water absorption capacity of the body of the tile can be more than 10 per cent of its weight and modulus of rupture only 17 MPa. Usually they are salt glazed. However, in stoneware tiles, which are produced from clay mixed with special silicon materials, the water absorption will not be more than 2.5 per cent and modulus of rupture can be as high as 25 MPa. Glazes used for earthenware tiles are technically difficult to be craze resistant, as earthenware unlike stoneware has only a small amount of glossy phase in the body. Similarly, because of the weak base, the glaze in earthenware does not stand up to much abrasion. It is also important to remember that the bright colours used for glazing require a body with a high expansion coefficient and this can be easily achieved only in a porous and relatively low strength ceramics. Hence, very brightly coloured tiles are not very suitable for heavy traffic areas such as shopping centres, transport terminal buildings, etc. One should be aware that until recently most glazed tiles were considered suitable for walls only, where there was no traffic. Only recently, with improved production techniques and glazes, tiles have been evolved for floors. Hence, it is very important to select the right brand of tiles for the type of wall or floor to be tiled. It is easy to select glazed tiles for walls but not so for floors.

The following may be used as a rough guide for the use of ceramic tiles in building. Suitable type of tiles should be selected for different areas as given below:

1. *Bathrooms.* Nonskid glazed ceramic floor tiles on floors and glossy wall tiles for walls can be used.
2. *Internal areas.* (areas not exposed to moisture and subjected to only very light traffic). Glazed ceramic tiles with colour can be used.
3. *Heavier duty floors like kitchen.* Only *stoneware* or *vitrified tiles* or *special glazed ceramic tiles* designed for heavy traffic are recommended.

However, in this chapter, we will concentrate only on the details of laying these tiles on floors and walls. A trial dry placing should be first made, if possible, before the final work.

17.3 LAYING CERAMIC TILE FLOORS

Ceramic tiles should be laid on a strong base concrete in ground floors as otherwise the flooring will not last long. The pattern to be used for laying has been dealt with in Chapter 14 (Fig. 14.2). The joints in floorwork are usually continuous in both directions (this is different from the tilework on roof terraces). The usual method of laying ceramic floors is to lay them on the base concrete in 1 : 3 cement mortar of average thickness 10 mm (but not less than 5 mm at any place) with neat cement slurry poured on the mortar as in terrazzo floors. The above mortar is first placed and allowed to harden sufficiently so that wooden planks can be placed for the mason to squat on it. Over this cement bedding, neat grey cement and water slurry (at the rate of 3.3 kg of cement per sq m and the consistency of honey) is spread. Only areas sufficient for laying 20 tiles are taken at one time. Tiles are soaked in water, washed clean and fixed in this grout one after the other by tapping each tile with a wooden mallet to fit it closely with the previous tile and properly embedded level with adjacent tiles. The joints should be as thin as possible and of the desired pattern. If necessary, for small areas to be covered, the tiles should be cut to size with a tile-cutting machine. Tiles adjacent to walls should enter at least 10 mm into the plaster of the walls. All surplus cement and mortar are removed from the tiles. Nowadays, tiles, specially with previously ground edges, are available to produce thin invisible joints between tiles.

The joints are then picked to remove the grey cement grout with wire brush 2 to 3 mm deep and flush pointed with white cement to which colour is added, if necessary, to match the surface. The floor is then cured by keeping it wet for seven days, after which the surface is washed clean. No part of the floor should sound hallow when sounded with a wooden mallet after fixing the tiles. Traffic should not be allowed on the floor till the tilework has been fully cured for at least seven days. (Nowadays, better tile adhesives as described in Section 17.6 are available for fixing these tiles. The bedding will also be thinner.)

17.4 LAYING CERAMIC TILES ON WALLS

A large number of failures in wall tiling due to lack of adhesion between backing materials and the tiles have occurred in practice. Hence great care should be taken in tilework on walls. Tiles are usually fixed to a bedding layer of cement plaster. To start with, the *joints of masonry walls should be raked* out to a depth of 15 mm while brick masonry is constructed. This raking out is very important to ensure a good bond between the brickwork and the plaster to be placed on it. Concrete walls should be hacked at close intervals and cleaned with brushes or specially treated with chemicals as in ceilings for good bonding. The tiles are fixed in one of the following methods:

Method 1: Cement Floating Method. A 12 mm thick plaster (not richer than 1 : 3 and not leaner than 1 : 4) is applied over the surface and allowed to harden for not less than 2 hours. This plaster is roughened with wire brushes or by scratching diagonally at close intervals. When the plaster can support the weight of the tiles, the tiles are soaked in water, washed clean and a coat of cement slurry applied liberally *at the back of the tiles* and set into the bedding layer, the tile being tapped into its proper position. The joints must be uniform, as fine as possible and truly vertical and horizontal, if so specified. The surface should be straight and level. This bedding is known as the *floating method* and is very successful if properly done by an experienced mason. This method is used in all superior works.

Method 2: Cement Bedding Method. In the second method of placing, it is done in one operation. The back of the tile is buttered with mortar to a thickness slightly in excess of the finished thickness required and the tile is then pressed directly to the moist wall and tapped back into position. This method is called *cement bedding method.*

Notes: 1. In these methods, the soaking of tile is very important. When fixing small glazed tiles on walls as in bathrooms, we may like to leave narrow spaces between the tiles. For this purpose, wooden spacer lugs (similar to matchsticks) are usually used to give a uniform and narrow joint pattern.

2. The above techniques are nowadays being replaced by a variety of tile adhesive techniques for decreasing the labour content of the work. This requires only a thin, special adhesive bedding. The tiles are fixed directly to the wall (see Section 17.6).

Method 3: The more efficient but costly method is to use tile fixing compound instead of cement and carry out the operation as in method 2 (see Section 17.6).

17.5 TERRACOTTA TILE FLOORING OVER CONCRETE

We have studied the properties of the various types of ceramic tiles in the book on *Building Materials*. Terracotta has a long history going back to the Greeks. They have fared very well and withstand the corrosive atmosphere of towns and cities. These tiles are available for sale with some reputed manufacturers. These tiles are laid on 1 : 3 cement mortar or 1 : 1 : 6 lime-surki mortar and grouted. The surface on which it is to be laid should be roughened, cleaned of all dust and wetted before the mortar is applied.

Laying terracotta tiles on floors. Terracotta is more porous than ceramic tiles and can be fixed in mortar. First the cement mortar 1 : 3 is spread in 12 mm layer over the surface evenly and to the required slope. *The tiles are soaked in water* for at least an hour before they are laid with open joints, flat on the mortar and lightly pressed to plane surface, true to slope, etc. using a trowel and a straight edge. In floors, these tiles are initially laid with continuous joints in both directions. The joints should not be more than 10 mm wide but in finer work, it should be as small as possible.

As soon as the paving is done, the joints may be raked and the open joints are grouted with cement mortar (1 : 3) or lime-surki mortar (1 : 1 : 6). The joints are finished flush with the surface and, if necessary, in the same colour as the tiles.

Laying terracotta tiles on walls. The same procedure as in floors is used when laying terracotta tiles on walls of small height. However, when laying in larger heights, where there is a danger of its falling down and causing damage to people and property, special care should be taken not only to embed them in mortar but also preferably to tie them to the wall at regular intervals like in stone veneering work described in Section 7.8. In walls, they are laid in a specified pattern, generally with their longitudinal line of joints perpendicular to the sloping wall surface. *The transverse joints may be continuous or break joints. (When tiles are laid on top of the roof as weatherproof/waterproof course, we lay the tiles breaking joints, but when tiles are laid as floors, the joints are usually continuous in both directions as in mosaic floor joints.)*

Note: Terracotta floors need polishing and good maintenance whereas terrazzo and ceramic floors need only regular cleaning with a wet cloth for its maintenance.

17.6 MODERN TILE ADHESIVES

Nowadays, special, ready-to-use *polymer-based hydraulically setting tile adhesives* (such as tilemate) are available in the market for fixing ceramic tiles. Whereas the old bedding with cement gives a thick bedding, these adhesives give a thin bedding. They have excellent bond strength and durability also. These adhesives are useful also for fixing natural stones, marble tiles, etc. also on any surfaces. They do not need water curing as when using cement.

In many places where we have to modify the floor and lay ceramic tiles over existing mosaic or other surfaces as in bathrooms, these adhesives can be used directly on the old surface. This procedure excludes the need to dismantle the old mosaic for fixing the ceramic tiles.

SUMMARY

Floors and walls laid with ceramic tiles are becoming very popular these days. Different types of ceramic tiles—from cheap glazed tiles to very expensive fully-vitrified tiles and porcelain tiles—are available in the market. They can be laid on thick cement mortar bed or laid on thin bed by using modern adhesives. We should select the right tile for different uses and they should be laid by experienced workmen for good results. As flooring can be an expensive item, much care should be taken in its selection and construction. Generally these tiles are laid with continuous joints in both directions as it will be easy to clean the joints.

REVIEW QUESTIONS

1. Enumerate the different types of ceramic tiles that are available in the market and indicate the places where they can be considered for use in a building.
2. Describe the methods of placing ceramic tiles (a) on the floor of a bedroom and (b) on a wall in a bathroom. What is the minimum curing period when tiles are laid in cement mortar?
3. Describe the method of constructing a floor with terracotta tiles.
4. Write short notes on:
   (a) Stoneware and glazed tiles
   (b) Floor tiles and wall tiles
   (c) Glazed tiles and fully-vitrified tiles
   (d) Joints in tilework in roofing and flooring

Chapter 18

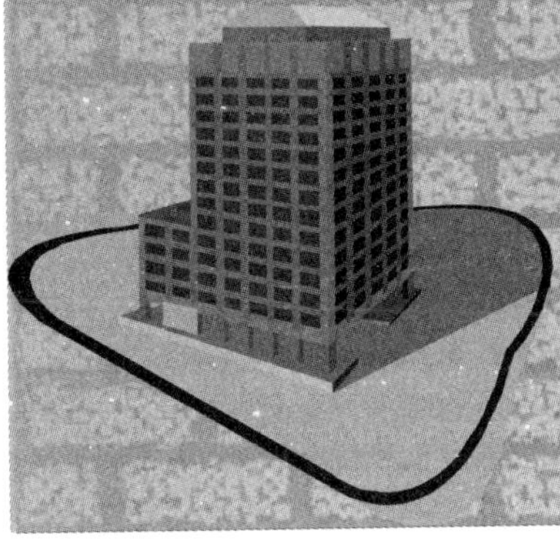

Resilient Floors

18.1 INTRODUCTION

Floors made from materials such as PVC, linoleum, rubber, etc. are called *resilient floors*. Resilient floors are generally laid in offices, computer rooms, show rooms, etc. Their composition, properties, etc. are dealt with in the book on *Building Materials*. A brief account of the laying of the following types of floors is given in this chapter.

1. PVC (vinyl floors)
2. Linoleum flooring
3. Rubber flooring

In *PVC (vinyl) floors,* the thermoplastic binder can be vinyl chloride polymer or vinyl chloride copolymer or both. The floor covering is backed with hessian or other woven fabric. Modern vinyl asbestos tiles have 40 per cent of chrysotile asbestos fibre together with powdered mineral fillers and pigments. It has greater resistance to grease for kitchen use even though it may not be wise to use it for industrial kitchens. The thicknesses of normal PVC floor coverings available in India are 1.5, 2.0, 2.5, 3.0 and 4.0 mm and the thicknesses of "backed floor coverings" come in 2.0, 2.5, 3.0, 4.0 and 5.0 mm. When supplied in rolls, the widths is usually 1.0, 1.5 or 2.0 m and length is 10 m. They may also be supplied as tiles.

Linoleum consists of oxidized or polymerized linseed oil and/or other suitable drying oils with the necessary driers and additives. Special grades are made by changing the type of drying oils and resins. The available thicknesses are 1.6, 2.0, 3.2, 4.5, 6.0 and 6.7 mm, width 2 m and lengths 3 to 5.5 m. The thicknesses of 2.0 to 3.2 mm are used for houses, 3.2 to 4.5 mm for offices and 6 to 6.7 for public buildings.

Rubber flooring is composed of natural rubber with various filling compounds. Because of the high cost of rubber, it is rather expensive. They are produced in sheets and also in tile form. An even subfloor is very essential for successful rubber flooring and the laying of these floors is a specialized job.

All these floors should be laid on a thoroughly damp-proof surface as otherwise moisture slowly rots the base and damages the adhesive resulting in the sheets/tiles being separated from the base. Then they tend to curl up. The subfloor drying can be determined accurately by electrical resistance test or hygrometer test. The moisture content should not be more than 5 to 6 per cent for good results. A qualitative determination can be made by sealing a sheet glass of 300 × 300 mm on the surface of the floor to be tested by putty (or any other suitable sealing material) and observing the surface after 24 hours. If the covered area of the glass is darker than its uncovered area, the floor is considered too wet and should be allowed to dry. Unless it is specifically required, it is better to avoid these materials for use in ground floors. PVC and linoleum are not used for skirting and dados. Wood is preferred for such works with these floors. The methods of laying these floors are dealt with in this chapter.

18.2 LAYING VINYL FLOORS

PVC flooring is usually laid on concrete sub-floors and the work should be taken only after plastering, painting, etc. is completed. Special adhesives such as rubber-based or polyvinyl acetate are to be used for sticking these materials to the base. As already explained, the subfloor should be dry when laying is taken up. When laying on ground floors, the *concrete subfloors must be laid in two layers* to make it level and moistureproof (section 15.2.1). The top of the lower layer concrete should be painted with two coats of bitumen at the rate of 1.5 kg/sq m to make it absolutely damp-proof. For this purpose, the surface is finished smooth and the bitumen is applied after the concrete has set and sufficiently hard and sanded. It should be noted that at the junction of floor and wall, the damp-proof course as above should also extend into the wall for at least 150 mm above the subfloor.

When laying in rooms that are to be always airconditioned as in computer rooms, it is ideal for the floor to laid on the subfloor under stable condition, i.e. the airconditioner working for several days with the temperature between 20–30°C before laying the floor. This condition should be maintained during laying and also for 2 days after laying. Laying of floors during monsoons on high humidity conditions should be avoided.

When using sheets, each sheet should be first laid out for trial without using adhesives according to the desired layout. The surface should be clean and dry. The adhesive is applied by using notched trowel (Fig. 18.1) to the subfloor and the back of PVC sheet or tile. It is allowed to set, free from dust and moisture for about half an hour when it will still be sticky but will not take finger marks. The flooring sheet is then slowly and carefully placed in position from one end onwards so that no air is entrapped between the sheet and the background surface. The sheet is then pressed by a roller to develop proper contact with the subfloor. The next sheet is then laid edge to edge with minimum gap between the joints.

When using tiles of these materials, they are usually laid from the centre of the area or other places as explained in Section 14.9. Care should be taken not to slide them but to lower them into position and roll with a wooden roller weighing 5 kg to ensure full contact between the tile and underlay. If any defect is noticed in laying, the tile should be relaid. The adhesive applied earlier to tile and subfloor should be thoroughly removed by using a proper solvent and the surfaces thoroughly cleaned before relaying.

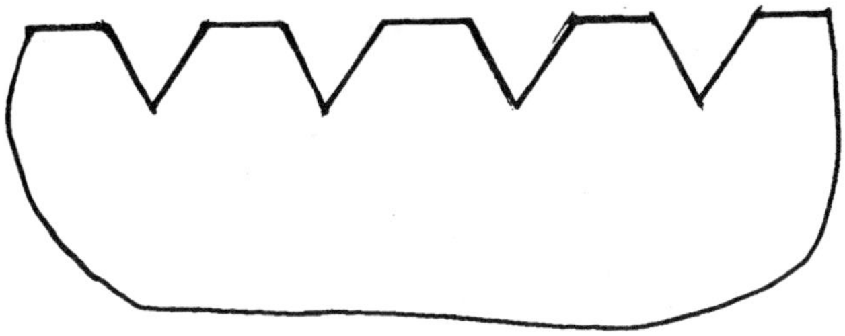

Fig. 18.1 Notched trowel for laying vinyl flooring.

Any adhesive squeezed between sheets or tiles should be wiped off immediately by a cloth. If any adhesive between joints has dried up, it should be cleaned by a solvent. A solution of one part of commercial butyl acetate and three parts of turpentine oil is suitable as a solvent. Where the edges of PVC sheets are exposed as on stair treads, suitable protection by means of metallic strips securely fastened to the subfloor should be used to avoid damage to the edges.

A minimum period of 24 hours should be allowed for development of the bond of the adhesive and for this time, the finished floor should not be put to any use. Thereafter, the floor is cleaned with a wet cloth soaked in dilute warm soap solution made with two spoons of soft soap in five litres of warm water.

18.3 LAYING LINOLEUM FLOORS

Linoleum floors also require a dry and damp-proof surface. It is important to brush away all dust and dirt from the floors before the flooring is laid. Linoleum, if used as sheets, must be unrolled and laid loose for 2 to 3 days before it is cut to size as it shrinks in length and expands in width after rolling. Special adhesive, as already mentioned, is required for sticking the linoleum tiles to the floor (about 1/2 kg of adhesive per sq m will be needed). The adhesive is brushed over the base and allowed to become tacky. The sheet is then firmly pressed down and rolled with a light cast iron roller to ensure that no air pockets remain under the sheet. If any air pockets are found, the sheet should be relaid after brushing more adhesive on the base and rolled. Sandbags are placed over the edges and joints to keep the sheet pressed down and prevent curling.

After laying, the excess adhesive must be removed with kerosene oil or spirit. The flooring should be cleaned with soap and wet cloth and then wiped dry. It may be wax polished to give a smooth surface. We must remember that linoleum perishes quickly under wet or moist conditions.

18.4 LAYING RUBBER FLOORS

There are many types of rubber flooring and skirting. They are available as plain sheets or with backings of fabric or foam or with fabric inset. These floors are costly and also deteriorate rapidly with direct sunlight especially in the presence of moisture. These are also fixed to the level base floors by means of special adhesives. Their laying is similar to vinyl and linoleum products.

SUMMARY

Resilient floors are very much used nowadays in places such as libraries, computer rooms, etc. They should be carefully laid by workmen experienced in such jobs. They are not as permanent as concrete, stone, terrazzo floors but are ideally suited for special situations.

REVIEW QUESTIONS

1. What are resilient floors? In what situations resilient floors are preferred to conventional floors?
2. Describe the method of laying vinyl floors.
3. Why is it not preferred to adopt resilient floors for ground floors? If they have to be used in ground floors, what precautions would you take?

Chapter 19

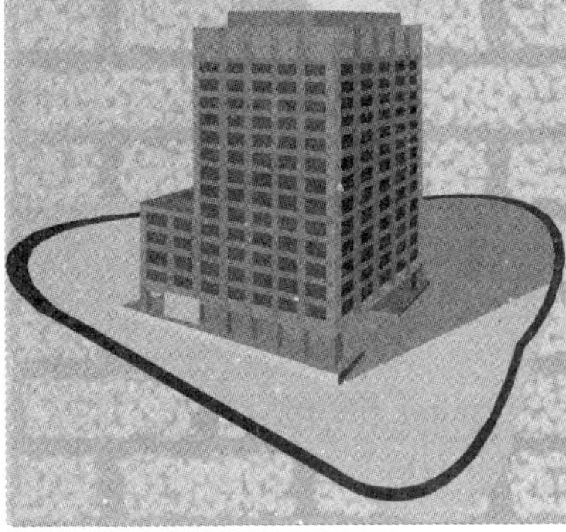

Woodblock and Parquet Flooring

19.1 INTRODUCTION

Wood floors can be divided into three groups—the traditional wooden floors, the woodblock floors and parquet floors. Traditional wooden floors are laid by wooden boards fixed on main beams and joists. As good timber is very costly and difficult to get, such floors are not common nowadays. Wood flooring on top of concrete flooring is quite often used in high class buildings for drawing rooms, reception halls, dancing halls, etc. The woodblocks can be thick or thin. Flooring using thick blocks is called *woodblock floors*. Wood flooring using thin (maximum thickness of 10 mm) pieces of treated wood is called *parquet flooring*. These types of work may also be carried out over plastered brick walls as wall panel, decorations. PVC flooring with woodfloor pattern on it is commonly used in many places as imitations of these floors. In this chapter, we will deal with the method of construction of woodblock and parquet floors.

19.2 WOODBLOCKS FOR FLOORS

The blocks of wood for woodblock floors must be of good and durable timber and regular in size. Teak and white Asian oak are the commonly used wood. These blocks are usually 250 to 300 mm in length and 75 mm in width (in plan). *The thickness is generally 38 mm.* In parquet floor, the wood is much thinner (maximum of 10 mm). The longitudinal edges of the woodblock floors are tongued and grooved near the bottom to assist fixing them firmly to the floor. (Another type of less popular locking system is the dowel and hole system.) The top and sides should be truly plane.

19.3 TRADITIONAL METHOD OF LAYING OF WOODBLOCK FLOORS

The subfloor is always made of cement concrete as described in Chapter 14. The final levelling layer of concrete would be 1 : 2 : 4 (10 mm) of 25 mm thickness. The cement concrete is placed and beaten with a wooden "*thappy*" till the cream comes up. The surface is then finished with a wooden float to give a sandpaper finish. The floor is also brought to the required level by a levelling course. It is very important for the success of the floor that its level is brought to a fine level with no undulation. This concrete is cured at least for a week and then allowed to dry thoroughly before laying the woodblocks on it.

19.3.1 Laying of Woodblocks

Woodblocks can be laid in different patterns. The usual patterns used are brick pattern, square basket pattern and herring bone pattern as shown in Fig. 19.1. After deciding on the pattern,

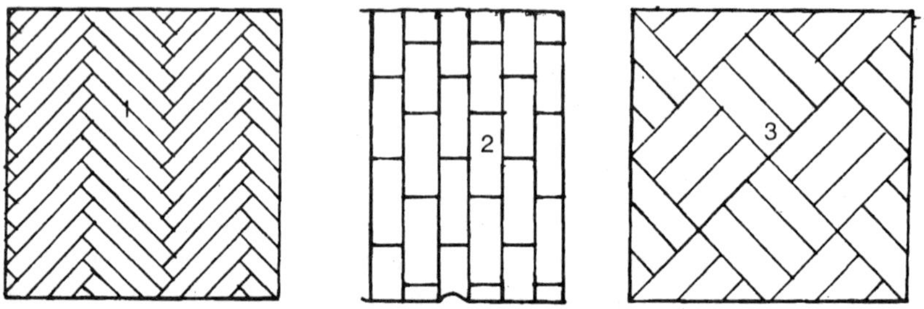

(a) Some patterns of laying wood floors

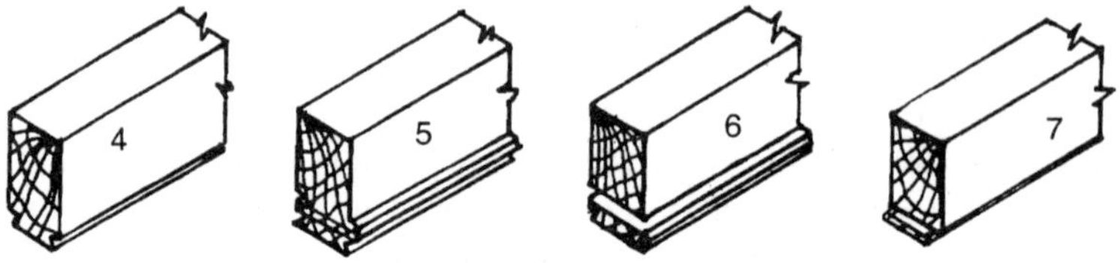

(b) Details of interlocking of blocks at base

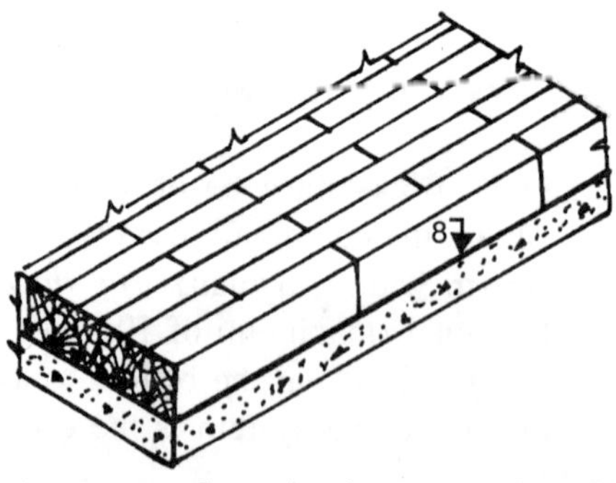

(c) Method of fixing woodblock to concrete floor showing penetration of bitumen into dovetail grooves

Fig. 19.1 Woodblock floors: 1. Herring bone pattern, 2. Brick pattern, 3. Square basket weave pattern, 4. Dovetailed groves at base, 5. Tangue and grooved above base, 6. Dovetail groove above base, 7. Tangue at the ends, (8) Hot bitumen @ 2.45 kg/m^2.

the wooden blocks are first laid dry for a trial fit of the blocks to suit the pattern. It is always recommended to make a dry placing in the desired form and then number of blocks serially. They are then removed and stacked in such a way as to facilitate laying in the same order and pattern.

The blocks are fixed to the concrete floors generally by hot blown bitumen of IS penetration grade 85/25 or equivalent. For this purpose, a small area is taken at a time and the surface is coated with a thin layer of hot bitumen melted to a temperature of not less than 180°C at the rate of 2.45 kg per sq m. The woodblocks are also taken one by one in turn serially and the base is dipped in the hot bitumen. It is dipped for about half their depth so as to thoroughly coat the bottom, fill the grooves on the block and also coat part of the sides. They are then quickly set and pressed into place in the required pattern on the previously coated concrete surface. When laid properly, the dovetailed grooves at the edges are filled with bitumen. The resultant joints must be very fine and thin. The surface is cleaned of any bitumen droppings and checked for level and smoothness. Bitumen on ground floor will also act as a DPC. Alternately, the block may be fixed by special adhesives that are now available.

19.3.2 Finishing of Woodblock Floors

When all the blocks have been laid, the surface is given a rubbing down with sandpaper to give a good finish. Running the special type of power-driven sanding machine with a drum surfaced with sandpaper will give a better and more uniform finish to the floor than using manual labour. After the sanding is completed, the floor is polished with beeswax or other well known brand of wax polish to give a smooth finish.

Nowadys, as the woodblock floors are mostly used for indoor sports (like badminton courts) they are called "sports floors".

19.4 PARQUET FLOORING

Parquet floors which are the floor mostly used in buildings nowadays, are made of thinner already polished wood pieces 10 mm or less thick. There are various types of parquet floors including very thin veneers of treated wood and plastics with wood like appearance. These are usually stuck to a concrete or any other dry surfaces by means of special glues in various patterns.

One of the prime requirements for sticking these thin pieces to the flooring material is that the surface on which it is to be stuck should be dead level with no level differences.

19.5 DISADVANTAGES OF WOODBLOCK FLOORS

The woodblock floors will sustain only light to moderate traffic. They cannot be washed and cleaned with water like stone, mosaic or ceramic floors. They need much careful maintenance. However they are light in weight compared to other types.

19.6 TIMBER FLOORS

True timber floors were commonly used in the past for dance floors and also in residences for the first and upper floors before the coming of reinforced concrete. They were made of *wooden planks* of good quality fixed on beams. Single joists are suitable for short spans, but for large spans, the layout will have to be more complex. As good wood is difficult to get, they are very expensive to build and are not much in use nowadays. We will not go into their construction details.

SUMMARY

Woodblock and parquet floors can be used for light traffic and are used only in special places. The work should be carried out by experienced workmen. The work using thick pieces of wood is called *woodblock flooring* and that using thin pieces of wood of maximum thickness of 10 mm is called *parquet flooring.*

REVIEW QUESTIONS

1. Explain the difference between woodblock floorıng and parquet flooring. In what situations, would you use them? How do they differ from timber floors?
2. Describe how woodblock floors are laid for a drawing room in the ground floor of a residence.
3. Explain why it is very important that the base concrete should be dead level for woodblock and parquet floorings.

Chapter 20

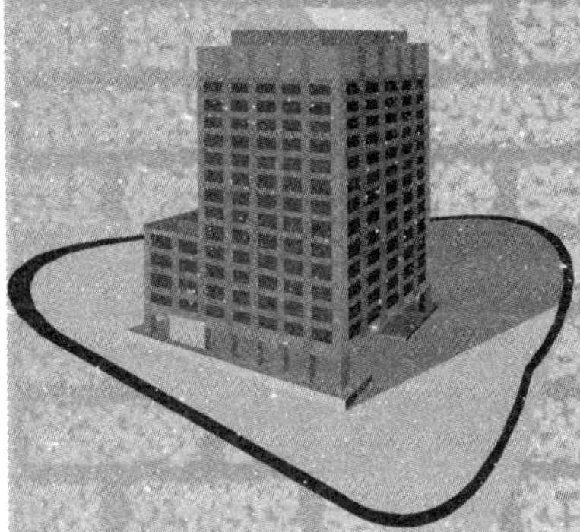

Terrazzo Work

20.1 INTRODUCTION

Mosaic work is the art of combining small pieces of stones, coloured glass, etc. into a pattern or picture. The early mosaics that humans made were simple patterns of pebbles found in Mesopotamia as early as 4000 B.C. Terrazzo work in contrast has no patterns. In modern usage, *terrazzo work* is also loosely called mosaic. Around 1970, terrazzo was the most popular type of floors for residences and general purposes in India because of its good appearance and easy maintenance. However, we should always remember that only with the *use of real marble chips and marble powder* (which will give the same wearing qualities as cement) that we can get good performance of terrazzo work. Terrazzo as a material is dealt with in the book on *Building Materials*. In this chapter, we will briefly deal with terrazzo floor constructions, as tiled and in-situ floors. Terrazzo floors are nowadays less commonly adopted only because of the difficulty of getting good tiles and also because it requires much more time and labour to construct good terrazzo floors compared to ceramic floors.

20.2 TERRAZZO

Terrazzo is also called *Venetian mosaic*. It is essentially a decorative concrete in which the aggregate is *marble chips* and the matrix is white coloured or grey cement. It was first used in Venice where there were plenty of marble chips resulting from marble statue work. It is used not only for floors but also for skirting, walls, door frames, staircases, counter tops, etc. There are two methods of laying the terrazzo flooring—"terrazzo tilework" and "terrazzo mix laid in situ". These are described below.

20.3 TERRAZZO TILEWORK

IS 1143–1959, Code of Practice for Laying and Finishing of Cement Concrete Floor Tiles, gives the method of laying these tiles. Terrazzo tiles are used for flooring, rises of steps, skirting dado, etc. The materials used and the method of preparation of terrazzo tiles are described in the book on *Building Materials*. The work is carried out as described further.

20.3.1 Laying of Tiles

Tiles laid as flooring. When tiles are laid as flooring, it can be laid on a subgrade of concrete in case of ground floor or on RCC slab in case of top floors. This subgrade should have a sufficiently rough surface for bonding. It is cleaned, wetted and mopped before the tiles are laid. Brushing with cement slurry will help the adhesive of mortar. The bonding of the tiles should be in mortar. The mortar can be lime surki mortar (1 : 1 : 2), lime mortar (1 : 3) or cement mortar (1 : 6). The thickness of the mortar can be maximum of 30 mm for 25–30 mm tile and minimum 10 mm for 20–25 mm thick tile. (The thickness of mortar is to be approximately equal to the thickness of the tile.)

If lime mortar is used, it is spread, tamped and corrected to proper levels. It is allowed to harden for a day before the tiles are set. If cement mortar is used, it is spread, tamped and screeded to proper levels. The mortar is allowed to harden but not completely and the tiles are laid on it while the mortar is still green. (Terrazzo tiles can be laid in the same way as we lay ceramic tiles as described in Section 17.3.) Over the mortar, a bedding of neat cement slurry of the consistency of honey is spread at the rate of 4.4 kg of cement per sq m over such an area as will accommodate about twenty tiles. The tiles are washed clean and fixed over the bedding one after the other, each tile being gently tapped with a wooden mallet till it is properly bedded and in line with the adjoining tile. The joint must be kept as thin as possible not exceeding 1.5 mm and in straight line or to suit the required pattern. The surface of the tiles must be frequently checked with a straight edge at least 2 metre long so as to obtain a true surface with the required slope.

The work should start as described in Section 14.9. Where full tiles cannot be placed, these must be cut by a tile cutting saw to required size and the edges are rubbed smooth to ensure a straight and true joint. Tiles, which are fixed in the floor adjoining the wall, must extend at least 12 mm under the plaster, skirting or walltiling with the junction is finished neat and straight. As the tiles are being laid, surplus cement grout that comes out of the joint should be wiped off with a damp cloth.

Tiles laid for skirting. When tiles are to be laid for skirting, risers of steps and for walling, a 12 mm thick plaster of cement mortar 1 : 3 is applied and allowed to harden. The plaster, while it is still green, is roughened with wire brushes or by scratching diagonal lines 2 mm deep at about 75 mm centres both ways. The back of the tiles is then buttered with a coat of ordinary cement slurry, the edges with cement slurry to which a pigment is added to match the shade of the tiles. The tile is set in the bedding mortar in the required pattern with butt joints. They are also corrected to proper levels. They may be level with the wall or projecting out by 6 mm. When flush with the wall, a groove 5 mm deep and 10 mm wide is usually provided above it. When used as risers in steps or as skirting, the tiles should rest on the tread or the flooring.

Note: It should be noted that a well manufactured and laid terrazzo tilework joint will be very thin and hardly visible after polishing. This is in contrast to laying of ordinary ceramic or other tiles where the joints cannot be avoided without great care and expense. This was considered as a great merit for terrazzo floors. However, with the advent of new types of vitrified tiles with its edges reground during the manufacturing, this problem has been minimized in the ceramic floors also.

20.3.2 Grouting, Curing, Grinding and Finishing of Tilework

Some factories deliver terrazo tiles to the site after an *initial first, one or even two, grindings in the factory*. Such terrazzo tiles will have the second grinding, then cured and finally ground for the third time and polished before the floor is put for final use.

Terrazzo work with unground tiles is carried out as follows. The day after laying the tiles, all joints are cleaned of cement grout to a depth of 5mm by a wire brush. All dust and loose mortar are removed and tiles cleaned. The joints are then grouted with grey or white cement slurry, if necessary, mixed with necessary pigment to match the colour of the surface. The *same cement slurry is also applied to the whole surface of the tiles as a thin coat.* This protects the surface from abrasive damage and also to fill any pinholes that may be present on the surface. The floor is then cured by keeping it wet for a *minimum period of 7 days* and the second and third grinding carried out as described below.

As already stated, many factories send tiles as unground and in such cases, they have to be first ground evenly after laying. In such works, the floor is first ground using a machine with coarse grade grit block (No. 60). Water is used profusely during grinding. After this *first grinding,* the surface is thoroughly washed with water (to remove all grinding mud), cleaned and mopped. It is then covered again with the same cement slurry as used before and cured for 3 to 5 days.

The *second machine grinding* is then carried out with a fine grade No.120 to 150 grit block. The day after the second grinding, a *third grinding* is made with a finer No. 320 to 400 grit block. If necessary, an intermediate grinding with No. 80 grit block may be done after the second grinding. In some places (as in restricted areas), hand grinding may have to be resorted to. For such cases, the carborundum stones used are first grinding – coarse grade No. 60, second grinding – machine grade No. 80, and final grinding – fine grade No. 120.

If the tiles are received from the factory as already ground (with the first two grinding), only final grinding is done at the site. This considerably saves the time for laying these tiles. (Time taken for laying terrazzo tiles is considered as a factor against its selection in many places.)

Final finishing. For the final polishing of terrazzo floors, *oxalic acid is dusted over* the surface at the rate of 33 gm per sq m. The surface is then sprinkled with water and rubbed hard with a pad of woolen rag. On the next day, the floor is wiped clean with a moist rag. After laying the floor, no tile should be loose and no part of the floor should sound hollow when tamped with a wooden mallet.

20.4 TERRAZZO LAID IN SITU

IS 2114–1962, code of practice for laying in-situ terrazzo floor finish, gives the procedure for laying terrazzo in situ. In situ terrazzo consists of a terrazzo topping and a cement concrete under layer 1 : 2 : 4 (10 mm) with a combined thickness of at least 40 mm. The minimum thickness of the terrazzo topping will depend on the size of chips used as given in Table 20.1 (IS 2114–1962). The chips should be preferably of good quality marble as described under terrazzo in the book on *Building Materials*.

Table 20.1 Recommended minimum thickness of topping

Grade No.	*Size of chips* (mm)	*Minimum thickness of topping* (mm)
00	1 to 2	6
0	2 to 4	6
1	4 to 7	9
2	7 to 10	12
3	10 to 15	(Above 10 mm thickness or 1.5 times size of chips)
4	15 to 20	
5	20 to 25	

Table 20.2 Shows recommended proportions of the aggregates for in-situ laying of terrazzo tiles.

Table 20.2 Recommended proportions for in-situ terrazzo

Grade No.	*Size of chips* (mm)	*Aggregate for 1 part of cement mix*	*Minimum thickness of top layer* (mm)
00	1 to 2	1.75	6
0	2 to 4	1.75	9
1	4 to 7	1.75	9
2	7 to 10	1.50	9
3	10 to 15	1.50	12
4	15 to 20	1.25	1.5 times maximum size of the aggregate
5	20 to 25	1.25	
6	Mixed	1.50	

The underlayer is laid on the sub-base in case of ground slab and a cushioning layer over the RCC slab in the top floors as described in Chapter 14. The topping consists of terrazzo mix of cement (with or without pigments) and facing aggregate (marble chips) marble powder and water. The cement and marble powder are mixed in the proportion of 3 parts of cement to 1 part of marble powder. For every part of this cement mix, the proportion of aggregate (marble chips) and the thickness of the top layer to be laid are as follows.

Where large chips are used they should be flat shaped so that they can be hand placed in the top layer. The full quality of dry mix necessary for one room should be prepared in one lot to ensure uniform colour. This should be stored in a dry place and kept well covered. It is to be mixed with water only at the time of laying. The in situ laying is carried out as follows.

Laying of underlayer. When we lay terrazzo in situ, as different from laying of tiles, we introduce an underlayer of 1 : 2 : 4 concrete 25 mm thick on top of the base concrete in the ground floor and the cushioning layer on top floors. The concrete for the underlayer should preferably be laid in panels made by 4 mm to 6 mm glass strips or 2 mm aluminium strips of not more than 2 m in length for inside situations. For exposed situations, the panels should not be more than 1.25 m. (Though the work can also be carried out by casting in alternate panels, the introduction of separation strips and laying in one operation is usually preferred in practice.)

When the underlayer concrete is laid on RCC slab, cement slurry at the rate of 2.00 kg per sq m is applied before laying of underlayer. The separators (glass, aluminium, brass, etc.) must have their top at necessary level and slope corresponding to the terrazzo floor.

Laying of topping. The topping of mix given in Table 20.2 is laid while the underlayer is still plastic (normally 18 to 24 hours after laying). To start with, a cement slurry, preferably of the same colour as the topping, is brushed on the underlayer surface immediately before starting of laying. The terrazzo layer is laid to a uniform thickness, slightly higher than the required, so that when finished, it will have the required thickness. The top surface is then rammed with a smooth marble stone of 150 × 150 × 250 and trowelled over, pressed and brought to the level by a straight edge and steel float in such a manner that the maximum amount of marble chips come up to the surface and are exposed uniformly on the surface.

Curing, grinding polishing. The surface is air cured for 12 to 18 hours and then underwater for at least four days and then ground thrice and polished as follows. After curing the top layer, the surface is ground evenly with a machine using grade No. 60 carborundum stone with copious sprinkling of water (first grinding). Then the surface is thoroughly washed and cleaned and covered with a grout of the same colour as the top to fill all the pin holes. The surface is allowed to be cured for at least 7 days and then ground with fine grit block No. 120 to 150 (second grinding). The surface is again cleaned and repaired as before and cured for 3 to 5 days. It is ground a third time with grit block No. 320 to 400 to get a smooth surface without pin holes. (This is the same procedure as in tiles sent unpolished from the factory.)

The finished surface will show the marble chips exposed evenly on the surface. If necessary, an intermediate grinding with grit size No. 80 may also be adopted. It is then polished by hand or machine with oxalic acid as described under laying of terrazzo tiles. In accessible places, grinding, etc. have to be done by hand. The first manual grinding can be taken up two days after completion of laying.

Terrazzo on vertical faces. Terrazzo finish on vertical surfaces such as skirting and dadoes should have an underlayer of stiff cement mortar (1 : 3) finished rough to provide a key to the topping. The combined thickness of underlayer and topping should not be less than 20 mm, the minimum thickness of topping assumed as 6 mm.

Items requiring special attention. When taking up in-situ terrazzo work, special attention should be paid to the following items:

(i) General condition of base and adequate bond between layers

(ii) Correct level of base

(iii) Suitability of terrazzo aggregates

(iv) Correct proportioning and proper mixing

(v) Suitable size of bay and quality of dividing strips

(vi) Correct level of the top of strips and gradient of floor

(vii) Sufficient and right type of consolidation to bring chips to the top.

(viii) Correct curing, grinding and polishing.

SUMMARY

Terrazzo work in floors can be carried out by means of tiles or by laying it in situ. When laying in situ, proper care should be taken to lay it in panels to allow for shrinkage of the cement mixes. If large areas of in-situ terrazzo are not laid in panels, they are liable to crack due to shrinkage. In both cases, the work should be polished by the specified methods (by machine or manual labour). It is also worthwhile to note that the chips used for these terrazzo floors should have the same hardness as cement, such as marble chips. Too strong chips such as coloured quartzite chips will not wear uniformly with the cement mortar, thus leading to serious maintenance problems such as pitting.

REVIEW QUESTIONS

1. Describe the method of laying a terrazzo tile floor with factory unground tiles.

2. Describe the method of laying in-situ terrazzo (a) for the ground floor and (b) for the upper floors.

3. Describe the method of grinding and polishing an in-situ terrazzo floor. How is this operation different from finishing stone floors?

4. (a) Explain the terms mosaic and terrazzo.

(b) How can you reduce the time taken at the site for laying terrazzo floor for a residential building?

(c) Why do we want an underlayer of concrete for in-situ terrazzo work on upper floors as different from tilework?

Chapter 21

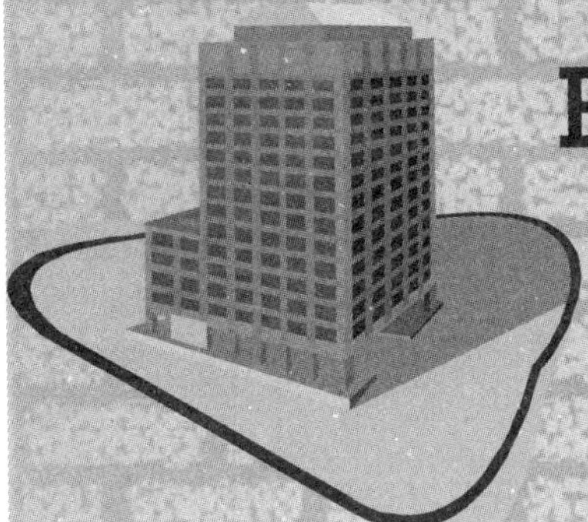

Flat-floor and Flat-roof Constructions

21.1 INTRODUCTION

Flat upper floors and roofs were used in buildings from long time back. In the past, flat floors were made of wooden planks resting on beams on walls. In temples, stone slabs were extensively used for roofs. Floors and roofs were also made using brick arches. The Madras roof terrace for floors and roofs was traditionally constructed with bricks on closely supported wood joists. However, with the advent of reinforced concrete, the upper floors and flat roofs are now generally made of reinforced concrete. In this chapter, we will first briefly review the construction of jack arches and Madras roof terrace (for floors and roofs) as a matter of historical interest and then deal with the construction of modern reinforced concrete floors and roofs in more detail.

21.2 JACK ARCH FLOORS WITH STEEL JOISTS

Jack arch construction with steel joists is shown in Fig. 21.1. They were very popular in India during the British rule. These floors or roofs were constructed with short span arches between I beams (rolled steel joists). The reaction of the arch has horizontal components which balance each other in the interior spans. The unbalanced horizontal forces in the *end arches* are compensated by steel tie rods of 20 to 25 mm diameter placed at suitable spacing (usually 1.8 to 2.4 m centres) with ends bolted outside the joists as shown in Fig. 21.1. The I beams are placed at 80 to 120 cm centres and encased in concrete. The arches can be constructed with bricks in lime or bricks cement mortar or in cement concrete and are given a rise of about 1/12th the span at the centre.

21.2.1 Construction of Arches with Bricks

The arches can be made of bricks set on formwork. Care should be taken to see that the

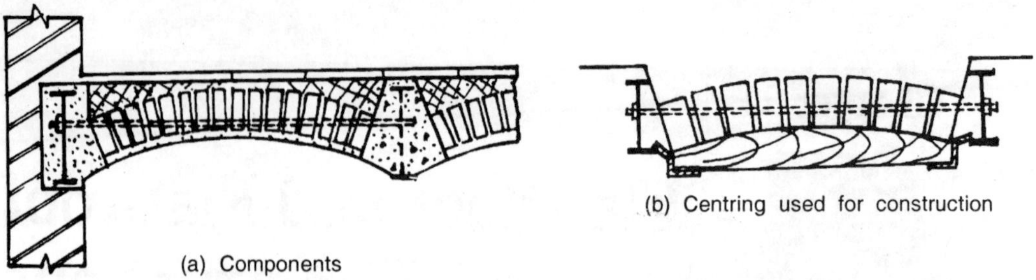

(a) Components

(b) Centring used for construction

Fig. 21.1 Floor construction by jack arches with rolled steel joists and brick arches.

encasing of the joists is always in cement concrete and not in lime concrete as steel in contact with lime gets rusted quickly.

When the arches are to be thick, they can be built ring by ring. Otherwise to break joints, the first ring can be built with alternate bricks of 20 cm and 10 cm lengths as shown in Fig. 8.3 (Chapter 8) to maintain bond with successive rings of arches and thus break joints. The space above the bricks can be of lime concrete as shown in Fig. 21.1. These types of construction were popular in olden days when cement was in short supply and not freely available in India. Some of the heritage buildings (such as the main hall of Freemason's Lodge in Chennai) built in India around 1920 are of this type.

21.2.2 Construction of Arches with Cement Concrete

Construction of jack arches with cement concrete is of later origin. In this case, the concreting can be completed in one operation. The type of centring used for forming these arches is shown in Fig. 21.2.

(a) Construction method

(b) Details of centring

Fig. 21.2 Cement concrete jack arch flooring with steel joists.

21.3 MADRAS ROOF TERRACING

Another type of flat-roof construction that has been popular for a long time in Tamil Nadu is the Madras roof terracing. Here again, a large number of old buildings with these floors and roofs exist as heritage buildings in Tamil Nadu. Their restoration should be done with care and knowledge of its original construction. These floors are made of "terrace bricks", 150 × 75 × 25 mm in size. The Madras roof terrace floor or roof consists of the following constructions:

(a) A layer of terrace brickwork laid on edge are jointed together with lime mortar, supported on *closely spaced wooden joists* (nowadays when repairing these floors, we can use reinforced concrete joints instead of wood joints) with a suitable ceiling finish applied to the soffit of the brickwork.

(b) A course of lime brick jelly concrete is laid above the bricks and the required top finish is given for floors and roofs.

For constructing this type of roof, the joists are placed first on the supporting walls to receive the bricks. The bricks are laid on them as follows.

Laying of terrace bricks. After keeping the terrace bricks immersed in water for at least four hours, the bricks are *laid on edge in diagonal rows spanning over the closely spaced joists*. The laying is to start from one corner and proceed towards the opposite corner. Each row is completed before the next one, adjacent to it, is started. The bricks are jointed in lime mortar 1 : 1½ (lime putty to sand by volume) breaking joints. The thickness of joints should not be more than 10 mm.

Over this brickwork, brick jelly concrete 1 : 2½ (slaked lime to brick jelly by volume) is spread over a thickness of 100 mm. This is consolidated by ramming by a wooden rammer (not more than 2 kg in weight) to a thickness of 75 mm. For floorwork, it is finished with the required finishing. *For roofwork*, it is completed with lime concrete roof terracing. The ceiling is also finished with a rich plaster.

21.4 REINFORCED CONCRETE FLOORS AND ROOFS

With the advent of reinforced concrete, flat floors and roofs of all modern buildings (residences, offices, commercial buildings) are nowadays made of reinforced concrete. They may be one of the following types as shown in Fig. 21.3:

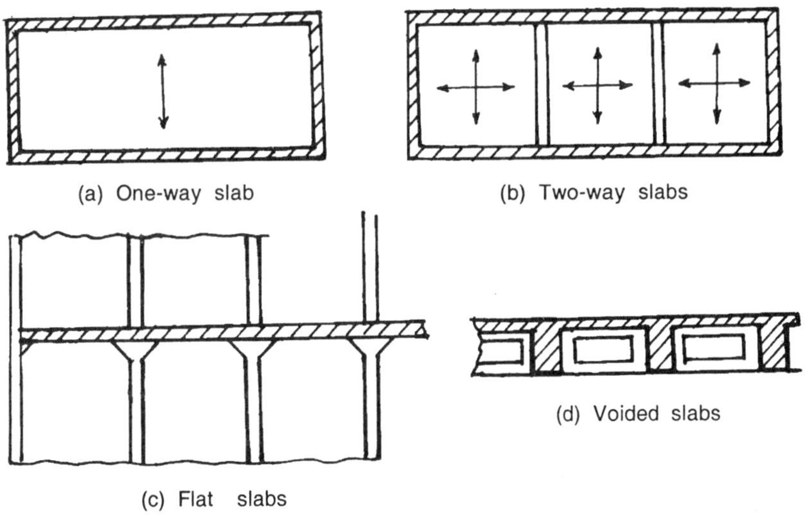

Fig. 21.3 Common types of reinforced concrete floors and roofs.

1. Simple slab type
2. Beam and slab type
3. Flat slab and flat plate type
4. Ribbed hollow block or voided slabs
5. Other types

Note: Even though *reinforced brick floors* made out of ordinary bricks with steel rods placed in between the bricks were once popular in India, their long-term performance *has not been found to be very good* as extreme care has to be taken to see that the steel does not come in contact with the stock bricks. With steel in contact with bricks, corrosion of steel will take place at a very rapid rate. Bigger special ceramic hollow blocks, which considerably reduce the weight of the roof, also has taken the place of such reinforced brickwork.

The above-mentioned reinforced concrete floors and roofs are discussed further.

Simple RC slab floors. When the breadth of the room to be spanned is not large (not more than 5 m), as happens in most residential buildings, the floor can be a simple reinforced concrete slab. If the length of the room is more than twice the width, the slab can be assumed to bend only in the direction of the breadth and it is called a *one-way slab* [Fig. 21.3(a)]. If the length is only 1.5 to 2 times the width, the slab tends to bend like a dish in both directions and it is called a *two-way slab* [Fig. 21.3(b)]. The ends of these slabs can rest on walls or reinforced concrete beams.

Beams and slab construction. When these slabs rest on reinforced concrete beams, part of the slab can be assumed to act with the beams. Such beams are called *T* and *L beams*. Thus the slabs can span between beams instead of walls. In reinforced frame construction, these beams span between columns.

Flat slab and flat plate construction. If beams are to be avoided, the slabs can rest directly on columns. If the columns are enlarged at the junction of columns and slabs, they are called *flat slabs* [Fig. 21.3(c)]. If there is no enlargement, they are called *flat plates*.

Ribbed, hollow block or voided slabs construction. These types of floors and roofs can be one of the following types. According to IS 456 – 2000, they can be cast in situ or made of precast units as follows.

(a) A series of concrete ribs with *topping cast on forms* which are removed after the concrete has set.

(b) A series of concrete ribs between *precast blocks* of concrete or ceramics for lighter weight construction which remain as part of the completed structure, the topping of concrete is of the same strength as that used in the rib.

(c) With a continuous top and bottom face but containing voids of rectangular, oval or other shape.

Voided slabs are shown in Fig. 21.3(d).

According to IS 456 for ribbed slabs, the in-situ ribs are to be not less than 65 mm wide and spaced at not greater than 1.5 m. The depth of the ribs excluding the topping is to be not more than four times their width. Generally ribs are also formed along each edge parallel to the span of one-way slabs. In addition, when the edges are built into the wall or rests on beams, a rib at least as wide as the bearing should be formed along the edges also. Ribbed floors with ceramic hollow blocks, considerably reduce the dead weight. These voided slabs can be cast in situ or made of precast units. Some of the precast type of floors are shown in Fig. 21.4.

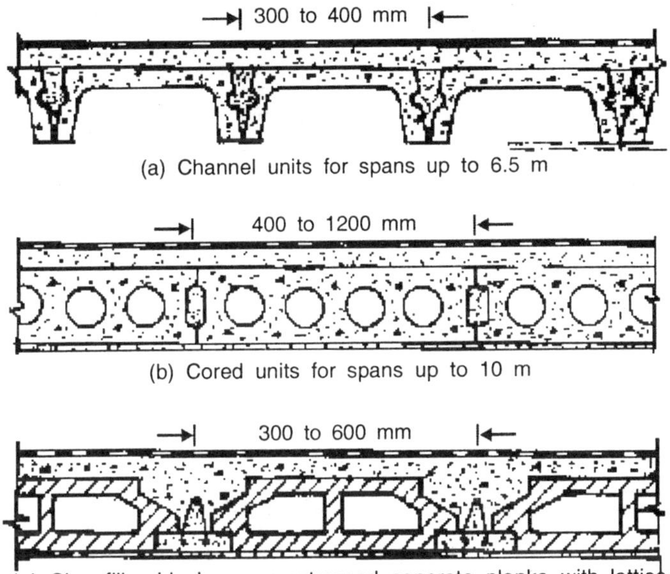

(a) Channel units for spans up to 6.5 m

(b) Cored units for spans up to 10 m

(c) Clay filler blocks on prestressed concrete planks with lattice reinforcements

Fig. 21.4 Preeast concrete floors and roofs.

As design and construction of reinforced concrete structures is a subject to be studied in detail in courses in Reinforced Concrete Design in civil engineering, we will not go further into the details of these constructions.

Other types of roofs. These can be shell roofs as described in Section 22.12.

21.5 PROVISIONS FOR EXPANSION

The average coefficients of expansion of concrete and steel are more or less the same—about 12×10^{-6} per °C. This property enables the steel and concrete to expand or contrast together with changes in temperature. The tensile strain capacity of mass concrete is only of the order of 90×10^{-6} and that of reinforced concrete will depend on the amount of steel in the concrete. The IS code recommends that all reinforced concrete framed structures more than 45 m (150 ft) in length (or with sudden change in dimensions) should be *provided with one or more expansion* joists. Usually a 25 mm gap is provided at expansion joints as shown in Fig. 27.3 in Chapter 27.

Expansion joints between non-continuous slabs resting on intermediate supports. Where expansion in joints is to be provided, a gap of 25 mm is *provided on roofs and* a gap of 12 mm between the slab is to be provided on floors. It is filled with bitumen filler or bitumen impregnated fibre. Waterproofing construction is carried out over this joint (See Figures 27.3 and 27.4 in Chapter 27).

Special attention should also be given to the following.

Treatment of top of walls on which RC slab rests. The top of these walls, on which simple RC *roof slab* rests, should be laid on craft paper placed over brickwork *plastered smooth with* 1 part of cement to 3 parts of sand plaster and finished with a floating coat of rich cement slurry and a thick coat of lime wash (CPWD specification). For *floor slabs,* the craft paper can be omitted. These are indicated in Fig. 21.5.

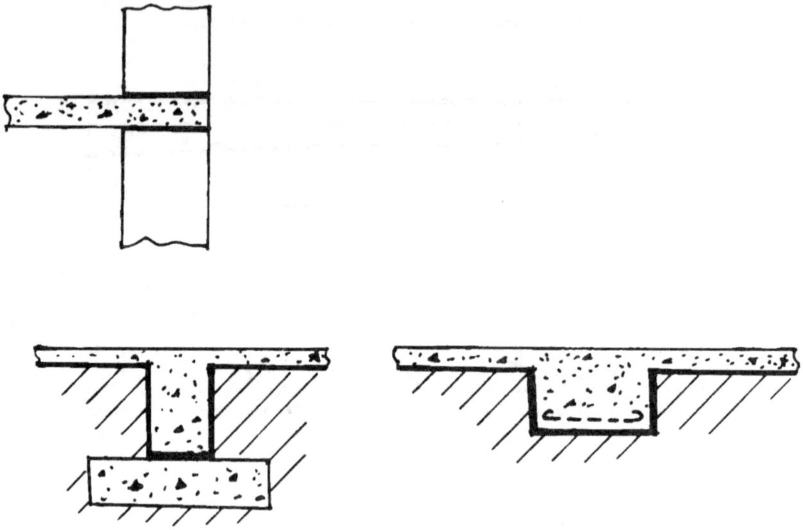

Fig. 21.5 Provision for expansion of roof slabs and beams resting on masonry by kraft paper placed on top of walls plastered with cement plaster with a coat of thick limewash on top of the plaster.

21.6 WATERPROOFING OF FLAT ROOFS

This topic is dealt with in detail in Chapter 27.

SUMMARY

Even though many types of floors such as jack arch were extensively used in the past, nowadays all suspended floors and roofs are generally made of reinforced concrete. However, a knowledge of the older types of construction will be useful when we have to repair or renovate old buildings. There are many types of modern reinforced concrete roofs and floors such as simple slab floors, slab and beam floors, flat slab, flat plates, voided slabs, etc. Their selection depends mainly on the use and layout of the building.

REVIEW QUESTIONS

1. Describe briefly the old method of construction of jack arch floors made of brick jack arches and rolled steel joists.
2. Explain the construction of Madras roof terraces.
3. Describe briefly different types of concrete floors used in buildings.
4. For a multistorey building, we have to adopt light weight floors by use of voided slabs. What types of floors would you recommend? Sketch the arrangements for such floors? What special features are to be provided at the four edges of such slabs?
5. Explain how expansion joints are to be provided in RC framed structures. What spacings would you adopt for these joints in such construction?

Chapter 22

Sloped Roofs

22.1 INTRODUCTION

We will deal with the sloped roofs in two parts:

1. Tiled and sheeted sloped roofs
2. Sloped concrete slab roofs and shell roofs

The framework for tiled and sheeted sloped roofs can be made of timber, steel or concrete. We will briefly examine the different types of frameworks used for sloped roofs and also the method of laying sheets or tiles over them. We will also examine the commonly used sloped and curved concrete roofs and other long span roofs.

22.2 TIMBER ROOFS

Owing to the introduction of concrete and steel and lack of good timber in India, modern roof constructions in large buildings are different from the traditional construction made from wood. Nowadays timber is generally used only for short spans, residential buildings and country houses. Large span roofs are mostly fabricated from steel sections. Also it is now possible to use light weight box sections in steel and simulate the appearance of wood. The construction of the wooden roof depends chiefly upon:

1. The plan of the building, and the style of the architecture
2. The climatic conditions
3. The covering to be used

22.2.1 Terms Used for Sloped Timber Roofs

Span. The *clear span* is defined as the horizontal distance between the walls, or the outer supports and the *effective span* is the horizontal distance between the centres of bearings of the supports.

Pitch. The slope of a roof is termed *the pitch* and may be expressed in terms of either the degrees of inclination to the horizontal or the rise to the span.

Eaves. In case of a sloping roof, the lowest course of the roof is called *eaves*. They are generally made to project beyond the external face of the wall (especially in places of high rainfall) to give protection to the walls from rains, for shading the walls from the sun and also for pleasing appearance. A projection 500 to 750 mm (1′ 9″ to 2′ 6″) is given for residences and 900 to 1500 mm (3′ to 5′) for large buildings, depending on the height of the wall to be protected. The ends of the rafters are usually cut in such a manner that the rain gutters can be fixed to these fascia by brackets.

Hipped end. Hip roof is formed by four sloping surfaces as shown in Fig. 22.1. The hipped end of a roof is the sloping end of a roof and is generally triangular in shape. The hipped side has the same slope as the main roof. The hip rafters with one end bearing on the wall plate and the other end cut and joined to the ridge board are used as the hips as shown in Fig. 22.1. The hip of a roof is the line of intersection of two roof planes.

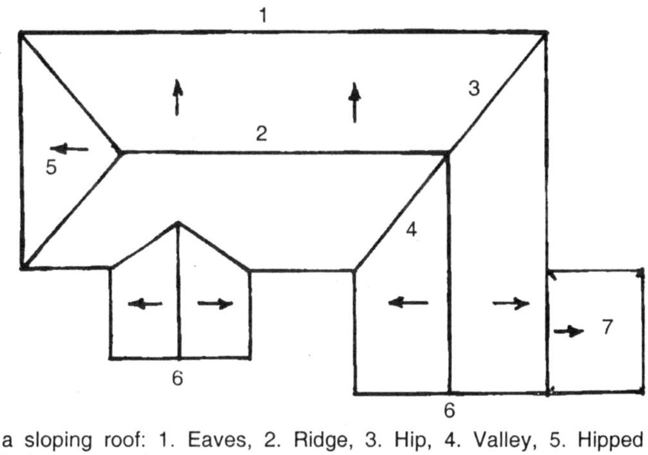

Fig 22.1 Parts of a sloping roof: 1. Eaves, 2. Ridge, 3. Hip, 4. Valley, 5. Hipped end, 6. Gabled ends (verge), 7. Lean-to roof.

Gable. Gable roof is a roof with slopes in two directions only. The two ends are vertical walls as shown in Fig. 22.1. Even though pitched roofs with two simple *gable ends* are cheap to construct, they look clumsy. Hipped sloped roofs are more pleasing and are generally preferred. The *verge* is the edge of the roof plane at the gable end.

Ridge. It is the name given to the upper edge, or the line formed by the intersection of the roof surfaces.

Valley. The line of intersection of two roof planes containing an angle of less than 180°.

Note: The main sloping member is called the *rafter*. Over these rafters, reepers are placed horizontally. If the horizontal members are large in size, they are called *purlins*.

22.2.2 Different Types of Sloping or Pitched Timber Roofs

The commonly used roofs are discussed further. They are shown in Fig. 22.2.

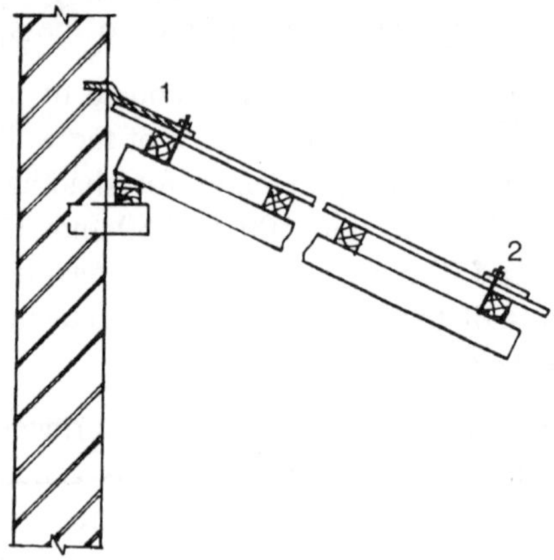

(a) Lean to roofs with flashing and wind tie: 1. Flashing of galvanised steel (400 × 1.25 mm) placed at least 50 mm into the wall, 2. MS flat (40 × 6 mm) fixed along eaves of sheets with hookbolts to purlins.

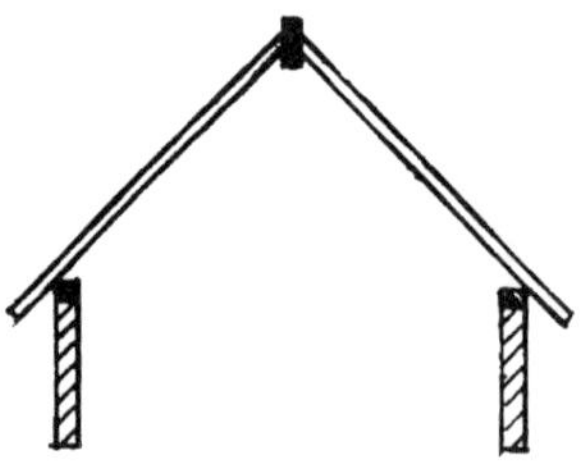

(b) Couple roof

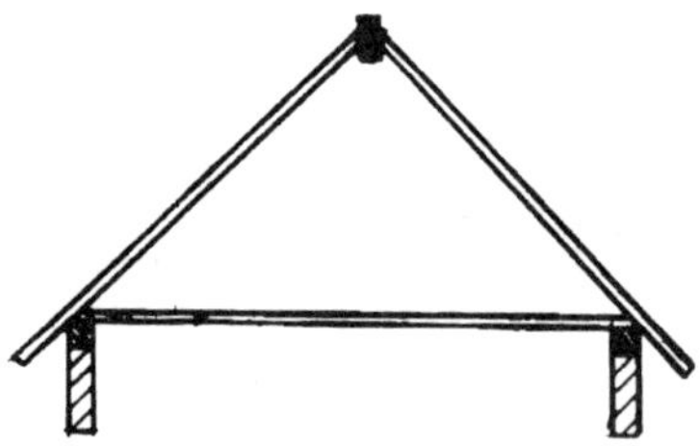

(c) Closed couple roof

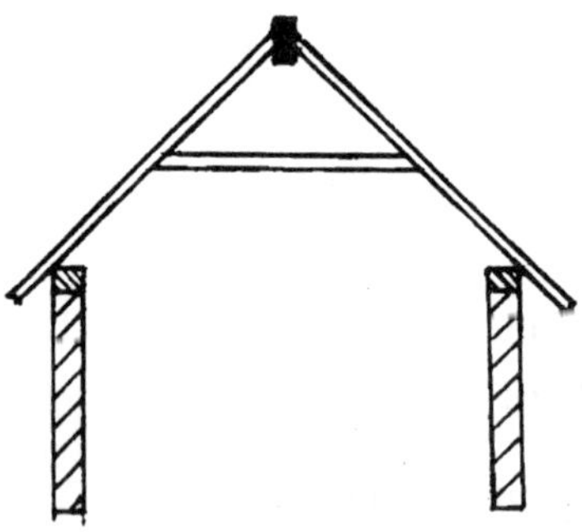

(d) Collar roof

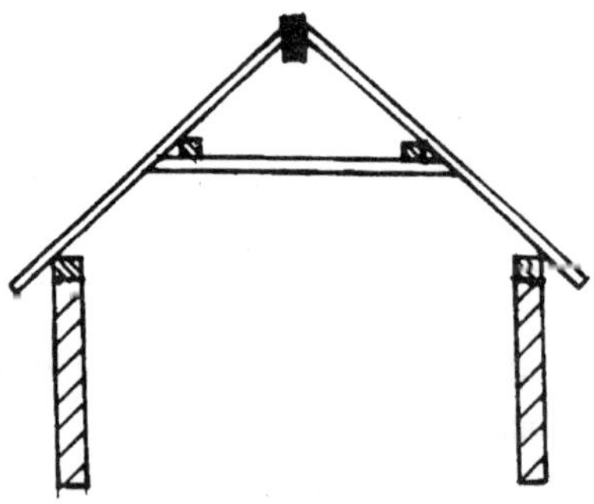

(e) Purlin roof (collar roof with a tie)

Fig. 22.2 Types of sloping roofs.

Lean-to roofs. The simplest form of roof is the lean-to or pent roof [Fig. 22.2(a)]. It is generally used for sheds and spaces which are attached to a main building. It is usually constructed as a mono-pitched roof. Maximum span usually is about 2.5 m.

Couple roof. Couple roofs, as the name implies, are roofs consisting of pairs of rafters fixed at their feet to the *wall plate* and pitching against the ridge piece [Fig. 22.2(b)]. These are suitable for spans up to about 3.5 m only as they tend to spread out and overturn the walls with larger spans.

Closed couple roofs. These roofs are suitable for spans up to about 4.5 m and are similar to coupled roofs, but with a tie fixed to the feet of the rafters [Fig. 22.2(c)]. This tie prevents the feet of the rafters from spreading, and thus the danger of the walls overturning is avoided. The ties also act as ceiling joists.

Collar roofs. To increase the inside height of a room, a collar tie roof is used [Fig. 22.2(d)]. The ceiling is obtained by ceiling the underside of the collar and part way down the roof rafters to the wall. They usually cover spans up to 5 m.

Purlin roofs. When the roof exceeds the span of the previously described roofs, we can use a collar roof with a tie running horizontally between the collars [Fig. 22.2(e)]. Advantage is also taken of any partition or wall to support the ceiling joints and rafters by struts and ties. These roofs are suitable for total spans (with cross walls) up to about 9 m.

Wooden roof truss. When the span exceeds about 5 m and when there are no cross walls, partitions or other support for ties for purlin roof, *trusses* are introduced for economy and strength. The advantage of a truss is that the total weight of the roof is *carried vertically* on the walls. Timber trusses consist of timbers framed together (triangulated) in such a manner that the frame, when subjected to the loads will not alter in shape. The trusses should not be placed more than 3 m apart. This distance will of course be influenced by the openings and piers in the wall. The tie beam at the base of the truss should be supported at every 4.5 m or less. This rule will determine the type of truss to be used. The common types of *wooden trusses* used are as follows:

King-post roof truss. These trusses may be used for spans up to about 9 m [Fig. 22.3(a)].

Queen-post roof trusses. Queen-post roof trusses may be used for spans up to about 13.5 m. However, they are seldom used nowadays, as steel roof trusses are and require less maintenance [Fig. 22.3(b)].

Laminated wood trusses. This form of truss is so called as its members are built up of thin wooden planks of timber instead of solid wood. Generally they are not used for tiled roofs but employed only when the covering material is light in weight, such as corrugated iron sheets. Roof construction of this type of truss is the most efficient and economical. These trusses are often used for farm buildings (stables, sheds, etc.) and huts, and also for so-called semi-permanent buildings, including school buildings.

In Fig. 22.4, a laminated roof truss for span up to 6 m is shown. It is made of wooden planks. The principal rafter consists of two planks, 120 mm × 30 mm in section, securely spiked at the joints. The main tie is made of two 120 mm × 30 mm planks. The other ties and strut are single plank 100 mm × 50 mm. The joint at the apex is made rigid by nailing the members to boards 25 mm thick acting as gusset plates. The feet are made rigid by

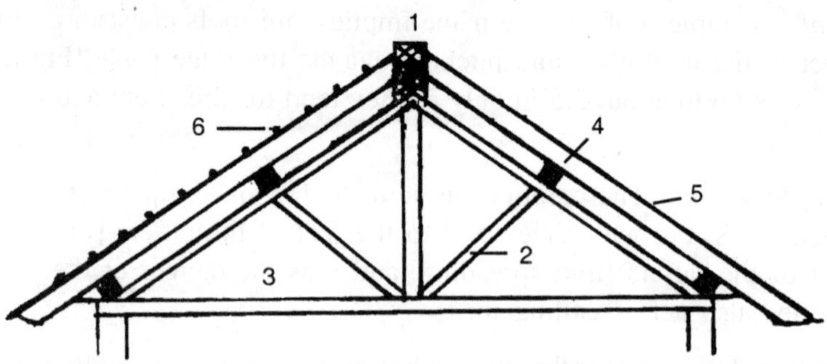

(a) King-post truss: 1. Ridge piece, 2. Strut, 3. Tie beam, 4. Purlin, 5. Common rafter, 6. Reepers (cladding not shown)

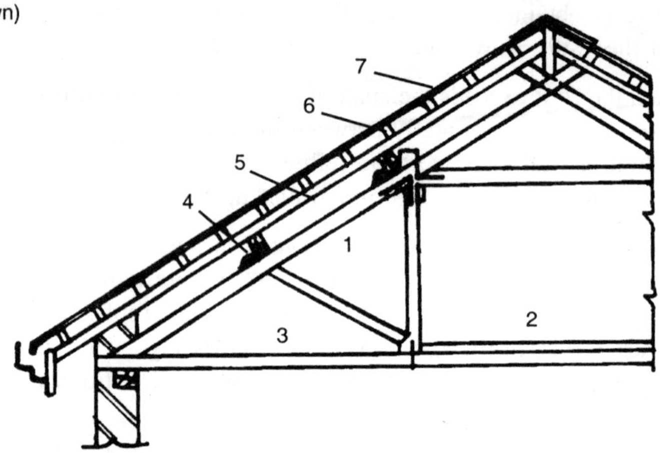

(b) Queen-post truss: 1. Principal rafter, 2. Straining sill, 3. Tie beam, 4. Purlin with cleats, 5. Common rafter, 6. Reepers, 7. Cladding

Fig. 22.3 Wooden trusses.

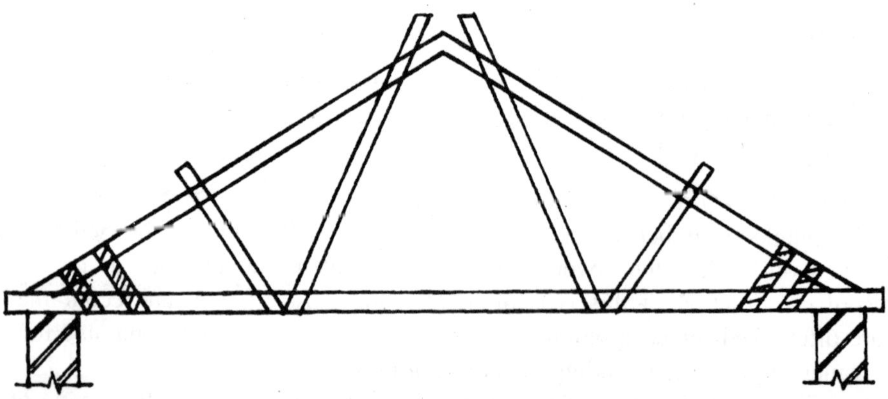

Fig. 22.4 Laminated roof made from wooden planks nailed together (members extended for seating of purlins).

100 mm × 50 mm boards nailed to the members as shown in Fig. 22.4. The struts are continued upwards to act as support for purlins. The spacing will depend on the weight of covering. For AC sheets, they can be spaced at 90 cm centres.

Trussed rafters. The majority of domestic roofs in UK are made of these triangulated plane laminated roof frames as shown in Fig. 22.4. They do not have purlins and the rafters are placed on wall plates at close centres at 60 cm (2 ft) centres and braced together by diagonal bracings. Lateral restraint is provided by the gable walls at the top and at the bottom chord level by mild steel straps over two members of trussed rafters. Reepers are directly fixed on these trussed rafters so that trussed rafters do not require any purlins or ridge board at top. For lateral stability, we use large reepers of size 50 mm × 25 mm. The basic idea of the trussed rafter roof is to combine every adjacent pair of rafter in a roof with the reepers to form a truss and thus remove the need for purlins. The joints of the trusses are made with plywood gusset plates and wood fasteners fixed on both sides.

22.3 STEEL AND CONCRETE TRUSSES

Trusses may be made of steel or concrete. Rolled steel sections such as angles, channels, ties, etc. are more often used for trusses than for concrete as steel sections will be lighter in weight. Types of steel trusses generally used are given in Fig. 22.5.

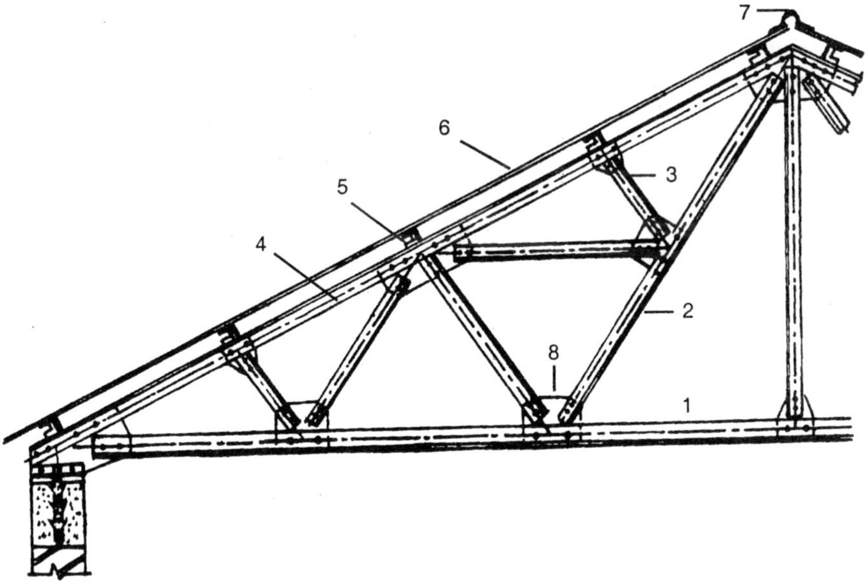

Fig. 22.5 Typical steel roof truss for sloping roofs for factory buildings: 1. Main tie, 2. Ties (members in tension), 3. Struts (members in compression), 4. Principal rafter, 5. Purlin with cleats, 6. AC sheets, 7. AC ridge, 8. Gusset plate.

22.4 COVERING SLOPED ROOFS

The materials used for covering a sloped roof are determined by the climatic conditions and by nature and importance of the building under construction. The roof covering must be as durable as the other parts of the building. Where the roof is prominent and can be seen on the elevation, the roofing material should be given due architectural and aesthetic consideration. Generally sloped roofs are covered with one or a combination of the following materials:

(i) Clay tiles
(ii) Corrugated asbestos sheets
(iii) Corrugated galvanized iron (GI) or aluminium sheets
(iv) Thatch
(v) Other tiles such as pan tiles, asbestos tiles, shingles

Poor conductors, such as tiles, make better coverings than good conductors such as corrugated iron in hot regions. Tiles tend to preserve a more equable temperature in the interior of the building. Roofing with clay tiles, AC sheets, GI sheets and aluminium sheets only are dealt with in this chapter. Roofing with other forms of tiles is similar in principle to roofing with Mangalore tiles.

In places where there is a good supply of coconut leaves such as Kerala and Tamil Nadu, a number of low-cost housings are built with sun-dried mud bricks, covered with bamboo or cashurina pole frames and thatched with palm leaves. However, they are not permanent and are highly combustible. Sprays made with boric acid or ammonium phosphates can be used to increase fire resistance of these palm leaves houses.

22.5 RECOMMENDED SLOPES OF ROOFS

The slope of a roof is to be determined by the material it is to be covered with and environment. It may vary from being almost in a horizontal position, as in case of a flat roof, to almost vertical, as in the lower part of a Mansard roof made of a combination of queen-post truss with nearly vertical sides at the bottom with a king-post truss resting on its top. Where the roof is in an exposed position or when the material is not leakproof as in case of tiles, the pitch should be steep as otherwise it will leak. This is specially important in tiled roofs, as joints between tiles are not impervious to passage of water. The roof is of a steep pitch in cold countries also because of the problem of snowfalls. In such cases, advantage may be taken in forming a room on top, light being obtained through a *dormer window* fixed in the roof as shown in Fig. 23.10.

In tropical countries, where only rain is to be taken care of, roof slopes can be smaller. For a given place (where the rainfall intensity is known), the slope depends on the material of the roof and its span. For lengths to be drained up to 4.5 m (15 feet) on either side, slopes of *10 degrees for AC sheet and 25 degrees for Mangalore tiles* are recommended for use. (A pitch of 26 degrees corresponds to a rise of one-fourth the span). For AC sheets with round tiles placed over AC sheets, a steeper slope of *about 15 degrees is necessary* for proper drainage. Slight adjustments may be made in the slope to adjust the length of the principal rafter to accommodate the *standard sizes of covering* (AC sheets, tiles, etc.) to be used. Table 22.1 gives the pitch usually adopted for roof covering for draining distances of up to 20 ft. For longer distances to be drained they may be suitably modified. (The length of the slope should also be adjusted with reference to the length of the sheets to be used.)

Table 22.1 Recommended roof slopes

Materials	*Angle of inclination (degrees)*	*Ratio of rise to span (recommended)*
Plain clay tiles (Mangalore pattern)	26 to 30	1/4
Pan tiles	26 to 30	1/4
Corrugated asbestos sheets	10 to 14	1/8
Corrugated iron and aluminium sheets	10 to 14	1/8
Thatch	45	1/2
Shingles	45	1/2

22.6 FLASHINGS

Flashings are devices fixed in roof junctions to prevent leakage. For example, when a lean-to roof starts from a vertical wall as shown in Fig. 22.2(a), the junction between the vertical wall and the horizontal edge of the sloping roof has to be made watertight. Roofing sheets, tiles, etc. can be continued and built into the walls but because of temperature effects, such junctions will not be fully leakproof in the long run. The sloping roof should be built close to the wall and the junction should be protected by a *flashing* or alternately by a projecting drip course. (The drip course should project at least 75 mm (3 inches) onto the sloping roof.) A flashing can be made of a plain GI or other flexible sheet (any flexible sheet as lead or copper can also be used), 1 to 1.25 mm thick and 40 cm overall width, built into the wall and spread over the sloping roof as shown in Fig. 22.2. It should penetrate the wall at least 50 mm and lap over the roof by at least 25 cm (10 inches).

Similarly when the vertical wall meets a sloping edge of the roof, we also need flashing. For this, we lay the overlap along the sloping roof and cut the end of the sheet that is to be set in the vertical wall brick joints in steps and use as stepped cover flashing the steps being fixed in the vertical wall as shown in Fig. 22.6.

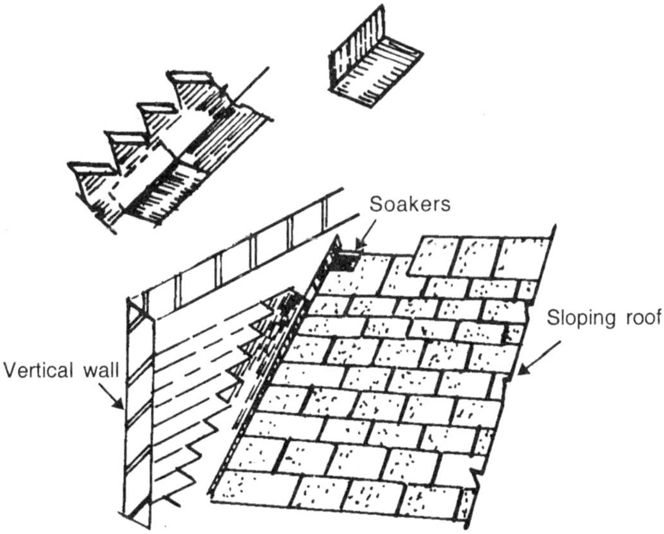

Fig. 22.6 Flashing between sloping roof and vertical wall-stepped flashing wedged and painted in brick joints.

Note: This principle of flashing can be used in many places in building construction such as in junctions of chimneys with roofs. Flashings are more important for sheeted roofs than for tiled roofs as the individual tiles are of smaller lengths and they can even be built into the walls. Junctions of tiles and walls are provided in many cases with mortar bands instead of metal flashings.

22.7 LAYING OF TILED ROOFS

Different types of tiles used for buildings have been studied in the book on *Building Materials*. The following are the common terms used in tiling work:

(i) ***Top*** of a tile is the exposed surface when laying.

(ii) ***Bed*** is the surface of a tile in contact with the actual roof timbers.

(iii) ***Head*** is the top edge of a tile.

(iv) ***Tail*** is the bottom edge of a tile.

(v) ***Course*** is a row of tiles.

(vi) ***Bond*** is the uniform arrangement of tiles whereby the edge joint between the tiles in any one course is in or near the centre of the tile immediately above and below them.

(vii) ***Lap*** is the amount which the tail of the tile covers the head of the course immediately below it.

(viii) ***Gauge*** is the distance between the tiles of each successive course. It is also the distance between the reepers. It depends on the length of the tile and the amount of the overlap.

22.7.1 Laying of Mangalore Tiles on Reepers

Different types of tiles are laid differently and the PWD specifications give details of their laying. As Mangalore tiled roofs are very popular, especially in places like Kerala and Tamil Nadu, we will examine it in detail. The following are the two types of commonly specified laying:

1. Mangalore tiles laid on reepers
2. Mangalore tiles set in mortar over flat tiles on reepers (Section 22.7)

In this section, we will examine the procedure of laying Mangalore tile on reepers. The main considerations in fixing Mangalore tiles on reepers are discussed further.

Fixing of reepers. Mangalore tiles are available in three sizes and the reeper spacing should depend on the readily available size of tile. The reepers for receiving the Mangalore tiles have to be prepared according to the type of the roof, whether it is a simple wooden roof or a steel truss.

If the main roof structure is of timber rafters, timber reepers 50 mm × 25 mm (2″ × 1″) are provided for 600 mm (2 ft) spacing of rafters. They are nailed by 50 mm (2 inches) long. French reeper nails across the slope at distances are required for the given gauge of tiles.

Usually the spacing is 310 mm (12.5 inches) but as different types of tiles are available in the market, care should be taken to see that the reeper spacing suits the size of the available chosen tiles. The reepers should be well seasoned and not shorter than the length which can cover at least four rafters. The underside of the reepers should be planed before fitting up. Joints are butt jointed over the rafters taking care that these are staggered. The top surface of the reepers is also usually coated with two coats of tar. To assist in fixing the reepers at the required gauge, a chalkline is first marked across the rafters at the required gauge intervals and the battens nailed in. For steel trusses, hollow steel sections similar to wooden reepers are now available for use as reepers. They can be spot welded to the steel truss at required intervals.

Tilting fillets. At the eaves, a special reeper, called a *tilting fillet*, is fixed near the ends. It is 20 to 25 mm (3/4 inch to 1 inch) thicker than the ordinary battens. It is used to tilt up the under or double eaves course (so that a proper bed may be formed for the tiles above). At the same time, it prevents the lifting effect of the wind, by decreasing the spacing of reepers at the edge of the roof near the eaves. Thus in addition to the increased thickness of reeper, the spacing of the lowest batten nearest to the eave is also not to be more than 250 mm (10 inches) from the one immediately above it. In some cases, the fascia board is utilized for the same purpose in place of the tilting fillet, the fascia board being placed higher than the battens by the amount required to tilt the tiles. It is also necessary to give a proper bed to the undertile and so prevent it from "riding". Alternately two reepers are fixed side by side at the eves to preserve the uniformity of slope instead of the special reeper mentioned above.

It is to be especially noted that in tiled roof design, the rainwater should be carried clear of the walls. This is especially true in places such as Kerala where the walls are usually made of laterite. This is also true when we use stabilized soil bricks, weak bricks, etc. The fascia board is provided for appearance also. In important buildings and in high roofs in places of heavy rains, gutters are also provided for proper drainage of rainwater away from the roof.

Laying the tiles. The tiles are laid from the eaves upwards the ridge with the rebates (holding devices) of the tiles resting fully against the reepers. The projection of the first row of tiles along the eaves is not to be more than 75 mm (3 inches). The tiles ending at the hips in a hipped roof should be suitably cut so that not to have any gaps when laid. A mortar band as described in section 22.6.2 is also to be provided at the hips.

If the tiles are to be abutted against a masonry wall at the sides of the roof, it is better to start work from the masonry wall end and work to the other end to ensure proper drainage. In these places, the tiles are to be taken into the walls to a depth of 50 mm (2 inches) and the joint grouted with 1 : 3 cement mortar. A band of mortar 100 mm × 100 mm is also formed along the junction above the roof surface with 1 : 1 : 5 lime-cement-sand mortar and finished smooth with 1 : 3 cement mortar. (A flashing can also be provided instead of mortar band.)

Ends over the gable should also be provided with a band of mortar 230 mm wide and not less than 50 mm in height in lime cement sand mortar to blend with the end tile along the gable and finished with 1 : 3 coloured cement mortar to match the colour of the tile.

After the sloping faces of the roof surfaces have been fully covered, the ridge is sealed with the "ridge tiles". They are laid last and the joints between themselves and the plain tiles are well grouted and set in cement-lime mortar. Usually a 1 : 1 : 5 mortar is used, pointed by 1 : 3 cement-sand mortar and suitably coloured. Sometimes pointing and finishing off with a mixture of red ochre and Portland cement is used to retain the red colour. All tiles to be set in cement mortar should be immersed in water for 3 to 4 hours before laying. After laying, the slope shall be uniform and the ridge and eave shall be horizontal and parallel to each other.

Wiring the tiles. In places where heavy winds are expected, tiles such as Mangalore tiles are fixed or wired by No.18 gauge galvanized softiron wire to the battens through the holes provided at the back of the tiles at every third, fourth or fifth course. This is quite satisfactory in most cases. But if the building is in very exposed position, every tile should be wired. Tiles at eaves, gable and those which project over the walls of a building must always be wired.

Ventilating tiles. If specially specified, ventilating tiles are provided at the rate of one per 100 sq feet of finished roof surface at convenient places.

22.7.2 Gutters and Flashings

Gutters and flashings are of prime importance in construction of sloping roofs. Brief discussion of these is given further.

Valley gutter. Valley gutters should be fixed before laying tiles near the valleys. Wooden planks 20 mm (3/4 inch) thick, called *valley boarding,* are fixed along the slope by nailing them to the rafters on both sides of the slopes. The top of this boarding is treated with two thick coats of hot tar and the underside of the board is treated with a wood preservative.

For valleys, galvanized iron gutters with at least 900 mm (36 inch) in width and of 18 BWG are used. The GI sheets are bent to shape and laid over the planks and dressed into the valley so that they extend equally on each side under the tiles. The depth of the trough is to be at least 100 mm (4 inches) at the centre. Laps along the slope must be at least 150–300 mm (6 inches to 1 foot). Two reepers, 50 mm × 25 mm (2″ × 25″), are fixed over the GI sheet 150 mm (6 inches) away from the centre line of the valley on either side to prevent the tiles from falling into the gutter.

At the sides of the gutter and for 9 inches on either side of the roof at valleys, cement plastering 1/2 inch thick in 1 : 3 cement mortar is provided to prevent the rainwater in the gutter from leaking through the sides of the valleys. This mortar is coloured to match the tiles if they are exposed. It is conventional to treat both faces of the sheet with two thick coats of hot tar. Other types of gutters for drainage of sloped roofs are described in Section 28.4 and shown in Fig. 28.3.

Flashings under the gable. When we have gabled roofs, we have to provide flashing on the gable to prevent leakage. For this purpose, at the gable of clay tiled roof end, the tiles are built into the gable wall and a flashing of suitable width and not less than 50 mm (2 in) in height is provided in 1 : 1 : 5 lime-cement-sand mortar to bind the end tiles to the gable. It is finished with 1 : 3 coloured cement mortar to match the tiles. The type of flashing shown in Fig. 22.6 can also be used for this purpose.

22.8 LAYING OF MANGALORE TILED ROOF SET ON FLAT TILES AND LIME MORTAR

To improve the appearance of the roof from below, the Mangalore tiles are sometimes set over flat tiles, which are available in plain or floral design on one side. In this work, the reepers are kept at central distances suited to the size of the flat tile which is usually only 150 mm × 150 mm (6 in × 6 in).

The flat tile is immersed in water for two hours before laying and allowed to dry out. Their undersides, if prescribed, are dipped in whitewash of the consistency of cream and allowed to dry. The tiles are to be laid dry on the reepers, the top surface of the reepers having been painted with two coats of tar. On the flat tiles, a layer of lime mortar 20 mm thick is laid over which the Mangalore tiles are laid.

22.9 LAYING OF AC SHEETS

Even though AC sheet is no longer used in many developed countries, it is still the cheapest material available in developing countries such as India for roofing. Hence it is still not forbidden in developing countries. Three types of AC sheets (corrugated, deep corrugated and profile sheets) are available in the market. Hence we must first decide which type of AC sheet we are going to use. The following points are important in laying of all types of AC sheets.

(i) Laying of AC sheets like GI sheets (described later) should be started from the point *which is away from the direction of heavy rains* in the region or site so that the overlaps do not face the direction of the winds and rains. Hence the work is usually started from the bottom leeward corner.

(ii) Holes in AC sheets for the hook bolts are drilled (not punched) through the ridge of the corrugation while the sheets are on the roof in their correct position. The holes are drilled with their diameter 1.5 to 2 mm larger than the diameter of fixing bolts.

(iii) AC roofing sheets are laid with its smooth side upwards. The slope of the roof should be at least 10 degrees or steeper according to the length of the roof to be drained.

(iv) The sheets are to be laid with a side lap of half to one corrugation and not more than one corrugation depending on the slope provided. The end lap is not to be less than 150 mm for roof pitches of 25 degrees and above and 225 mm for roof pitches below 25 degrees.

(v) The sheets are fixed with galvanized hook bolts not less than quarter inch diameter or by coach screws depending on nature of purlins. J bolts are generally used for angle iron purlins. L and J bolts are used for I beams concrete or wood purlins. In cyclonic areas, J bolts are preferred to L bolts. A rubber or bitumen "underwasher" is placed immediately above the asbestos sheet and over the washer, an iron limpet (metal) washer is put and over this metal washer, nut is screwed down tight. The length of these hook bolts should be 50 mm longer than the depth of purlin at points of single sheet fixing and 90 mm longer over lap points. The grip of the hook bolt shall not less than 25 mm [Fig. 22.7(a)]. The coach screws used for sheet fixing in timber purlins, they should be square headed and of length at least 120 mm. The bitumen washers are to be of 35 mm diameter and 1.5 mm thickness while the GI washers must be of 25 mm diameter and 1.6 mm thickness.

(vi) The hook bolts are preferably put on the 2nd and 6th corrugations in general roof areas of the roof where the effect of wind is minimal. The bolts are put on the 2nd, 4th and 6th corrugations in wind-sensitive areas of the roof such as gable ends and eaves overhangs and near ridges (see section 22.8.1).

(vii) Care should be taken to see that the bolts bear properly against the purlins and fit tight against it all along its length.

(viii) The maximum overhangs of sheets at the gable end should be limited to 150 mm (6 inches) while the overhang at the eaves (measured along the slope length from the lower end of the crown of bolt holes of sheets) must not be normally more than 300 mm (12 inches) for 6 mm thick sheets and 400 mm (16 inches) for thicker sheets.

The spacing of the purlins will depend on the size of the sheets. For normal cases, it should not exceed 1.2 to 1.4 m for 6 mm sheets and 1.6 m for 7 mm sheets. The ridge purlins should be fixed 75 to 115 mm from the apex of the roof.

(ix) Thus the maximum distance between the purlins for 6 mm sheets is recommended to be as follows depending on the wind conditions:

Areas subject to normal winds (zone 3):	120 cm (4′0″)
Intermediate areas (zone 2):	105 cm (3′6″)
Cyclonic areas (zone 1):	90 cm (3′0″)

These are to be reduced by 15 cm (6 inches) for very important buildings in zone 1 and zone 2 areas.

(x) The courses of the sheets are to be laid so that the corrugations are in continuous straight lines as shown in Fig. 22.7. (This is different from laying tiles where we break joints.)

(xi) Where the edges of four sheets meet and have to be lapped, the overlapping of two sheets have to be cut off neatly so that a snug fit can be obtained. This is called *metering of corners* and should always be carried out correctly on the ground and not on the roof for accuracy. A mitre cut is to be made so that the overlap effect will only be due to two sheets. (AC sheets can be cut by a wood saw.) A cat ladder may be used between purlins to adjust the sheets [Fig. 22.7(f)].

(xii) Ridges can be made of AC ridge with the same corrugations while the hips are to be plain. These may also be of Calicut pattern clay tiles and if they are specified in the drawings, these should be set as described under clay tile roofing.

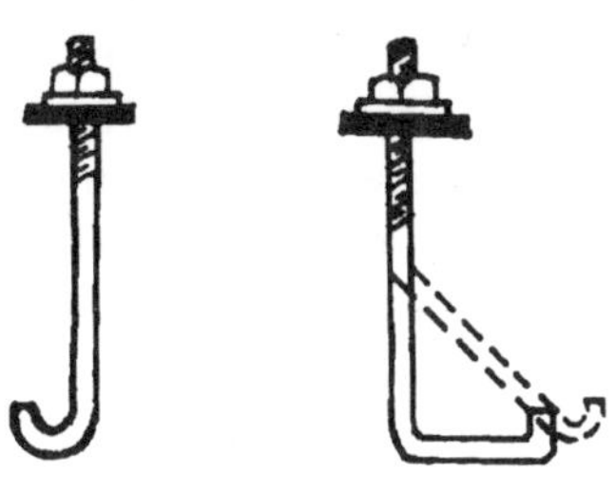

(a) Fixing bolts

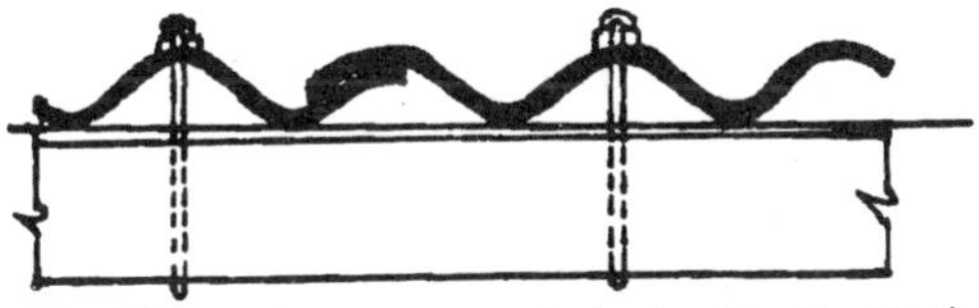

(b) Fixing deep corrugated sheets to steel angles

(c) Fixing ordinary corrugated sheets to wooden purlins

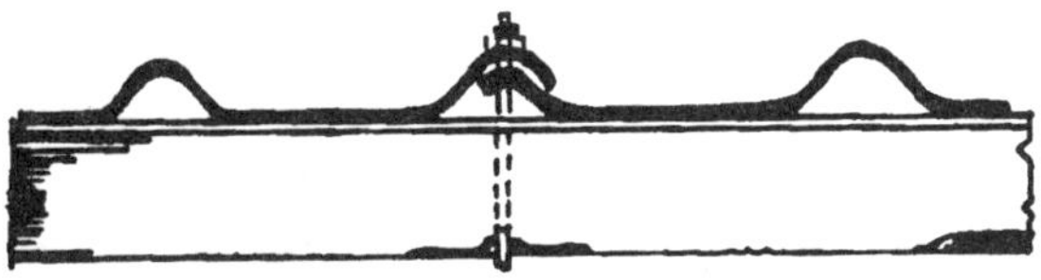

(d) Fixing Tafford sheets to steel angles

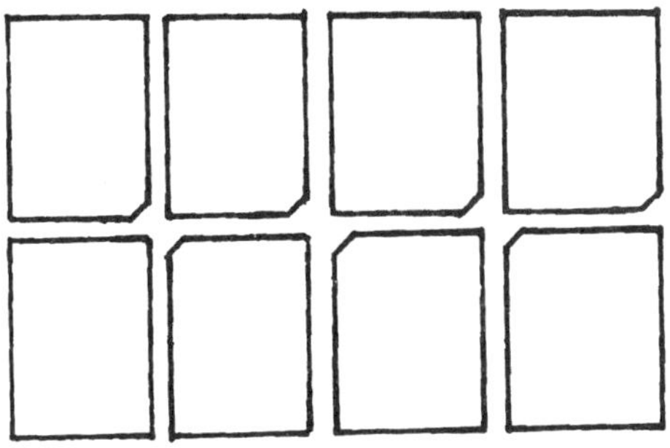

(e) Plan of mitering of corners

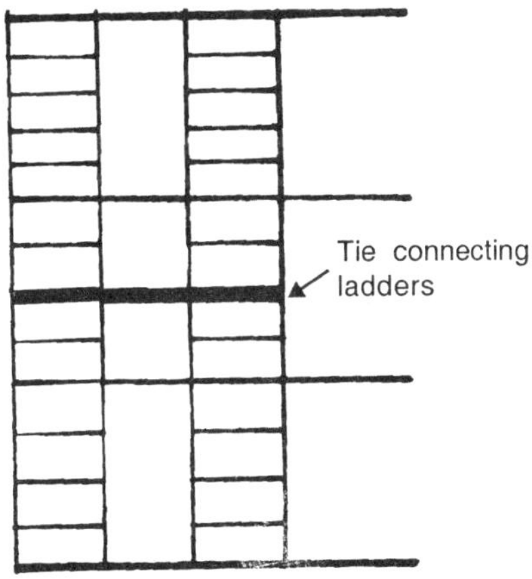

(f) Plan of cat ladder on roof top

Fig. 22.7 Laying of AC sheets on sloped roofs.

22.10 EFFECT OF HEAVY WINDS ON SHEETED ROOFS IN CYCLONIC AREAS

Sloped sheeted and tiled roofs are the worst affected by cyclones. Experience and tests show that the most severely affected parts of a sloped roof covered with sheets are the *eaves and the gables* and with slightly less intensity (but more than the average) *along the ridges.* We must resort to closer bolting or anchoring of sheets in these areas. Also projecting the sheets beyond the gable walls should be avoided as it can lead to considerable damage in heavy winds. Such projections are the first to be lifted up during a cyclone. Hence in cyclonic areas, the sheets should be stopped at the middle of the gable wall. In addition, a binding wall of 300 mm height should be constructed over the sheet at the gable to give it fixity at the gable end. In addition, care should be taken to see that the above constructed masonry wall is along the ridge of the corrugated sheets to prevent entry of rainwater into the masonry wall. In cyclonic areas, the fixing bolts are also provided at more points on the sheet than normally provided.

22.11 LAYING OF CORRUGATED GI SHEETS

Fixing of corrugated GI sheets is similar to that of AC sheets. The important points to be noted in the use of galvanized iron sheets are as follows:

1. The slopes of the roof should not be less than 10 degrees. A slope of 1 vertical to 2 horizontal corresponds to 26.5 degrees.
2. The minimum end lap should be 150 mm (6 inches) where the pitch is not less than 15 degrees. In locations less than 15 degrees, the lap has to be increased to 200 mm (8 inches) to 225 mm (9 inches). These lap ends are for sheets laid over purlins.
3. The sheets are to have a side lap of *half to one corrugation*. (CPWD recommends two ridges.) Lapping in CGS sheets shall be a coat of approved primer and two coats of point (such as chromate point) before the sheets are fixed in place.
4. They are to be fixed by 6 mm (1/4 inch) diameter galvanized hook bolts *with rubber washer* directly over the sheet and then a metal *washer screwed* down by a nut.
5. Alternately when fixing on wood, special roofing galvanized nails of 2¼ inch length and 1/8 inch diameter with twisted square shank are used. These nails should have self-sealing spring heads. In addition, a metal diamond-shaped GI washer is provided in all important buildings in cyclone areas (zones 1 and 2) when fixing with these nails.
6. In the general roof areas, in all zones, the fixing is to be made on the 1st, 4th and 8th corrugation (there are generally 11 corrugations per sheet and the first fixing being through the crown of the overlap corrugation). Special precautions should be taken in fixing of eave overhang and gable end areas. In these places, fixing is to be made in 1st, 3rd, 6th and 9th corrugations in zones 2 and 3 and in the 1st, 5th, 7th and 9th corrugations in zone 1.
7. The sheets are to be fixed commencing from the point away from the direction of heavy rains (leeward ends) so that water is not forced through the laps.
8. The ridge and hip capping as well as the junctions of ridge with hips are to be plain sheets of the same material. Special sections are also available. They are fixed with side laps of not less than 15 cm (6 inches) and the spacing of fixing is not to exceed 18 inches. If the roof framework does not permit this spacing, the capping sheet should be fixed with the above spacing by additional seam bolts or rivets.

9. The recommended spacing of purlins is the same as for AC sheets as shown in Table 22.2.
10. In case these sheets are to be covered with tiles 4 cm × 4 cm, battens are laid over the ridges and secured to purlins below by long screws or bolts and tiles are laid on reepers fixed to these battens.

Table 22.2 Recommended spacing of purlins for GI corrugated sheets

Thickness of GI sheet (mm)	*Max. spacing of purlines* (m)
1.60	2.80
1.25	2.40
1.00	2.00
0.80	1.80
0.63	1.60

The top surface of the purlins should be plane and the purlins should be painted before fixing the sheets. Embedded portions of wooden purlins should be coated with two coats of coal tar.

22.12 LAYING OF ALUMINIUM SHEETS

Aluminium steels are fixed to purlins as in the case of GI sheets by hook bolts or drive screws made of aluminium and not GI bolts. In contact with steel, aluminium produces galvanic action. Hence with steel purlins, the steel should be painted with two coats of a zinc chromate- or barium chromate-based paint, or with two coats of bitumen paint or two coats of zinc rich paint before the sheets are laid on them.

The end laps are to be 150 mm (6 inches) for pitches of 15 degrees or more and for more than 15 degrees, the lap should be 225 mm (9 inches) as with GI sheets. The side laps should be 1½ corrugations laid away from the direction of heavy rains. For vertical sheeting, the side laps are to be as above but end laps can be 4 inches. The overhang at the eaves is not to exceed one quarter the span between purlins and the overhang at the gable ends should not exceed 150 mm (6 inches) beyond the purlins. The sheets are to be so fixed as to close the ridges and hips. The details of ridge and hip capping and also that of junctions of ridge with hips should be as in roofing with GI sheets.

22.13 CEILINGS

Ceilings for sloped roofs and *false ceilings* for flat roofs are provided for the following reasons:

1. To reduce the heat transfer from the roof
2. To improve the appearance of rooms under the sloped roof
3. To conceal all the ducts, pipes, wiring and light fittings and for the room

4. To reduce the volume of the room and improve its air conditioning and acoustical quality.

In the case of flat roofs, the ceilings can be provided below the roof and are usually flat. In sloped roofs, they can also be built flat or sloping along the slope of the roof. In collar roofs, they can run along the rafters and the collar. Some roofs, such as the Mangalore tile roof, sit on flat tiles as described in Section 22.6 do not need another ceiling.

Standard false roofing materials are wood, gypsum board, plaster of Paris, thermocoal, extruded aluminium and acoustical board. Glass mirrors, metal sheets, canvas and many other materials are also used for special effects. The cost may range from a few rupees to hundreds of rupees. Generally, nowadays, boards of these materials are set in aluminium framework. Standard specifications give the following types of ceilings:

1. Wooden ceiling made with timber, boards and provided with tongue and grooved joints and beadings (for beauty)
2. Ceiling with insulating building boards
3. Plaster of Paris (anhydrous gypsum) tile ceiling
4. Ceiling with thermocoal (EPS) or polyurethane foam (PUF) slab with aluminium framing.

Details of these constructions are given in PWD specifications such as CPWD specification 77. In modern practice in RC buildings, the aluminium frame is fixed to the concrete roof and the insulating boards built into these frames.

22.14 SLOPED CONCRETE SLAB ROOFS WITH TILES ON TOP

Traditional sloped roofs with tiles are very pleasing in appearance. Because of the high cost, difficulty and considerable time required for such construction, more and more residential buildings are being built with *sloped concrete roofs on which we lay traditional or special tiles.* These tiles are laid by one of the following methods:

1. On reepers fixed to the concrete slab
2. By fixing the tiles to the concrete slab by cement or lime mortar
3. Only a few rows of tiles at eaves and a few tiles at other places laid with mortar and all other tiles placed directly on the slab as continuation of the eave tiles. This allows circulation of air below the tiles and less transmission of heat from the top to the concrete slab. (Special tiles to give these air gaps are available in the market.)

The most important consideration for leakproof construction of these roofs is the proper compaction of the concrete on these sloped roofs. (This is also true for folded plates and shells described later in this chapter). In flat roofs, it is easy to compact concrete in place. But in sloped roofs along the slope, and in valleys which also slope in one direction, it is *not easy* to compact concrete properly. If left untreated, it will lead to bad leakage. Hence it is advisable to treat the top surface of the concerete slab and especially the valleys with extra waterproofing layer over the concrete slab and also provide proper drainage (refer to Chapters 27 and 28).

Folded plates are also similar to sloped roofs, and because of the inclination of the slab, it is difficult to compact in situ concrete. So special attention should be given for waterproofing the slab surface and the valleys.

22.15 CONCRETE SHELL ROOFS

Concrete shell roofs are a special class of sloped roofs. They are commonly used for long span structures. Shells can be defined as very thin slabs curved in shape with their thicknesses small compared to their radius of curvature. They get their strength from their curved shape (called shell action) which is difficult from slab action. In shell action the forces are principally membrane forces (tension or compression) whereas in slabs the principal strength is obtained from bending action. Shells can be of single curvature or double curvature, the latter having more strength than the former. These shells are supported at its periphery by rigid members.

Geometrically, shells can be divided into the groups:

1. Rotational shells
2. Translational shells

Rotational shells are formed by a curve rotating about a central axis. The circular dome can be considered as a rotational shell. Translational shells are those produced by a geometric line or curve moving over another curve. Cylindrical shells and hyperbolic paraboloids belong to this group (see below). The following are the more popular shapes of shells used in building construction.

Domes. Domes are double curvature rotational shells obtained by a curve rotating about a central vertical axis [Fig. 22.8(a)]. It can also be viewed as a translational shell. (A circular dome can be produced by a circle riding over another circle.) Pendentive domes are formed from a hemispherical dome by cutting it by vertical planes on the four sides as shown in Fig. 22.8.

Cylindrical shell or barrel vault. This is a single curvature shell or a cut cylinder. It is a translational shell produced by a line moving over parts of two circles at the ends (or part of a circle moving along a line). They are classified as long and short shells. A long shell has its length along the axis large compared to its radius. The length of a short shell is small compared to its radius.

Conoids. These are single curvature translational shells generated by a straight line moving over two different curves at the ends or moving over a curve at one end and a vertical line at the other [Fig. 22.8(b)].

Hyperbolic paraboloids (saddle shells). These are double curvature translational shells produced by a vertical parabola with upward curvature moving over another parabola with downward curvature [Fig. 22.8(c)]. The generated surface is shaped like a horse saddle which is difficult to construct. This type of shell can also be formed by raising or lowering one or more of the corners of a rectangle. Thus the resulting shape can also be easily formed by raising or lowering the opposite corners (the diagonal) of a square, thus forming a warped parallelogram. For shell action, the rise of the diagonal should not be less than 1/15th the

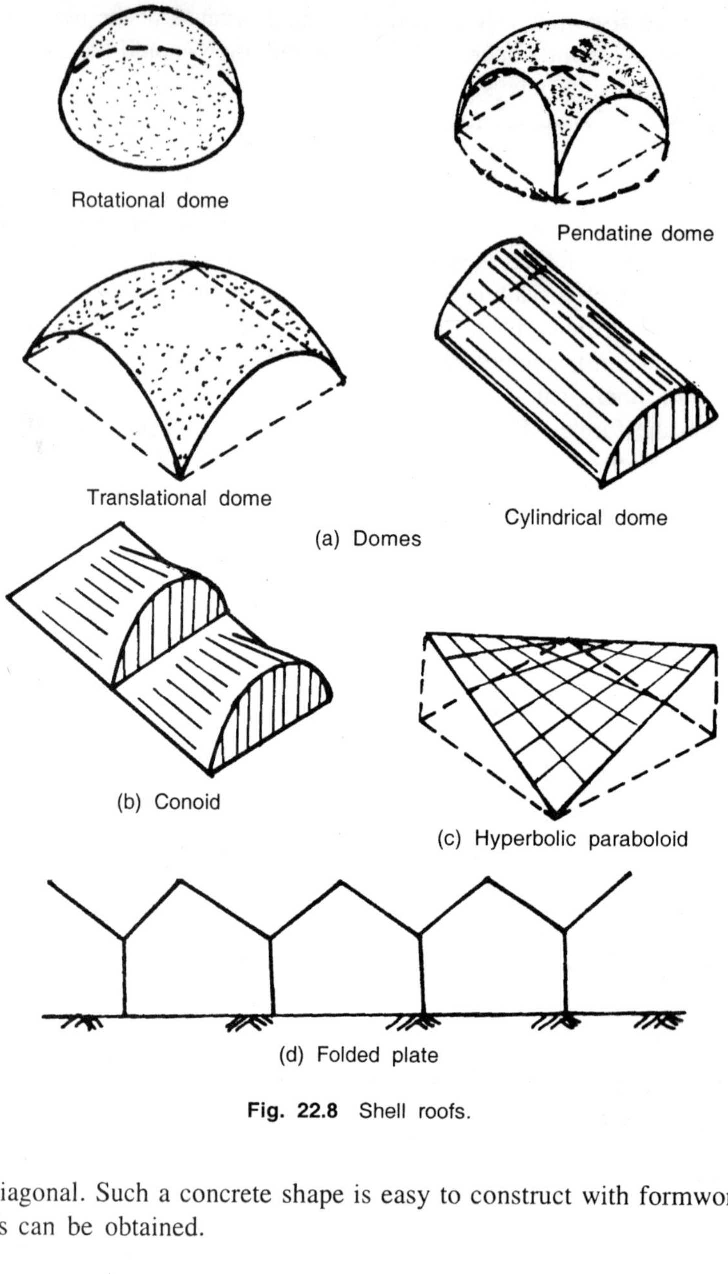

Fig. 22.8 Shell roofs.

length of the diagonal. Such a concrete shape is easy to construct with formwork and various shapes of roofs can be obtained.

22.16 LONG SPAN ROOFS

When we need long span structures such as sports stadium, exhibition halls, etc., we use special roofs. These can be tension cable structures or temporary tension membrane structures (made of special cloth such as membranes which can be stretched) or steel frames made from steel tubes with special joints as shown in Figures 22.9(a)–(c).

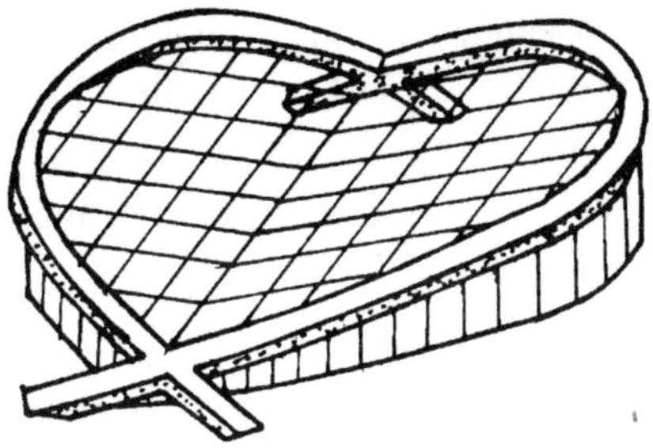

(a) Tension cable structure

(b) Tension membrane structure using special tensionable membrane

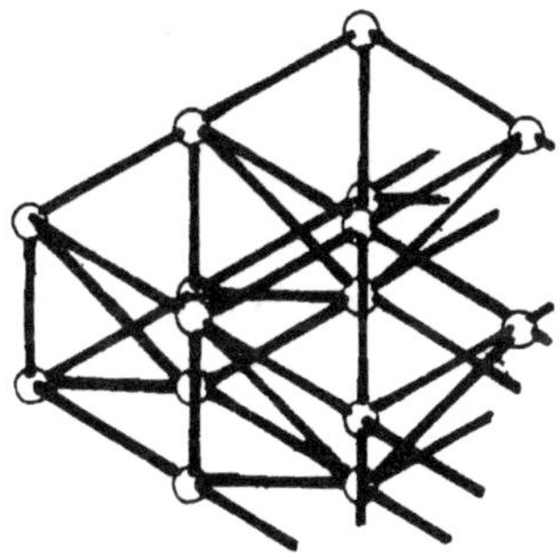

(c) Steel frame made from steel tubes and special joints

Fig. 22.9 Long span structures.

SUMMARY

Sloped roofs from wood are light and very popular for residences all over the world. Sloped roofs with AC, GI or aluminium sheetings are very commonly used in India for factories. Use of AC sheets is very much discouraged in developed countries. Roofs of low-cost buildings such as watchman's shed, storage sheds at a construction site, etc. can be made with wooden frames and sheetings so that they can be dismantled easily and reused in another place.

Sloped roofs and shells of various shapes made of thin concrete slabs are also very much used in building construction. Considerable care is necessary in compacting the concrete in these structures to make them waterproof. Large span structures made of tension cables, steel tube space frames and tension membranes are also becoming popular these days.

REVIEW QUESTIONS

1. Sketch and explain the various types of pitched roofs indicating the spans for which they are applicable.
2. What type of roof would you select for a large span workshop? What type of roof coverings would you choose? Explain the considerations.
3. Explain the method of laying Mangalore tiles on a pitched wooden roof.

4. Describe the method of laying AC sheet on a steel truss with purlins. Sketch the details of fixing AC sheets to (a) an angle purlin, (b) wooden purlin and (c) I beam purlin.
5. What are the precautions you will take when covering a grain storage building with AC sheets in a cyclonic area?
6. What are the purposes of ceilings? What materials are used for laying of ceilings?
7. Explain the use of flashings in sloped roof construction.
8. Write short notes on:
   (a) Mitering of AC sheets
   (b) Fixing of roofing sheets in cyclonic areas
   (c) False ceilings
   (d) Flashings in (i) vertical wall with a horizontal roofline of a beam to roof, and (ii) vertical walls with a sloping roofline of a lean to roof
   (e) Give a short account of the different types of concrete shell roofs.

Chapter 23

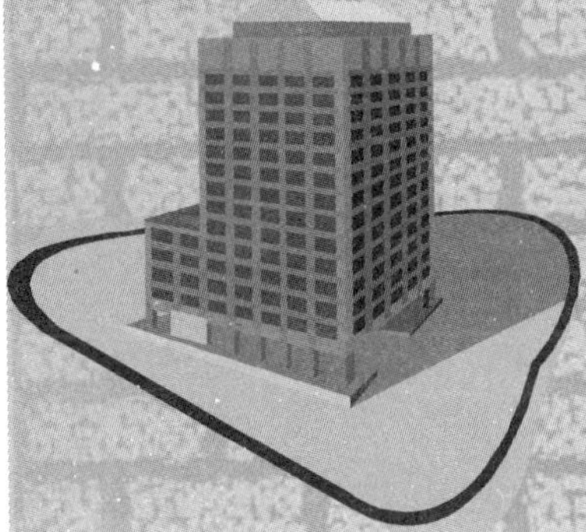

Doors, Windows and Ventilators

23.1 INTRODUCTION

The cost of providing doors, windows and ventilators in a building can work out to about 18 to 25 per cent of the total cost of civil works. This is due to the fact that expensive materials such as wood, fittings and skilled labour are involved in making them. There are many types of doors and windows and it is difficult to deal with all of them in one chapter. Hence in this chapter, we will examine the types of the generally used doors and windows and their fabrication. For more details, reference may be made to the CPWD specifications 77 and other PWD specifications such as Tamil Nadu Building Practice Vol II. One of the important details to be thought of in designing the width of shutters of windows and doors opening to the outside of the building is that the cantilevering portion of the *chajja* above these shutters should fully cover the shutter when opened and they do not become wet during rainy season.

23.2 DOORS, WINDOWS AND VENTILATORS

Doors are provided as external doors at the various entries and exits of the building and also as internal doors in the interior places where we want to divide the space into separate parts. External doors should be more robust than internal doors since they should give protection from pilferage and the elements. Windows are provided on outer walls for natural light and ventilation. Ventilators are necessary for air circulation in various places such as bathrooms, stores and where windows are not usually provided. The old practice to provide ventilators on top of doors and windows is not generally practiced today. Small ventilators at roof or underside of top floors in all the main rooms with conventional ventilators in stores, bathrooms etc. is the common practice followed nowadays.

Bedroom windows can be provided in two halves so that the lower halves can be closed for safety and convenience. We will briefly examine some of the popular types of doors, windows and ventilators.

23.2.1 Panelled and Glazed Doors

Panelled doors are those doors in which the shutters are made up of a number of panels as shown in Fig. 23.1. Glazed doors are those provided with glass panels.

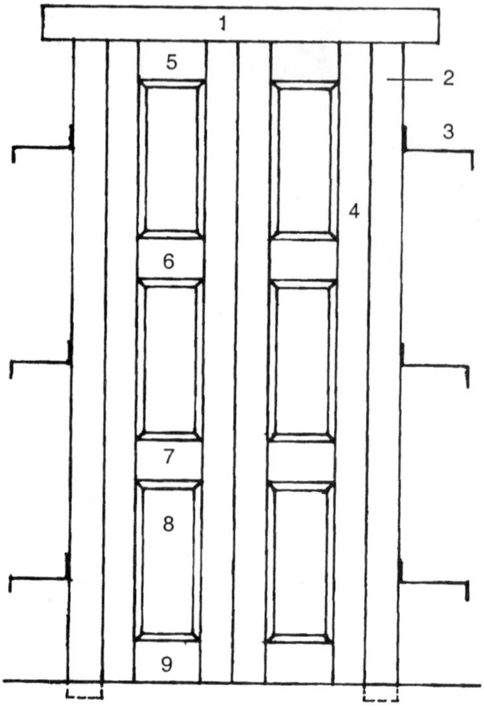

(a) Panelled door

(b) Fully glazed door (c) Glazed and panelled door

Fig. 23.1 Types of doors: 1. Head, 2. Frames, 3. Holdfast, 4. Style, 5. Top rail, 6. Frieze rail, 7. Lock rail, 8. Panel, 9. Bottom rail.

The important parts of a door are the following.

Door frame. It is the assembly that is fixed to the wall. It is fixed into the wall with hold fasts as shown in Fig. 23.1. The shutters are fixed to the frame by hinges. The top part of the frame is the *head* and the optional projecting part is the *horn*. The bottom part is the *sill* which nowadays is omitted for inside doors. However, for the doors in the *ground floor* leading to the outside of the building, the sill is also generally provided to prevent entry of insects. This part of the main entrance door may also be provided with a *threshold plate*. (Nowadays it may be in the form of a depression for door mat.)

Shutters. These are the opening parts of the door and fixed to the frames by hinges. The following are the constituents of the shutters:

(i) *Styles.* These are the vertical outer members

(ii) *Rails.* These are the horizontal members, divided into top rail (at the top), lock rail (in which locking arrangement is fixed) and the bottom rail. The bottom rail is broader than the other rails. The top rail is also called the freeze rail.

Panel. It is the area between the rails and styles consisting of thinner planks between the rails.

The masonry parts of the door opening of the wall are called by the following terms:

Jamb. It is the external part of the vertical wall face of the opening which supports the frame at right angles to the wall face. (see Fig. 1.2 of Chapter 1).

Reveal. It is the revealed part of the jamb on the sides of door and window frames (Fig. 1.2).

Threshold. The wooden plank or slab of concrete or stone usually provided at the bottom of an entrance door is called the *threshold*. (In general sense, it also means the entrance to the building.)

23.2.2 Windows

In hot and humid regions, the window area should be 15 to 20% of the floor area for good ventilations, even though only 10% of the floor area is needed for lighting. As a rule, each room must have at least two openings in two different walls for cross-ventilation.

The various parts of a typical window are shown in Fig. 23.2.

The following special terms with reference to the window frames are of interest:

Mullion. It is the vertical member in the middle of a window frame which subdivides the window. There will be one mullion in a window with two halves. (It also refers to doors if provided.) Mullions are nowadays generally provided in many windows with two shutters.

Transom. It is the horizontal member in a window frame used to subdivide the frame opening by a horizontal member. In many cases, it may not be necessary.

Rebate. It is the recess made inside the door or window frame to receive the shutter.

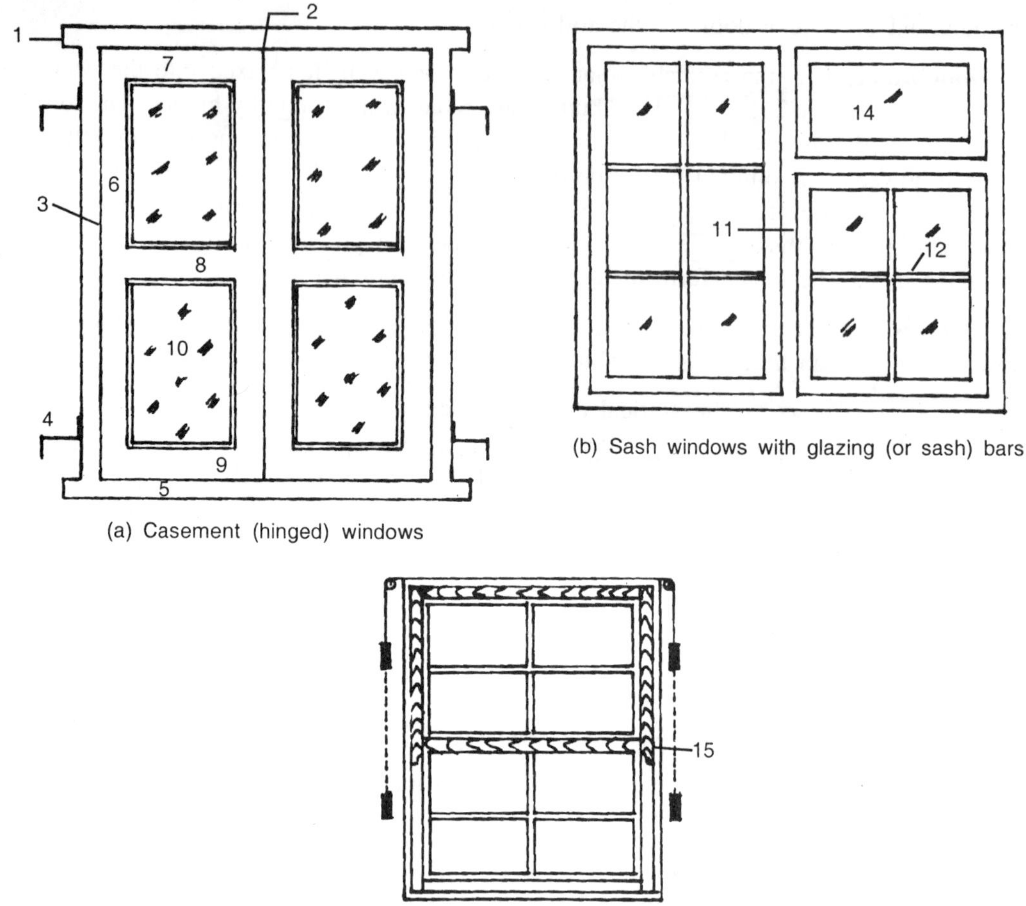

(a) Casement (hinged) windows

(b) Sash windows with glazing (or sash) bars

(c) Hung windows

Fig. 23.2 Windows: 1. Frame, 2. Head, 3. Jamb post, 4. Holdfast, 5. Sill, 6. Style, 7. Top rail, 8. Intermediate rail, 9. Bottom rail, 10. Panel, 11. Mullion, 12. Glazing (sash) bar, 13. Transom, 14. Ventilator, 15. Meeting rail.

23.2.3 Ventilators

Ventilators are small openings, its ideal location in general rooms and kitchen is just below the roof in the top storey or below the upper floor in the room. In bathrooms and lavatories, it is generally fixed above small windows. Simple brick or cement concrete jaliwork can be used as ventilators. Generally ventilators consist of a frame and shutter (generally glazed). The shutter ventilators are usually horizontally pivoted as shown in Fig. 23.3 so that it can be tilted with the *top edge of the shutter opening downwards in the inside of the building* and the bottom edge opening outside of the building to facilitate rainwater falling outside when closed. Ventilators may also be in the form of simple openings left in the masonry work under the roof slab and made with concrete jaliwork. In bathrooms, where the windows are small and are fixed at higher levels, ventilators can be combined with windows.

Other methods of fixing ventilators are shown in Fig. 23.11.

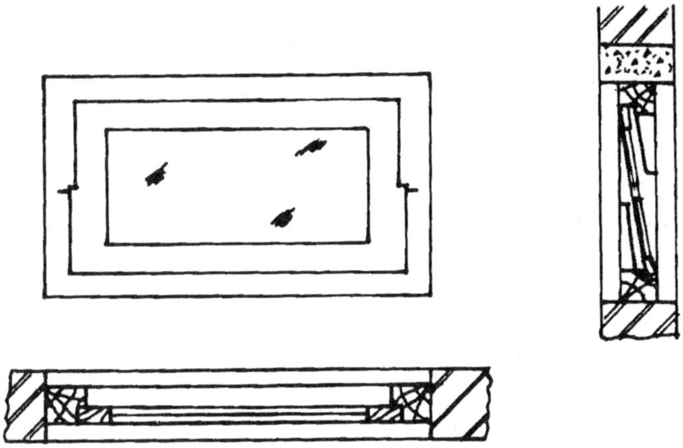

Fig. 23.3 Ventilator (centre hung type).

23.3 STANDARD SIZES OF DOORS, WINDOWS AND VENTILATORS

The dimensions of doors, windows and ventilators are measured *inside to inside of openings* and designated by letters such as 8DS20, 10 WT12 and 6V6. In this designation, the *first number* indicates the width of door opening in modulus of 100 mm (10 cm). Thus 8 means 800 mm. The first letter D denotes door, W windows and V ventilator. The second letter S denotes single and T double shutters. The last number denotes the height of the opening in modulus of 100 mm. Thus 20 means 2000 mm (2 m). The standard sizes of openings according to IS are given in Tables 23.1 to 23.3.

In a building, the top of the doors and windows are placed at same level. Hence the height of a window will depend on the height where we want to fix the window sill to ensure circulation inside the building. The width of windows will depend on the openings required. In places such as kitchen windows are generally provided above the level of the kitchen platform. Fixing the height of doors and windows is explained in Section 23.4.

Table 23.1 Indian standard dimensions of timber door frames

S.No.	*Designation*	*Size of opening*	*Size of wood frame (width and height)*	*Size of door shutters*
1.	8 DS 20	800 × 2000	790 × 1990	700 × 1905
2.	8 DS 21	800 × 2100	790 × 2090	700 × 2005
3.	9 DS 20	900 × 2000	890 × 1990	800 × 1905
4.	9 DS 21	900 × 2100	890 × 2090	800 × 2005
5.	10 DT 20	1000 × 2000	990 × 1990	900 × 1905
6.	10 DT 21	1000 × 2100	990 × 2090	900 × 2005
7.	12 DT 20	1200 × 2000	1190 × 1990	*1100 × 1905
8.	12 DT 21	1200 × 2100	1190 × 2090	*1100 × 2005

* Doors with two shutters 560 mm each shutter and 20 mm overlap when closed.

Table 23.2 Standard dimensions of timber window frames (mm)

S.No.	Designation	Size of opening	Size of wood frame (width and height)	Size of window shutters (wood)
1.	6 WS 12	600 × 1200	590 × 1190	500 × 1100
2.	10 WT 12	1000 × 1200	990 × 1190	460 × 1100
3.	12 WT 12	1200 × 1200	1190 × 1190	560 × 1100
4.	6 WS 13	600 × 1300	590 × 1290	500 × 1200
5.	10 WT 13	1000 × 1300	990 × 1290	460 × 1200
6.	12 WT 13	1200 × 1300	1190 × 1290	560 × 1200

Table 23.3 Standard dimensions of timber ventilator frames

S.No.	Designation	Size of opening (L × H)	Size of wood frame	Size of shutter
1.	6V6	600 × 600	590 × 590	500 × 500
2.	10V6	1000 × 600	990 × 590	900 × 500
3.	12V6	1200 × 600	1190 × 590	1100 × 500

Note: The thickness of the shutters are 20, 25 or 30 mm depending on size of the opening.

23.4 FIXING SIZES AND HEIGHTS OF DOORS AND WINDOWS

Generally the height of doors (as defined in Section 23.3) is kept not less than 1950 mm (6 ft 6 in) in residences and 2.1 m (7 ft) in public buildings *from the floor level.* We have already seen that the top levels of doors and windows are always kept at the same height. The sill or bottom part of windows is usually kept at 700 mm (2 ft 4 in) to 800 mm (2 ft 8 in) above the floor level. In some places such as drawing rooms, window sill may also be fixed at lower levels. Bay window sills are usually kept at a low level—at about 500 mm (1 ft 8 in). The size of windows will, therefore, depend on these dimensions.

Considering width of doors, we note that as all modern furniture has the minimum dimension of 750 mm (2 ft 6 in), the minimum width of door openings should be at least 800 mm (2 ft 8 in). Thus the minimum size of main doors should be 8DS20 doors to stores, bathrooms, etc. may be smaller in width 675 mm (2 ft 3 in), which is less than the standards specified.

When we plan a building such as a residence, we try to keep the width of the various door openings of the main building to be not more than the following three sizes:

(i) Entrance doors: 1200 mm × 2100 mm (4.0′ × 7.0′) or 1000 × 1950 mm (4′ × 6′6″)

(ii) Internal doors: 900 × 2100 (3.0′ × 7.0′) or 900 × 1950 mm (3′ × 6′6″)

(iii) Doors for bathroom, stores, etc.: 750 × 2100 (2′6″ × 7.0′) or 750 × 1950 mm (2′6″ × 6′6″)

It is also very important to remember that if we are using flush doors, the openings and the frames we are planning must suit the commercial flush door shutters that are available in the market.

23.5 SIZES OF TIMBER TO BE USED FOR DOORS AND WINDOWS

Timber doors and windows are more popular than other types for residences. The sizes of timber scantlings adopted for doors and windows should be of reasonable size as otherwise they will quickly get out of shape. The following sizes of timber are recommended for doors of 2.1 m in height.

23.5.1 Doors

The sizes of timber to be used for the various parts of a panel door are given in Table 23.4.

Table 23.4 Timber sizes for panelled doors

Part of panel door	*Timber sizes (sections)*
Frames	For single shutter frame: 100 mm × 65 mm For double shutter frame: 130 mm × 65 mm
Panelled shutters	For styles: 75 mm × 50 mm For top and intermediate rails: 75 mm × 40 mm For lock rail: 125 mm × 40 mm For bottom rail: 150 mm × 40 mm
Panels (planks)	For single leaf: (1.0 m × 2.1 m) 15 mm thick For small panel: 12 mm thick

(Panels can also be moulded from planks of wood or made from plywood.)

23.5.2 Windows and Ventilators

The sizes of timber to be used for the various parts of windows and ventilators are given in Table 23.5.

Table 23.5 Timber sizes for windows and ventilators

Part of window	*Timber sizes*
Frames (windows and ventilators)	For up to 1.2 m width: 75 mm × 60 mm For more than 1.2 m width: 100 mm × 60 mm
Panels (styles and rails)	For window styles and rails: 75 mm × 35 mm For ventilator styles and rails: 60 mm × 35 mm
Panels (planks)	For planks for panel: 12 mm thick For glazing bars: 40 mm × 35 mm

23.6 METHOD OF FIXING DOOR AND WINDOW FRAMES

23.6.1 Important Considerations

The fixing place of door and window frames needs special attention. It will depend on the way the shutters are to open out. Usually the front doors open to inside the building. Usually the door of shutters are planned to swing 90° or 180° and be parallel to a wall when fully opened. Similarly ventilators can be fixed to be flush with the inside walls of the room or the outside the wall. When we fix window frames flush with the outside walls in a one-brick (9 inch) wall, we get a ledge on the inside of the room. If we fix the frames on the inside, they will be flushed with the inside walls. We must decide these positions well in advance of the construction. The following methods are used to fix frames to the walls of the building:

In modern building construction, where concrete lintels are used over openings, door and window frames are fixed after roof construction and before plastering of masonry is taken. Before fixing all the surfaces of the wood of the door and window frames that come into contact with masonry, they should be treated by painting with coal tar mixed with anti-termite solution. The door frames are fixed by three holdfasts on each side whereas windows are usually provided only two holdfasts on each side as shown in Figures 23.1 and 23.2. Small sized ventilators need only one holdfast on each side.

23.6.2 Fixing to Masonry

There are many methods to fix door and window frames to the wall masonry. Broadly they can be in two ways—prepared opening method and built-in method. These methods are discussed further.

Method 1: Prepared opening method of fixing in masonry. We use this method for *door opening with lintels*. In this method, the doors are fixed after the masonry is fully built but before the plastering. This method is preferred for superior type of doors and also for other modern type of doors. It can also be carried out by any of the following procedures:

(a) In the conventional method, holes are left or made in the masonry by Z-type holdfasts. These are 30 mm wide and 5 mm thick flats 200 mm in length bent into Z shape. They are inserted into the holes and the holes are filled with concrete. The portion of the holdfast inside the wall should point downwards and the portion on the frame upwards. For fixing a door, three holdfasts are used. One holdfast is fixed at the centre of the frame and the other two at 30 cm from top and bottom of the frame. For a window, two holdfasts are used—each at 30 cm from its top and bottom.

(b) Hard wood plugs can be embedded in the masonry at appropriate points with their ends flush with the face of the opening. The frames are then screwed into the plugs by means of 75 mm galvanized wood screws for door frames and 50 mm screws for windows. This type of installation makes replacement of door frames later on easy.

(c) Instead of wood plugs, we may install a cast in situ concrete block with a nut and greased bolt. The door frame can then be fixed to this nut by a bolt.

(d) Where the window or door frame has to be placed adjacent to a concrete column, prior planning should be made to cast a 15 mm GI pipe in the column at predetermined positions (where the frames are to be fixed). A corresponding hole can be made in the window or door frame when it is to be fixed by bolt head on the frame side and a nut on the other side of the column. The countersunk hole in the frame is filled up with the same type of wood as that of the frame.

Before fixing the frames, we plaster the jamb with fine plaster and when the mortar is still wet, we force the frame into the opening covered with mortar so that a close fit can be obtained.

Method 2: Built-in method of fixing in masonry. This is the old practice where the modern RCC lintels were not used. In this method, the doors are fixed to the wall as construction proceeds. This ensures a good and clear joint without any space between the wall and the frame. Two crossbattens are provided diagonally to keep the frame from deforming during construction. However, as it takes very long time to complete the roof, till then the frames are subjected to heat and moisture leading to warping of wood. Hence this practice is not recommended to be followed nowadays.

23.7 TYPES OF DOORS

The following are the main types of doors. Some of them are shown in Fig. 23.1 and Figures 23.4 to 23.8.

1. Panelled doors (Fig. 23.1)
2. Glazed doors (Fig. 23.1)
3. Louvered doors (Fig. 23.4)
4. Flush doors (Fig. 23.5)
5. Battened doors (Fig. 23.6)
6. Wire-gauged doors
7. Revolving doors, sliding doors and swing doors
8. Collapsible steel gates (Fig. 23.7)
9. Rolling steel shutters (Fig. 23.8)
10. Metal doors
11. Others

Glazed doors are the same as panelled doors except that in glazed doors, instead of wooden panels between frames, we use glass. The panels can be fully or partly glazed. In general, panelled doors are considered more durable than other types of doors. Flush door shutters are made in the factory. They can be of various types such as block board core, cellular core, hollow core and particle board core or medium density fibre (MDF) fibreboard. The most common specification for most of the city flats in India is "hardwood door frames with flush doors". Specially moisture-resistant flush doors are needed for bathroom doors. These flush doors should be resistant to water. If dipped in 30 cm of water for 24 hours and allowed to dry for 24 hours and this cycle is repeated 8 times, there should be no delamination at the end of the test. Flush doors of different thicknesses are available and a suitable size should be selected for use. (Types of flush door shutters are explained in the book on *Building Materials*.) Sliding doors and folding doors are useful in places where the door leaves cannot be accommodated.

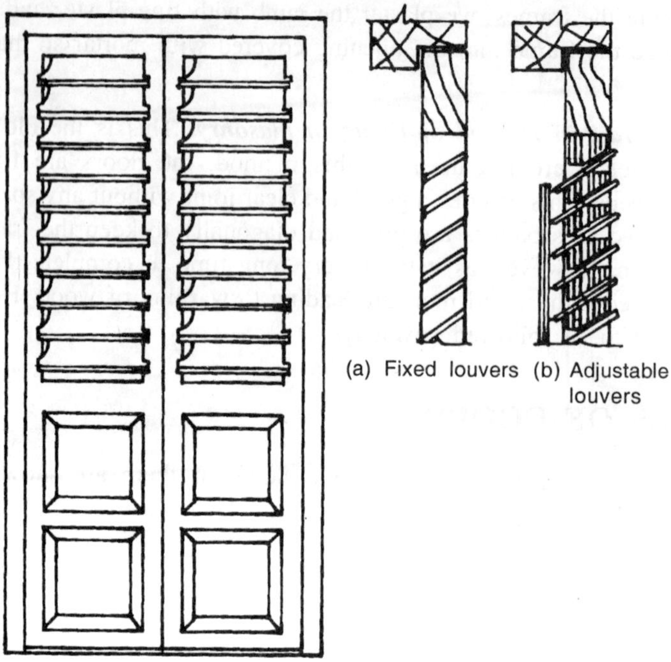

Fig. 23.4 Louvered doors.

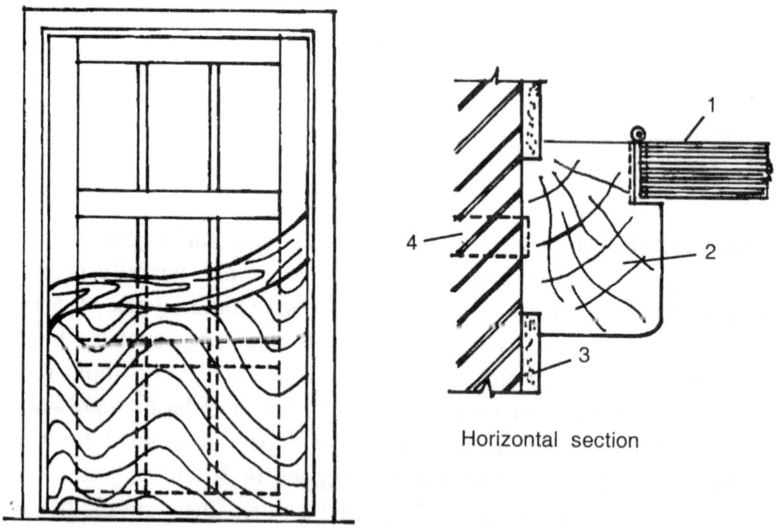

Fig. 23.5 Flush doors: 1. Shutter, 2. Door frame, 3. Plaster, 4. Holdfast.

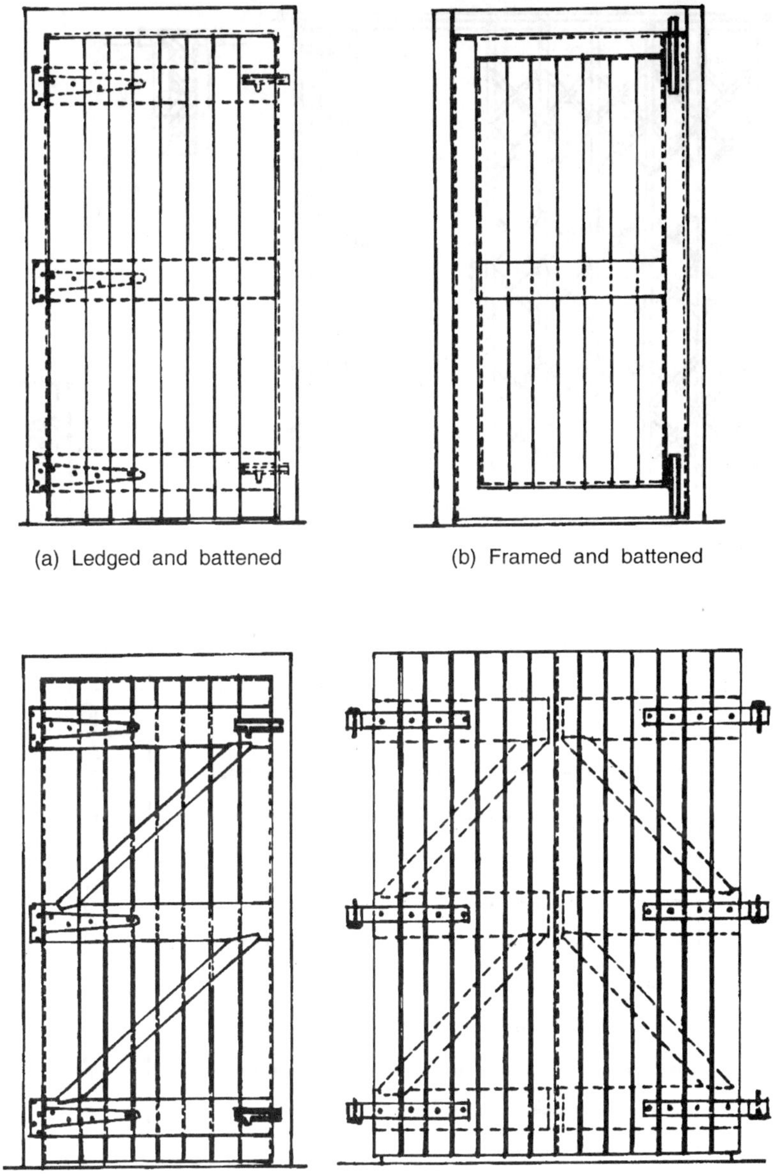

(a) Ledged and battened

(b) Framed and battened

(c) Ledged, braced and battened

(d) Garage door

Fig. 23.6 Battened doors.

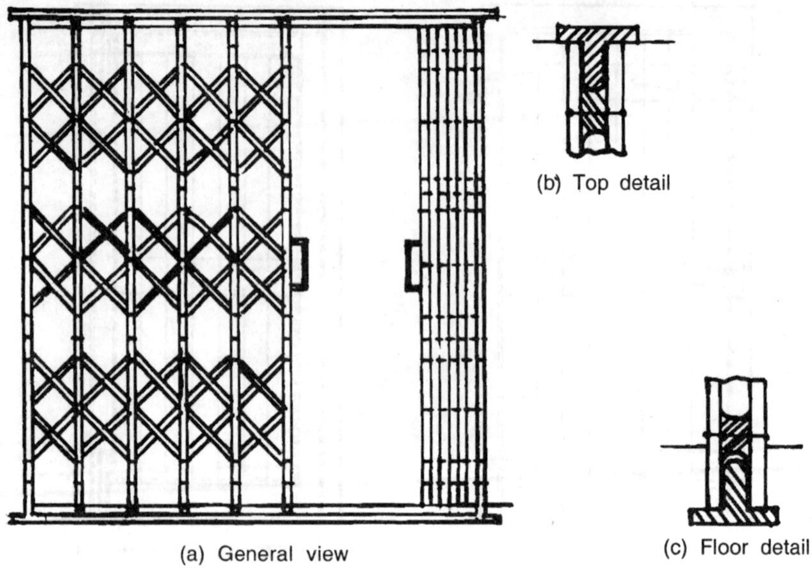

Fig. 23.7 Collapsible steel door.

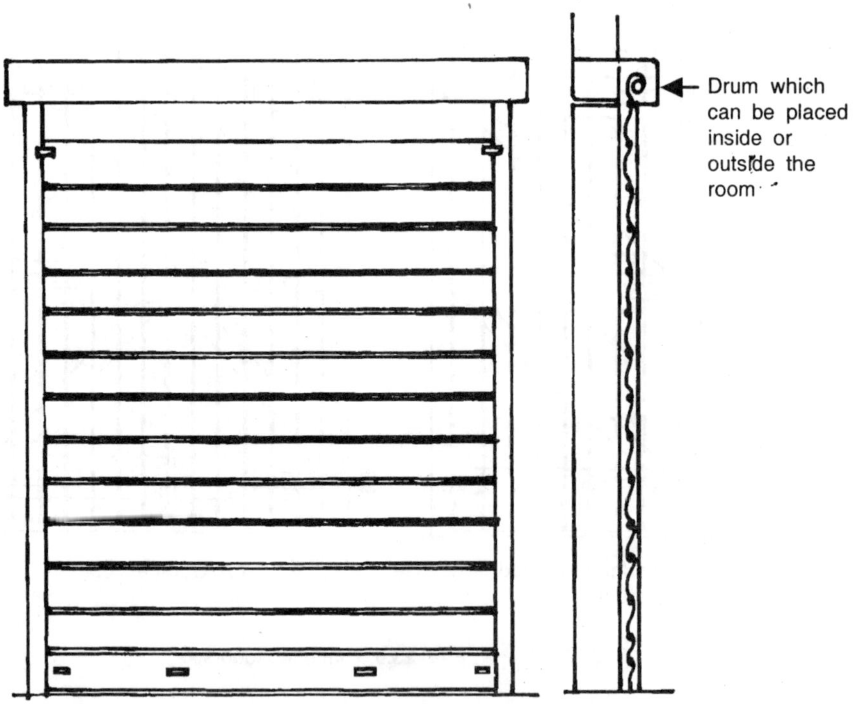

Fig. 23.8 Rolling shutter.

Battened doors need special mention as they can be fabricated from small sized timber and can be used in many locations of a building. These are shown in Fig. 23.6.

23.8 FITTINGS FOR DOORS

The following are the main *fittings for doors:*

1. Z holdfasts (3 on each side of frame as described in section 23.6.1)
2. Hinges of minimum size 80 (3 on each shutter)
3. Top tower bolt with eye
4. Bottom tower bolt
5. Horizontal tower bolt
6. Locking arrangement
7. Door-stoppers (optional)
8. Handle (optional)
9. Swing damper such as hydraulic closure (optional)
10. Push plate (optional)
11. Name plate (optional)
12. Kick plate (optional)

Note: There are also a large number of fittings such as magnetic door holding devices that are available nowadays in the market.

For external doors, the travel of tower bolt must be at least 45 mm for top and bottom bolts. For internal doors, it can be shorter. Hinges should be at least 100 mm long and width should suit the thickness of door. It is fixed with steel "wood screws". Hinges can be steel (anodized), brass or aluminium with plastic washers. For heavy doors, we should not use aluminium hinges as the plastic washers tend to wear out soon and they do not last long. (It is better to use oxidized steel hinges.)

23.9 WINDOWS

Windows are necessary for ventilation and lighting. These are usually glazed with clear or opaque glass. As already stated not less than 10 to 15 per cent of the floor area of a room is given to windows opening to the outside. The smaller the floor area, the larger will be the percentage. Windows can be of many types as shown in Figures 23.2, 23.9 and 23.10. Some of them are as follows:

1. Casement windows (those opening outside) [Fig. 23.2(a)]
2. Double hung windows (those moving up and down) [Fig. 23.2(c)]
3. Louvered windows (Fig. 23.9)
4. Bay windows [Fig. 23.10(a)]
5. Dormer window [Fig. 23.10(b)]
6. Gable window [Fig. 23.10(b)]
7. Sliding windows (those which slide horizontally)

There are many other types of windows such as sash windows (fully glazed window divided into square or rectangular portions by sash bars) Fig. 23.2(b). Bay windows are described in Section 23.11. As already stated, windows are fixed at 750 mm to 900 mm above the floor level. Bathroom windows (if provided) are fixed at 1.2 m height (with ventilators placed on top of the windows).

Fig. 23.9 Louvered windows.

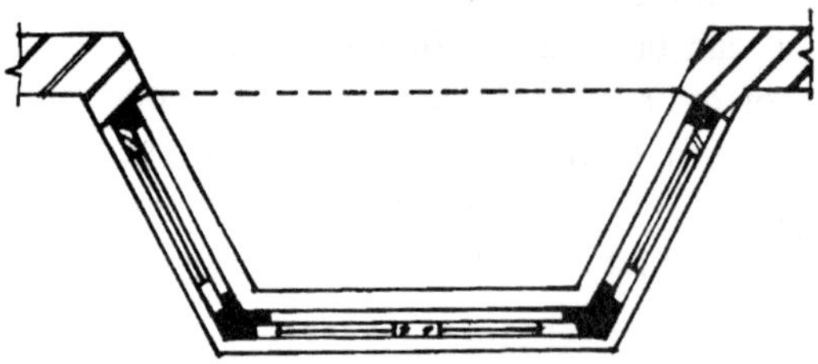

(a) Bay windows

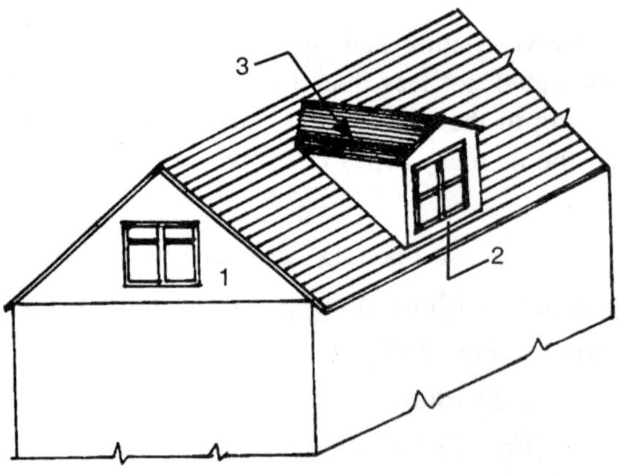

(b) Other types: 1. Gable windows, 2. Dormer window, 3. Dormer roof

Fig. 23.10 Special windows.

23.9.1 Window Fittings

The most common type of windows is the casement window and the fittings of this window are as follows:

1. Two Z holdfasts at quarter and three fourths heights, 50 mm × 60 mm × 200 mm, fixed with 50 mm screws. In total, four numbers of holdfasts are used for windows of height less than 1.7 m and six numbers for windows of height more than 1.7 m. The portion in the masonry should point downwards as described for doors.
2. 50 mm hinges, 175 mm wide and 100 mm deep fixed with 40 mm wood screws — 2 numbers for single leaf and 4 numbers for double leaf (2 on each leaf).
3. Top and bottom barrel bolts
4. 200 mm cabin hook and eyes (of 6 mm round steel)
5. Ring or handle
6. Glass panes to proper dimensions

23.9.2 Fixing Grills to Windows and Ventilators

In many cases, especially for ground floor windows, simple iron bars or welded iron grills can be provided for protection from intruders. To reduce the cost of grills, iron bars can be fixed to an aesthetic spacing to the window frame itself. These bars should be 15 to 20 mm in diameter (18 mm bars are generally used), round or square. Otherwise separate *fabricated grills* made out of steel flats can be fixed to the window frame. These grillworks are measured in kg/m^2 or kg/sq ft (on an average, the grillwork works out to about 2 kg/sq ft). The procedures to fix these grills are described in Section 37.13).

23.10 METAL DOORS AND WINDOWS

Due to difficulty in availability of good wood and easiness of maintenance, there is a tendency nowadays to use metal (especially aluminium) and reinforced plastics for doors and window frames. "Fully-glazed anodized aluminium window frames, with sliding shutters and MS grills" is the most common specification for most of the residential flats now being built in India in its principal cities. Aluminium can be powder coated to give different finishes, even wood finish.

Steel doors and windows as shown in Fig. 23.11 are used in industrial houses where regular maintenance is available. Without good maintenance, steel fittings corrode very quickly, especially near sea coasts.

23.10.1 Fixing of Metal Windows

Metal windows, such as aluminium windows, can be fixed to the walls in two ways.

Method 1. Metal windows can be fixed to wooden frames which are fixed to masonry wall by conventional methods.

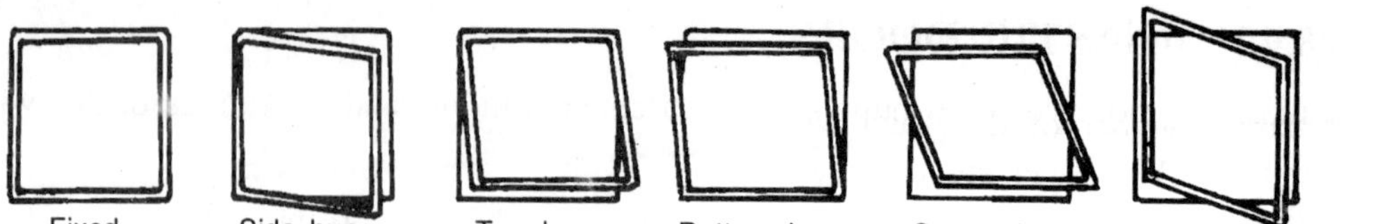

(a) Ventilators

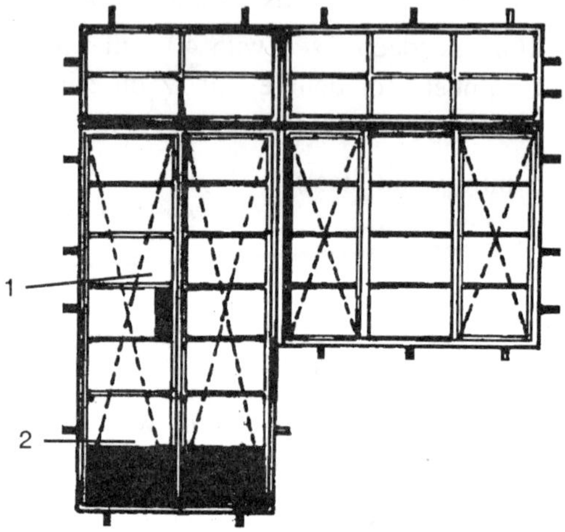

(b) Glazed steel door with window: 1. Lock plate, 2. Kick plate

(c) Glazing with clips

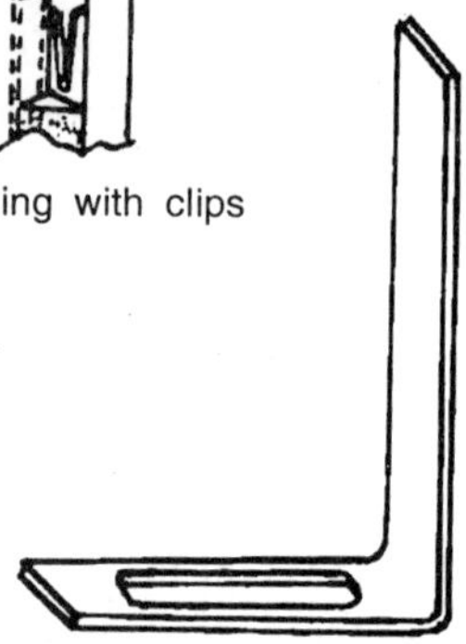

(d) Slotted log to be cast in concrete blocks for fixing steel doors and windows by countersank screw and nuts

Fig. 23.11 Steel ventilators and door with windows (Ref. 70).

Method 2. Metal windows can directly be fixed to the walls by first making 5 cm square and 5 to 10 cm deep holes in the jams of the masonry. Wooden plugs are then inserted in these holes and well grouted in the wall. The metal frames are screwed on to these plugs.

Steel doors and windows are usually fixed by slotted lugs as shown in Fig. 23.11.

23.11 SPECIAL TYPES OF WINDOWS

There are many types of special windows of which the bay, gable and dormer windows are important. These are shown in Fig. 23.10.

Bay windows. Bay windows need special mention. These windows project outside the external walls of a room and are extensively used in upper floors in housing projects as they provide increased floor area on top storeys. They can also be provided in ground floors for ordinary houses. They admit more light and air into the room. The most common type is the splayed windows as shown in Fig. 23.10(a). They can also be semi-square, semi-hexagonal, semi-octagonal, semi- circular, etc. in shape. Usually they start at window sill level with a cupboard underneath. A loft can also be built on the top. If the shutters are fixed and the window is meant only for lighting, the cost of the frames can be reduced by making the frames with reinforced concrete, with the shutters fixed to the frame by wooden plugs.

23.12 PLASTIC DOORS AND WINDOWS

A large number of door and window frames and shutters made of plastics and plastics reinforced with steel are nowadays available in the market. They are useful for doors for bathrooms and other special situations. Due to shortage of wood all over the world, such fixtures are nowadays becoming more and more popular. However, their service life may be limited.

23.13 COLLAPSIBLE STEEL GATES

Typical collapsible steel is shown in Fig. 23.7. They act as a steel curtain and are useful in many places such as workshops and public places. They are fabricated out of steel double channels (20 mm × 10 mm × 2 mm) joined together to form hollow sections or square hollow sections. They move along T sections at the top and bottom (see also Section 37.5).

23.14 ROLLING STEEL SHUTTERS

A typical rolling shutter is shown in Fig. 23.8. They are used instead of large doors when we have to enclose the room from outside as in garages, godowns and retail shops. It consists of a frame, a drum with shaft and spring arrangements and shutter made of thin steel plates (laths or slats) joined together. The diameter of the drum varies from 200 to 300 mm depending on the size of the doors. They are available with simple push-pull arrangement for small doors and with mechanical gear arrangement for large shutters for closing and opening them (see also Section 37.4).

23.15 SLIDING DOORS

For covering large size openings such as garages, sliding doors are very useful. They are usually hung by rollers running on steel sections fixed to a timber head as shown in Fig. 23.12. There are special arrangements for these doors to be folded up to the side when the opening is to be kept open.

23.16 COST REDUCTION IN DOOR AND WINDOW CONSTRUCTION

As the cost of doors and windows is high, any reduction in their cost will be a substantial reduction in total cost of construction. This is especially true in low-cost housing projects financed by agencies. At the same time, external doors and windows should be safe against intruders. Hence we must also be conservative in our design of external doors. However, much savings can be made in the design of *interval doors*. Some of them are as follows:

1. Many interior doors need not be there at all as an opening in the wall with arrangements for hanging a curtain is all that is needed. Hence if a door is not necessary, do not put it.

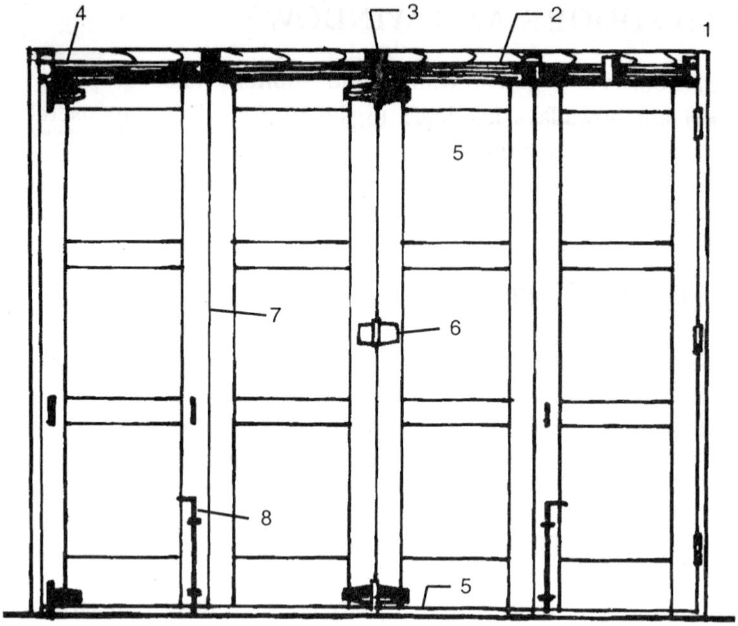

Fig. 23.12 Sliding (or folding) doors: 1. Fixed end, 2. Timber head, 3. Centre hanger, 4. End hanger, 5. Door guide or bottom rail, 6. Backflap hinges, 7. Butt hinges, 8. Drop bolts.

2. Any doorway traditionally consists of the door frame and the shutter. In many cases, the costly door frame for *internal doors* can be omitted or substituted by a thinner plank fixed to the side wall of opening at close centres, this plank being used to fix the shutter by hinges. Alternately, the door shutter itself can be fixed onto the wooden plugs cast in concrete and built in the masonry wall. The masonry opening for doors can be finished straight and square so that it will look aesthetic.
3. Traditionally in many temples, the doors have no frames and hinges. They are fixed at top and bottom in pin and socket arrangements. If we design a large door opening more than 1.5 m (5 ft) in width, one of the ways to support it is to fix the shutter off centre and balance it at the top and bottom with pins.
4. In low-cost housing projects, the wooden door frames can be substituted by reinforced concrete door frames, cast in steel moulds to give it a fine finish (see Section 23.17).
5. If wooden doors are to be selected, we should go in for cheaper treated timber products for the frames and shutters.

Similarly cost reduction in windows can be made as follows:

1. Provide jalis instead of window wherever possible. Brick jalis are easy to make and are attractive.
2. Use top and bottom pin and socket or other arrangements instead of costly hinges for the shutters.
3. When using aluminium or other metal windows, they can be directly fixed to the brick walls with wooden plugs without wooden frames.

23.17 REINFORCED CONCRETE DOORS AND WINDOW FRAMES

As an alternative to timber door and window frames, RCC frames can be used, especially in all temporary buildings. The chief objection for their extensive use lies in the difficulty in obtaining a suitable fixing arrangement for the shutters and lack of good finish. The increasing scarcity and cost of timber have revived the interest in RCC frames for low-cost and temporary housings as it provides a durable and economical alternative. RC frames up to 2.1 m in height may be made as described below. (These frames are made according to IS 6523 – 1983 and are nowadays available in the market.)

Concrete mix and size of reinforcement. The mix of concrete recommended is 1 : 2 : 4 with 10 mm graded aggregate. As observed from tests under simulated conditions of use, section of 34″ × 24″ reinforced with at least 3 numbers of 6 mm diameter bars (2 nos. at the face) is recommended for adoption as reinforcement for all openings up to 2.1 m.

Mould. The mould (or formwork) may be of either timber or steel. When timber is used, it is advisable to line it with GI sheets for longer life and better finish of the product. Lining also helps in reducing undue distortion of the mould due to warping and shrinkage. For large scale production, steel moulds are preferred. If the frames are assembled from members, the moulds can also be used for producing frames of various sizes. Provision is made in the mould to accommodate the fixing devices and holdfast. If required, a rebate may also be provided to act as a plaster groove.

Fixing hinges for shutters. Various methods of fixing the hinges to RCC frames are in vogue. Two cheap methods recommended by the Building Research Station Roorkee are as follows:

(i) *Aluminium pipe fixture.* This consists of a 5 mm 3/16″ aluminium tube. The rear end is pressed flat and bent. The threading at the other end can be readily done at site by holding the tube in a vice. The fittings are inserted in the mould from the inner side. Flattened ends of adjacent fixtures are kept facing different directions for better bond. Screws are inserted in the tubes from outside of the mould to keep the fitting in position during pouring of concrete as also to prevent concrete from getting into the holes of the fittings.

(ii) *Wire fixture.* The second alternative method for fixing shutters consists of embedding a helically wound wire of 22 SWG wound on a steel wood screw. The screw is held in a vice and the wire wound around it, till it covers the screw and trails out by about 4″. The wire and screw are embedded in the concrete frame and the screw withdrawn before the concrete sets. Pull out tests show that the pipe fixture is superior to the wire fixture in its performance.

Casting. The frame may be fully precast single element or assembled at the site out of more than one precast members. The former has the advantage of reducing work but introduce difficulties in transport and handling. Since damage to even one part of the frame can result in rejection of the whole frame, the latter method is relatively more suitable and economic. When cast in separate parts (the vertical and the horizontal), one of the bars of the vertical member of the frame is kept projecting so as to tenon into corresponding holes in the horizontal members. The holes in the horizontal members should be somewhat larger to facilitate easy insertion of the projecting bars. It is advisable to use a vibrating table for compacting the members or alternatively a shutter vibrator may be used. Hand compaction can provide the required strength but the finish is likely to suffer.

Removal of members and curing. The cast products should be removed from the mould after 24 hours. They should be allowed to cure in a water tank for a fortnight. Finally, they should be removed to a covered place and preferably stored for a month before use.

Finish. A good finish can be obtained by employing smooth surfaced mould and by vibration. The defects, if any, can be readily made good by rubbing the frame with carborandum stone before erection. As far as possible, plastering or touching up should not be permitted.

Various decorative treatments such as painting and terrazzo finish can be given to members. When terrazzo finish is to be given, the mix used should be of 1 cement—2 chips laid on the exposed faces of the members while casting the frame. Some care is necessary to ensure that the inner edges do not suffer damage. If it is painted, the paints used should normally be cement paint of an approved brand. When an enamel paint finish is desired, it should be done in three stages as follows:

(a) Priming coat of 2 coats of alkali fast primer should be applied first.

(b) Undercoat should consist of synthetic resin paint conforming to IS 520 – 1954.

(c) Finishing coat should consist of 2 coats of synthetic enamel paint conforming to IS 520 – 1954 (colour as per choice).

Erection. The RCC frames may be assembled and erected as in case of timber frames. They, however, require some care in handling, as they are heavier and the joints are liable to give way if not handled carefully. Alternatively if they are made of separate pieces, the vertical members may be held in position and the top member placed over the vertical members. The whole frame is then plumbed and supported temporarily till the hold fasts are embedded in jambs. Cement sand slurry of mix 1 : 3 should be used in grouting the joints between the vertical and top members. In cases where four members are used such as in window or doors having sill, the bottom members are first placed in position and the other pieces erected as described.

23.18 FLYPROOF DOORS AND WINDOWS

It is quite common nowadays to provide flyproof and mosquitoproof doors and windows to residential buildings. In doors, separate additional shutters with flyproof netting (metal or nylon) have to be provided. For windows, we have the option of either providing fixed netting or additional shutters with flyproof netting. The latter is more costly but it allows greater freedom to close and open the windows. Nylon mosquitoproof nets with Velcro fixing devices that can be fixed inside and easily opened or closing and opening the shutters are also available in the market. In any case, prior planning is needed for the type of arrangements to be used.

SUMMARY

Cost of doors and windows forms a major part of the total cost of the building. Hence this work needs careful planning. There are many types of doors and windows to choose from. Nowadays as cost of good wood is extremely high and it is also difficult to get good timber, alternative materials such as aluminium, steel reinforced plastics and steel are being thought of alternative materials for these fixtures. Use of RC door and window frames, at least in temporary buildings, will considerably reduce the cost of construction.

REVIEW QUESTIONS

1. (a) How are doors and windows designated by Indian Standards?
(b) Discuss how you will select the sizes of doors and windows you will adopt for a middle class residential house. Specify the door sizes you will adopt.

2. Name the important components of a panelled door and illustrate your answer with the help of a sketch. Why would you recommend the bottom sill for external doors, especially in villages?

3. Name the important components of a casement window and illustrate your answer with the help of a sketch.

4. In which places would you install ventilators in a residential building? Illustrate the typical ventilators you would install in each of these places in the building.

5. Explain the procedures for fixing wooden door frames and window frames to (a) masonry walls and (b) concrete columns. How do you fix aluminium windows in position?

6. Write short notes on:
(a) Flush doors
(b) Aluminium windows
(c) Battened doors
(d) Ventilators
(e) Mosquitoproofing of residences

7. Explain how you can reduce the cost of doors and windows in a residential building.

8. Explain briefly the following types of windows:
(a) Bay windows
(b) Dormer windows
(c) Gable windows

9. Write short notes on:
(a) Threshold plates
(b) Doors for bathrooms
(c) Steel doors and windows
(d) Rolling shutters

Chapter 24

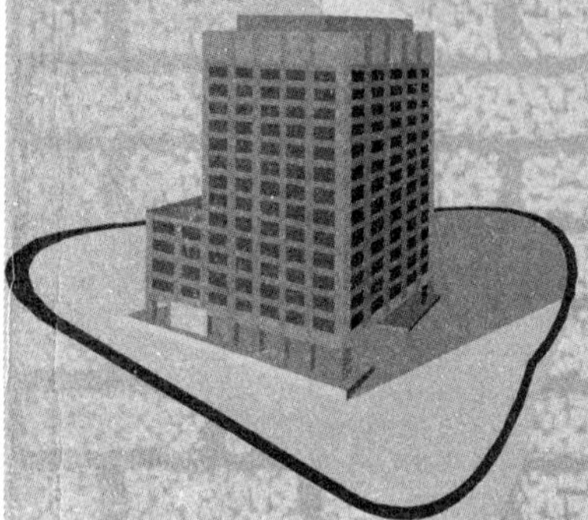

Timber Joints and Glazing

24.1 INTRODUCTION

Timber joining and glazing are important works in fabrication of doors and windows. Joining of timber is also important in furniture making. In this chapter, we will briefly deal with the following:

(a) Timber joints
(b) Timber fasteners
(c) Glazing of doors and windows

24.2 TIMBER JOINTS

24.2.1 Types of Timber Joints

Timber joints can be classified into the following:

1. Lengthening joints
2. Widening joints
3. Bearing joints
4. Framing joints
5. Angle or corner joints
6. Oblique shouldered joints

There are many forms of each of the above-mentioned joints, but we will examine only the important ones.

Lengthening joints. These are used to lengthen members. The important types of lengthening joints are given below.

1. Lapped joints [Fig. 24.1(a)]

2. Fished joints (with fish plates on each side) [Fig. 24.1(b)]
3. Scarfed or spliced joints [Fig. 24.1(c)]
4. Tabled joints [Fig. 24.1(d)]. (These are similar to scarf joints. They join surfaces that are horizontal and vertical to take both tension and compression.)

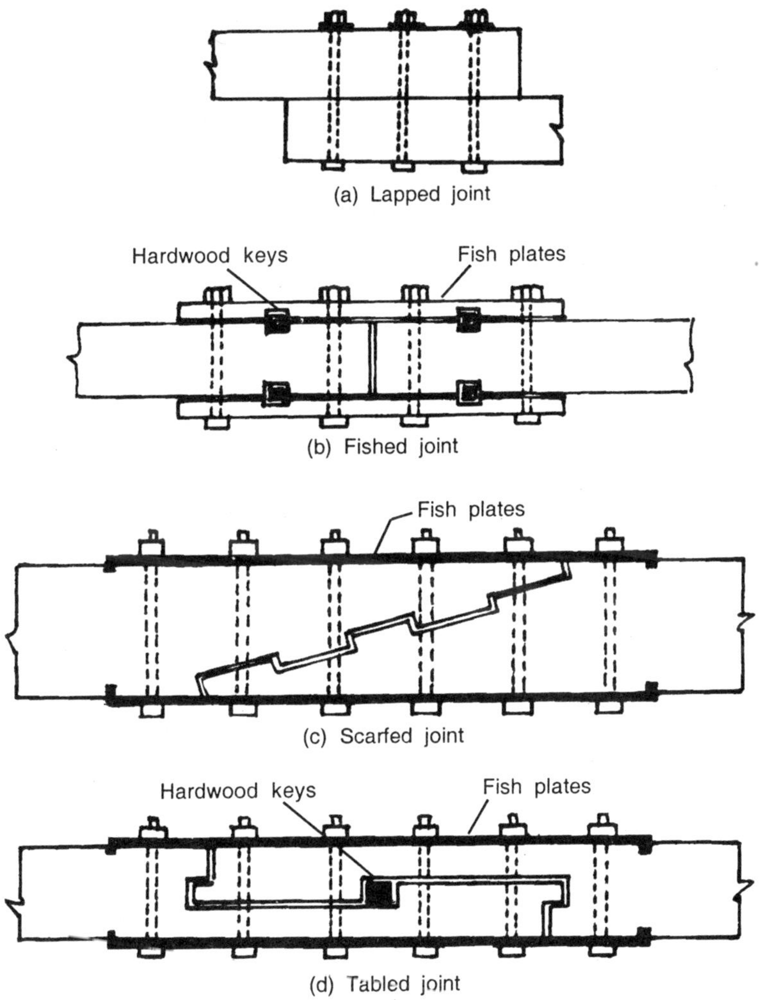

Fig. 24.1 Lengthening joints in timber.

Widening joints. There are many joints used to make a wide piece of wood out of small pieces. The commonly used ones are as follows:

1. Butt joint [Fig. 24.2(a)]
2. Rebated joint [Fig. 24.2(b)]
3. Tongued and grooved joint [Fig. 24.2(c)]
4. Dowelled joint [Fig. 24.2(d)]

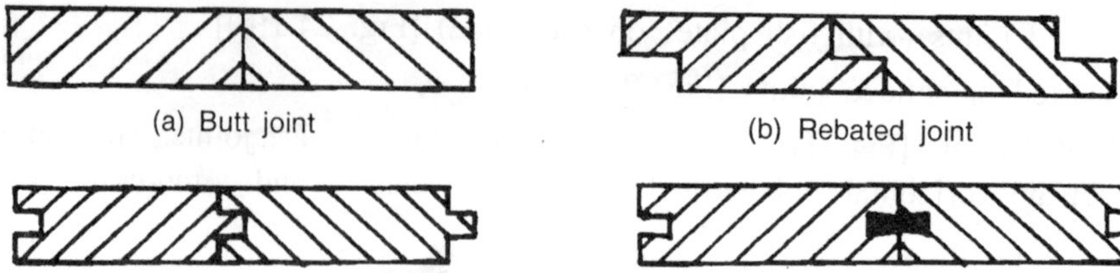

(a) Butt joint

(b) Rebated joint

(c) Tongued and grooved joint

(d) Dowelled joint with dowel (similar to dovetail joint)

Fig. 24.2 Widening timber joints.

Bearing joints. These joints are used to join two members that meet at right angles. Commonly used joints under this group are as follows:

1. Halved joint [Fig. 24.3(a)]
2. Notched joint [Fig. 24.3(b)]
3. Cogged joints [Fig. 24.3(c)]
4. Dovetailed joints [Fig. 24.3(d)]
5. Mortice and tenon joint [Fig. 24.3(e)]
6. Stub-tenon joint [Fig. 24.3(f)]

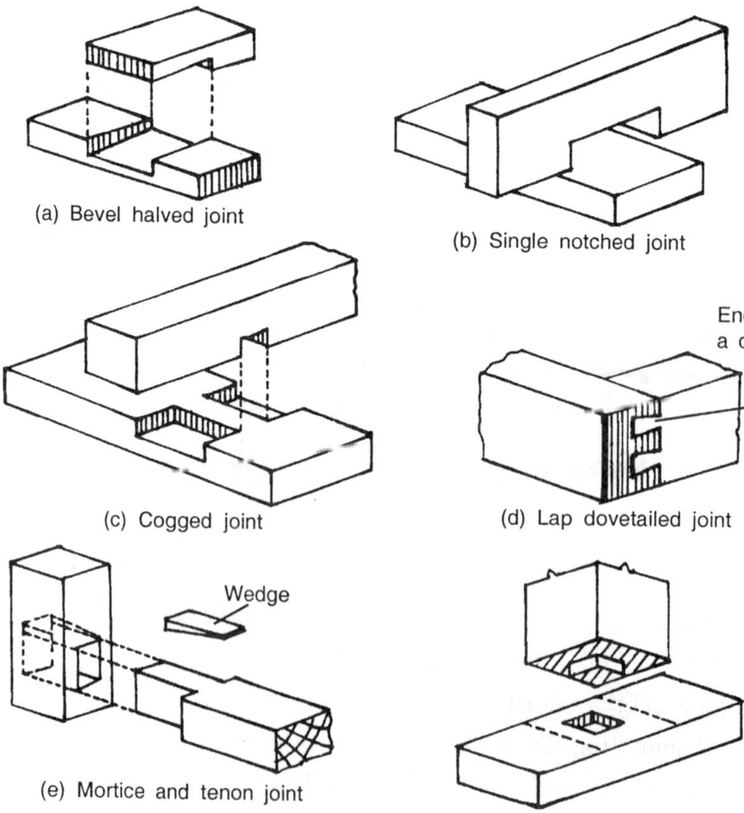

(a) Bevel halved joint

(b) Single notched joint

(c) Cogged joint

(d) Lap dovetailed joint

(e) Mortice and tenon joint

(f) Stub-tenon joint

Fig. 24.3 Bearing joints.

Joints in door and window frames. The most common types of joints used for door and window frames are as follows (Fig 24.4):

1. Butt joints
2. Tongued and grooved joint

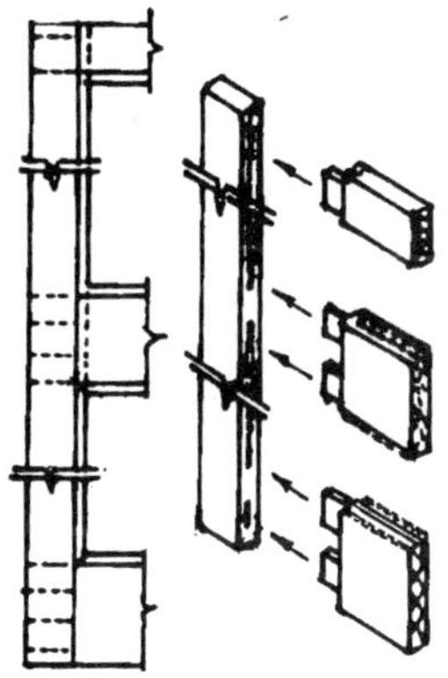

Fig. 24.4 Joints in a panelled door frame.

Joints in wooden trusses. As failure of these joints will lead to serious consequences, the important joints of timber roof trusses are usually strengthened by steel straps with through bolts as shown in Fig. 24.5.

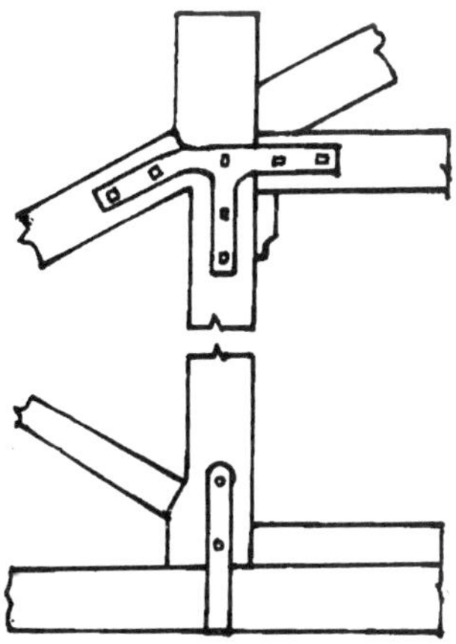

Fig. 24.5 Joints in a queen-post truss with metal straps.

24.2.2 Details of Joining

When carrying out the joining work, we can use glue and also steel nails or wood pins. For example, in a mortise tenon work, the contact surfaces are treated with glue (polyvinyl dispersions such as *Fevicol*) before we put the pieces together. We also use pins (10 mm diameter of hardwood or bamboo pins) after gluing the frames together. The members are pressed into position by means of a press.

24.3 TIMBER FASTENERS

Timber pieces can also be joined by timber fasteners instead of the joints described above. In addition to the traditional devices such as nails, screws and bolts, special connectors are also nowadays used to connect solid or laminated timber. The selection of fasteners depends on the forces in the member. The fasteners are galvanized when exposed to atmospheric weather, but for connections used in chemical industries, non-corrodible dowel joints with specially protected fastening devices should be used to withstand the corrosive atmosphere. When technical facilities exist, glued joints may also be used in such cases. Description of the *traditional type of joints* is given further.

24.3.1 Nailed Joints

Nail is the most commonly used fastener for timber. It is used to prevent separation of one member from another as in wooden boxes (where resistance to withdrawal is most important requirement) or to prevent movement of one member past another as in a joint in a roof truss (where the nail is principally in bending but some resistance to withdrawal is also required).

The diameter of a nail is to be determined by the load it has to carry. Large diameter nails are not to be used in thin members for danger of splitting and too small a diameter may cause the nail to bend while being driven. Usually the diameter of a nail is one-sixth to one-eleventh the least thickness of members to be joined.

Simple butt or lap nail joints are found to be sufficient for most purposes. The nails generally used are of plain head diamond pointed or blunt tapered made from mild steel wire.

Recommendations for spacing of nails. These can be taken as follows:

End distance : 10 d minimum along the grain
Intermediate distance : 5 d minimum along the grain
Edge distance : 5 d perpendicular to grain
Row spacing : 5 d across the grain
where d is the shank diameter of nails.

Recommendations regarding penetration of nails. In lap joints, where nails are in single shear and two members of different thicknesses are joined, the nails should always be driven from the side of the thinner member into the thicker member. The minimum penetration to thicker member should be 1½ times of that in the thinner member, i.e. total penetration should be 2½ times of the thickness of the thinner member.

In case of butt joint, where only one cover (gusset) plate of lesser thickness is used and the penetration into the parts jointed is partial, and nail is in single shear, it is recommended that the length of penetration should be at least equal to the thickness of the thinner (outer) member. In case of butt joints with double cover plates, where the nails are to be in double shear, the nail should penetrate all the members completely.

Recommendations regarding gusset plates. The following are the rules:

(a) In *lap joints,* the area available for nails can be increased if they are connected through gusset plates. These plates when used should be at least two-thirds the thickness of the thicker member.

(b) In *butt joints,* the gusset plates should be three-fourths the thickness of the central member.

Recommendations for driving nails. These are as follows:

(a) Adjacent nails in a joint should be driven from opposite faces.

(b) Nails should be so arranged that there should be more vertical rows than horizontal.

(c) Minimum number of nails in any row should be two.

(d) Preboring to 4/5th of diameter of the nail in hardwood and 3/5th of diameter of the nail in softwood is allowed.

24.3.2 Wood Screw Joints

In many situations, screws are used to fix fasteners to wood. There are many types of screws including self-tapping screws. Wood screws are generally used in light works such as cabinet work and are seldom used in structural work. Such joints can be easily dismantled and reassembled. The length of the screw should be so chosen that the penetration into the second piece of timber is at least seven times the screw diameter. In some hardwoods, it may be necessary to initially prebore a hole of smaller diameter to fix the screws in.

24.3.3 Coach Screw Joints

Coach screws (called lag screws in USA) are special types of wood screws up to 1 inch in diameter and 12 inches in length with hexagonal bolt-type head. They have threads for part of its length and ends in a point. Their normal use is in structural work and they are galvanized for use in places exposed to weather.

24.3.4 Bolted Joints

Bolted joints are extensively used in structural work. They are used in situations which require holding parts firmly together or for ease in dismantling later. The sizes of bolts in structural work will depend on the loading of the joints. When bolts are used in wood not seasoned, they should be tightened after some months of installation to take up the size due to shrinkage of the timber. The following two types of bolts are in general use:

(a) Machine or engine bolts with hexagonal or square heads and cut threads.

(b) Coach or carriage bolts with mushroom-shaped heads (under which there is a short square section as shank) with cut or rolled threads.

The square section of the coach bolt fits into the prebored hole and prevents the bolt from turning during the tightening. Square or round washers should always be used under the nut and also under the head of an engine bolt. The following recommendations for spacing of bolts are given in British Code of Practice 112:

1. The centre to centre spacing of bolts in the direction of load should not be less than four times the bolt diameter.
2. For loading perpendicular to the grain, the spacing at right angles should be 2½ times the bolt diameter where the ratio of thickness of timber (t) to bolt diameter (d) is unity and the spacing should be 5 times the diameter when the ratio is 3 or more. For values between 1 and 3, it is to be linearly interpolated.
3. For loading parallel to the grains, the distance between the edge of the timber and centre of the bolt is to be not less than 1½ times the bolt diameter.
4. The edge distance should not be less than 7 times the bolt diameter in tension and 4 times the bolt diameter when the loads are perpendicular to the grains.

24.4 GLAZING

Glazing is an important indoor and window work. Manufacture and properties of glass have been dealt with in the book on *Building Materials*. The work of fixing glass, plastic or similar material on a framework or opening to admit light is called *glazing*. Glass is fixed to a door or window shutter in a *rebate* or depression made inside the panels especially for fixing the glass.

The clearance between the bottom edge of the glass and the rebate is called *edge clearance* and the clearance between the inside face of the glass and the rebate face is called *back clearance*. Any frame sash or casement into which glass is glazed is called the *surround*. *Sprigs* are the small headless nail or triangular strips used for securing glass in wood surrounds while the putty hardens. Clips are generally used to secure glass in steel frames until the front putty hardens [see Fig. 23.11(d)].

Unless otherwise specified, all glass used in first class buildings should be patent flattened sheet glass of the following specifications.

(i) Up to 600 mm × 600 mm in size: 2.5 mm thick at 6.3 kg/m^2
(ii) 600 mm × 600 mm to 750 mm × 750 mm in size: 3 mm thick at 7.5 kg/m^2
(iii) Above 750 mm × 750 mm in size: 4 mm thick at 10.0 kg/m^2

Fixing of glass to frames depends on the type of frame (whether wood or aluminium) and the appearance required. The commonly used methods are as follows:

(i) Glazing with traditional putty for wood and metal frames
(ii) Glazing with beads for wood and for metal frames with special beads
(iii) Glazing with sealants
(iv) Glazing into grooves with gaskets
(v) Other patent systems

Glazing in which either one side or both sides of the glass are exposed to the outside of the building is known as *external glazing*. Glazing from inside is called *internal glazing*. It should be remembered that the side of the glazing exposed to the outside of the building should be weather resistant and waterproof.

One should be careful to choose the fixing system suitable to the probable degree of movement of the frame and the glass due to various factors.

In this chapter, only the common type of glazing used in ordinary buildings will be dealt with. For more details for fixing large openings, etc., specialized literature should be consulted.

24.4.1 Fixing Devices for Glazing

Glazing putty. The traditional linseed oil putty is used for traditional internal glazing in softwood or absorbent hardwood. It should not be used on non-absorbent hardwoods such as teak.

Metal casement putty differs from linseed oil putty and adheres to non-porous surfaces after setting. It is suitable for steel frames, sealed timber, sealed concrete surfaces but not suitable for aluminium, stainless steel, plastics, etc. (Glass is fixed in aluminium frames with snap on glazing clips and gaskets as described later in this chapter.)

Linseed oil putty can be made by mixing by hand one part of white lead with three parts of finely powdered whiting and adding raw linseed oil to make a stiff paste. Varnish is then added to this paste at the rate of 1 litre for every 18 kg of mixture and again mixed to make the final putty. (Boiled linseed oil will make the putty friable.) When mixed, it should not adhere to the hands. After being well kneaded, it should be left for a week to soften and worked up in small pieces till it becomes smooth and then used while it is quite fresh. It can be used for a period not less than 6 months from the date of manufacturing. If the putty becomes dry, it should be restored by heating and working it again while it is hot. If the putty is to be exposed, it can be coloured to suit the colour of the frame and given a coat of similar coloured paint to protect it from the environment.

Glazing beads. Glazing beads are strips of wood metal or other suitable material used to retain the glass in the frames *as an alternative to putty*. It is also used where conditions are more severe than can be withstood by putty. The beads are fixed to the frame by nails, screws, clips, etc., so that they are sufficiently rigid to prevent movement of the fixing under wind or other loads. For hardwoods such as teak, it is better to fix glass with glazing beads. Bead glazing gives a good appearance and should always be used for front doors.

Sealants. These are comparable to flexible glazing compounds. These are applied generally by a gun and used for capping and bedding of glass to the frames.

Preformed compression gaskets. Gasket glazing is usually adopted for any type of frames. In India, they are principally used with steel and aluminium frames. The glass is fitted into the grooves with these gaskets. They have several advantages over conventional glazing materials as they are easy to install and reduce wastage. There are different types of gasket that suit the frame. They may be made of rubber, plastics, etc. The principal types are the H-type, grooved type and press lock single sided type, which do not require a filler strip. The H type mounts onto the flange or nib in the frame. A lock strip is inserted along one side of the gasket, which affects the seal. The grooved type is also called the 'tongue and groove' gasket. Data about these can be obtained from the manufacturers.

24.4.2 Storing Glass at Site and Maintenance of Glass in Buildings

Glass should be stored in racks in near vertical position secured against wind. Stacked glass cases should not be exposed to direct heat or sun as their expansion can cause breakage. Paint, plaster, etc. that may fall on glass should be removed when wet as they harden when dry and cause damage to glass on removal. Alkali water from cement finishes or concrete from above should not be allowed to drip onto glass as the alkali will chemically affect the glass.

For maintenance, glass should be cleaned regularly first with warm water and soap or mild detergent and then with clean water. Washleather or cloth is good enough for plain glass but for a textured surface, a stiff plastic or bristle brush may have to be used to remove the dirt. Methylated spirit may be used to remove very bad stains; otherwise use pumice powder and rub it down (see also Section 40.4).

24.4.3 Preparation of Frame for Glazing

Before glazing, all the rebates and grooves must be cleaned as well as dried. All beads and rebates in woodwork made of softwood should be given a coat of wood primer. In case of hardwood, which is non-absorbent, a 'metal casement putty' should be used.

After the rebates are sealed with primer, at least one undercoat of paint should be applied before the putty is placed. In case of metal surrounds, it should be cleaned of all dirt, grease, etc. and a specified primer should be applied to the surround and painted before glazing is commenced. Stone/brick/concrete surrounds, rebates or grooves must be sealed with at least two coats of alkali-resistant sealer before application of glazing compounds.

24.4.4 Cutting of Glass

Glass panes should be cut to fit the rebates of the sashes leaving an edge clearance of 3 mm all around. For panes exceeding 0.2 m^2 in area, the glass should be set on small blocks of resilient material spaced not more than 75 mm from the corners to locate the pane properly within the surround.

If putty is used in timber or metal frames for fixing, the pane should be placed on a bed of putty so that the glass does not touch the surround at any point. We will now examine in more detail the three common methods used for glazing (see Fig. 24.6):

(a) Glazing by putty

(b) Glazing by beads (dry glazing) and with beads and putty

(c) Glazing by clips and putty

They are briefly described as follows.

24.4.5 Glazing with Putty

The first and conventional method is by putty. A rebate of 10 mm should be available for fixing glass as shown in Fig. 24.6. The bedding putty is applied uniformly over the rebate and the glass pressed into position and secured with glaziers sprigs spaced at 450 mm around the frame. (Thin flat slightly headed nails called *brads* can also be used for securing the glass.)

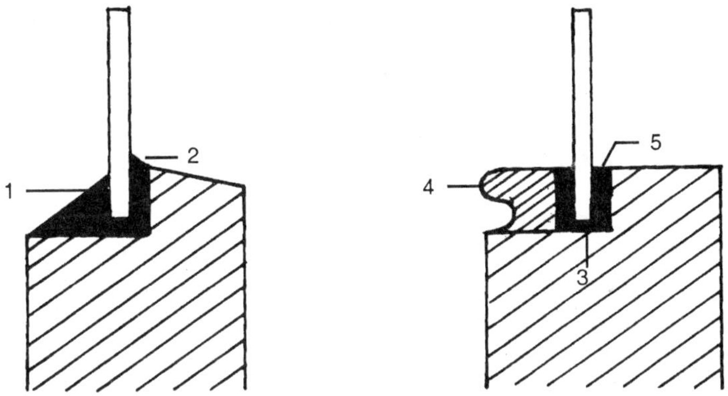

(a) Simple glazing with putty (b) Glazing with putty and beads

Fig. 24.6 Glazing: 1. Front putty, 2. Back putty, 3. Selting block and putty bed, 4. Glazing bead, 5. 1.5 mm thick putty.

On pressing in the glass, a certain amount of bedding putty is pressed out and the remaining back putty should be at least 2 mm thick between the glass and the rebate. The glass is then puttied with 'front putty' to about 2 to 3 mm the sight line of the rebate so that when paint is applied, it extends to the sight line and also seals the edge of the putty to the glass. The front putty is cut to a straight line and the back putty stripped at an angle at the sill level as shown in Fig. 24.6. The quantity of putty used should not be less than 185 gm/metre of glass perimeter.

Linseed oil putty should be painted within 2 to 3 weeks after fixing of glazing to avoid its cracking. Otherwise special formulations should be used as putty.

24.4.6 Glazing with Beads (Dry Glazing) and Glazing with Beads and Putty

The second method of glazing is by beads. Moulded wooden fillets are called *beads*. Large glasspanes or plate glasses are fixed in the rebate of wooden frames by beads fixed all-round the rebate with rust-proof brass countersunk screws, nails or springs. A strip of felt or rubber is first inserted in the rebate under the glass to act as a cushion. The bead should be fixed by nails or screws and should not be more than 75 mm from the corners and at spacings not more than 200 mm apart. The glazing operation is to be as follows.

Where beads are proposed, drill holes for screws should be made in advance in the frames. Rebates and beads are preferably first sealed with a proprietary sealing compound applied by brush. The glass is then bedded in glazing compound and set in position using setting blocks and distance pieces to restrain its movement. The bead is then bedded with glaziers putty and screwed or pinned in position. For internal bead glazing, the bedding of the bead can be omitted. Strip of asbestos tape or felt can also be used to cushion between the glass edge and the bead.

For internal glazing (not exposed to outside), dry bead glazing can be specified. The materials used can be self-adhesive glazing tapes or self-adhesive black velvet or washleather, etc. passed round the edge of the glass and trimmed off flush on both sides according to the instruction of the manufacturers of these glazing tapes. This is then fixed with beads (or putty as specified).

24.4.7 Glazing with Clips and Putty for Large Sized Works

Glazing clips are special clips, one end of which is secured on the frame and the other to the glass, thus holding glass to the rebate as shown in Fig. 23.11(d). They are used to fix glass to the frame when the glass sizes are more than 300 mm × 600 mm.

Glazing sprigs or clips need not be provided for glass sizes up to 300 mm × 600 mm in glazing with putty. Four clips are provided per glass pane for a size larger than the above. For sizes exceeding 800 mm × 2000 mm, not less than six glazing clips should be used. Their spacings should not exceed 300 mm. Glazing clips may also have to be provided for lesser sized glass in very heavily-exposed situations. In all these cases, holes for these clips should be preformed during the fabrication itself. Front putty is applied after fixing these clips. Thus in this method, we use putty and also clips.

24.4.8 Glazing of Aluminium Doors and Windows

Glazing of aluminium doors and windows is generally carried out by special sealing compounds.

24.5 USE OF PLASTICS FOR WINDOWS

In some cases, as in windows of showcases, it may be advisable to use unbreakable transparent materials other than glass. Polycarbide is a modern material which is transparent and very strong. Such materials are being used more and more in places such as roofs of railway platforms, shop windows, protective cover for valuable paintings, etc.

SUMMARY

Timber joining and glazing form an important part of construction of doors and windows. The beauty of these fixtures is very much affected by the details with which these items are made. They should be carefully planned by the building engineer.

REVIEW QUESTIONS

1. (a) Enumerate the various types of joints in woodwork.
   (b) Sketch the different types of joints used in the frame and shutter of a panelled door.
2. Sketch the joint you will use for the following:
   (a) Tie member of a queen post truss
   (b) Corner of a door frame

(c) For making a measuring box for aggregates in making concrete
(d) The junction between the king post and tie of a king post truss

3. What are the principles to be followed in nail jointing two wooden members?

4. (a) What are the methods used for glazing of doors and windows?
(b) Explain glazing with beads used in a teakwood glazed door shutter.

5. (a) Explain glazing with putty of doors and windows.
(b) What transparent material would you use instead of glass for windows to make them safe against breakage by vandals?

Chapter 25

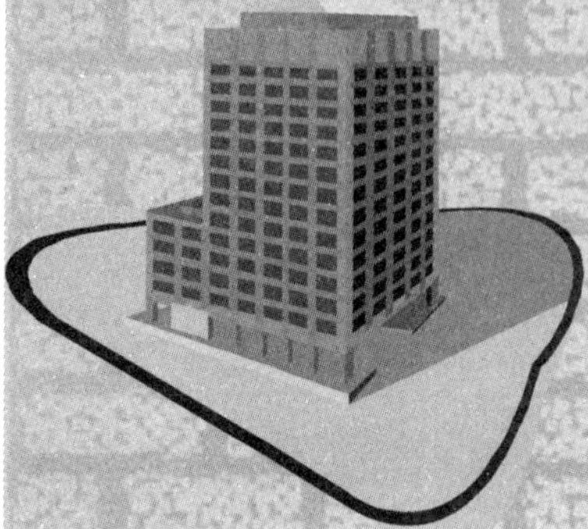

Stairs and Lifts

25.1 INTRODUCTION

Stairs, electric lifts, ramps and escalators are the commonly used means of vertical transportation in buildings. For buildings up to three floors, only stairs are usually provided but for buildings more than three floors, both electric lifts and stairs are to be provided. There are many types of stairs, some of them such as helical and circular stairs are very ornamental in nature. In this chapter, we will examine only the simple types of stairs to understand their important features. We will also deal briefly with lifts (which are commonly used nowadays in offices and flats) as well as ramps and escalators built in places where the traffic is heavy.

25.2 TERMINOLOGY USED IN STAIRS

The common terms used in simple stairs are as follows (see also Fig. 25.1):

Going. It is the horizontal distance between faces of two successive risers.

Tread. It is the horizontal portion of the steps on which we put our steps to climb the staircase.

Rise. It is the vertical distance between two successive treads.

Riser. It is the vertical portion of a step that supports the tread.

Handrail. It is the member placed on top of baluster to hold our hands while climbing the stairs.

Baluster. It is the member supporting the handrail.

Balustrade. It is the system consisting of balusters and the handrail.

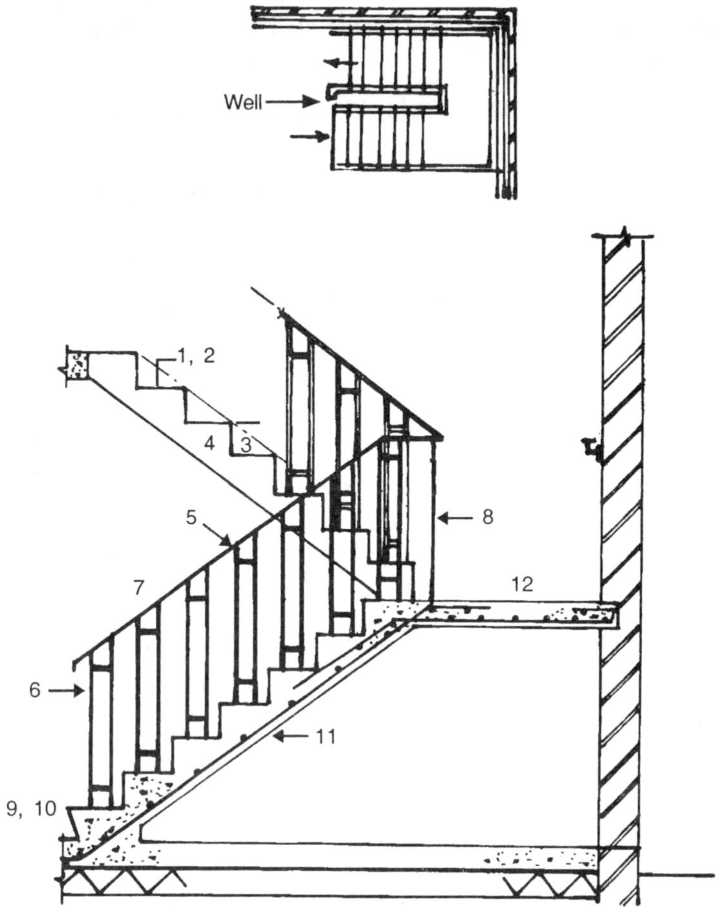

Fig. 25.1 Cast-in-situ reinforced concrete doglegged half-turn stairs: 1. Tread, 2. Going, 3. Riser, 4. Rise, 5. Handrail, 6. Baluster, 7. Balustrade (baluster and handrail), 8. Newel, 9. Nosing, 10. Scotia (moulding under nosing), 11. Soffit, 12. Landing.

Newel. It is the post usually provided at the beginning and end of the flights supporting the handrail. It gives stability to the handrail and should be properly anchored.

Nosing. It is the projection of the tread beyond the face of the riser to provide as wide a space for the tread as practicable. It is usually rounded off beyond the face of the riser to avoid a sharp edge. It is also customary to provide nosing by sloping the riser.

Scotia. It is the moulding provided under the nosing to improve the elevation of steps and, in some cases, to strengthen the nosing.

Soffit. It is the underside of a staircase.

Flight. A flight of steps is the uninterrupted series of steps between landings.

Headroom. It is the minimum clear vertical distance between the tread and overhead ceiling or floor.

Pitch or slope. It is the angle of rise of the stair to the horizontal. It can be defined by the line joining the nosings.

Stairwell or well. It is the space provided between the flights of a half turn or quarter turn staircase.

Strings or stringers. These are the sloping members provided in wooden staircases to support the steps in the stairs.

Winders. These are the tapered treads provided at the turnings of the landing space to reduce the number of steps required in the other straight portions of the stairs and thus economize the length required for the staircase [Fig. 25.3(b)].

25.3 TYPES OF STAIRCASES

Stairs can have parallel treads in rectangular layout or curved treads in curved layout (circular, helical, etc.) as shown in Figures 25.1 to 25.4. In many buildings, the staircase is built as an architectural feature. The following are some of the arrangements commonly used and are shown in Figures 25.2 to 25.5.

Straight flight stairs. This is a straight run with or without landing in between as shown in Fig. 25.2.

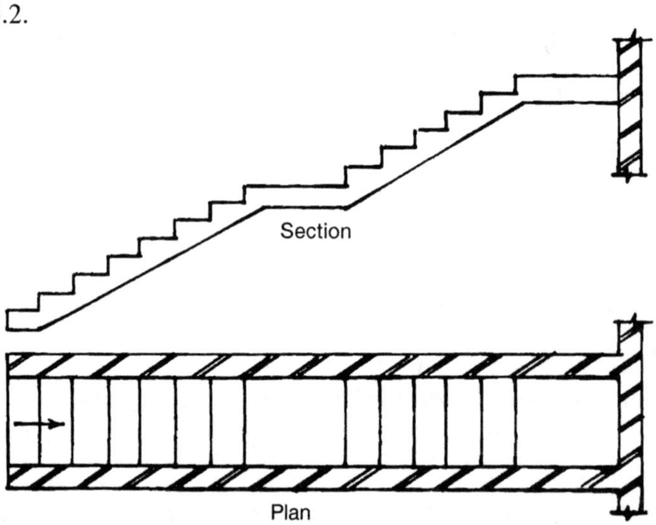

Fig. 25.2 Straight stairs.

Quarter-turn stairs. In this arrangement, the turning at the landing is only 90 degrees. The landing is known as *quarter-space landing*. Quarter-turn stairs can be of three types (Fig. 25.3)

(a) Open well quarter-turn stairs: These have two turns with three flights and go around a well.

(b) Newel quarter-turn stairs with winders: These have only one turn and two flights.

(c) Geometrical quarter-turn stairs

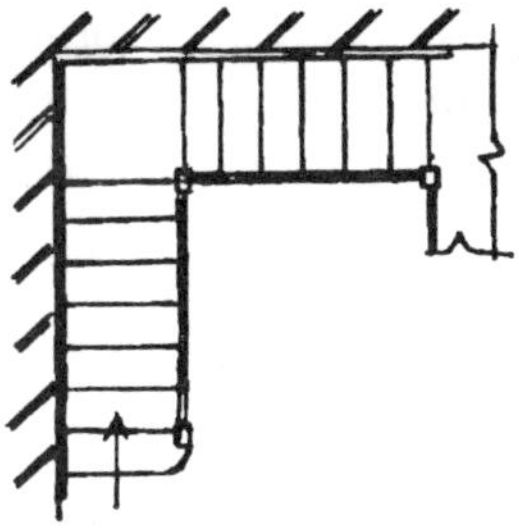

(a) Newel quarter-turn stairs with quarter space landing (with prominent nowel posts)

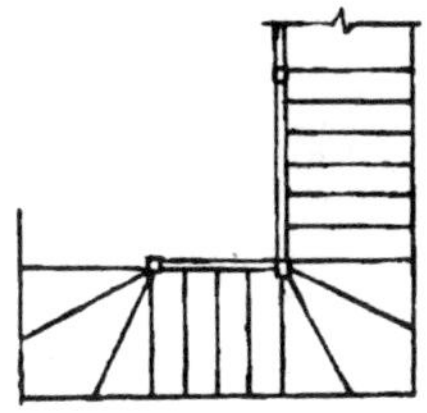

(b) Newel quarter-turn stairs with winders

(i)

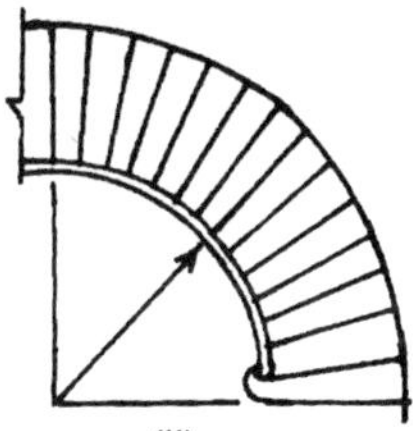

(ii)

(c) Geometric stairs with landing and continuous type

Fig. 25.3 Various layouts of quarter-turn stairs.

Half-turn stairs. In this system, the direction of climbing at the landing is reversed through 180°. This type of landing is called *half-space landing*. There can be three types of such stairs (Fig. 25.4):

(a) Doglegged (or newel) half-turn stairs with no well
(b) Open well half-turn stairs with a well
(c) Geometrical half-turn stairs

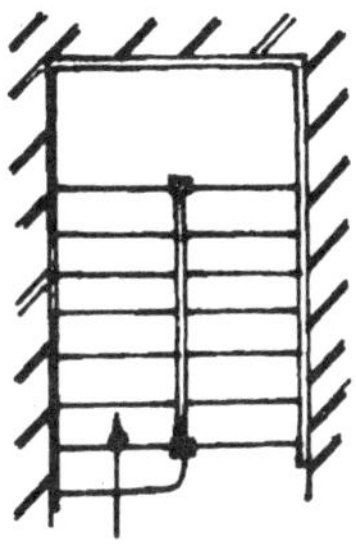

(a) Doglegged half-turn stairs (with half space landing)

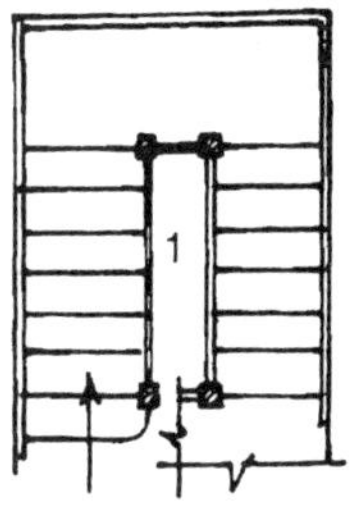

(b) Open well half-turn stairs

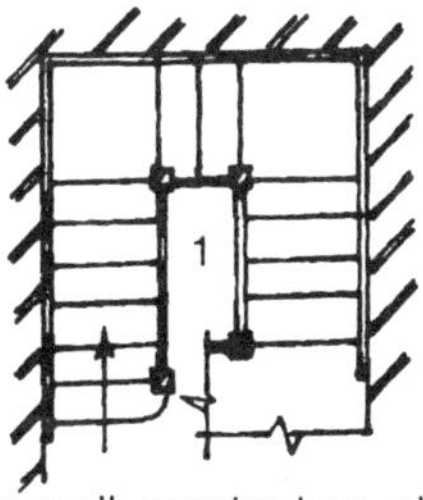

(c) Open well quarter-turn stairs

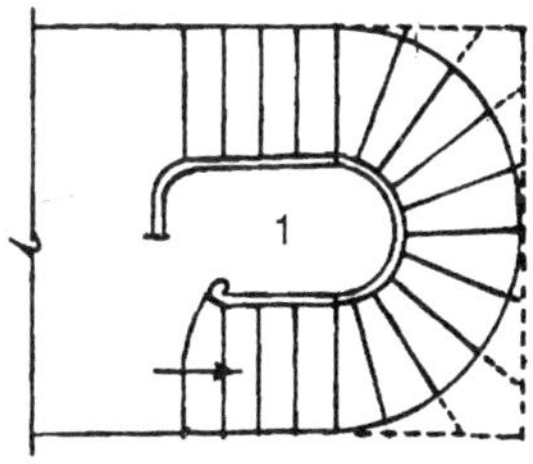

(d) Geometrical half-turn stairs

Fig. 25.4 Other types of stairs: Stairs with wells: 1. Well.

Note: Half-turn stairs can be built in many ways. It can be in the form of concrete slabs or they can be slabs supported on a central beam or individual steps may cantilever from walls of the staircase room. Thus we can have a very large number of types of staircases to choose from.

Other types. In addition to the above common types, there are many other special types of staircases. Some of them are given below:

(a) Circular staircase as shown in Fig. 25.3c(ii).

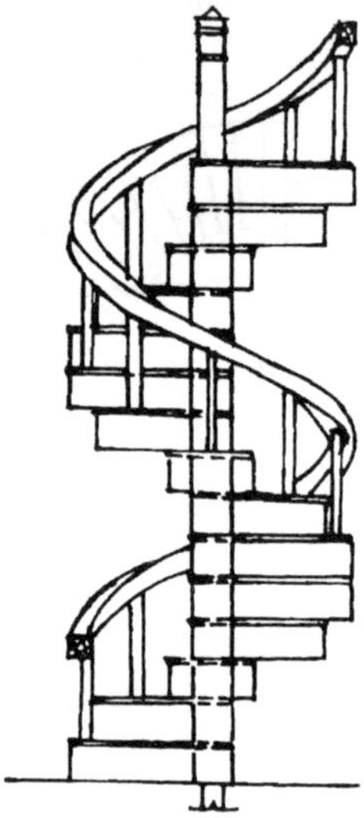

(a) Spiral stairs

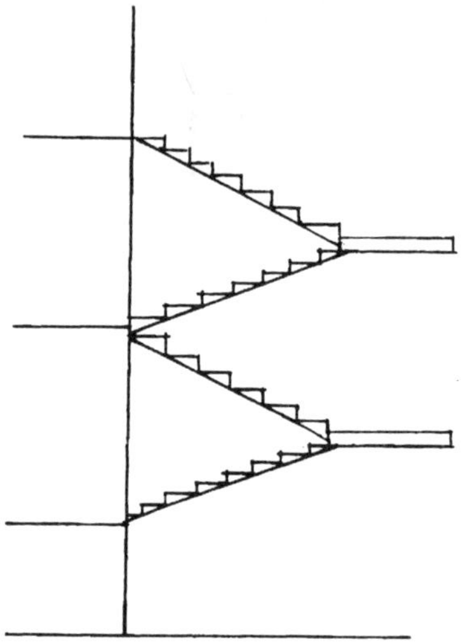

(b) Cantilever or Jackknife stairs

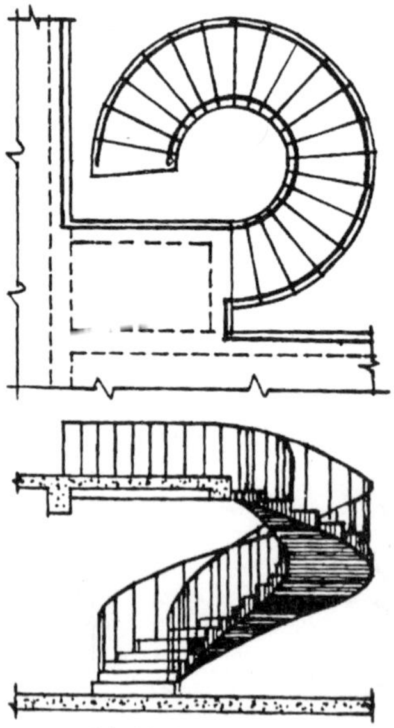

(c) Helical stairs

Fig. 25.5 Special staircases.

(b) Spiral staircase *rising around a central core* with separate horizontal steps [Fig. 25.5(a)].

(c) Cantilever (or jackknife) staircase which are usually doglegged staircases that fully cantilever from the building without support [Fig. 25.5(b)].

(d) Helical (helicoidal) stairs *with an open well at the centre* which is a continuous slab staircase without a central core [Fig. 25.5(c)].

There are many other types of staircases such as the stairs with a central spine or stringer beam supporting waist slabs cantilevering both sides which are used for wide staircases in public buildings. However, we will not go into their details.

Spiral stairs have separate steps describing a helix around a central column whereas a helical stair has an open well. A helicoid is a *warped surface* generated by a line wrapped around a central imaginary core. It will be like the curve of a screw. The slab is like an inclined plane. It can be circular or elliptical (it has no central structure). Spiral stairs are generally provided at the back of buildings as emergency exit and as access to working people to the various floors. Helical staircases are very aesthetic, more difficult to design and are generally provided as an ornamental feature in libraries, assembly halls and also in very luxury buildings.

25.4 REINFORCED CONCRETE (RC) STAIRS

Nowadays reinforced concrete is the preferred material for stairs in residential as well as in office and other public buildings. For ornamentation, the concrete structures are sometimes covered with wood. Steel staircases are common in factories especially in chemical plants. We will study in a little more detail the simple doglegged reinforced concrete stairs, which are used in most residential buildings.

25.4.1 Layout Requirements of RC Stairs

The following are the general requirements to be met when we plan the layout of a staircase.

Width of stair. It depends on its use. The recommended values are a minimum of 90 cm (3 ft) in residences and 1.5 to 2 m in public buildings.

Length of flight. Generally the number of steps in one flight (to the landing) should not exceed 12 to 16 and not be less than 3.

Pitch of stair. The pitch or slope of the stair depends on the rise and tread adopted. They should depend on the use of the building. Public buildings should have larger treads or going and smaller rise than in dwelling houses. The values usually recommended for tread and rise are as follows:

(a) In residences, we give a tread or going of 250 mm (9 to 10 inches) and a rise of 160 to 175 mm (6½ to 7 inch) approximately.

(b) Public buildings should have longer treads and smaller rise. Treads of 270 mm to 300 mm and rises of 100 to 150 mm are usually given.

(c) Rises and treads of all the steps should be the same. It is very important that we should not change the dimensions of tread and rises from the start to the finish of the stairs. Sudden changes in dimensions can lead to accidents.

The following empirical formula between going and rise is usually used:

$$(2R + G) > 550 \text{ mm but} < 700 \text{ to } 600 \text{ mm (approximately)}$$

Head room. The clear distance between the tread and the soffit of a flight immediately above or between the tread and floor above should not be less than 2.1 to 2.3 m (say 7 feet 4 inch) so that a person can use the stairs with a luggage on his head. This provision of head room is very important.

Height of handrail. The height of the top of the handrail from the tread should be between 850 to 900 mm (about 3 ft) to make it easy for a person of average height to hold on to it by hand.

Stairs to open terrace. Where the staircase leads to an open terrace, the level of the upper landing slab should be 30 cm (1 ft) higher than top of roof slab so that there will be clear difference in height of about 15 cm after the weathering course is laid. (This need not be so if another storey is envisaged in future.)

Staircase room dimension. The minimum clear width of staircase room in residential buildings should be 2.1 m (7 ft) so that there will be a clear width of staircase of 90 cm with enough width for the ballustrade of 15 cm and a well of 15 cm (90 + 90 + 15 + 15 = 210 cm). In public buildings, the minimum width of staircase room should be 2.85 m (9′6″).

Note: The height to the bottom of the upper floor of most residences is about 3.2 m. If we provide 19 steps, the rise will be about 170 mm (7 inches). In many cases, we may also provide a store, washbasin or bathroom under the landing of the half space landing. This may require about 12 to 13 steps in the first flight of the stairs to reach a height to provide at least 2 m under the landing slab.

25.4.2 Construction of Simple Concrete Stairs

There are many ways in which the simple concrete staircase can be constructed. Two of them—inclined slab construction and cantilever slab construction—are discussed further and are shown in Fig. 25.6.

Inclined slab construction. These types of stairs can be built in two ways. Firstly the inclined slab and steps can be built together with reinforced concrete. Steps are built with proper shuttering. Alternately, in cheaper construction, only the inclined slab is first built in concrete and the steps are later constructed with brickwork as shown in Fig. 25.6. This latter procedure considerably reduces shuttering costs and is commonly used for residences.

Cantilever slabs. It was remarked in Section 25.3 that the stairs can be built in many ways. In residences, where the traffic is light, the individual steps can be cantilevered from the surrounding walls of a staircase room. Otherwise for very wide stairs, the individual steps can be centrally supported and cantilevered from a central cast-in-situ spine beams specially

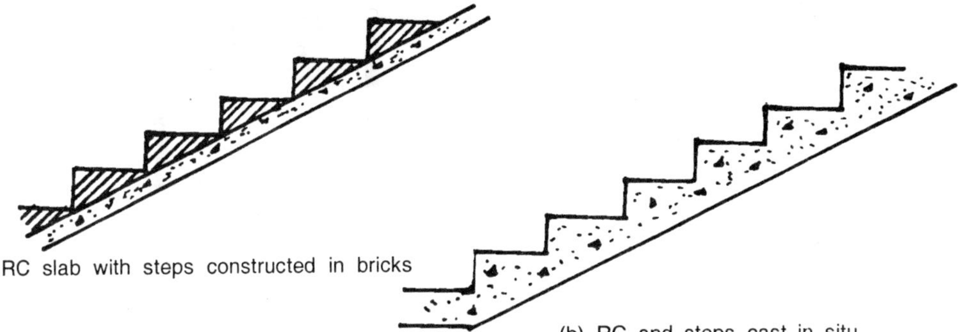

(a) RC slab with steps constructed in bricks

(b) RC and steps cast in situ

Fig. 25.6 Construction of steps in stairs.

built as part of the stairs. The latter type of construction is very common in office buildings where a wide staircase is planned. When they are cantilevered from walls as in residences, it will be desirable to have a concrete beam in the wall connecting all the ends of the slabs of the stairs to improve stability with long term use. Otherwise the fixing ends may get loose due to vibrations and long term use.

25.4.3 Finishes to Concrete Steps

Concrete stairs surfaces are usually given a finish which is easy to clean and give a good appearance. Terrazzo and chequered tiles, and ceramics woodwork are quite often used. It may also be carpeted or a given woodblock paving as described under woodblock flooring.

25.5 PLANNING A SIMPLE STAIRCASE

The method of planning a simple staircase for a residence is shown in Example 25.1 given at the end of the chapter.

25.6 LIFTS

According to the present building regulations, it is mandatory to provide ramps or lifts in all public buildings for the convenience of the elderly and the invalids (physically challenged).

The main components of lifts are the following (Fig. 25.7):

1. The lift car moving on guide rails
2. Machine room with winding machine with electric motor and other memory mechanisms
3. Suspension ropes (steel wire ropes with factor of safety 12 to 20)
4. Counterweight on pulleys to balance the car with 40 to 50% of maximum live load. This is provided usually at the back of the car.

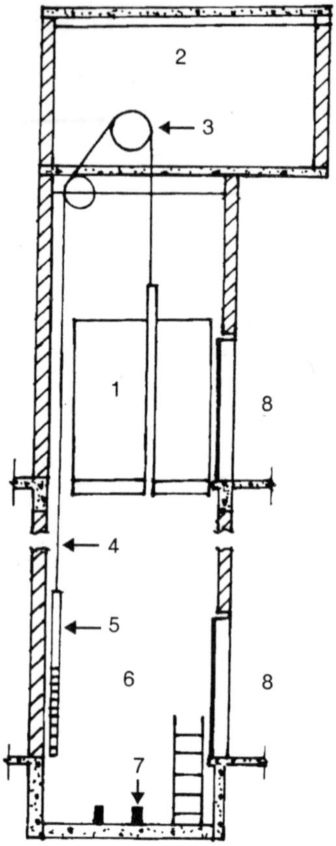

(a) Section

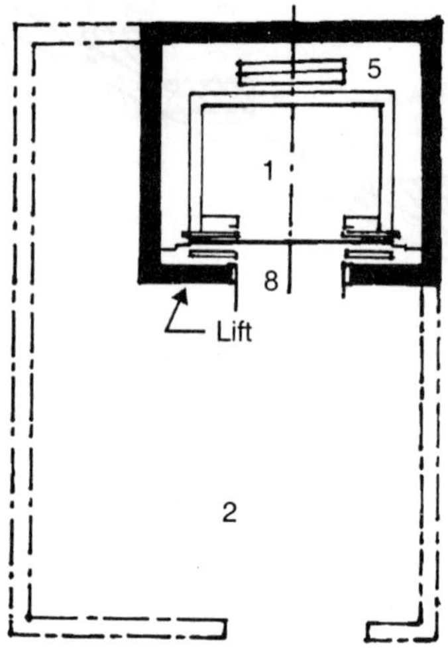

(b) Plan

Fig. 25.7 Components of passenger lifts: 1. Lift car moving in guide rails, 2. Machine room, 3. Rope winding machine, 4. Suspension ropes, 5. Counter weights, 6. Lift pit, 7. Buffer, 8. Landing doors.

5. Buffers for the car in the lift pit floor. Spring buffers are used for low velocity lifts and oil buffers for speed in excess of 1.5 metre per second.
6. Specially operating landing for entry and exit
7. The passenger capacity of a lift is usually rated assuming the weight of a person as 68 kg.

25.6.1 Structural Components of Lifts

The civil work components necessary to accommodate a lift are usually made of RCC and are as follows:

1. A lift well of suitable size, usually extending up to 1600 to 2600 mm, below the bottom landing.
2. Openings of height of 2 metres for entry of people at every floor level. The breadth of opening will depend on the width of the lift well and number of passengers to be handled. (Hospital lifts, for example, will be wide.)

3. An upper machine room on top of the lift to suit the lift size chosen and according the specification of the supplier of the lift. IS 14665, Parts I to IV, can be used for their design.

25.6.2 Specifications of Lifts to be Provided

Lifts are usually classified as follows:

1. Passenger lifts
2. Goods lifts
3. Hospital lifts
4. Service lifts
5. Fireman's lifts

Each type of lift has its own specifications. Passenger lifts are provided in flats, residences, hotels, assembly halls, etc. The number of passenger lifts to be provided, their capacity and travel speed required for a building, etc. are governed by the number of floors to be served, the number of passengers to be handled, floor area, floor heights and other considerations. IS 14665, Parts I to IV, which deals with electric traction lifts, can be used as a guidance for these purpose. Lifts are available in the market with capacity for 4 to 20 persons (with average weight of a person considered as 68 kg).

Lifts for other purposes such as goods are provided in many places and these lifts come under a different class. The relevant IS code should be consulted for further information on the subject.

25.7 RAMPS AND ESCALATORS

Ramps. Ramps are provided where large numbers of persons or vehicles have to be moved from floor to floor. It is also provided for the convenience of the old and the invalids in places wherever they are necessary. Multistorey car parks that are generally provided in the heart of the cities are usually provided with ramps or lifts for taking cars to and from the upper floors. Ramp is a uniformly sloping surface or inclined plane. However, they occupy much larger space than stairs and lifts for construction. In India, many cinema halls and low rise hospitals are provided with ramps instead of stairs. It is easier for sick people to climb a ramp then a series of steps

Escalators. Escalators are power driven, inclined, continuous stairways used for transporting passengers in large departmental stores, airports, exhibition halls, etc. They operate continuously without operators. They transport persons between two levels with speeds not more than 38 metres per minute and its capacity may vary from 3200 to 6400 passengers per hour depending on the width of the escalators. There are comb plates at the entrance and exit. They should mesh with the slots in the tread surface so that clothings of passengers should not be caught between them. Special care should be given in India in this aspect in all escalators, as people in India wear loose clothes, covering up to their feet. The rise of the step is rather large. It is around 220 mm (16 inches). They are provided with balustrades with a handrail moving along with the escalators in the same direction, the handrail being also extended not less than 300 mm (1 foot) beyond the beginning and end of the escalator. Escalators are always provided with emergency stop bottoms near the top and the bottom landings to stop the escalator if there is an emergency such as the dress of a passenger being caught in the comb plate of the escalator.

25.8 BRICK STAIRS

In ancient buildings, stairs were made from masonry. Brick stairs instead of RC stairs are nowadays commonly used in village single storey houses as single straight flights outside the building for access to the flat roof. They are made of either solid wall attached to the outside wall of the building or with arch openings used for storage spaces as shown in Fig. 25.8.

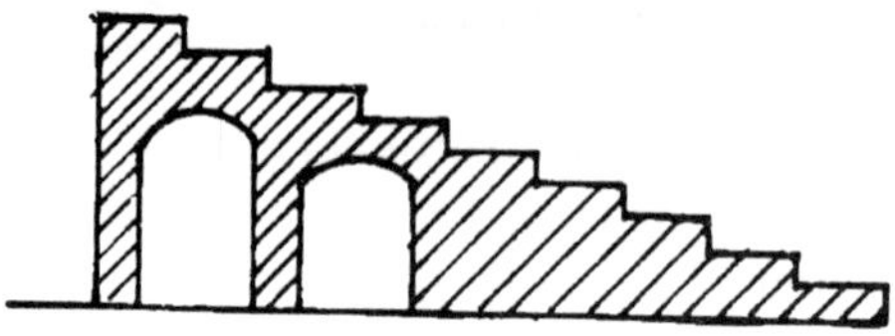

(a) Staircase with arches adjacent to building built in masonry

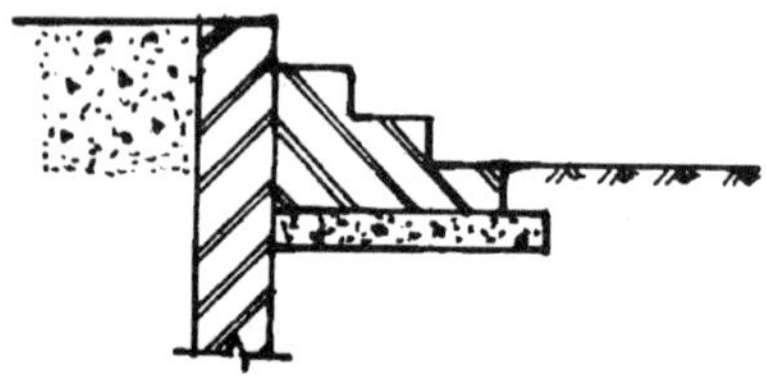

(b) Entrance steps to buildings built on concrete foundation

Fig. 25.8 Brick stairs.

Example 25.1: Layout of a simple staircase

A doglegged concrete staircase is to be provided for a two-storey residence. The floor to ceiling height of the ground floor is 3 m. The top floor structural slab is 125 mm thick and the finishing and flooring on the top floor is 40 mm thick. Design a doglegged staircase and determine the dimensions of the staircase room needed. Sketch the layout of the staircase.

Steps	*Calculations*
1	*Find total height (top of GF to top of FF).* Height = 3000 + 125 + 40 = 3165 mm Assume width = 900 mm as it is a residence
2	*Adopt rise and going.* (a) Rise: Assume a rise of 165 mm (usual for residences) No. of steps $= \dfrac{3165}{165} = 19.18$ (assume 19 steps) (19 is an odd number which some people, as in Tamil Nadu, prefer; others as in Kerala prefer 18 steps in residences)

(b) Tread = (going + nosing of 30 mm)
Assume going of 250 mm usually used in residences.
Check: (2R + G) = 2 × 165 + 250 = 580 > 550

3 *Sketch the arrangement (see Figure).*
First flight = Landing space + 9 goings + landing (equal to width of stairs)
Second flight = Landing + 9 goings + landing

4 *Find total length and width required.*
Length required = 900 + 9 × 250 + 900 = 405 cm = (13′6″)
Width required = 2 × width of stairs + Newel
(Assume newel of 75 mm square)
Width required = 2 × 900 + 75 = 1875 mm (6.25ft)
Required staircase room dimension = 4.05 m × 1.875 m.
[Sketch the stairs as in Fig. 25.1 and a layout in plan as in Fig. 25.4(a) with dimensions.]

SUMMARY

Means for vertical transport of persons and also materials are needed in multistorey buildings. Only stairs are provided in buildings up to 3 floors. With increasing heights of buildings, both lifts and staircases near the lift have to be provided. Emergency staircases are also to be provided for exit in case of fire. Escalators and ramps are normally provided for rapid transport of persons as in airports, offices, theatres, etc. Stairs and ramps and the structures necessary for lifts and escalators are included in civil works. The installation of lifts and escalators are carried out by specialized agencies.

REVIEW QUESTIONS

1. (a) Enumerate and sketch the common types of quarter-turn and half-turn staircases used in buildings.
(b) Draw the plan and section of a doglegged stair connecting two floors.

2. Design a doglegged slab staircase for an office in which the floor to floor height is 3.5 mm. Assume other dimensions as required.

3. (a) Indicate the situation where you would recommend (i) a ramp and (ii) an escalator.
(b) Distinguish between a spiral staircase and a helical staircase. In what places, would you recommend each of them?
(c) What are the structural components to be provided in a building where a lift is to be installed?

Chapter 26

Painting

26.1 INTRODUCTION

Painting industry caters to both industrial painting (as in automobiles, ships, etc.) and decorative painting. In building construction, we are interested only in decorative painting. The materials used for painting of buildings are usually one of the following:

1. Lime (fat lime) for whitewashing.
2. Materials for colour washing: They contain fat lime with colouring materials.
3. Cement paints: These are made from cement and additives. They can be used on plastered brick, concrete, AC sheets, etc.
4. Dry distemper is a water-based natural substance such as chalk or acrylic resin materials in powder form, as vehicles.
5. Oil bound distemper: Oil emulsion type (with natural or acrylic materials as base) comes in thick paint form. It is generally called OBD.
6. Oil paints: These are cheap paints consisting of oil as base, pigments and a drier.
7. Plastic emulsion paints are water-based paints with acrylic binders.
8. Enamel paints: These contain resinous matter which forms a smooth hard film. The resins may be ordinary natural resins obtained from materials such as turpentine or may be plastic alkyd resins as in *synthetic enamel paint*. These are used on metal, wood, etc.
9. Exterior paints for walls are acrylic or cement paints specially made for exterior use.
10. Primers for paints: There are different primers that have to be applied to surfaces before applications of the paint. They provide the base for a good paint job. We have different primers as follows:
    (a) White cement primer is used as a primer for plastered surface.

(b) Metal primers: Red oxide zinc chrome primer (IS 2074–1979) is used for steel as primers. Simple zinc chromite primer is used for aluminium and it can also be used for steel. (These primers are more effective than the simple red oxide primer paints used in former days.)

(c) Wood primers: Pink wood primer, aluminium paint or boiled linseed oil can be also used as primer for wood (refer to section 26.2.2).

11. Varnishes and polishes for woodwork.

These materials and their covering capacities are described in the book on *Building Materials*. In a middle class single- or two-storey residence, we may choose cement paints (especially those not requiring watercuring) for the exterior and interior walls. For interior walls, we may also use the more expensive distempers, enamel paints or acrylic emulsion paints. In this chapter, we will deal only with the important considerations in choosing a paint and carrying out the work in building construction.

It is also of interest to note that certain architectural features can also help maintain the beauty of painted external walls for a longer period. Protection of painted external surfaces from rain by means of ample overhang at eaves in sloped roofs and continuous projections such as sunshades or other features in RC roofed buildings can protect the walls from discolourization and fungus growth, especially in regions of heavy rainfall. These should be incorporated at the planning stage.

26.2 GENERAL CONSIDERATIONS

Paints can be classified into "breathing and non-breathing" types depending on their ability to let moisture pass through the painted surfaces. Oil-based paints leave an impermeable film on the surface after drying. Hence if moisture is entrapped behind the film, the surface blisters. This is the non-breathing type of paint. Cement paint after drying allows moisture to pass through the painted surface. They are the breathing type of paints. In places where moisture tends to accumulate in RC slabs (under bathrooms and wet areas of upper floors), it is preferable to use the breathing type of paints. Before we go into the details of these works, we should remember the following general precautions to be taken.

1. Cement paint has to be applied to wet surfaces. Lime wash and colour wash can also be applied to wet surfaces, but other superior paints such as OBD and plastic emulsion paints should be applied only after the moisture in the walls have dried out. In case of new walls, these special paints can be applied only about six months after plastering has been completed and the moisture in the wall has completely dried out. (In many cases, cement paint is given initially and after a period of time, the final superior paint is applied.)
2. *New surfaces should be given at least three coatings of paint.* The first coat is the *priming coat* with a suitable primer. There are different primers for wood, plaster and metal. The second coat, called *undercoat*, covers the primer coat and acts as bond to subsequent coats. In many cases, it is only a thinned paint of the finishing

coat. The third coat is the *final finishing coat.* The final finish can be as follows in order of smoothness:

(a) Mat finish or flat finish
(b) Eggshell
(c) Oil gloss
(d) High gloss such as enamel

However, a high gloss finish will highlight the irregularity of the surface and are generally not preferred except in high class works.

3. When painting outside walls, it is advisable to paint, if possible, the surfaces which are on the shady side of the building so that the direct rays of the sun are avoided.

26.2.1 Preparation of Newly Plastered Walls for Painting

After the wall has dried and matured for a few months, we first apply a coat of cement primer (white). Irregularities of the surfaces should be then rectified by putty before painting is started. For example, in ordinary house construction, we can use plaster of Paris as putty. For superior work as in showrooms, special copolymer emulsion putty, suitable for exterior and interior, *containing the primer coat* is also available in the market. They are, however, costly. (It is more convenient to first apply white primer on cement plastered walls before applying putty. The white background will show up the irregularities and the putty can be applied more effectively. This procedure will also reduce the amount of putty required.) Rub putty with sandpaper to smoothen the surface. Apply one more coat of primer and allow it to dry for 10 to 12 hours before applying the chosen paint as undercoat. The following points are important:

1. Oil and acrylic paints and distempers do not fare well if there is alkalinity in the surface to be plaster or if the wall has been previously whitewashed with lime. Hence it is essential to apply alkali-resistant cement primer on these whitewashed surfaces before applying these paints.
2. When painting exterior walls all moulds, algae, moss, etc. should be removed by scrapping. The organic growth can be arrested by ammoniacal copper solution containing 7 gm of copper carbonate dissolved in 80 cc of liquid ammonia diluted in one litre of water. Alternately 2.5% magnesium silicon fluoride solution may be used. When cleaning with these solutions, the surface should be allowed to dry before painting. Special exterior paints are more long lasting than normal paints. They also resist growth of algae and fungus.
3. Where efflorescence is seen in brick wall, water should not be used to wash out the efflorescence. If water is used, the salt may be absorbed back into the pores and will be precipitated again when it dries out. The deposit should be brushed and cleaned in the dry state itself.

26.2.2 Types of Primer Coat

The primer paints to be used for various building works is given in Table 26.1.

Table 26.1 Primers for painting

S.No.	Surface	Type of primer
1.	Hard and soft work	Pink wood primer (IS 3536)
2.	Resinuous and plywood	Aluminium primer (IS 3585)
3.	(a) Aluminium and light alloys	Zinc chromate primer (IS 104)
	(b) Iron, steel and GI	Red oxide zinc chromate primer (IS 2074)
4.	Cement concrete, plastered surface, asbestos	Cement primer (IS 109) (white)

(The covering capacity of these primers is 18 to 20 m^2 per litre per coat.)

26.2.3 Types of Paints to be Used

The approximate breakdown of the present consumption of paints in *urban buildings* is reported as follows.

Cement paint	17%
Dry distemper and oil bound distemper	22%
Emulsion paints	5%
Enamel paints (for steel and wood)	56%

Exterior paints. As labour costs are increasing and as painting of urban high rise buildings is cumbersome, there is a trend to go in for long life exterior paints, *especially in high rise flats and office buildings.* As exterior painting of high rise buildings has become very important, special exterior paints (100% acrylic emulsion based paints and also cement paints) are now used more and more. Many paint manufacturers specially advertise only their exterior paints.

Choosing paint. In rural housing which forms more than 65% of the homes in India, the low cost, colour wash distemper and cement paints are quite popular. The possible alternatives are discussed further.

1. *Painting of interior plastered brickwork.* We can use whitewashing, colourwashing, cement paint, dry distemper, oil bound distemper, plastic emulsion paint or enamel interior paint depending on our affordability.
2. *Painting of exterior plastered brickwork.* We can use whitewashing, colourwashing, special waterproof cement paint, or special exterior acrylic paints. (Distempers are not used for exterior surfaces.)
3. *Painting of plastered concrete surfaces.* (Same as plastered surfaces above)
4. *Painting of metal surfaces.* Pretreatment with metal primer (red oxide or zinc chromate primer) and exterior enamel paint for exterior surfaces and interior enamel paint for interior surfaces.
5. *Painting of woodwork.* Good wood in which we want the grains to be seen is not painted but varnished. If we want to expose the gains as in furniture, we first rub it down to expose the grains, then apply a sealer coat and finally apply clear the varnish. We can also stain the wood with wood stains before we apply varnish. If we want to paint wood with opaque paints, pink wood primer is applied first and the

wood can be painted by oil, synthetic enamel or plastic emulsion paints. Clear paints are also now available for superior wood as given in section 24.4.7.

6. *Painting AC sheets and boards.* We apply two coats of alkali-resistant primer and then the acrylic paints or synthetic enamel flat paints.
7. *Painting GI sheets.* We apply priming coat of red oxide or zinc chromate paint followed by synthetic enamel paint. (New GI sheets should not be painted till they have weathered for about an year in contact with atmosphere and the protective oxides are formed on the surface.)

26.2.4 Choosing Colours

Colours have a positive effect on our mood and emotions. The choice of colour should also depend on the size of the room. Colours that might look good in a large auditorium foyer will look gaudy in a small flat. Colours can create optical illusion of increasing or decreasing the size of a room. A ceiling painted white seems higher than a ceiling coated darker than the walls. When choosing colours for not very well lighted rooms as in block of flats, the colour should be chosen with respect to the effect of sunlight and the effect of artificial light during nights. The reflective values of colours vary from 70 to 90% in white to only 10 to 20% for deep colours. In any case, light colours are always safe in residences and give good reflective surfaces. There are more than hundred shades of paints in the market to choose from.

26.3 DESCRIPTIONS OF PAINTWORK IN BUILDINGS

Next we will consider in detail the following paintworks that are usually carried out in building construction:

1. Whitewashing and colourwashing
2. Painting with cement paint (Ordinary cement paint requires curing. Supercem does not need curing)
3. Painting with dry distemper
4. Painting with oil bound distemper
5. Painting with plastic emulsion paint
6. Painting iron and steel with enamel paints
7. Painting new woodwork with enamel paints
8. Polishing woodwork with attractive grains

Paintwork is to be carried out under the following three steps:

(a) Preparation of material
(b) Preparation of surface to be painted
(c) Application of paint

We will principally deal with the first operation as the second and third operation have different types of paints. Preparation of a newly plastered surface for painting has already been described in section 26.2.1. It is understood that the paint is applied by brush. Applying paint by spraying and by special rollers can also be carried out, but brushing is the most common and economical method for application of paint.

One stroke from top downwards, another from bottom upwards over the first stroke, followed by horizontal strokes from left to right and right to left, constitute one coat. No brushmarks should be visible in a well painted surface. In thicker paints, the horizontal strokes may precede the vertical strokes.

26.4 WHITEWASHING AND COLOURWASHING OF PLASTERED WALLS

26.4.1 Preparation of Materials for White and Colour Washing

As these paints are used for low cost buildings, preparation of surface is not given much importance. Whitewashing is carried out using fat lime and colourwashing is done by adding approved pigments to fat lime. Alternatively colourwashing powders consisting of readymade mixtures of fat lime and colouring matter are also available in the market. For preparing whitewash, thoroughly slaked fat lime is mixed with water and made into a thin cream. This is allowed to stand for a period of 24 hours after which it is screened through a coarse cloth. For colourwash, it is advisable to buy the ready mixed coloured paint material. If we are to mix the ingredients, we should thoroughly mix sufficient quantity of the mix in one operation to avoid shade differences. Some of the pigments used are ultramarine or indigo, copper sulphate (blue) and yellow and red ochre. To the above basic materials, we also add the following materials:

(a) One kg of gum dissolved in hot water for each cubic metre of lime cream. (Alternately rice or *kanji* water is added.)

(b) 1.3 kg of common salt dissolved in hot water for every 10 kg of lime powder. (Salt quickens carbonation of calcium hydroxide and makes the coating hard and resistant to abrasion.)

(c) Indigo (neel) up to 3 gm per kg of lime used for the mix is usually added to make the paint.

The thoroughly mixed solution prepared as indicated above is applied with a brush or a *moonj*, generally in three coats to a new surface and two coats to an old surface.

26.5 PAINTING OF PLASTERED WALLS WITH CEMENT PAINTS

The surfaces to be painted are prepared as already indicated. The cement paint comes in powder form. It is manufactured from white cement with pigments and special additives. The paint has also waterproofing qualities. Cement paint can be applied to concrete, plaster, asbestos cement, etc. but not on gypsum plaster, metal, wood surfaces, etc. which are hard surfaces devoid of a key. These surfaces are to be painted with distemper.

For painting with cement paint, the surface to be painted is first cleaned with clean water. Before application of the paint, the surface should be thoroughly wetted to take care of the suction effect of dry surfaces. Pitting, if present in plaster, should be made good by an initial coating of cement paint as primer.

Preparation of cement paints for painting. The cement paint is prepared as follows. Two parts by volume of cement is slowly added to one part of water to obtain a *creamy consistency*. Allow it to stand for 5 to 10 minutes. Then again add one more part of water to make the mix ready for painting. (The quantity of water added for further mixes is kept the same to get constant colour from the paint.) Only enough paint is to be mixed each time to last only for an hour or more depending on temperature. The paint should be kept stirred during use. *The covering capacity of cement paint is about 4* m^2*/kg per coat.*

Curing. When using *ordinary cement paint*, the painted surface must be cured. The surface should be sprinkled with water several times a day in between the coats and at least two days following the final coat. Generally three coats for new works and two coats for old works are prescribed. Hence, this is not suited for rooms with electric connections.

As curing interior work is very difficult and it çan damage electrical wiring, a new superior cement paint, called *Unique*, which does not need any curing, has been introduced into the market by *Super Cem company*. It is claimed that even initial wetting of surface is not needed for its use and it can be used as an internal or external paint without curing after paintings. The cost of supercem is higher than the cost of ordinary cement paints and is to be used where curing can not be done.

26.6 PAINTING OF PLASTERED WALLS WITH DRY DISTEMPER

The surfaces to be painted should be prepared as described in section 26.2.1. New surfaces should be allowed to dry for two months before painting. Dry distemper contains a ghee like vehicle and looks like tinted powdered chalk. It is marketed as a powder. The paint is prepared for application by mixing one kg of powder with 600 cc (0.6 litres) of clean warm water. It is allowed to stand 30 minutes before use. Whatever is mixed should be used in that day itself. They are generally used for plastered walls and are considered only just superior to whitewashing. Work is carried out as in cement painting but does not need curing with water. *Its covering capacity is about 10* m^2*/kg per coat.*

26.7 PAINTING OF PLASTERED WALLS WITH OIL BOUND WASHABLE DISTEMPER (OBD)

Oil bound distemper (OBD) comes as a thick paste. It is used as an intermediate costly paint for *interior walls* and ceiling where moisture is not present. It is an oil emulsion type of paint. Synthetic washable type distemper wears itself to a smooth, durable matt finish that can be wiped clean with water and mild soap.

Preparing surface for painting with OBD. Surface should be prepared as given in section 26.2.1. The new plastered surfaces should be allowed to mature for a minimum period of 6 months. In previously painted buildings, remove all lime or colourwash and powder distemper from the wall by scraping. Cracked or flaking paint should be scraped off completely and the surface should be clean and dry before application of paint. Earlier, similar coatings of oil based or emulsion based paints need not be removed if it is in good condition. However, the sheen on such surfaces must be removed by sandpapering.

For the best results in new walls, a coat of alkali-resistant cement primer (also called wall neutralizing cement primer) is applied and the dents, if present, are filled with suitable putty applied in the layers. One more coat of cement primer is applied over the places where putty was applied and OBD is applied only after it has dried over two days.

Preparation of paints for application. Water is to be added slowly to the thick OBD paste while stirring to bring it to a brushable consistency. Normally 500 to 800 ml of water is needed per kg of distemper, i.e. 50 to 80% dilution is made. Generally three coats are prescribed over the cement primer for new surfaces and two coats on old surfaces. These are applied only at intervals of at least 24 hours between successive coats.

OBD is available over thirty different colours and a suitable colour of distemper should be chosen for the work. OBD should not be used in places where moisture will penetrate. In such places, it peels of as the paint cannot breathe. A large number of such failures of OBD can be seen around us. *The covering capacity of OBD is about 10 m^2 per litre per coat.*

26.8 PAINTING OF PLASTERED WALLS WITH EMULSION PAINTS LIKE PLASTIC EMULSION PAINT

Plastered walls and concrete surfaces in high class buildings are commonly painted with emulsion paints. The term *emulsion* simply signifies that the pigment and medium are dispersed as small globules in water. *Many emulsions such as oil, synthetic resin and bitumen emulsions are available*. Thus emulsions can be ordinary emulsions or plastic emulsion. Plastic emulsion paints have plastics in them and can give different types of finishes such as flat to egg shell.

Acrylic plastic emulsion paints, which are very popular, are luxury flat finish paint for interiors and exteriors of plastered walls. They can also be applied to fibre board, asbestos, wood, metal etc. after applying suitable primer. It is quick drying (touch dry in one hour) and is washable after one month of self-curing (when stains, grease mark and finger marks appear on it) with water. It also has a low odour level.

Preparation of surface. The surface should be prepared very carefully for these paints. An old wall surface previously treated with paint or oil bound distemper should be cleaned and rubbed down with sandpaper to ensure adhesion and these may not require further application of primer. Where lime wash and dry distemper has been used, it is necessary to wash surface with water to remove traces of alkali and loose distemper and apply an alkali-resistant cement primer. For new surfaces of plaster, brick, asbestos, fibre board, etc., the acrylic paint thinned with an equal volume of water can be used as the first primer coat. For new surfaces of wood, we should use a wood primer and for new iron and steel, we use chromate primer.

Preparation of emulsion paints for painting. The paint is prepared by 100% dilution with water for the first coat (priming coat) and 75 to 35% dilution with water for subsequent coats, depending on the nature of surface and the number of coats to be applied. For the best results, we usually apply the undercoat and the finishing coat, allowing at least 2 to 4 hours between successive coats. The paint comes in over 100 different shades to choose from so that a proper choice of readymade mix is possible. *Plastic emulsion paints have a covering capacity of about 20 m^2/litre per coat.*

26.9 PAINTING IRON AND STEEL WITH ENAMEL PAINTS

Enamel paints are available as ordinary enamel and synthetic enamel paints. Ordinary enamel paints are made from natural rosins (solid residue got after distillation of oil of turpentine from crude turpentine. They are also loosely called *resin*). Synthetic enamel is based on synthetic alkyd resins. Enamel paints are available as interior and special exterior grades. In buildings, these paints are used for painting iron and steel items such as iron grills, steel windows, etc. They are available as ready mixed paints.

Method of Painting of iron and steel with enamel paint. Painting should be started only on clear sunny days. Painting is carried out in the following five operations:

1. Cleaning and sandpapering
2. Application of primer coat
3. Stopping and filling
4. Application first undercoat
5. Application of finishing coats

In the *first operation,* the surface is thoroughly cleaned by mechanical means of scraping, brushing with wire brushes. It is then smoothened with sandpaper.

In the *second operation,* the primer coat red oxide zinc chromate primer (IS 104) or preferably ready mixed red oxide zinc chromate primer (IS 2074) is applied by brushes. After the first coat has completely dried, a second primer coat is applied, so that a film fully covering the metal is obtained. It is important that within the shortest period after cleaning, the first primer should be applied. This is particularly important in places with high humidity.

The third operation is stopping and filling after the primer coat has hardened. We apply the paste filler coat. For this purpose, we first roughen the primer coat with emery paper (not sandpaper) without scratching and without damaging the primer coat. We free the surface from dust. Fill the deep dents with paste filler (as used in body rebuilding of motor cars) using a good putty knife pressing firmly into the dents and applying only the optimum pressure. Allow as many layers of paste filler as needed to get a smooth surface allowing each layer to hard dry. Then wet rub to a smooth finish using waterproof emery paper 280/320. After the last coat of paste filler is hard and dry, it is wet rubbed down where necessary. We then apply a coat of readymade filler (R/M filler) by brush. Allow it to dry and wet rub to a smooth finish.

The fourth operation is to apply the first undercoat by brush with minimum brushmarks. The brush is held at about 45° to the vertical. Paint is first applied in vertical strokes until the surface is covered and *then is applied crosswise with light strokes.* Finally it is laid off with vertical strokes. This constitutes one coat.

The *fifth operation* is to apply two finishing coats. The first finishing coat is applied (by brush or spray) and it is allowed to dry. We then gently remove the gloss for the entire surface by wet cut with waterproof emery paper 320/400. After the surface is dry and free from dust, we apply the second coat.

Notes: 1. Repainting of old iron surfaces is done first by cleaning the surface with turpentine oil to remove grease and perspiration of handmarks, etc. The surface is allowed to dry and then repainted.

2. If exterior surfaces of iron has developed fungi, it should be removed by ammoniacal copper solution or 2.5% magnesium silicon fluoride and allowed to dry (see section 26.2.1).

3. The covering capacity of enamel paints is 18 to 20 m^2/litre per coat.

26.10 PAINTING NEW WOODWORK WITH SYNTHETIC ENAMEL PAINT

Woodworks such as doors and windows can be painted or polished. In this section, we will examine painting and in the next section, we will deal with polishing of woodworks. Quality painting of new woodwork must be carried by the following five preparations:

1. Removal of knots and preparation of surface
2. Applying primer coat
3. Stopping and filling (filling hole with putty)
4. Applying undercoat
5. Applying topcoats

These steps are described below.

Removal of knots and preparation of surface. *The first operation* is the removal of knots and preparation of surface. Wood to be painted such as doors and windows should be well seasoned and free from discoloured sapwood and from resinous or loose knots. The surface is prepared as follows. It should be first smoothened with abrasive paper and projecting fibres removed. Any knots, resinous streaks or bluish sapwood should be treated with pure shellac knotting thinly applied around it extending to about 25 mm (1 inch) beyond the knot area. If the area is small and the wood is not highly resinous instead of applying two coats of knotting, we apply one coat of shellac slightly pigmented with aluminium powder. The above method of knotting or killing knot is called *patent knotting*. (It can also be done by *ordinary* or *size knotting*, where we use a paste of red lead ground in water mixed with strong glue size applied hot on the knots. *Lime knotting* consists of covering the knot with lime for 24 hours after which it is treated with ordinary knotting. Another easy method now widely practiced is to fill the knot with *Fevicol* mixed with fine woodpowder scraped from wood.)

Applying primer coat. *After knotting, the second operation is to rub down the surface with sandpaper and apply the wood primer coat* or sealer coat to fill the pores of wood. This priming coat is generally applied before the woodwork is placed in position and is to be applied on all portions of the wood.

Stopping and filling. *The third step is stopping and filling*. It consists of rubbing down the primed surface and filling the holes with putty. (For high class work, we use a putty with one-third white lead and two-thirds ordinary putty). It is then rubbed down with sandpaper to a smooth surface.

Applying the undercoat and finishing coats. *The fourth operation is applying the undercoat.* After drying of the undercoat, the surface should be carefully rubbed down with sandpaper and wiped clean before the fifth operation of applying the finishing coats of paint. Generally only two coats of paints (undercoat and finishing coats) are prescribed in ordinary works.

26.10.1 Repainting of Old Wood Work

If the old paint is cracked and is in a perished condition, it can be removed by readymade paint removers available in the market or by traditional mixtures prescribed by PWD specifications given below. Thick layers are removed by blow lamp or a burning stove. Then it is painted as described above for new work.

Note: Some of the traditional prescriptions to remove old paint from painted wood are as follows:

(a) Mix 1 part of soft soap and 2 parts of potash in boiling water. Add to it 1 part of quick lime and apply this hot mixture on old paint and leave it for 24 hours. Wash with hot water to remove the old paint.

(b) Prepare a solution of 1 kg of sodium hydroxide (caustic soda) in 5 litres of water. Rub this solution with a rag fixed on a stick (caustic soda attacks human skin). The paint gets dissolved and can be washed off.

(c) Make a creamy solution with 2 parts of quicklime and 1 part of washing soda. Apply it on the old paint and leave for an hour after which the paint can be easily removed.

26.11 POLISHING WOOD WHICH HAS ATTRACTIVE GRAINS

Even though exterior woodwork needs painting for durability, interior surfaces and furniture which are made of wood with attractive grains (as in teak wood) can be given finer polish with the following clear finishes:

1. Clear and shellac varnish (varnishing)
2. Spirit of French polish (French spirit polishing)
3. Melamine based finish (melamine finishing)
4. Polyurethane-based finish (polyurethane polishing)

26.11.1 Method of Polishing Wood

The polishing of wood is carried out by the following four operations:

(a) Knot filling, glue sizing and filling of holes

(b) Staining

(c) Sealing

(d) Finishing by varnish, spirit polishing or clear paint

Knot filling. The first operation is knot filling, glue sizing and filling of holes. All the knots are first treated as already described in section 24.4.6. Glue size is a glue-like substance that stiffens with time. We first rub down the wood with fine emery paper to expose the grains. Then we apply glue size to seal all the pores in the wood to prevent the absorption of moisture and the oil from the varnish. First a thin layer of size is applied all over the surface. When dry, it is rubbed down smooth with sandpaper and cleaned. A second coat is also usually applied and rubbed down with fine sandpaper and cleaned. Alternately a transparent wood filler coat may be used. Any holes, etc. are also be filled with metal paste (used for automobile repair) and rubbed down. The filler coat can also be prepared by mixing 1.5 kg of whiting (ground chalk) with 1 litre of methylated spirit.

Staining. This second operation, staining, enhances the beauty of wood. Desired stains can be applied evenly by brushing. Staining of furniture is usually carried out by dissolving staining colour powder in clear varnish and applying it on the wood.

Sealing. Sealing is the third operation. A suitable readymade sealer available in the market is applied on the stained surface to seal it from subsequent coats.

Finishing. The fourth operation is the final finish. It is carried out by one of the following finishes.

(i) *Finishing by varnish.* It is obtained by applying two or three coats of varnish. Usually *Copal* varnish obtained by dissolving *Copal* resin in denatured spirit is used. Each coat should be *applied with a full brush* and allowed to dry. Special fine hard brushes, and not ordinary paint brushes, are to be used for this procedure. The surface is then rubbed down and flattened after each coat with fine sandpaper. The final finished surface should be uniform and glossy. (Types of varnishes have been covered in the book on *Building Materials.*)

(ii) *Finishing by French spirit polishing.* French polish is a product obtained by dissolving shellac in commercial alcohol. The surface for French polish is prepared as already described by first sizing the knots and coating it with a filler compound such as that made of mixing 1.5 kg of whiting (ground chalk) with 1 litre of methylated spirit. This process is to fill the pores. It is allowed to dry. Then it is rubbed down with sandpaper and cleaned. On this prepared surface, French polish is *applied with a polishing pad of absorbent woolen cloth covered with a fine cloth* (and not by brush). The pad is moistened with polish and then rubbed hard with quick light strokes along the grains of the wood. On drying, the surface is rubbed down with fine sandpaper and cleaned. Several coats are applied to get the described finish.

(iii) *Finishing by melamine-based and polyurethane-based finishes.* These special paints are available as clear paints. Melamine-based paints are available in transparent and opaque finishes and polyurethane paints are available as a clear finish. Polyurethane paint is cheaper and is recommended for clear painting of external doors and windows. They also have a good tolerance to climatic change and humidity when compared to French polish.

Note: French polish is cheaper than varnish but the surface gets easily scratched and stained. It does not withstand weathering and hence is suitable only for inside work, for polishing furniture, etc. It is not suitable for external work.

26.12 DEFECTS IN PAINTWORK

The common defects that should be avoided in painting are as follows:

Blistering. These are formed by water vapour trapped inside non-breathing types of paints.

Bloom or flashing. These are formation of dull patches usually due to the defect in paint or bad ventilation.

Brush marks. These occur due to defective work.

Cracking. It occurs due to defect of paint and fast drying.

Crawling or sagging. It occurs due to application of too thick a paint.

Flaking. It occurs due to poor adhesion of paint to the surface.

Lack of opacity or body. It happens due to overthinning of paint or inadequate stirring of paint during its application.

Pin holes. These are formed when there are small holes present in the surfaces such as walls even before painting. The air from these holes can burst forth and create holes. Surface should be levelled with putty before painting.

Slow drying. It can occur due to a moist unhardened undercoat, bad quality of paint or painting in damp weather or on a greasy surface.

26.13 ROUGH ESTIMATION OF QUANTITY OF PAINT

The following approximation based on plinth area may be used to estimate *quantity of paint required* for painting an ordinary residential building. (It should not be used for estimation of cost of painting.) The covering capacity of various paints has been already indicated in the various sections of this chapter.

Ceiling work = 1 × Plinth area

Interior walls = 2 × Plinth area

Outer walls = 1 × Plinth area

Doors = 15% of the Plinth area

Windows = 15% of the Plinth area

The coefficients used to estimate the equivalent plan areas of uneven surfaces and the labour required for payment of paintwork are given in books on estimating.

SUMMARY

Painting and finishing of walls, wood walls, grillwork, etc. is important items in building works. Very cheap to very expensive materials are available for these works. Similarly the materials used for exterior works can be different from those used for interior decoration. Hence proper care and attention should be taken in the selection and application of paints in various types of building works. A simple 2 coat lime wash or 2 coat white cement or dry distemper finish will be enough for low-cost buildings.

REVIEW QUESTIONS

1. What are whitewashing and colourwashing? How are these carried out in a building?
2. (a) What type of paint would you use for the walls of a newly built low-cost house?
   (b) What is OBD? How do you apply OBD on a plastered surface? What are the precautions to be taken in its use? How is it different from dry distemper?
3. What are "breathing" and "non-breathing" types of paints? What are the considerations in application of these paints to the ceiling under the wet areas and wall surfaces?
4. (a) Which paints will you use for painting the steel grill for a window?
   (b) Describe the procedure of painting a metal surface.
5. Describe how you will
   (a) polish an interior teakwood door
   (b) paint an exterior door
6. (a) What are the differences between plastic emulsion paints and plastic enamel paints? Where would you use these materials?
   (b) Which paint would you choose for the exterior surface of a high rise building consisting of residential flats? Give reasons for your choice.
7. Specify what type of paint would you use for the following:
   (a) Doors and windows made of ordinary wood
   (b) Window grills
   (c) Exterior wall surfaces
   (d) Interior wall surfaces
   (e) Ceiling of rooms
8. (a) What are the differences between varnishing and French polishing? Which of these would you use for (i) the exterior teakwood door of a building and (ii) the teakwood furniture in the drawing room?
   (b) What are polyurethane-based finishes? Where would you use these in a building?

Chapter 27

Waterproofing and Weatherproofing RC Roofs and Waterproofing Wet Areas

27.1 INTRODUCTION

In tropical countries, it is very important to detail roofs of building properly and make them waterproof. The most important point to note in waterproofing of roofs is to give the necessary slopes for drainage. The detailing of the roof junctions and the construction of waterproofing systems to be used are also important They depend on the climatic conditions of the region. Detailing of roof for drainage of rainwater is dealt with in Chapter 28. In this chapter, we will deal with methods of *waterproofing of roofs and wet areas of the floor*. We will also examine methods of insulation of roof from transmission of heat through the roof slabs.

27.2 DETAILING OF BEARING OF FLOOR AND ROOF SLABS ON WALLS

Firstly detailing of the bearings, especially the bearing of the RC *roof slabs* on the outer walls, is very important in waterproofing of buildings. The bearings of simple slab roofs (which are different from slab and beam construction) should allow expansion and also prevent ingress of rainwater through the slab/wall joints. In old MDSS specification for *ordinary wall and slab construction* (non-framed buildings), the top of the roof slab bearing walls should be plastered smooth with 1 : 3 cement mortar and whitewashed with lime and the roof slab should rest on bituminous kraft paper to allow expansion of roof slab. For structural reasons, the bearing *should not be* less than 15 cm or equal to the depth of the slab resting on the wall, whichever is greater. If the wall is thicker than one brick and the slab does not extend the full width, then the space for expansion of the slab should be left between the slab and the end of wall. This space should be filled with bituminous material or joint filler as shown in Fig. 27.1. This will allow expansion and contraction without leaving a gap. The slabs otherwise are to be carried over the full width as continuous slabs or used as a *chajja* (sunshade). It is also useful

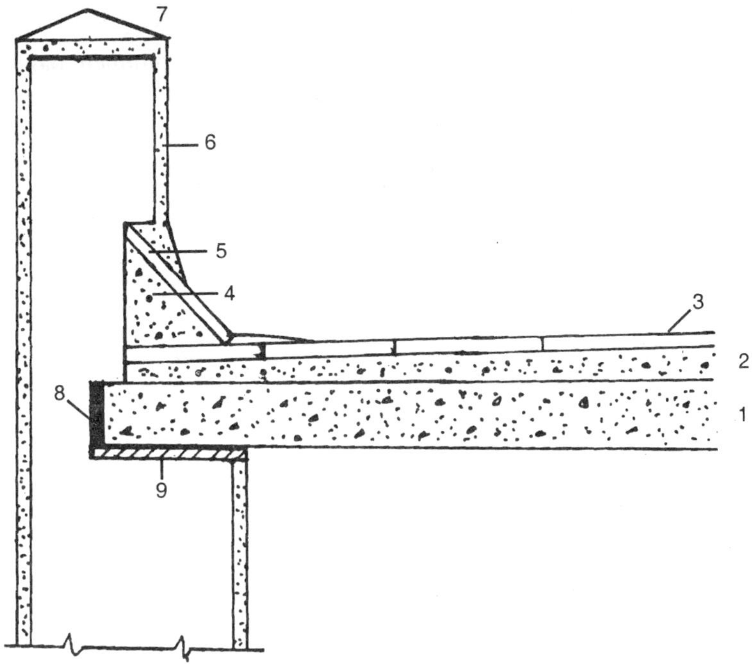

Fig. 27.1 Details of waterproofing junction between parapet and brickjelly concrete terracing: 1. RCC structural slab with bitumen coat on top, 2. Brickjelly concrete, 3. Pressed tiles on roof, 4. Fillet of brickjelly concrete, 5. End tiles laid at 45°, 6. External rendering, 7. Coping (preferably with inward slope), 8. Joint filler, 9. Brickwork finished in 1:1 cement mortar whitewashed and laid with kraft paper.

to detail the junction with a drip course (by providing a projection or a recess) if the slab *runs through the wall without a chajja* so that water does not run along the junction of wall and slab but is carried off the joint. *Leakage through these outer walls and simple RC slab has been a source of trouble in many buildings.*

As regards floor slabs, which do not undergo much expansion, it is considered enough if the bearing part of the wall is plastered smooth and painted with cement grout and also a coat of lime wash is applied before placing the steel bars for the slab. The bituminous paper is optional (Fig. 28.2).

The second detail is the junction between the top roof slab surface and any vertical face such as the parapet wall as shown in Figures 27.1 and 27.2. This aspect is dealt with in Sections 27.6 and 28.3.2.

27.3 WATERPROOFING FLAT ROOF SLABS

A number of methods are available for waterproofing of roofs in India. Their uses depend on the climatic condition of the region. Thus, for example, the system used in the dry regions of Rajasthan will not work in rainy regions such as Bengal or Kerala. The following are *some of the systems* used in India. Combinations of these can also be used depending on its cost. (Waterproofing materials are described in the book on *Building Materials*.)

1. Mud phuska terracing covered with paving brick tiles.
2. Waterproofing with bituminous products such as bitumen paint or ready-to-use bituminous membranes (felts) which can be stuck to the roof by simple torch-on process or hot applications of bitumen. (As bitumen produces a dark surface, they have to be painted with aluminium paint or covered with light coloured pebbles to reduce heat absorption.)
3. Lime concrete terracing with or without tiles laid above this terracing.
4. Application of modern elastomeric paints directly on concrete or on tiles laid over the concrete.
5. Waterproofing by *special chemical slurry compound applied to concrete* as soon as roof is laid and covering it with tiles has been found very effective on sloped roofs where gutters have to be specially treated for leakages. Two component materials which can be mixed with water to get a waterproofing slurry for application on concrete are available in shops dealing with concrete chemicals (see the book on *Building Materials*). For good performance of surface, coats not stable under direct sunlight should be covered with tiles, fibre glass or other fibre reinforcement.

We should also remember that in a hot country such as India (in both hot humid and hot arid regions), it is desirable not only to waterproof the roof but also try to prevent the roof as much as possible from being heated up in the summer. Hence black surfaces on roof should be avoided. In addition, it is also desirable introduce an insulating layer over or under the roof to reduce the heat.

27.4 DESCRIPTION OF METHODS OF WATERPROOFING FLAT ROOFS

Details of various types of conventional waterproofing treatments can be found in PWD specifications. We will, however, examine only the following five popular specifications to understand the principles involved. The type to be used depends on the rainfall at the place.

(i) Brickjelly lime concrete terracing with or without tiles for flat roofs
(ii) Membrane waterproofing with bituminous membranes
(iii) Thermal insulation combined with waterproofing for flat concrete roofs
(iv) Waterproofing and insulating roofs by elastomeric membranes. (This method is more recent.)
(v) Waterproofing and weatherproofing by mud puska terracing.

Variation in the actual execution can be made depending on the requirements of the case. Sloping roof requirements are described in Section 27.5.

27.4.1 Method 1: Brickjelly Lime Concrete Terracing with or without Tiles

The steps in preparation of this roof finish with lime concrete are as follows (Fig. 27.1):

Application of bitumen. The RC roof is first thoroughly cleaned with a rag soaked in kerosine oil and a layer of *bitumen primer* is applied with brush at the rate of 0.24 kg per sq m to assist adhesion of bitumen to concrete. Three coats of hot blown grade 85/25 are then applied uniformly over the surface at 1.45 kg/m^2. This can be done either by applying raw blown bitumen heated at the site or by using its readymade paint preparations available in the market. A layer of coarse sand at the rate of 0.6 cm of sand per 10 sq m of roof area is spread over the bitumen and pressed into the bitumen.

Making of lime concrete. The lime concrete is made of brickbat aggregates lime and surki. It is prepared as follows. Firstly we make *lime mortar*. For this purpose, *one part* of slaked lime and *two parts* of surki by volume (instead of sand) are mixed in a watertight platform. It is sprinkled with the required quantity of water and ground well using a mechanical grinder. It may also be mixed thoroughly by manual labour.

The necessary lime concrete is then obtained by thoroughly mixing the brick aggregate of 20 mm size (well burnt brick ballast) and the above lime-surki mortar in the proportion 2½ : 1 by volume (The mix will approximate to 1 part of lime, 2 parts of surki and 7 parts of aggregate.) The brick aggregate should be soaked thoroughly in water for a period of not less than six hours before use in the concrete. The brick dust obtained during breaking of the brick can also be used along with surki. The lime mortar with surki should be used within 36 hours of preparation of the lime mortar. Where surki is not available, mix 2 parts of lime and 2½ parts of brick ballast to get lime concrete (12 kg of bar soap and 4 kg of alum dissolved may be added or every m^3 of concrete for improving the waterproofing quality).

Laying of lime concrete. The laying of the lime concrete is carried out as follows. The average thickness of finished lime concrete should be 10 cm with a minimum of 7.5 cm laid to slope. *The slope should be not less than 1 in 60 when top tiles are used and 1 in 50 otherwise*. The mixed concrete is first placed and rammed with rammers weighing not more than 2 kg. Further consolidation is made by wooden thappies by persons sitting in a row and the gang moving forward gradually. The beating is normally carried out for seven days till the *thapi* makes no impression on the surface and rebounds on beating. During beating, the surface is sprinkled liberally with lime water mixed with a water extract obtained by soaking for 24 hours Kadukai nuts in jaggery water (500 g of jaggery, 150 g of Kadukkai and 100 litres of water for 25 sq m^2 of work is used. Alternately, a solution obtained by mixing *3 kg of jaggery and 1½ kg of Bael* fruit in 100 litres of water may also be used).

Curing. The lime concrete is cured after compaction for 6 days or until it hardens by covering it with a thin layer of grass or straw which is to be kept wet continuously.

Finishing. To reduce the expansion and contraction of the roof with variation of temperature, one layer of pressed clay tiles 230 mm × 230 mm or two layers of smaller country tiles are laid on top of the lime concrete. The tiles should be laid in an impervious mortar prepared as follows. First a 1 : 3 cement-sand and mortar is mixed thoroughly. To this crude oil equal to ten per cent of the weight of cement is added and mixed to the necessary consistency. (This may be substituted by a similar mortar with waterproofing chemicals). The tiles are bedded in this mortar with joints not wider than 6 mm. Before the work dries up, the joints should be raked out and again pointed with the above mortar. Proper slopes and roof drainage (one 100 mm diameter pipe for every 40 sq m of roof area) is also to be provided. In earlier days, the tiles laid over lime concrete were small country tiles laid in two courseworks. The first course was laid diagonally and the second square to the parapet, thus breaking joints. The tiles were immersed for at least two hours in water before use. Nowadays one layer of large-size pressed tiles (150 mm × 150 mm) is used instead. These tiles are usually *laid breaking joints* (differently from laying floor tiles which are laid with joints) in line. The joints should be finished with the crude oil or waterproofed mortar described above.

27.4.2 Method 2: Membrane Waterproofing with Bituminous Membranes

As described in the book on *Building Materials*, there are different systems of bituminous membranes for waterproofing. In the old systems, the membranes were made from organic materials such as jute and were stuck to the roof by painting the roof with hot bitumen. This is called the *pour and roll system*. In the latest torch-on system, plastics are used instead of felts for the membranes. By heating the concrete surface and the plastic underlayer of the membrane by a blow torch working from a gas cylinder, the membrane can be stuck onto the roof slab. This system is called the *torch-on system*. (Refer the book on *Building Materials.*)

In the old practice, a "four-course" treatment with bituminous felt will consist of the following:

1. Intial application of hot blown asphalt as bonding meterial
2. Application of bituminous felt
3. A second layer of bonding material
4. Final layer of stone grit of pea gravel

A six-course treatment will have two layers of bitumen felt, sandwiched between the bituminous paint and grit. The felt should be laid at right angle to the slope of the roof and the laying should commence at the lowest level nearest to the rainwater outlet.

27.4.3 Method 3: Thermal Insulation Combined with Waterproofing for Flat Concrete Roof

In hot climates, it is desirable to cut off, as much as possible, the heat from the sun penetrating the roof. One of the methods commonly used for this purpose is given step by step below (see Fig. 27.2):

1. Apply blown bitumen on concrete surface by hot painting.
2. Before the bitumen dries out, thermal insulation in the form of an insulating material is fixed on the roof. It then sticks to the bitumen paint. These insulating materials can be cellular concrete, vermiculite, thermocole, fibreglass, mineral wood, etc.—25 to 50 mm thick. The joints should properly filled and sealed with bitumen.
3. A sheet of polyethylene (25 micron) is laid over the insulating surface to act as the vapour barrier for the insulating material (see the description of vapour barriers given below).

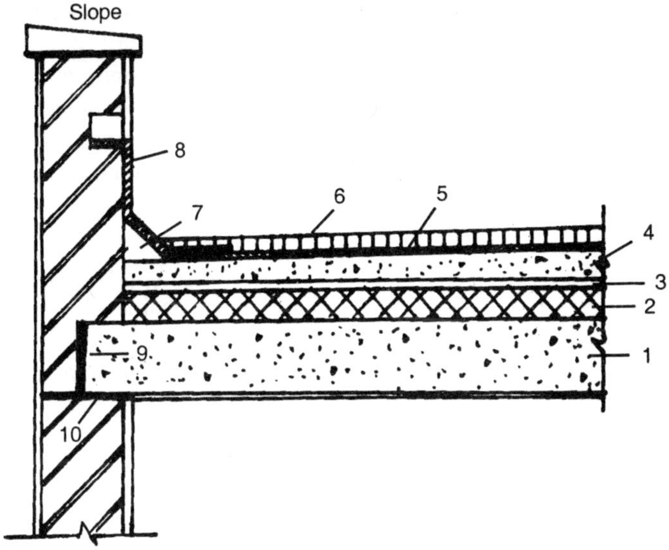

Fig. 27.2 Details of waterproofing of roof with thermal insulation: 1. RCC slab coated with 85/25 grade bitumen coat for adhesion of insulation, 2. Insulation EPS/PUF slabs, 3. 250 micron polythene sheet as vapour barrier, 4. 1 : 2 : 4 concrete with wire mesh laid to slope, 5. Membrane waterproofing, 6. Surface finish, 7. Fillet in cement or brickjelly concrete in corners, 8. Membrane waterproofing taken into parapet, 9. Joint filler, 10. Finish in 1 : 1 cement mortar whitewash and kraft paper (see also Fig. 28.2).

4. A layer of hexagonal wiremesh of galvanized iron 22 gauge with 20 mm (3/4 inch) openings is fixed on the insulating material (thermocole slab) by an adhesive so that the membrane is continuous without holes.
5. 1 : 2 : 4 concrete of average thickness 10 cm is applied on the above layer to the necessary slope (1 in 100).
6. Two layers of terrace tiles are laid over the concrete after curing with necessary slope to facilitate drainage.
7. Detail the junctions of walls properly as described in Section 27.7.

Vapour barriers. Insulating materials are effective against transfer heat only because they contain still air between the fibres in minute spaces. If hot humid air is allowed to get in these air spaces, the vapour will condense later when cooled and the pores will get filled with water. Then the materials will loose their insulating property as water in the pores is not a good insulator. Hence a barrier for the hot humid air should be *provided at the hot side of the insulator* which is called *a* ***vapour barrier***. It can be a sheet of bitumen, polythene (as described above) or aluminium sheet impermeable to moisture. PVC membrane also acts as a good vapour barrier. Similar principles apply in air conditioning also.

27.4.4 Method 4: Waterproofing and Insulating Roofs by Elastomeric Membranes

There are a large number of modern elastomeric paints, *which are white in colour*, used for waterproofing. They are made of high quality polymers and pigments that are solar reflective. It is claimed that this coating system has been specially developed to be durable and able to reflect the heat waves. Upon curing, it forms a seamless, joint free, watertight *elastic membrane* (as opposite to plastic bitumen) making the coated surface impermeable. There paints bond well with most building materials such as concrete, bricks, tiles, asbestos, bitumen, aspect coatings, etc. It can even be used to repair old asphaltic surfaces and can be painted over bituminous materials after using the specified primer coat. Separate primer coatings are available for coating this paint on asphalt and concrete surfaces (see waterproofing materials in the book on *Building Materials*). These paints, being white, also give insulation for the roof.

27.4.5 Method 5: Waterproofing and Insulating by Mud Phuska Terracing with tilebrick paving

Step 1: Coating with bitumen. This method of waterproofing and heat insulation is used in North India. The concrete roof is first coated with bitumen as described in section 27.4.1.

Step 2: Mud phuska application. Special soil, suitable for brickmaking (P.I. value of 10 to 15%) is mixed with just enough water so that the soil can be made into a stable ball. The mud is applied on the bitumen laid surface to a slope 1 : 50 with and about 25 mm thickness near the water outlets.

Step 3: Mud plastering. After laying the above mud phuska, the surface is coated with 25 mm mud plaster (a mixture of clay soil, wood shavings and water).

Step 4: Gobri leaping. After the plaster has dried, the surface should be coated with *gobri* (cowdung) leaping so as to fill all the hair cracks that may be in the mud plaster. The paste is made with a mixture of fresh cowdung and clay sieved through fine sieve and water. Cutback bitumen thinned with kerosine (penetration 80/100) can also be added to make the mud plaster waterproof.

Step 5: Paving with tiles. Flat tile bricks are laid using a minimum of plain mud mortar as bedding for correct slope. The mud mortar should not rise in the vertical joints by more than 12 mm. The tiles should be laid close so that the joints are 6 to 12 mm thick. The tiles are allowed to set and the mortar is alloured to dry up without workmen walking on it.

Step 6: Grouting joints. After the bedding mortar has dried up, the joints of the tiles are grouted with cement mortar made with waterproofing compound at 1 kg per bag of cement.

Step 7: Water curing. As soon as the cement sets (after 8 to 12 hrs), the surface should be covered with wet gunnybags and later the surface should be cured with frequent sprinkling of water for seven days.

Note: Good performance of this surface depends on prevention of cracks in the tile joints so that soaking of the soil below from rainwater does not occur.

27.5 WATERPROOFING OF CONCRETE SLOPED ROOFS AND THEIR VALLEYS WITH WATERPROOFING SLURRY COATS

When concreting sloped roofs, it is difficult to place concrete on slopes of sloped roofs properly. This is especially true for valleys in sloped roofs. In many cases, the concrete tends to be of low slump and hence porous after it is placed. In such places, it is the usual practice to treat the leaking sloping surfaces of the roof with cement grout *as soon as the formwork is removed*. First we test the roof for leakage by pouring water over it and wherever there is leakage, we pour a thin grout and then successively increase the amount of cement and thus grout the porous surface thoroughly. A better method is to treat it with waterproofing slurry as mentioned in Section 27.3 (item 5). This slurry is applied before tiles are placed on these roofs. It consists of two components to be mixed with a slurry and then applied. As already stated, these slurries are found to perform better if they are not exposed to sunrays but are covered with tiles or a layer of mortar. Otherwise it should be reinforced with glass fibre cloth.

27.6 DETAILING OF ROOF AND PARAPET WALL JUNCTION

This is a very important subject as improper detailing of junctions of roof slab with parapet or wall is a major source of leakage in many buildings. It is important that the prescribed details should be incorporated for a leakproof concrete roof. The detailing will depend on the type of waterproofing used. We will study two cases—that of lime concrete and bituminous felt waterproofing—to understand the basic principles involved.

Junction between roof and parapet when laying lime concrete terracing with tiles. The following details are important:

(a) It is usually specified that in the case of a flat concrete roof, a brick parapet wall, *one brick thick, must be built* for a minimum height of 30 cm. The lime jelly concrete and the roof tiles should extend into the wall at least 50 mm as shown in the Fig. 27.1.

(b) In addition, along the junction of the parapet and roof, we should lay additional tiles at 45° to the face of the parapet wall after laying a backing of brick jelly concrete at the junction as shown in Fig. 27.1.

(c) Another arrangement of building khura is shown in Fig. 28.2.

Detailing junction when waterproofing with bitumen felt. A similar treatment with bitumen felt or bitumen mastic can be made by taking the felt or mastic layer up to the parapet and tucking it into a chase (hole) 65 mm × 75 mm placed at a minimum height of 150 mm above the finished roof level as shown in Fig. 27.2. This fillet should be filled with 1 : 4 cement mortar or 1 : 2 : 4 baby jelly concrete.

27.7 DETAILING JOINTS IN SLABS AND BEAMS

As explained in Chapter 10, expansion joints are provided in structures. Joints in slabs, beams, etc. at roof level should be properly treated with waterstops, joint filler, etc. as shown in Figures 27.3 and 27.4. Similarly, if expansion joints are not provided in long sunshades, it will crack at intervals and cause leakage of rainwater.

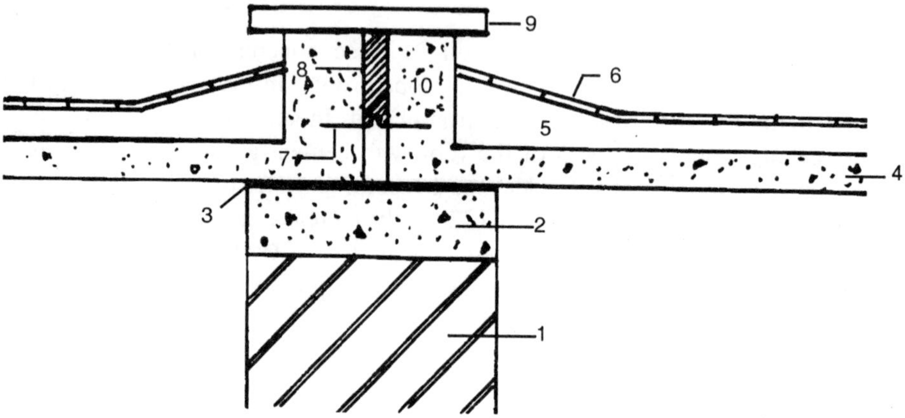

Fig. 27.3 Expansion joint between beams supported on brickout at its ends: 1. Brickwork, 2. Concrete bed block, 3. Kraft paper, 4. Slab, 5. Brickjelly concrete, 6. Membrane waterproofing, 7. Waterstop, 8. Joint filler, 9. Two courses of concrete slab laid breaking joints, 10. Beams.

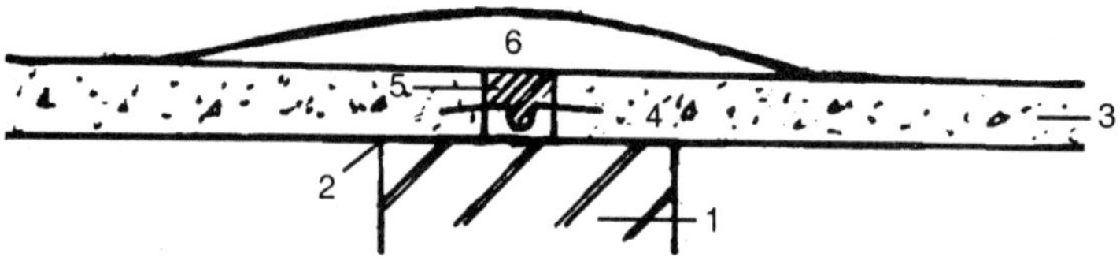

Fig. 27.4 Expansion joint in slabs on walls: 1. Brickwork, 2. Kraft paper, 3. Slab, 4. Waterstop, 5. Joint filler, 6. Waterproofing.

27.8 REDUCTION OF HEAT TRANSMISSION THROUGH ROOFS

The methods used for the transmission of heat from roofs are the following:

1. Use heat insulation materials along with waterproofing as described in sections 27.4.3 and 27.4.5.
2. Build a false ceiling with an air gap under the roof.
3. Apply reflective materials such as white paint, glazed tiles, etc. on the roof.
4. A cheap method is to whitewash the roof during every summer.
5. Arrange a sprinkler system to sprinkle water intermittently on roof.
6. Sloped concrete roofs are generally insulated from the direct rays of the sun by tiles fixed to the roofs on wooden reepers fixed to the top of the roof or by lean lime/ cement mortar (Section 22.12).

27.9 WATERPROOFING OF WET AREAS IN A BUILDING

Toilets, baths, area below sinks, etc. are called *wet areas* as these areas tend to remain wet most of the time of the day. Porticos exposed to rain also come in this category. Leakage from these wet areas from upper floors above the ground floor is a common problem in multistorey buildings. Leaking toilets and baths as well as presence of wet patches under the slab, peeling of paints under these rooms, etc. are caused by seepage of water in these areas. This is especially important when parts of the water supply and drainage pipe system in toilets, are buried or concealed under the floor as is the practice in modern buildings. Areas near floor traps in bathrooms and washbasins are also potential places of leakage. Great care should be taken in waterproofing these areas *during construction itself* as any repair after construction will be costly and tedious. In Section 27.10, we will deal with the basic principles to be considered in these areas.

27.10 BASIC PRINCIPLES OF WATERPROOFING WET AREAS

Usually the floor slab of toilets and bathrooms is sunk for easy drainage of floor, for concealing of water supply and sanitary pipes and to accommodate the depth of *nahni trap* and water closet (WC) drainage systems. Generally sunken depth of 200 mm (8 inches) is enough for *nahni* traps. Indian WC requires a minimum depth of 500 mm (20 inches). (It can also be built on a raised platform.) After laying the water closet pipes, etc., the space is filled with brickjelly concrete in lime *without sand* (sand will increase density of concrete). The top 100 mm can be the base concrete for the floor over the brickjelly. There are two factors that enter into the problem. First is the water tightness of the *structure* (floor, wall, etc.) and second is the water tightness of the various pipe and *sanitaryware joints*. Both these aspects should be looked into for an efficient solution of the problem. Taking care of one aspect without taking care of the other will not give a satisfactory solution.

27.11 METHOD OF WATERPROOFING WET AREAS

If the tile joints of the wet areas (as in bathrooms) are treated in the same way as other dry areas, in the course of four to five years, lots of tiny cracks can be formed at the various joints of tiles (due to drying shrinkage of cement). Water can seep through these joints and saturate the brickbat koba in the sunken area. This can also happen in the pipe joints after a few years of completion of the work if the various pipe joints have been joined only with cement mortar (and not by special joint fillers that do not shrink). Moreover, if *the underside of the brick koba* (the concrete floor) is not properly treated for water tightness, all the moisture in the sunken space tends to flow down through the lower slab.

Hence we must take care of both the following points:

(a) Special treatment of the floor and surrounding area.

(b) Carefully carry out the various joints of the concealed sanitary connections.

27.12 TREATMENT OF SUNKEN FLOORS IN WET AREAS FOR LEAKAGE

The principles to be followed in detailing the sunken and bathroom areas are described below (see Fig. 27.5):

1. The top surface of the sunken slab corresponding to the wet area and the depression given in the slab for Indian water closet or concealment of the discharge pipes should be rendered waterproof by plastering floor concrete with cement mortar 1 : 3 mixed with integral waterproofing compound. The surface of the plaster *should be further painted by* bituminous coat or a slurry coat using the two mix components available in the market. (Alternately the Indian WC can be built on a raised platform in the room so that there is no need for a depression in the floor.)
2. In a depressed floor, it is advisable that the wall portion of the depression is also covered with 50 mm thick 1 : 2 : 4 concrete with nominal 6 mm MS reinforcement. The concrete should be cast leaving the exit holes on the walls for a 25 mm pipes to pass through at the predetermined positions as shown in Fig. 27.5.

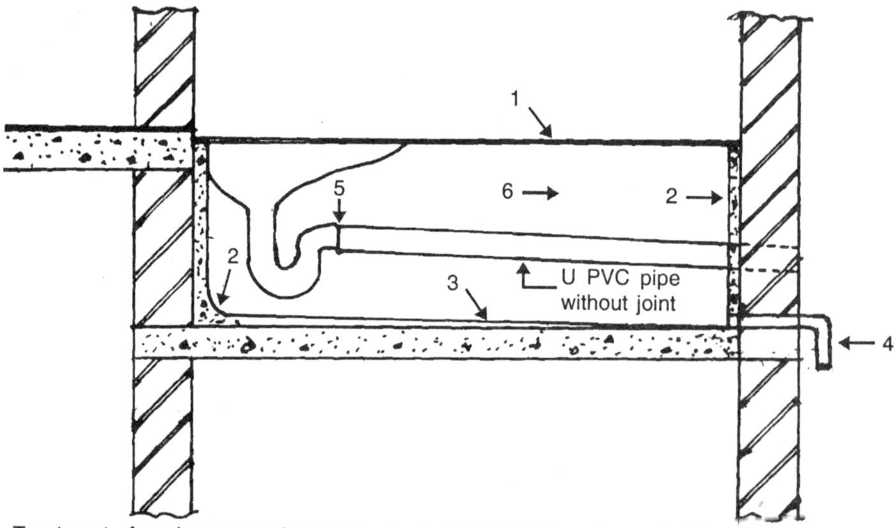

Fig. 27.5 Treatment of sunken areas for water closets (WC) in upper floors: 1. Floor tiles with special joint filler, 2. 50 mm of 1 : 2 : 4 reinforced cement concrete layer coated with britumen or special waterproofing slurry, 3. Waterproofing plaster coated with bitumen as special waterproofing slurry coat, and give a slope of 1/40, 4. 25 mm exit pipe, 5. Pipe joints with special plastic sealant to prevent shrinkage and also allow some movement, 6. Brickjelly concrete (without sand) filling.

Note. Nowadays to make repairwork easy we use a level floor (instead of sunken floor) with fittings below step. A ceiling is provided at the level of the old sunken floor which can be taken out for repairs.

3. A 25 mm PVC exit pipe should be laid from the bottom of the sunken floor projecting 275 mm (9 inch) outside the wall so that any possible leak from the joints will escape through the pipe. The pipe should be located away from any ventilator or window openings below.

4. In case of the European type WC situated far away from the outer walls, it is customary to take the WC connections to the outer face through a sunken floor. In such case, the fittings and pipes should always be tested by a water test for water tightness after they are fixed and before they are covered. For this test, the exit end of the pipe can be closed and water allowed to stagnate up to the top surface of the fixture and checked for leakage. Alternately a smoke test using a suitable apparatus should be made to check for leakage of joints. These tests should be made compulsory for all concealed pipes.
5. After checking the joints, the space around the unit (Indian W/C) and the connecting pipes should be filled with *brickjelly lime concrete without sand* (as the use of sand will result in denser and less drainable concrete). The top 100 mm (4 inches) of the filling can be in cement concrete forming the base concrete for the flooring.
6. If the floor tiles are fixed with cement mortar (which may shrink and cause leakage), the joints should be raked and filled with nonshrink, waterproof, tile adhesive or suitable plastic filler such as epoxy putty.
7. All the pipe joints inside the encasement should be made as stated below in Section 27.12.1.

27.12.1 Taking Care of Joints of Pipe and Sanitary Fittings

It is not enough if we make the structure waterproof. All the pipes and sanitary joints should also be watertight. For this purpose, we should proceed as follows:

1. Sanitary pipe joints, trap joints, etc., that are to be concealed under the floor, should be sealed with multipurpose epoxy putty especially manufactured for that purpose. These joints should not shrink and will be flexible to allow slight movements. Most cement joints(such as jute and cement joints) may be satisfactory to start with but will give trouble a few years after their installation as it is rigid. Any slight movement tends to produce small cracks which can lead to leakage. Hence for *these concealed joints,* we should not be satisfied with the ordinary cement joints that shrink with time.
2. Bituminous or asbestos rope coverings should be preferably provided to these concealed portions of the pipes and joints of fittings.
3. In all cases, the concealed pipe fittings should finally be tested for leakage before they are buried in mortar (see Chapter 38).

Note: Water-supply pipes that are always filled with water under pressure, if concealed, can give more leakage and trouble than drainage pipes. In general, water supply lines should not be laid in floor concrete.

27.12.2 Providing Vapour Escape Holes on Sides

Most of the wet areas are exposed to the outside air on one side. Hence in portions such as sunken floors filled with brickbats, we may also think of providing an exit for the moisture that may be trapped in the brickkoba through these external walls. A 25 mm PVC pipe as shown in Fig. 27.5 can be provided for this purpose.

SUMMARY

The subject of waterproofing and weatherproofing of concrete roofs and wet areas is an important item of work in modern building construction. There are many more ways than those given in this chapter for waterproofing depending on the location of the building with respect to the different regions in India. It is also very necessary to detail the joints and junctions of roofs according to modern practices so that there will be no leakage through these joints during rains. Only a very brief account of these aspects has been discussed in this chapter. Waterproofing of wet areas of the floor such as bathrooms is also important. It is a special item to be considered especially when pipes and sanitary fittings are concealed.

REVIEW QUESTIONS

1. (a) How would you detail the top of the load-bearing walls bearing the (i) floor slabs and (ii) roof slabs?
   (b) Sketch the detail of the end bearing of a 100 mm RC roof slab on a 1½ brick load bearing wall.
2. Sketch the detail of the junction between the parapet wall and the RC slab provided
   (a) with lime concrete terracing with two layers of tiles laid for waterproofing and
   (b) with bituminous left.
3. Describe briefly the traditional and the modern methods of membrane waterproofing by bituminous membranes.
4. Describe briefly the following:
   (a) Waterproofing with lime concrete terracing and one layer of pressed tiles
   (b) Thermal insulation of concrete roofs
   (c) Waterproofing of a valley of sloped concrete roofs
   (d) Slopes to be provided for drainage for:
      (i) Lime concrete terracing without tiles
      (ii) Lime concrete terracing with tiles
      (iii) Mud phuska terracing
      (iv) Waterproofing with bituminous membranes
5. What are the wet areas in a building? Give the general principles of waterproofing of wet areas in a building.
6. (a) Sketch and describe the arrangements for waterproofing the depressed area in a bathroom in the upper floors of a building through which the drainpipes of a wet water closet are taken to the stack pipe.
   (b) If a depressed area is not provided in the first floor of a building, how would you install an Indian type water closet?
7. The OBD paint used in the ceiling below a bathroom of a flat was found blistering. What can be the reason for this trouble? What remedial measures would you recommend (a) if the trouble is only marginal and the peeling of paint is not severe and (b) if the trouble is acute?

Chapter 28

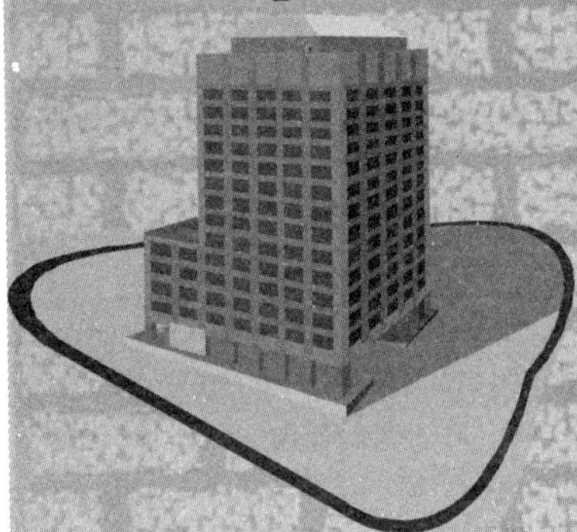

Roof Drainage and Repair of Leakage

28.1 INTRODUCTION

Even though roof drainage is one of the last pieces of work to be carried out in the construction of a building, it is a very important item of work, especially in places of high rainfall. The drain systems in flat roofs and sloping roofs are dealt with differently. Only some of the basic principles of *roof drainage* are briefly dealt with in this chapter. Waterproofing of roofs and wet areas are separately dealt with in Chapter 27. In this chapter, we will deal with drainage of roofs and the usual causes and remedy for leakage in buildings from rains.

28.2 MATERIALS FOR RAINWATER PIPES

Cast iron, asbestos cement and unplasticized PVC (UPVC) pipes are commonly used as rainwater pipes. Of these, UPVC pipes are the most used nowadays, as they are light, unbreakable and easy to install. The basic material for these special pipes is PVC. Other materials such as stabilizers lubricants, fillers and pigments are added to it. The stabilizer protects it from ultraviolet radiation of sunlight. Since the quantity of plasticizers is kept less than 4%, it is called UPVC (Unplasticised rigid PVC) pipes. However, AC pipes are much cheaper than PVC pipes and are used to reduce total cost of construction. However, maintenance of work with AC pipes will be costlier than that with PVC pipes. The following points should be remembered when fixing rainwater pipes:

1. The rainwater pipes should be fixed to the outside of external walls and if possible, especially in case of PVC pipe in special recesses, should be built in external walls. With this arrangement, we can shield plastic pipes from direct sunlight.
2. A bell mouth inlet at the roof surface in a flat roof is effective only if proper slopes of the roof are provided towards the inlet. Good planning should be made in providing the necessary slopes.

3. The rainwater pipes should *run through the parapet wall* so that the rainwater does not come in direct contact with the parapet wall during its discharge into the drainage pipe. It should not be fixed to the outer side of the parapet wall.
4. Nowadays, it is recommended that the rainwater pipes should discharge into devices to facilitate "rainwater harvesting". For example, the water on the roof may go directly to fresh water sump or to rainwater collecting wells or pits as desired.

28.3 DRAINAGE OF FLAT ROOFS

In planning drainage of roofs, the roof area should be conveniently divided into areas and sloped by lime terracing or other means to the pipe entry points as shown in Fig. 28.1. Stagnation of water on flat roof is one of the common reasons for leakage and dampness in flat roofs. Hence the slope of the roof terrace should be such as to drain off all the rainwater before it has a chance to seep into the roof. The minimum recommended slope for a roof with roof tile is 1 in 60. For a non-treated rough surface, it is preferable to keep it as 1 in 40 to 1 in 50 so that the water is drained quickly. The size and spacing of the rainwater pipes also require special care. In general, the diameter of rainwater pipe should depend on the rainfall in the given area, but it should not be less than 100 mm (4 inches). Atleast one 100-mm pipe should be provided for every 40 sq m of the roof area to be drained. (Theoretically, for an average rainfall intensity of 50 mm per hour, a 100-mm pipe should drain 85 sq m of the roof area.)

28.3.1 Method of Fixing Drainpipes in Flat Roofs

Firstly the flat roof area should be divided into areas suitable for drainage of each pipe and sloped in such a way that rainwater will flow easily to the point where the drainage pipes are located. The level of the drain pipes should be so fixed in relation to the roof slab that the rainwater completely drains with no water collecting even near the drain. The entry to the pipe should be at the lower level of the slab. The division into areas and some of the methods of fixing the rainwater pipes at the entry point on flat roofs are shown in Fig. 28.1.

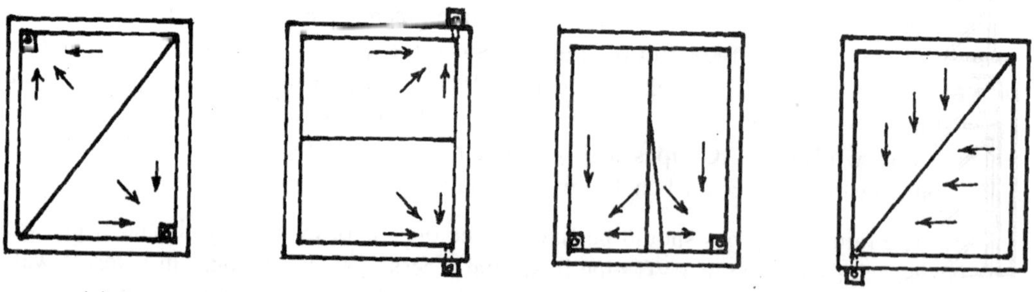

(a) (b) (c) (d)

Fig. 28.1 Drainage of roof: various slope arrangements and positions of rainwater pipes for roof drainage.

28.3.2 Construction of Concrete Khurras and Golas

CPWD specification gives the details of construction of khurras for fixing drainage pipes and golas in parapet walls in flat roof terraces.

Construction of Khurras. Khurras are spaces constructed near the rainwater outlet to guide rainwater from roof terrace to the spout of the roof drainpipe [see Fig. 28.2(a)]. It is constructed before the masonry work in parapet wall is taken up. A PVC sheet, 1 m × 1 m × 400 microns, is first laid on the concrete roof slab and a 1 : 2 : 4 concrete slab, 3 cm thick, is laid over it. The finished plastered top surface should be lower than the finished level of the roof by the following values to facilitate drainage:

(i) 20 mm in roof finished with lime roof terrace
(ii) 50 mm in roof finished with mud phuska with brick tile covering
(iii) 70 mm in roof finished with lime concrete with brick tile covering
(iv) The khurra and the sides of outlet and junctions are plastered with waterproof 1 : 3 12-mm thick cement mortar. This plastering should be done when the concrete is still green. It should be 45 cm × 45 cm in size when finished.
(v) The size of the 3 cm concrete slab should be larger than 45 cm × 45 cm so that the topping of the lime concrete should overlap it by at least 7.5 cm on all the three sides.

Construction of golas. The junction of the terrace with any vertical surface such as parapet is highly vulnerable for leakage [see Fig. 28.2(b)]. Hence we form a concrete projection with 1 : 2 : 4 concrete in a chase 75 mm wide and 75 mm deep cut in the parapet with a slope at the top to drain off the water. The top surface is plastered with 1 : 3 waterproof cement mortar. To allow for expansion, the joints should be provided in these projections at 3.5 to 4.5 m length, which should be filled with bitumen mastic (of proportions 80 kg bitumen, 1 kg of cement and 0.25 cm of sand). Other methods of detailing these junctions are shown in Figures 27.1 and 27.2.

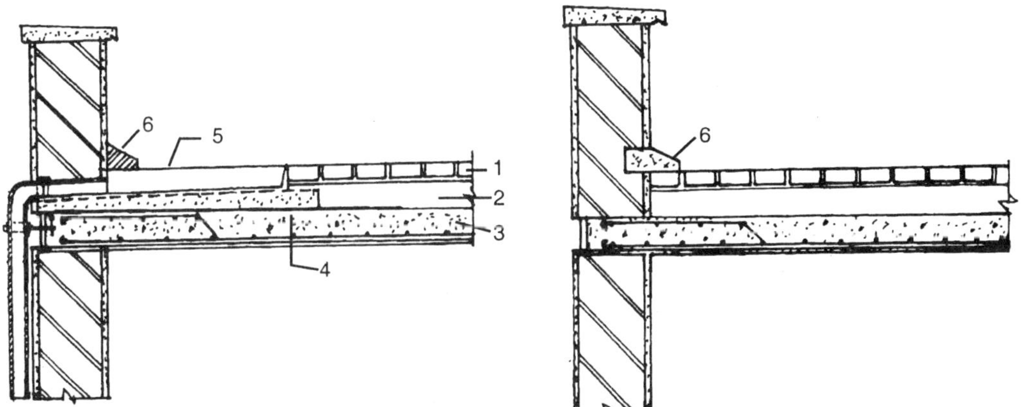

(a) Khurra for installation of drainage pipe (b) Gola

Fig. 28.2 Construction of khurra and gola for good drainage and prevention of leakage of junctions in roof: 1. Brick roof tiles, 2. Lime concrete or mud phuska, 3. RCC slab, 4. Polythene sheet covered with concrete, 5. Drainage space, 6. Gola construction at junction of parapet and roof.

28.4 DRAINAGE OF SLOPED ROOFS

In sloped roof construction, we use *valley gutters, eave gutters* and *boundary wall gutters* for drainage and disposal of rainwater. The minimum slope to be provided for these gutters is 1 in 150. These are shown in Fig. 28.3.

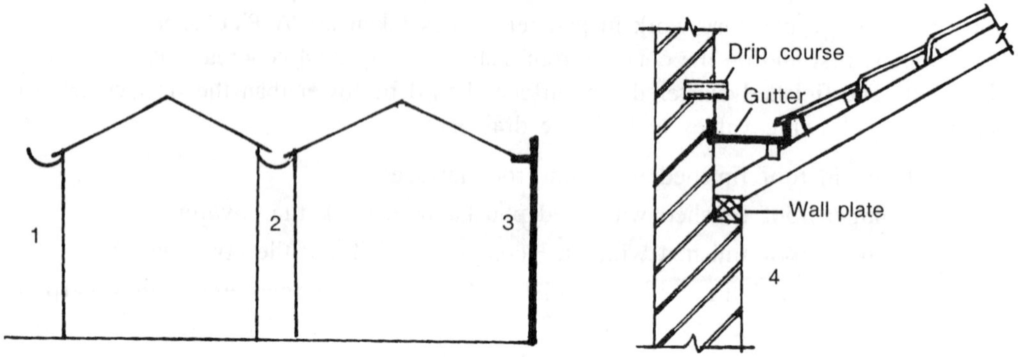

Fig. 28.3 Gutters in sloped roofs: 1. Eave gutter, 2. Valley gutter, 3. and 4. Boundary wall gutters.

Valley gutters. These are provided at the valleys or junctions of two slopes. In traditional tiled roof construction, these were made of galvanized iron sheets. Nowadays plastic gutters are available. They rest on the valley rafter and the reapers.

Boundary wall gutter. In industrial buildings, where steel trusses and AC sheets are used, and when the roof slopes to the end of the property line (or boundary line), special gutters are provided to carry rainwater off the boundary. These are called ***boundary wall gutters*** as they are formed with boundary walls.

Eave gutters. These are provided at the eaves of sloping roofs to collect rainwater and dispose it of through drainpipes. In single-storey buildings, as the height of fall of rainwater from roof is not high, gutters may not be provided. In ordinary buildings, they are laid on brackets fixed to the fascia at the edge of the roof. In industrial buildings, they may also be in the form of "boundary wall gutters".

28.5 CONSTRUCTION OF OPEN SURFACE DRAINS

The size of the open drain built in the ground to carry rainwater is specified by the width of the drain at the top. Narrow drains of 15 or 25 cm width with proper slope and depth are enough to carry the rainwater. The performance of these surface drains can be improved by providing "egg shape" for the lower portion of the drain. It should correspond to the acute angle end of an egg in order to provide good flow even at minimum discharge. This type of egg-shaped drains can be built by a mason using a template. The surface should be plastered smooth with 1 : 3 cement plaster of 18 mm thickness. The slope should be sufficient to give a velocity of flow of about 0.6 to 1 m/s when running half full. The maximum speed of water should not exceed 3 m/s to avoid erosion due to sand and other gritty materials that may be in the water. They are usually built with a 1 : 2 : 4 cement-concrete slab of 100 mm thickness as foundation on the ground and half brick walls built on top of the slab. The inside width can vary from 150 to 250 mm and height can be 200 mm to 300 mm depending on the required discharge.

28.6 SOAK PITS

In granular soils, where the groundwater level is low and well below the ground level, the rainwater or even waste water can be collected in soak pits so that it can seep into surrounding soil. They are described in Section 31.7 and shown in Fig. 31.5. When the groundwater level is high, we resort to dispersion trenches which are narrow channels filled with coarse materials such as stone or gravel just below the ground so that the water percolates down through the top soil as shown in Fig. 31.4.

28.7 REPAIR OF RAINWATER LEAKAGE IN BUILDINGS

We have already discussed the methods of waterproofing of buildings in Chapter 27. However, leakage can occur with time and due to lack of proper maintenance. Leakage of rainwater in buildings is very annoying and also damaging. It spoils materials and equipments stored in the rooms. Electrical circuits with earth leakage circuit brakers (ELCB) will trip with wet walls. Otherwise the electric leakage in wet walls can give shocks and is dangerous to the occupants. The most important factors in repair of water leakage are the following:

1. The material that is used for repair is suitable for the repair.
2. The source of leakage should be carefully identified and rectified.

28.7.1 Materials Used for Waterproofing

As explained in the book on *Building Materials*, some of the many available waterproofing materials used in building construction are as follows:

1. Integral waterproofing compounds.
2. Waterproof coatings on concrete with special coatings which with capillary action forms compounds inside the capillaries in concrete.
3. Waterproof coatings such as hot blown asphalt.
4. Membranes with blown asphalt as in tar felt.
5. Elastomeric coatings with elastomeric paints.
6. Elastomeric coatings with reinforced fibre of polyester or glass fibre fabrics for crack bridging.
7. Epoxy compounds (such as araldite).

28.7.2 Rainwater Leakages and Their Treatment

The commonly seen leakages and their treatments are discussed further.

Leakage in flat terrace roofs. The following are the important considerations:

(a) All concrete flat roofs during construction should be thoroughly cured by pounding water over it. Any leakage noticed at this stage can be rectified easily by cement grouting (by pouring grout of consecutive higher cement) till the leakage stops. Alternately, all the surfaces can be given a coat of hot blown asphalt after curing.

(b) The necessary (minimum 1/50) slope for drainage should be provided to the drainpipe points and water should not stagnate at any place on the roof during rains.

(c) The necessary fillets, where the horizontal surface meets a vertical or inclined face (Figures 27.1, 27.2 and 28.2), should be provided and waterproofed. This item is very important and many leakages along walls are due to defeats at this junction.

(d) All the entry points of the rainwater drainage pipes should pass through the parapet walls as shown in Fig. 28.2. If they are laid beside parapet walls they should be properly detailed. Entry points should not be blocked with leaves, etc.

(e) The waterproofing of the roof should be intact and any cracks of exposed surfaces of tiles joints in tiles, joints with other surfaces such as roof and parapet should be properly grouted. If necessary, blown bitumen or elastomeric paints with fibre glass, polyester or other reinforcement to give gap to the crack bridging strength to the coatings.

(f) Very porous concrete slabs may have to be grouted under moderate pressure to make it free of voids.

Leakage in sloped roofs and shells. It is quite common that these sloped surfaces are not compacted properly during construction due to difficulty in compaction. As the surfaces are sloped and the concrete is dense, they tend to drain well during normal rains. However, if the concrete is porous, water can seep through during heavy rains and they tend to leak. This is especially true of valleys of sloping roofs where, unless concrete is well placed, it can become porous (see Section 22.12).

Hence all parts of the sloping roof and in any case, all the valleys should be checked for leakage and grouted at the construction stage itself. As the roofs are usually covered with tiles, it is difficult to repair the roof after construction. It is advisable at the construction stage itself to apply a waterproofing coating in the form of (plastomeric) bituminous coat or special coating which work on crystallization technique or elastomeric paint coat with fibre reinforcement, if necessary (depending on the state of the concrete). A large number of slope roofed buildings have been found to leak in heavy rains due to lack of this precaution.

In cases where the roof is very porous, grouting with cement grout under low pressure may be necessary. However, these are extreme cases.

Leakage through junctions of walls and sunshades. This is very common in many buildings and the following items should be checked:

(a) Check whether proper slope and drainage have been given at the construction stage. Water should drain away from walls and should not become stagnant in the sunshade. In any case, the architectural features over sunshades should not cause blocking of rainwater flow from the sunshade. (This is a common mistake made in many buildings). A throating (or drip course) at the bottom end of sunshades will make the draining water to drip properly (Figures 8.7 and 10.4).

(b) Continuous long sunshades without joints tend to crack at intervals. Water can flow through these cracks and to the walls. Fill the cracks with crack-filling bituminous or elastomeric compounds (see also Fig. 27.4). All leaves, etc. collected over the sunshades should be removed and sunshades should be kept clean before the rainy season.

Leakage through wall cracks. The following items should be attended to (see also Section 39.3):

(a) All external cracks in walls should be deepened to 5 mm and filled with paintable sealant. It is also good to paint external walls with special waterproof cement or external paint which can keep the walls waterproof. Silicon paints in walls containing airconditioning or other electrical connections wills prevent wetting of walls during rains.

Seepage through junction of wall and roof and floor slab. We have dealt with the top of roof slab under 'leakage in flat terrace roofs'. Another important place to check for leakage is the bearing on the junction between the roof/floor slab and the wall—in simple wall and slab buildings as well as in framed buildings. There is a possibility of crack due to expansion of concrete and movement over the wall which will be seen more under roofs than under floors. If there is a crack at this junction at the external face, the water that flows down the wall tends to flow into the building. This can be avoided if during the construction stage itself, a drip course is provided (by extending the slab or by constructing a separate drip coarse in framed buildings in the form of a band at that junction). The old practice of providing bituminous craft paper under the slab on walls is to avoid this trouble. If such leakage occurs in an existing building, the joint has to be sealed with a water proofing sealant or by application of elastomeric paint with glass fibre or polyester fabric reinforcement.

Leakage from external electric points. The conduits laid for wiring for external light fixtures can also form source of leakage of rainwater if they are not sealed properly. Though leakage through these conduits can happen in many unexpected places of the building if they are connected together. All external light points and fittings should be sealed with good *sealants* and made waterproof.

Joints of slabs and beams. Joints provided by the designer in slabs and beams should be properly filled with *approved joint fillers*. As described in the book on *Building Materials*, joint fillers should be properly installed to allow expansion and contraction of the structure. (Joint fillers are different from sealants.)

SUMMARY

Roof drainages of flat and sloped roofs are carried out differently. Drainage of a flat roof requires more attention than that of a sloped roof as any faulty construction can flood the flat roof and lead to leakage and corrosion of steel in the roof slabs in the long run. Hence attention should be given to proper drainage of these roofs. Similarly drainage of large sloped areas, as in industrial sheds, should be properly detailed. There are many case histories of very large damages occurring in factories during heavy rains due to faulty design and blocking of drainage pipes due to bad maintenance of roof drainage systems. When repairing for a rainwater leakage, it is the most important consideration to find the exact source of the leakage and then treat it efficiently using the right materials. Rebuilding the structure in the system to prevent leakage is rather costly and in many cases, unnecessary.

REVIEW QUESTIONS

1. (a) What slope would you give for flat RC roofs for drainage of rainwater? How is this slope provided?
(b) Sketch the methods of fixing rainwater drainage pipes (khurra) to a flat roof at the entry point. Describe how it is constructed.
(c) What is a 'gola'? Describe how it is constructed and indicate its use in flat roof construction.

2. (a) What types of gutters are used for drainage of rainwater in sloped roofs?
(b) What methods would you recommend to dispose of the rainwater collected on a roof by the roof drainage system?

3. What are soak pits and dispersion trenches? What are the types of soak pits and dispersion trenches used in practice? *(See also Chapter 31.)*

4. What are the usual types of leakages of rainwater found in buildings and how would you rectify them?

Chapter 29

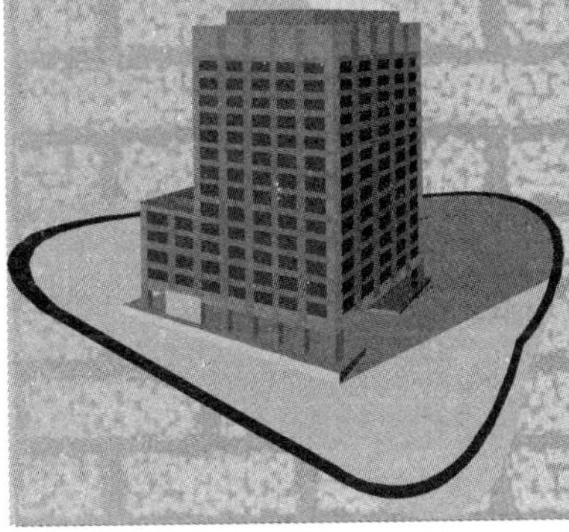

Water Supply in Buildings

29.1 INTRODUCTION

We can divide the plumbing systems in a building into the following four groups:

1. The rainwater drainage system *above the ground level* (which is covered in Chapter 28)
2. The water supply system inside the building (which is covered in this chapter)
3. The soil and wastewater drainage system *above the ground* (which is covered in Chapter 30)
4. The drainage *system below the ground* carrying the foul water to septic tank or the public sewer (which is covered in Chapter 31)

In this chapter, we will only deal with the water supply system in a building.

29.2 WATER SUPPLY SYSTEM

In villages and small towns, water is obtained from private wells. The usual diameter of domestic wells can be 1.2 m (4 ft) and that of agricultural wells can be as much as 3 m. Water is usually pumped up to an elevated storage tank from where it is distributed in the building. However, in large towns and cities, filtered and treated water is supplied by the public agencies through street mains. The commonly used connection from the municipal street mains to the buildings is carried out as shown in Fig. 29.1.

Note. If a metre is introduced in the supply line, it should be in a lower position so that it is always filled with water. If left moist for a long time, it is likely to get out of order.

The brass or bronze ferrule shown in Fig. 29.1 can be inserted into the mains which may be under pressure. The flexible goose neck, 40 to 50 cm long, is made of brass, copper or plastics and is provided for flexibility of the system. The main inlet service pipe is nowadays

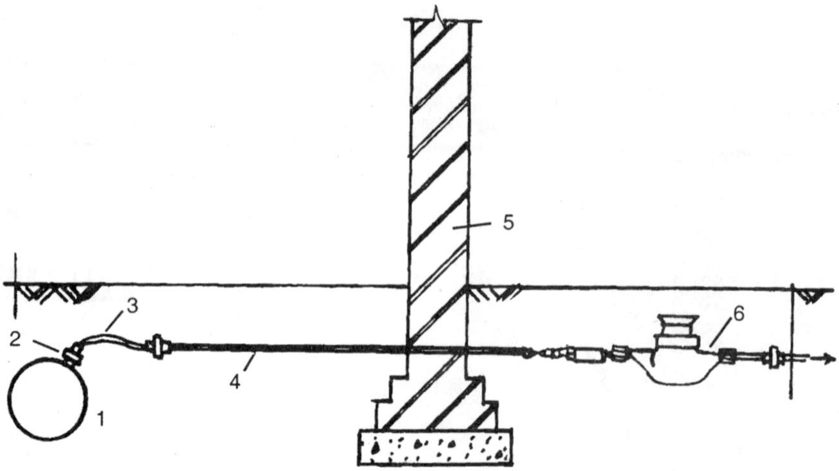

Fig. 29.1 House water supply from street mains: 1. Street mains, 2. Ferrule, 3. Goose neck, 4. PVC (noncorrodable) connecting pipe, 5. Boundary wall, 6. Watermeter.

made of plastics so that it does not corrode when buried in the earth. Generally, the diameter of the service pipe to buildings is fixed on the basis of the occupants. The diameter varies from 12.5 mm (for 4 persons) to 30 mm (for 60 occupants).

If there is enough pressure, the municipal water supply can be directly connected to an overhead tank. However, in most towns and cities in India, the pressure is too low, so a sump (water tank below the ground level) has to be provided within the house property, preferably as near to the street water mains as possible. The inlet to the sump should be at sufficient depth so that water from the municipal supply line can flow into the tank. The depth of the tank should also be such that there should be enough storage capacity below the level of the incoming pipe. It can happen that there may be no supply of water through the mains in summer, and then, we may be forced to buy water. Hence, *the total capacity* of the low level storage tank (sump) should be, if possible, a little more than the capacity of the water lorry that supplies water which is usually 9000 to 12,000 litres.

Water is pumped from this sump at regular intervals to an overhead tank from where the water is distributed to various parts of the building. The capacity of the overhead storage tank will depend on the number of persons using it. It usually varies from 200 to 500 litres. In any case, if it is more than 5000 litres in large buildings, it is advisable to arrange it in a series of interconnected tanks so that they can be isolated for easy cleaning. This tank can be of brick, concrete or readymade PVC tanks. Its outlet pipe is fixed at 50 to 75 mm above the bottom of the tank preferably with copper strainers. A drainpipe is also provided to the tank to drain and clean it.

The underground and overhead tanks should be cleaned at regular intervals. They should also be disinfected after cleaning by using chemicals. For this purpose, we should give the water a dose of 50 parts of chlorine in one million parts of water. If ordinary bleaching powder is used for this purpose, the proportion should be 150 gm of powder for every 1000 litres of water in the tank. The powder is mixed with water to a creamy consistency and then suitably added to the tank before filling. The pipe can also be disinfected by allowing the above water to run through the pipes—one pipe at a time.

In some layout of water supply in buildings, a common but separate flushing storage tank is provided exclusively for flushing water closets, etc., as the quantity of water used for this purpose is large and this water can be of low quality recycled or untreated water such as bore well water. The water to this tank is usually supplied separately and not directly from overhead tank. Table 29.1 gives the recommended capacities of water tanks to be provided for these separate flushing storage tanks.

Table 29.1 Flushing storage capacities

S.No.	*Classification of buildings*	*Storage capacity*
1	Residential houses	270 litres for one WC seat and 180 litres for every additional WC seat
2	Tenements having common convenience	900 litres per WC seat
3	Factories and workshops	900 litres per WC and 180 litres per urinal seat
4	Cinemas, assembly halls	900 litres per WC seat and 180 litres per urinal seat

Note: 1. One flushing will need 5 to 15 litres depending on the type of the tank.
2. For water supply in buildings with no municipal connection (see Section 31.11).

29.3 DESIGN OF WATER DISTRIBUTION SYSTEM

Design of water supply in multistorey buildings needs special layouts and many factors are to be considered in their design. Here we will only consider water supply in ordinary low-rise residences and deal with the following items:

- Estimation of water requirements
- Determination of size of pipes to be used
- Principles of the pipe layout

29.3.1 Estimation of Water Requirement

IS 1172–1983 and SP 7–1983 recommend 135 litres of water per head per day to be used for design of water supply system for residences. (A supply of 540 litres per day is considered enough for a family of four. The rated capacity of the common large plastic bucket is 20 litres.) The consumption is estimated as shown in Table 29.2. In case of buildings such as hospitals with beds, it will be larger (of the order of 340 litres per bed per day), and for a day school, it can be as low as 45 litres per head per day. In many towns and cities in India, the supply of treated water from the mains is much less, and it has to be supplemented from other sources such as bore wells inside the property coupled with rainwater harvesting.

Table 29.2 Approximate pattern of domestic consumption of water

S.No.	*Use*	*Litres*	*Percentage*
1.	Drinking (water, tea, etc.)	5	3.7
2.	Cooking	10	7.4
3.	Bathwash (hygiene)	35	26.0
4.	Washing clothes	15	11.0
5.	Toilet flushing, etc.	55	40.8
6.	House cleaning	15	11.1
	Total	135	100

29.3.2 Size of Distribution System

For an average middle class house, where there will be only 5 to 10 taps, pipes of minimum nominal size of 20 mm (3/4 inch) are adequate. When water closet flushes are directly operated from the overhead tanks, a 25 mm size is usually recommended for these pipes so as to get sufficient flow. The minimum permissible sizes should be as given in Table 29.3.

Table 29.3 Recommended size of water supply pipes

Details of pipe	*Size of pipe* (mm)	*Size of pipe* (inch)
Service pipe	20	3/4
Bath tub and shower stand	12	1/2
Lavatory	10	3/8
Water closet (tank type)	10	3/8
WC valve type	25	1
Kitchen sink	12	1/2
Drinking fountain	10	3/8

The sizes of the various pipes can also be estimated as follows:

Riser from underground tank to overhead tank is to be 25 mm.

Separate pipes from overhead watertank should be laid for supply of water to each floor (say ground and first floors).

Sizes of outlet pipes should depend on the number of connections.

To control the delivery from overhead tanks, a stop valve or gate valve must be fixed *in an accessible place* to stop the water supply during repairs.

A thumb rule for sizes of pipes which supply two or more branch pipes is as follows:

(a) Up to 3 numbers of 10 mm (3/8 inch) branches can be supplied by a 12 mm (1/2 inch) pipe.

(b) Up to 3 numbers of 12 mm (1/2 inch) branches can be supplied by a 20 mm (3/4 inch) pipe.

(c) Up to 3 numbers of 20 mm branches can be supplied by a 25 mm (1 inch) pipe.

More complex arrangements are needed for a multistorey building with many flats at each level.

29.3.3 Principles of Pipe Layout

The following four basic rules should be followed in laying out of water supply lines:

1. There should be *no cross connection and backflow*. Backflow, as shown in Fig. 29.2, is the flow of polluted water into the water supply system. For example, where the water supply outlet is located below the flood level of a cistern, the outlet might submerge when the cistern is filled. In such cases, reverse flow can happen if the pressure in the water supply line drops. This is called backflow or *back siphonage*. Special care should be taken to see that backflow does not happen from WC flush tanks. Backflows can be eliminated by ensuring the following:
   - (i) A suitable air gap is always provided between the inlet and final flood level in all places such as washbasins.
   - (ii) A positive pressure at the pipe outlet points should always be assured.
   - (iii) In cisterns connected to WC, the inlet and float valve should be suitably fixed to prevent backflow.

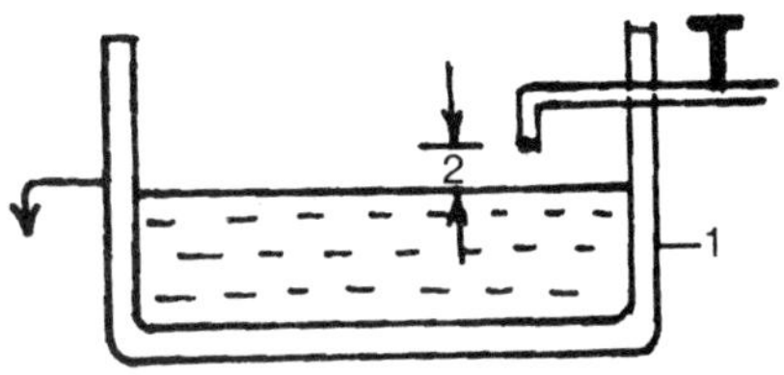

Fig. 29.2 Prevention of backflow in water supply fittings: 1. Washbasin with overflow device, 2. Air gap (25 mm minimum).

   The air gap for a washbasin should not be less than 25 mm (1 inch) and that for a lavatory supply should not be less than 40 mm (1.5 inch). For kitchens and laundry, the air gap should be increased by 50% of the latter.

2. Water supply pipes and drainage pipes should not be laid very close to each other under the ground or above the ground. There should be no chance, under any circumstances, of the drainage pipe leaking on the water supply pipe. Drainage pipes, especially stoneware pipes which are liable to leak if damaged, should never be laid above the water supply pipes. Similarly, water supply pipes should not be laid alongside the pipes carrying foul water (sewage water and waste water).
3. All pipework should be planned so that it is accessible for inspection, replacement and repairs. Burying GI water supply pipes in the ground is not a good practice as they tend to corrode in contact with soils. GI pipes can run along walls.
4. All pipes, including PVC pipes, laid inside the building as concealed pipes should always be tested under pressure for leakage before the final concealment.

29.3.4 Caution in Concealed Piping

It should be emphasized that as in most modern buildings, where the plumbing work inside the building is generally concealed, sufficient safeguards and precautions should be taken to prevent leakage. GI pipes tend to corrode when encased in brickwork without concrete cover. Hence it is advisable that we use inert plastic pipes. All pipes meant to be concealed should be first tested for leakage wrapped with bituminous material and tested again before concealment (see Section 38.11). The CPWD recommended pressure for testing of pipes in ordinary buildings is 6 kg/cm^2 equivalent to 60 m of water.)

29.4 WATER SUPPLY FITTINGS AND FIXTURES

As water supply is a separate subject of study for civil engineers and architects, we will not go into details of the subject. There are many types of fittings (different from pipe fittings such as sockets, reducers, etc.) which are to be used for water supply. Some important ones are discussed here and shown in Fig. 29.3.

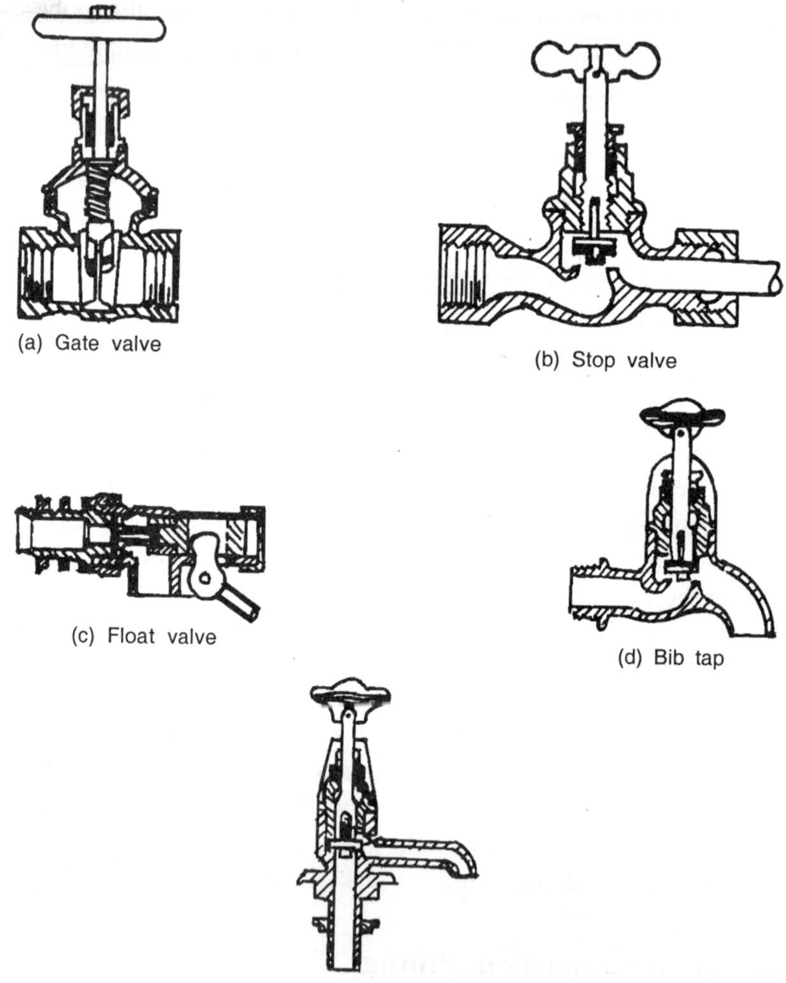

(a) Gate valve

(b) Stop valve

(c) Float valve

(d) Bib tap

(e) Pillar tap

Fig. 29.3 Common types of valves and taps.

1. *Stop valves (IS 781–1984).* These are fixed in an accessible place *for each outlet* such as WC, washbasin, etc. to control flow and facilitate repair [Fig. 29.3(b)]. There are various types of stop valves available in the market.
2. *Gate valves (IS 778–1984).* These operate a gate in the pipe, and when fully open, it offers very little resistance to flow [Fig. 29.3(a)]. They are used for closing the *main pipes from tanks.*
3. *Ball or float valves (IS 1703–1989).* These are provided in storage tanks, flushing cistern, etc. to control the level of storage of water [Fig. 29.3(c)]. When water reaches a designed level, its supply is cut off.
4. *Bib taps or Bib cocks (IS 781–1984).* These are the common types of fittings used in washbasins [Fig. 29.3(d)].
5. *Pillar tap (IS 1795–1982).* It is the outlet provided from a vertical inlet and horizontal outlet in washbasins in kitchens, laboratories, etc. [Fig. 29.3(d)]. Various types of pillar tops available in the market.
6. *Self-closing taps (IS 7711–1970).* These are provided in railway stations, hotels and remain open as long as the valve *is pressed down* and closes when the pressure is released.
7. *Magnetic taps.* These are installed in airports, hotels, hospitals, etc. Such type of tap operates as soon as the hand is placed below the tap (as it cuts off the magnetic effect) and closes automatically as soon as the hand is removed. This is a very hygienic tap where one does not have to touch the tap to operate it.

There are many other fancy water supply fittings used in buildings which can be seen in any sanitary shop dealing with these devices.

Note. The above list of valves are those used inside buildings. There are many other types of valves used in a pipe line conveying water from reservoirs or tanks. The following are some of them and they are different from those used inside buildings. They are (a) *Sluice valves* for isolating the flow used in irrigation work, (b) *Air inlet valves* used when the flow in the pipe is gravity flow and (c) *Check valves* used in places where reverse flow can happen.

29.5 MATERIALS TO BE USED FOR WATER PIPES

The type, sizes and pipe materials to be used in buildings should be carefully selected. GI pipes are excellent if the quality of water running through the pipes is good and the pipes do not come in contact with the ground or bricks. These pipes available in three grades (or classes)—A, B and C. Grade C pipes are the costliest and meant for heavy duty purposes. Grade B pipes are generally used for water supply in buildings. Colour bands painted at every 2 m spacing can be used to identify these pipes. The colours are yellow for grade A, blue for grade B and red for grade C. Where the quality of water is good, the distribution pipes can be laid above the ground (not in contact with bricks, etc.) concealed in concrete or laid over continuous sunshades of the building.

Unplasticized PVC (UPVC) pipes are nowadays preferred as water supply pipes.

However, they do not fare well if exposed to sunlight. Moreover, as the coefficient of expansion of UPVC pipes is about eight times higher than that of GI pipes, the supports or clamps should allow free longitudinal thermal movement and should not grip the pipes tightly. UPVC pipes also require closer supports. The support spacing should be about 700 mm for 20 mm pipes (35 d) to 975 mm for 50 mm pipes (20 d). More details of layout of plumbing come under the subject Building Services under public health engineering.

SUMMARY

Only a very brief survey of the water supply to a residence is given in this chapter. Supply of water in very large multistorey projects involves a number of other considerations and is an important subject in civil engineering. It will be learned in greater detail under the subject Environmental Engineering.

REVIEW QUESTIONS

1. Sketch and explain the connection to be made for supply of water from the street water mains to a building. Explain how you will decide on the size of the water supply sump to be built for a residential building.
2. Describe briefly how water taken from water mains is supplied to different parts of a building in a residential house situated in a city.
3. List the various appliances used for water supply in a residential building.
4. Discuss briefly what material and sizes of pipes would you select for the internal water supply in a residential building for a medium size family.
5. Explain how to select roughly the sizes of various pipes to be used for water supply in a residence. What are the four basic rules to be followed in laying water supply pipes in a building?
6. What is meant by backflow? How is it prevented in the layout of water supply pipes in a residence?
7. Write short notes on the following types of fittings, indicating their nature and use:
   (a) Magnetic taps
   (b) Ball valves
   (c) Gate valves
   (d) Stop valves
   (e) Bib taps
   (f) Pillar taps

Chapter 30

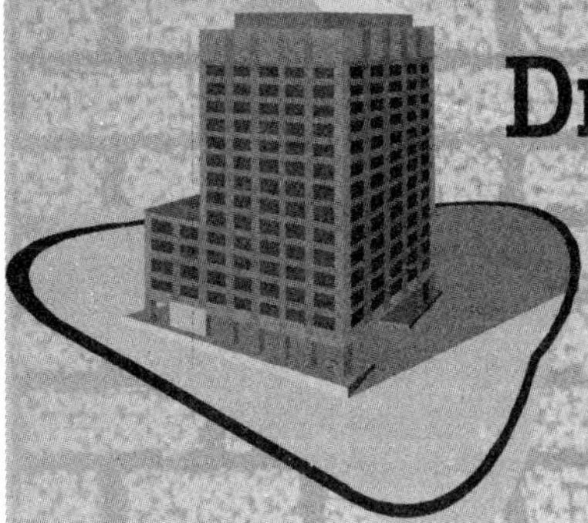

Drainage of Wastewater and Sewage above Ground

30.1 INTRODUCTION

We have dealt with drainage of rainwater in Chapter 25. The next problem is the drainage of foul water. Water discharged from waste appliances (such as washbasins, sinks, showers, etc.) is called *wastewater or grey water*. Water discharged from soil appliances (such as water closets, urinals, etc. is called *sewage*. The wastewater from sewage is called *foul water*. The wastewater and foul water have to be collected safely and disposed of from a building. There can be mainly three types of arrangements or sewerage systems that can be designed to collect the foul water above the ground—combined system, partially separate system and separate system.

Combined system. In this system, all the foul water (waste and sewage) is carried away by the same drain.

Partially separate system. In this system, the whole of the sewage and a part of the wastewater are carried off by the same system. The rest of the wastewater is carried off separately.

Separate system. In this system, the wastewater and the sewage are carried away separately. The wastewater is carried off and recycled or seeped into the ground. Sewage is to be drained to the sewer or to the septic tank. (Many believe that wastewater which contains disinfectants, chemicals, detergents, etc. should not be led to septic tank not only because it is bad for bacterial digestion but also because it unnecessarily increases the needed capacity of the septic tank. In an economic layout, only sewage need be drained and digested in the septic tanks. This consideration need not apply *when the drain is connected to a public sewer.*)

The collection and disposal of foul water is treated in two ways—drainage above ground and drainage below ground level. In this chapter, we will deal principally with the system of drainage of waste and sewage water above the ground. Drainage system for foul water *below the ground* (about sewers, manholes, etc.) will be discussed in Chapter 31.

30.2 PIPES TO BE USED FOR DRAINAGE OF FOUL WATER

The pipes used for drainage of foul water above the ground are as follows (see Figures 30.1 and 30.4):

Waste pipes from sinks, etc. These carry wastewater from sinks, washbasins, etc. The sizes used are 30 to 50 mm (generally 40 mm) initially where it is horizontal and 50 to 75 mm (50 mm usual) when the horizontal pipes join as vertical down pipes.

Soil pipes. These carry human waste. Usually a minimum of 100 mm diameter pipes are used as soil pipes for WC which is also of the same diameter as the main down pipes.

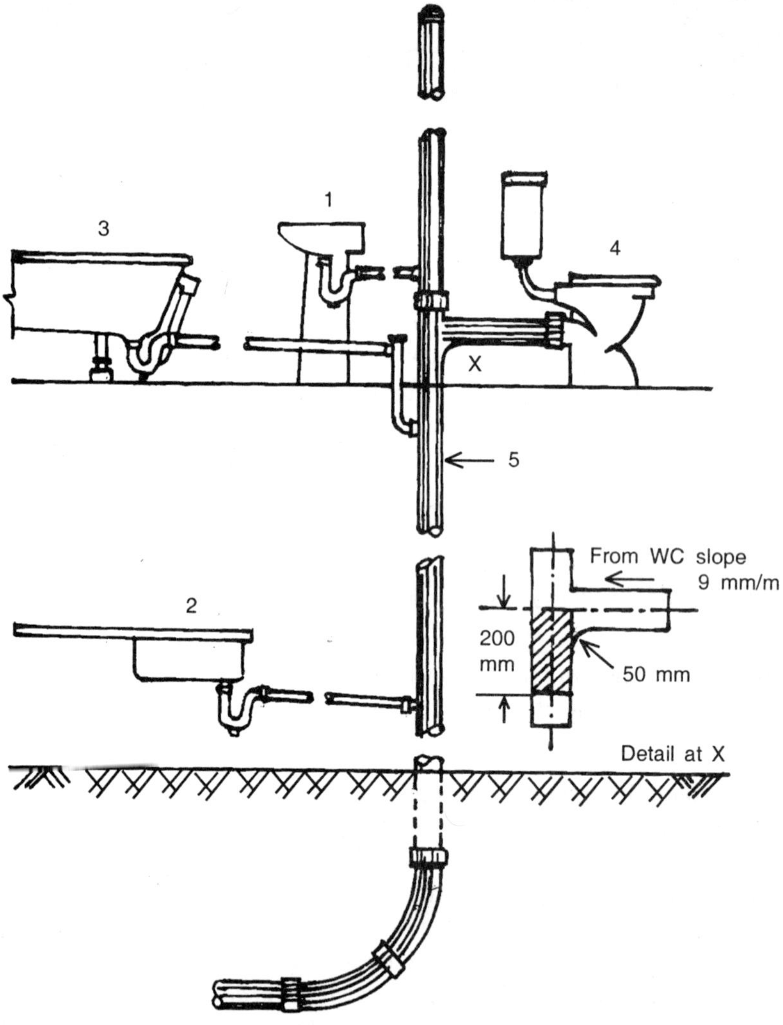

Fig. 30.1 Single-stack drainage system without vent pipes: 1. Washbasin, 2. Kitchen sink, 3. Bath, 4. Watercloset 5. Stack pipe.
(*Note.* Required diameters of drainpipes for each appliance is shown in Table 30.1. Connections with vent pipes are shown in Fig. 30.4.)

Vent pipe (ventilating stack). These are the main vertical pipes open at both top and bottom carried at least 1 m higher than the roof level. This is used for ventilation or escape of foul air. When exclusively used for venting, it can be of 50 mm diameter. But in single stock system described below, the soil and vent pipes are combined and then the minimum diameter should be 100 mm.

Anti-siphon pipes. These pipes are of smaller diameter and are meant to prevent siphonage of water from the traps described in the next section. The anti-siphon pipes start from the traps and are connected to the vent pipe (Fig. 30.4). Nowadays, in single stack systems (as shown in Fig. 30.1), anti-siphon pipes need to be used only in special cases as described in section 30.6.1, item (1).

30.3 TRAPS

One of the import devices we use in buildings for drainage of waste and sewage water is the *trap*. Traps are fittings or parts of various appliances used to retain water seal which is meant to prevent the passage of foul air and insects such as cockroaches from drains to the building. They should always be introduced in salient points along the drainage system. Traps may form an integral part of the appliances such as a water closet or it may be a separate fitting (attached trap) which is to be separately installed such as the *floor trap* and *Nahani trap* in a bathroom. A good trap should be self-cleaning and should form an efficient water seal under all conditions of flow as well as in the absence of flow. In water closets and floor traps, a minimum water seal of 50 mm is to be maintained. However, in 'hand poured flushed' type rural (Indian) water closets, it may be specified as 20 mm only for manual flushing. In general, these traps can be classified as P, Q or S types as shown in Fig. 30.2. Some of commonly used traps and their various locations are described further.

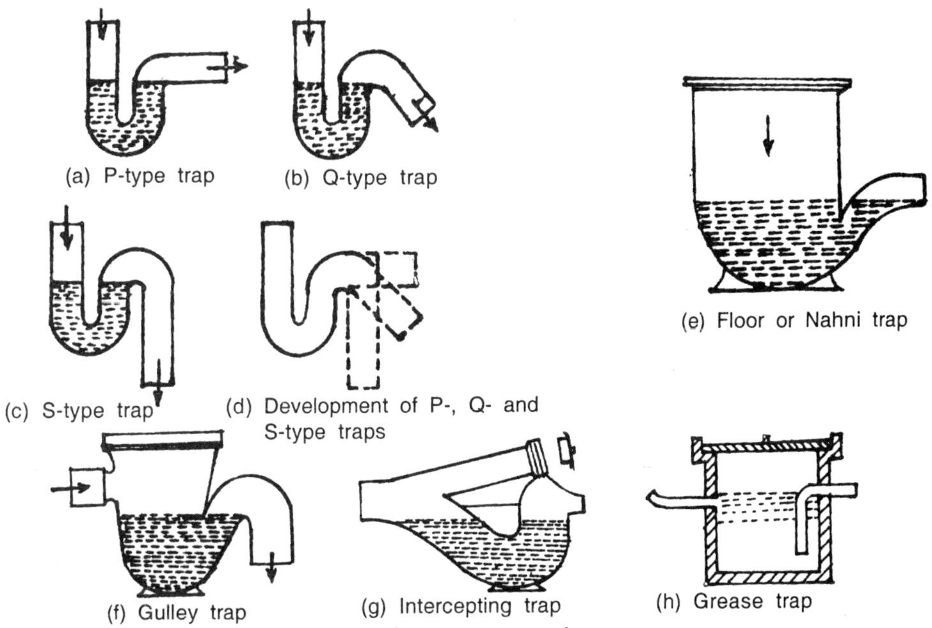

Fig. 30.2 Common types of traps.

Traps integrated with appliances. These are P, Q or S type of traps that come with appliances such as water closets [see Figures 30.2(a)–(d)].

Floor traps. These are provided in the floor of wet areas (bathroom, washroom, etc.) to collect wastewater from washbasin, bathroom, etc. These are usually made of cast iron or PVC and are provided with a grating at the top of the floor level to prevent large particles, hair, etc. to fall into the trap and clog it. It generally gives a 50 mm water seal. PVC floor traps are the most popular traps nowadays. We should always provide a floor trap in bathrooms, under washbasins, sinks, etc. as it will prevent cockroaches and bad odour entering the room from the drains. A special type of floortrap is the *Nahani trap* shown in Fig. 30.2(e). These traps with water seal will have a minimum depth of 100 mm (4 inches). If they are to be provided in upper floors, the thicknesses at their planned locations should be at least 150 mm. Alternately, in areas such as bathrooms, the floor is usually depressed by a suitable amount. (In case of Indian type squatting pans, the trap should have a depth of 320 mm to 430 mm depending on the make.) In case the floor of bathrooms on upper floors is not depressed or thickened to install the traps, for washbasins, etc. we have to be content with shallow cast iron or plastic devices without water seal. Usually, a removable perforated cup with a grating is provided in these devices to prevent the entry of cockroaches and other insects. These cups can be taken out for cleaning.

Gulley traps. The term *gulley* means an opening for drainage. When we want to discharge rainwater or wastewater to the sewer drain, we discharge it through a gulley trap so that the foul air and insects from the sewer do not escape to the outside [see Fig. 30.2(f)]. It is usually made of stoneware or cast iron. A cast iron grating is also usually provided at the top. It can be an S- or P-type trap with a water seal of 50 mm.

Intercepting trap. This type of trap as shown in Fig. 30.2(g) is provided at the end of the house drain before the street sewer to prevent entry of sewer gas into the house drain. This trap is used in drains below the ground.

Other traps. There are also other traps such as grease trap for collecting grease from kitchen water or food processing unit [Fig. 30.2(h)] and sand trap to collect the grit when grit is used for washing purposes.

30.4 SANITARY APPLIANCES

There are many types of appliances used for sanitary works. We will only deal with the following two principal appliances to understand the plumbing system used for soil and waste water disposal:

- Soil appliances (water closets)
- Waste appliances (washwater appliances such as washbasins, kitchen sink, bath tubs, etc.)

30.4.1 Water Closets

The following four types of water closets are mainly available in India as shown in Fig. 30.3:

1. Oriental design or squatting type—washout system
2. Anglo-Indian type (combined)—washout system
3. European design or sitting type
4. Siphonic type

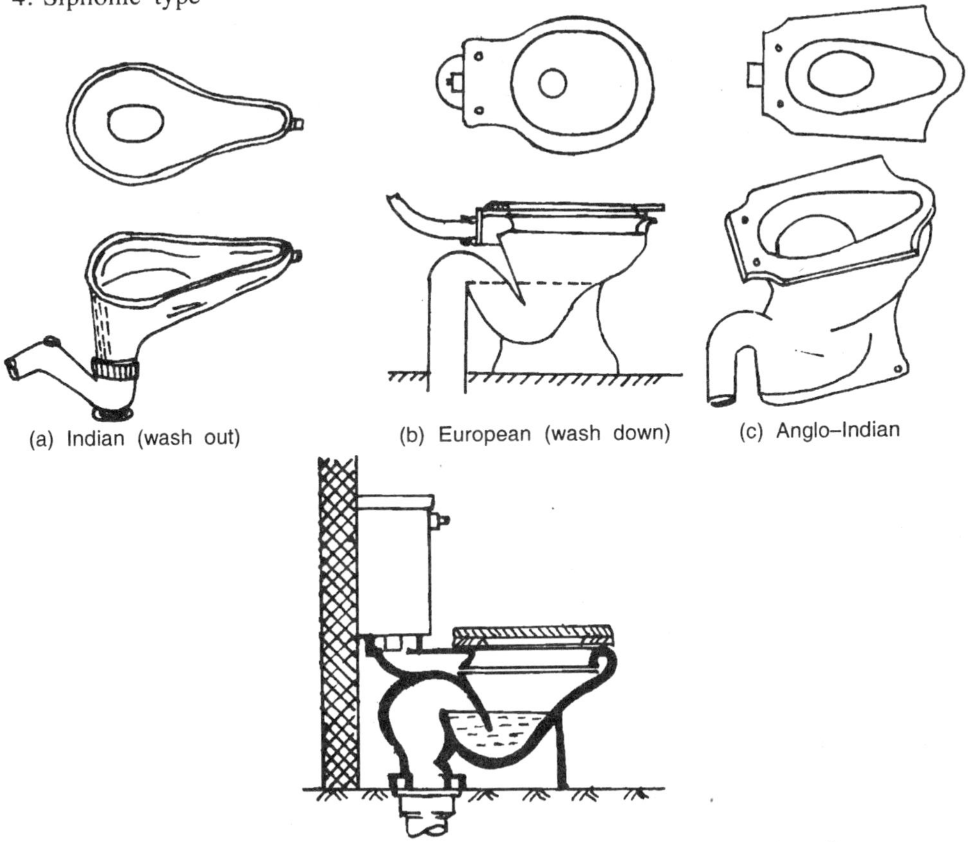

(a) Indian (wash out) (b) European (wash down) (c) Anglo–Indian

(d) Siphonic low level pistern (can be in single or double trap format)

Fig. 30.3 Types of water closets.

The disadvantage of the Indian WC design is that the human excreta falls in the pan and has to be washed out. In the European design, the excreta falls directly into the trap water and hence, it requires less water to flush the system than in the squatting type WC. In the European type WC itself, there are two categories, namely:

(i) The washdown closet which relies on velocity of the flush water to wash away the excreta.

(ii) The siphonic water closet which works on siphonic action and is quiet and efficient in its working, but has higher blockage risk if not used properly.

The common flushing arrangement for water closet is the siphonic type. The tank also has a float valve arrangement to cut off water supply when the tank is full. Central Building Research Institute (CBRI), Roorkee, has developed devices to reduce the consumption of water. By incorporation of these devices, flushing tanks for partial or full flushing as is necessary can be made. Mechanical arrangements by which the flushing system is connected to a high level tank as well as low level flushing tanks operated by a valve with a handle are commonly used in many public buildings and residences in India.

30.4.2 Washwater Appliances

There are many types of washwater appliances such as washbasins, kitchen sinks, etc. They are fitted in bathrooms, kitchen and other places. As they can be seen in any building, they do not need further description.

30.5 PREVENTION OF SELF-DRAINAGE OF TRAPS

It is very important that the water in various traps should not be drained off. The trap water can be drained off by one of the three ways, namely *siphonage, momentum* and *evaporation*. Siphonage can be caused by induced siphonage or self-siphonage. *Induced siphonage* is caused by water from adjacent fittings running full bore in the main pipe bearing a vacuum in front of the trap. This can be avoided by connecting the trap to a ventilator shaft. Such a connecting pipe is called an *anti-siphonage* or *ventilating pipe*. (Even though ventilating pipes are nowadays not used for ordinary buildings, for very tall buildings, these are still used.) *Self-siphonage* can occur when the discharge from the waste pipe runs full bore in a long pipe. It causes suction at the back of the trap. The possibility of such siphonage can be reduced by *limiting the length of the connection of the trap to the main pipe and also by making the radius of the bend of the connection from the trap to the main pipe easy.* On an average, if the length from inlet to discharge of discharge pipes is large (more than 3 metres), a ventilating pipe connection has to be introduced. In addition, the minimum radius of the bend should be kept 50 mm for smooth flow. Evaporation loss in traps happens only when the system is not used for a long time in hot weather.

30.6 SINGLE- AND TWO-STACK SYSTEMS

The waste and sewage water (foul water) can be drained separately by two pipes or alternately collected in one pipe and disposed of into the sewer. If we are to use the single-stack system, we have to provide a discharge pipe of minimum diameter 100 mm for the foul water drains or stack.

Theoretically, as is done in practice, we have to provide another *ventilating pipe* of smaller diameter to ventilate the traps. The function of these pipes is to ventilate the traps to atmospheric pressure so that the water in the traps is not siphoned off along with the dischange of the foul water. Thus there have to be two stacks, one of which is the *discharge stack* and the other is called cross connection or *ventilating stack* as shown in Fig. 30.4.

However, research by the Building Research Establishment in UK has shown that practically for buildings up to 15 storey high, a single stack for collection of all the foul water (without another stack for ventilation) is enough for proper drainage, if the conditions given in section 30.6.1 further are satisfied. They can be met when the sanitary appliances are grouped together. Such a system without a ventilating stack is called a *single-stack system*. In *two-stack system*, the seals are fully ventilated. Hence, we have two separate stacks—the *discharge stack* and the *ventilating stack*. Another modification of the two-stack system is meant for cases where the limitation of distances of 3 m of the single-stack system (as given in Table 30.1) cannot be followed. In such cases, only the branch pipes which are more than 3 m in length are provided with ventilation and the others are not ventilated. This, is the *partially ventilated system*. Thus, we can have three different systems of drainage (namely single stack, fully ventilated and partially ventilated) with or without ventilating stacks. The various details of the above disposal systems will be studied under the subject Building Services. To make their study simple here, we will examine only the commonly used single-stack system and that too very briefly.

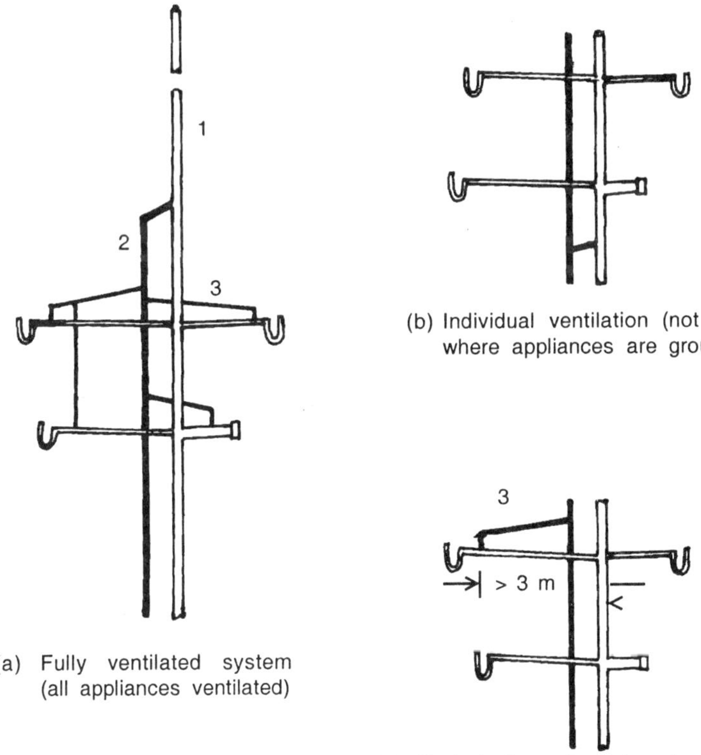

(a) Fully ventilated system (all appliances ventilated)

(b) Individual ventilation (not provided where appliances are grouped)

(c) Modified single-stack system (ventilation provided only for special cases as explained in Sections 30.5 and 30.6.1).

Fig. 30.4 Ventilated stack systems: 1. Stack pipe, 2. Ventilating pipe, 3. Anti-siphonage pipe.

30.6.1 Conditions to be Satisfied for Single-stack System in Multistorey Buildings

The single-stack system, where the ventilating stack is not provided, is the most economical system for buildings of moderate height, very safely up to 10 storeys. For the main vertical pipe, a combined soil waste pipe (SWP) of 100 mm for buildings up to 4 storeys and a pipe of 150 mm for buildings up to 10 storeys in height are recommended. Ventilation is provided through the same pipe open at the top. The following limitations should be followed when using single-stack system (see also Fig. 30.1). If these rules are kept, anti-siphon pipes are not neçessary.

The minimum sizes and distances given in Table 30.1 should be kept to limit siphonage in single-stack systems.

Table 30.1 Diameters and permitted lenghths of branch pipes from appliances to stack pipe in single-stack system

S.No.	*Appliance*	*Diameter of branch pipe*	*Permitted maximum length of the branch pipe from trap to stack pipe*
1	Washbasin	32 mm	1.7 m
2	Sink	40 mm	2.3 m
3	Bath	40 mm	2.3 m
4	Water closet	100 mm	1.5 m

1. The distance of the branch connection to the main stack pipe should never be more than 3 metres and preferably, it should be as given in Table 30.1.
2. A slope 18 to 90 mm per metre (depending on length of the connection) should be provided to the smaller branch connecting pipes for easy discharge.
3. Provide bends of large radius at joints of pipes. When joining pipes to the main stack, the bends should be given a minimum radius of 50 mm.
4. Water closet discharge should be connected to the stack pipe by a branch pipe of the same diameter in which the flow takes place downwards (with a slope of 9 mm/m) in the direction of the flow in the stack. The minum size to be used is 100 mm.
5. Not more than two numbers of toilet units should discharge into the same point of the single stack at each floor.
6. Bath waste should not be connected to the main down pipe at distances less than 200 mm (8 inches) below the WC branch.
7. If the distance of the connecting pipe is more than 3 m, then an anti-siphon pipe should be provided for it as shown in Fig. 30.4(c).

30.6.2 Connection for Single-storey Buildings

In a single-storey building, there are no stack pipes. The units are directly connected to the underground system through manholes described in Chapter 31. However, a vent pipe is usually provided for the passage of foul air.

30.6.3 Provision of Mosquito-proof Cover

All stack pipes are to be extended to a height 600 mm (2 ft) above the roof of the building (for prevention of bad odour) and provided with mosquito-proof nets to prevent breeding of mosquitoes in the drains as shown in Fig. 30.5. If ventilating pipes are used they need not be taken up but can be cross connected to the stack soil pipe at a point which is above the spillover level of the highest appliance in the building as shown in Fig. 30.4(a). Alternatively, the small diameter ventilating stack can also be taken up above the building for the sake of appearance.

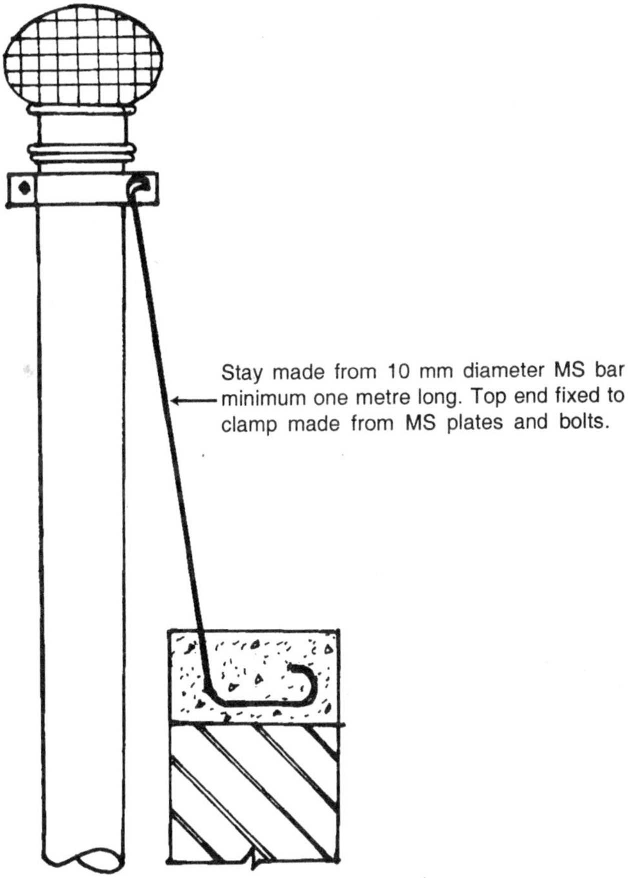

Fig. 30.5 Fixing stack pipe with mosquito proof wire mesh cage to top of parapet wall using 100 × 100 mm size 1 : 2 : 4 concrete block.

SUMMARY

Theoretically, we should have to ventilate all the drainpipes to reduce the chances of breaking the water seals of the various traps. However, in practice, for ordinary buildings, we can adopt a single-stack system if we obey the required restrictions regarding length, slope and number of discharge connections at a point in a single-stack system. Only elementary principles of the subject applicable to a residence are explained in this chapter. The subject will be more fully studied in civil engineering under Environmental Engineering.

REVIEW QUESTIONS

1. (a) Explain briefly the use of traps in drainage of water from a building.
   (b) What are the various types of traps used for drainage of a building?
   (c) How can self-drainage of the traps occur?
   (d) How would you install a floor trap in the first floor of a building?
2. What are the types of water closets available in the Indian market? Enumerate the advantages and disadvantages of each type.
3. Explain what is meant by siphonage in foul water drainage. Also explain how this can be prevented in single-stack systems without ventilating pipes.
4. (a) Explain what is meant by a single-stack system.
   (b) What are the rules to be kept in adopting single-stack system?
   (c) For buildings of how many storey high can the single-stack system be used?
   (d) What are ventilating pipes and when do you need them in the foul water drainage system?

Chapter 31

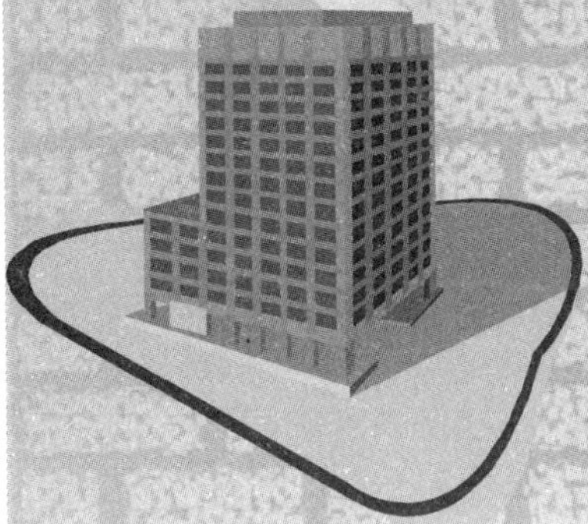

Drainage of Foul Water below the Ground Level

31.1 INTRODUCTION

In this chapter, we will consider the drainage of foul water (wastewater and sewage) below the ground level. All the foul water from various sources is to be collected together and disposed of through pipes laid below the ground level. All rooms such as lavatories, urinals, bathrooms, kitchen which produce foul water should be planned, if possible, to be grouped together and also located in such a way that the walls of one of these rooms form the periphery of the building. This facilitates an easy collection of the foul water and its conveyance to the house drain which usually runs around the perimeter of the building. In multistorey buildings, the areas that use water should be located one above the other for easy collection of discharge through the same pipe running vertically up on the side of the building. In this chapter, we will briefly deal with the common underground drainage system used in ordinary buildings. As already stated, the *wastewater* can be collected separately and recycled but the sewage is usually disposed of in street sewers or septic tanks. We will also briefly deal with septic tanks.

31.2 GENERAL LAYOUT OF DRAINAGE SYSTEM

The house drains (sewers) are usually laid around the perimeter of the building as shown in Fig. 31.1. Inspection chambers (manholes) (see Section 31.3) are provided at the various points of discharge of foul water and also at changes of direction of flow. They are also placed at intermediate points of the drain if they are too long so that the spacing of the inspection chambers is not more than 6 m in ordinary buildings (and not more than 30 m in street sewers). These chambers allow access to the underground drainage system for its inspection as well as clearing of any blockage that may happen. The layout of a house drain is illustrated in Fig. 31.1. The gradient to be given is given in Section 31.4.

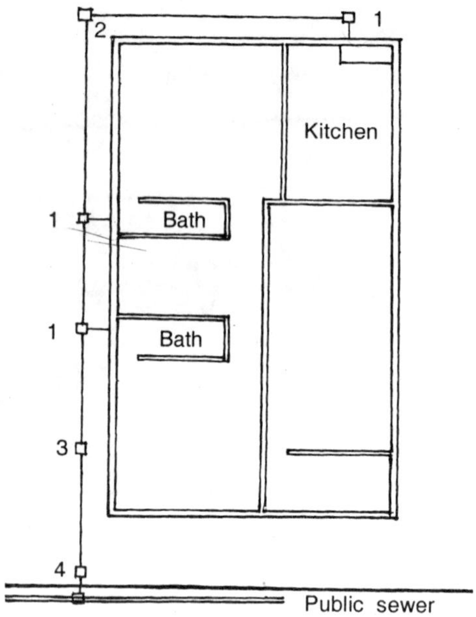

Fig. 31.1 Layout of house drains and inspection chambers for a residential building: 1. Inspection chambers at pipe junctions, 2. Inspection chamber at corner, 3. Inspection chamber at intervals, 4. Inspection chamber at end of property discharging into muncipal sewer.

31.3 INSPECTION CHAMBERS AND MANHOLES

The drainage below the ground mainly consists of inspection chambers connected by pipes. A chamber the depth of whose invert level does not exceed one metre is called an *inspection chamber*. When the depth of the invert level of a chamber exceeds one metre, it is referred to as a *manhole. Thus, manhole is a deep inspection chamber. There should be a distance of at least one metre between the layouts of the inspection chambers and the outer walls of the building*.

The inspection chambers are usually constructed in bricks as shown in Fig. 31.2. No drain from the house fitting (such as soil pipe, gulley trap, etc.) to the inspection chamber should be more than 6 metres in length (the maximum distance between the manholes being 30 m). If the drain from the building to the inspection chamber is more than 6 m in length, we provide inspection chambers in between to fulfil the 6 m rule. The inspection chambers up to 0.8 m in depth should have internal dimensions of 700 mm (in the direction of flow) by 500 mm i.e. minimum size of 700 mm × 500 mm. Usually a size of 750 mm × 750 mm is provided. Manholes of 0.8 to 1 m depth should be at least of size 1200 mm × 900 mm. Details of a manhole are shown in Fig. 31.2. For deeper manholes, circular chambers with a minimum diameter of 1400 mm are usually provided.

All soil-pipe connections (which have traps built into the system), after passing vertically down along the side of the building must be below the ground level and shall be led by an easy bend preferably into the adjacent manhole as shown in Fig. 31.2. Alternately, it should be connected by an easy bend into the house sewer pipe. Similarly, all waste pipes from the

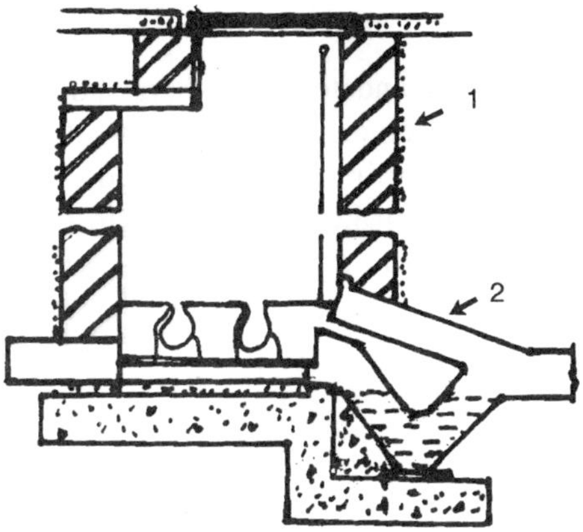

(a) Section of an inspection chamber with intercepting trap: 1. Chamber, 2. Incepting trap

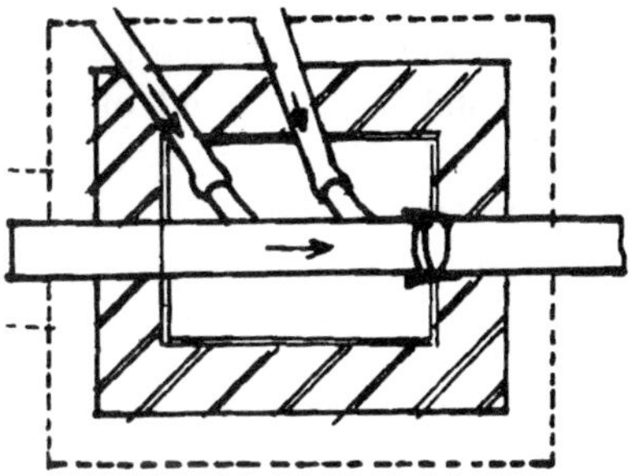

(b) Plan showing layout pipe

Fig. 31.2 Inspection chamber in drainage system.

ground level should preferably be provided with a *floor trap* and connected to the manhole. If a floor trap is not provided, the connection should be through a *gulley trap*.

31.4 TYPES OF DRAINPIPES USED

The pipes laid between the manholes are buried in the ground. Cast iron pipes are durable but costly. Stoneware pipes are cheap and were popular once. They are still used for cost reduction, but it requires experienced workmen for laying. Where there are a large number of trees around the pipes, the roots migrate to the leaky joints and damage them. Stoneware pipes come with spigot (the plain end) at one end and socket (the enlarged end) at the other. They are laid with the *plain end downstream*. Stoneware pipes are joined with two rings of hemp yarn soaked in thick cement slurry, chalked with hardwood tools and finished with 2 : 1 cement-sand mortar. The pipes are usually *laid on a bed of concrete* in good soil condition or covered with concrete in very bad situations (refer to Chapter 37).

Nowadays in many middle class residences, UPVC pipes are used extensively as drainage pipes. They cost more but the cost of labour for their installation is small. The following rules for sizes of pipes and slopes are usually recommended. The minimum size of sewage pipes should be 100 mm.

1. For less than 20 dwellings: 100 mm with gradient 1 in 40 to 1 in 50
2. For more than 20 dwellings: 150 mm with gradient 1 in 60
3. In places where public sewers are provided: The slopes should be carefully worked out so that there will be a good fall from the house drain outlet to the public sewer. There should be no backflow from the public sewer to the house drain during the monsoon.

31.5 CONNECTION TO SEWER

If the house sewer is connected to a municipal sewer, it is mandatory that there should be an inspection chamber at the end of the property before the drain discharges into the public sewer. It may also be required by regulations to provide an intercepting trap described in Section 30.3 and shown in Fig. 30.2 so that no foul gas from the sewer gets into the house sewer or drain. In many cases, this trap may be optional as in a well laid out sewer, there should be no foul gas.

31.6 SEPTIC TANKS

31.6.1 Use of Septic Tanks

Where there is no public sewerage system, the sewerage has to be led into a septic tank for treatment. A septic tank is a combined sedimentation cum digestion tank. It acts on the principle that *aerobic bacteria* thrive in the presence of oxygen of the air and *anaerobic* bacteria thrive in water or under saturated conditions. In septic tank, the sewage is retained for the specified retention time so that there will be maximum digestion of organic solid matter in the sewage and the heavy solids settle down. *It works on anaerobic digestion of sewage*. In a properly constructed septic tank, millions of anaerobic microorganisms *break down human waste* into gases, liquids, etc. The solid matter settle to the bottom and is known as *sledge*. Usually, in the tropics a storage period of 24 to 36 hours is desirable for the breakdown. In colder places it will be move up to 192 hours. We estimate the needed capacity by providing a storage of 180 to 200 litres per person using the tank if all the water used is led into the septic tank. If waste water is excluded, it will be much less. Capacity of the tank for less than five persons is not recommended. A free board of 300 mm (12 inches) is usually provided. The length is usually made three times the width. The details of the layouts of small and large septic tanks are shown in Figures 31.3(a) and 31.3(b). The details and levels of the inlet and outlet of a septic tank are very important and should be carefully laid out. Inspection covers and ventilating pipes with mosquito netting are also provided for the septic tanks.

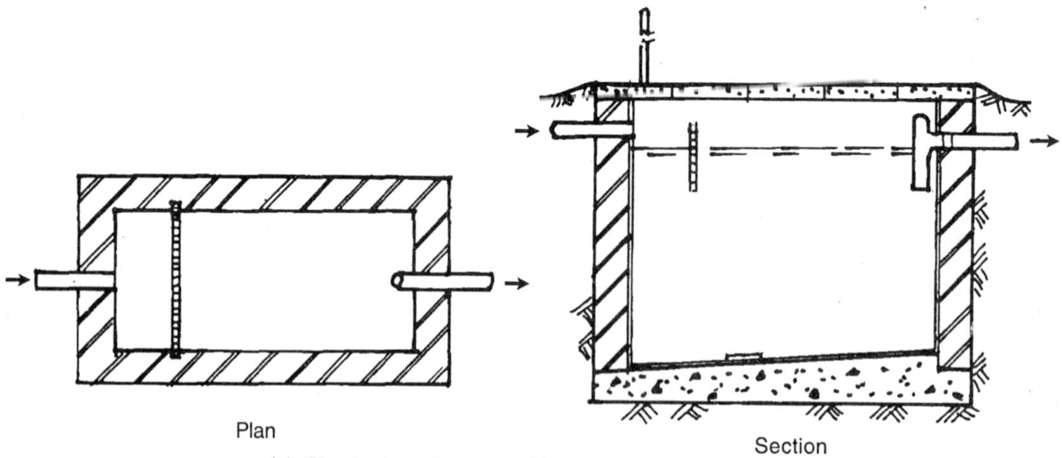

(a) Simple type for a residence with two chambers

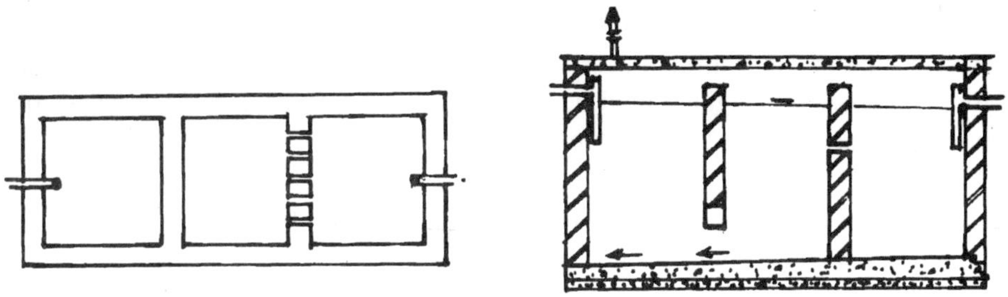

Plan Section

(b) Ideal septic tank with three chambers for large installations

Fig. 31.3 Septic tanks.

31.6.2 Principles of Design of Septic Tanks

The following are the general principles to be followed in construction of large sized septic tanks:

1. The capacity to be provided is to be a minimum of 100 litres per user and the minimum number for design should be 5 persons. The minimum retention period should be 24 to 36 hours.
2. The minimum inner width should be 75 cm (2.5 ft). The length should be three times the width. It is customary to divide a medium to large tank into 3 equal chambers. For small tanks, as in ordinary individual residences with 4 to 5 persons, two divisions may be adequate.
3. The liquid depth should be 1 to 2 m only (normally 1.5 m or 5 ft) and the top level of the tank should be 30 cm above the liquid level, thus forming an air space.
4. There will be two cross walls inside the tank with three chambers. The first wall is provided near to the entry point and should be constructed on a R.C.C. (1 : 1½ : 3) lintel placed 30 cm *above the bottom of the tank.* It should extend 15 cm above the liquid level. The first chamber is sometimes referred to as the *grit chamber.*

   The second cross wall should be built from the floor level to 15 cm above the liquid level. Holes with honeycombed brickwork are to be provided *at a depth of 40 cm from the liquid level* in this wall. The water will pass through the holes to the third chamber and the scum will be floating on the top in the second chamber. [In a small two-chamber tank, the second cross wall is omitted as shown in Fig. 31.3(a).]
5. The base slab is to be R.C.C. (1 : 2 : 4) and the walls can be one brick walls for tanks up to 3.6 m (12 feet) in length. (For large tanks, it can be made as twin tanks side by side with a middle common wall.) For easiness in the cleaning, the top cover slab can be precast slabs which can be easily lifted up and covered with earth to make it airtight. If the top is a continuous slab, manholes should be provided for cleaning.
6. The inside and outside should be plastered with 1 : 3 cement mortar (12 mm thick). The bottom concrete floor is given a 1-in-20 slope towards the inlet side so that the sludge will get collected there. The sludge collected should not be spread on the ground. It should be removed from the site or should be buried in a pit.

7. The inlet level should be *higher by 50 mm than the outlet level.*
8. A vent pipe of 100 mm diameter with a wire mesh cowl is to be provided above *the first chamber*. Its height should be 2 m when the septic tank is 15 m away from the building. If the tank is nearer the vent pipe should extend to 2 m above the top of the building.
9. The outlet should have a dispersion trench so that the effluent can be led to a dispersion trench provided at least 15 cm below the ground level. The stoneware pipes laid in the dispersion trench should be laid with loose joints *with sockets at the downstream direction* (different from the normal) to allow easy dispersion of the outflowing water (Fig. 31.4).

(As a thumb rule, a tank with inside dimension 1.5 m × 0.75 m in plan and 1.8 m deep with two divisions can serve 5 persons and a tank with inside dimension 2.3 m × 1.1 m in plan and 1.8 m deep with three divisions can serve 20 persons.)

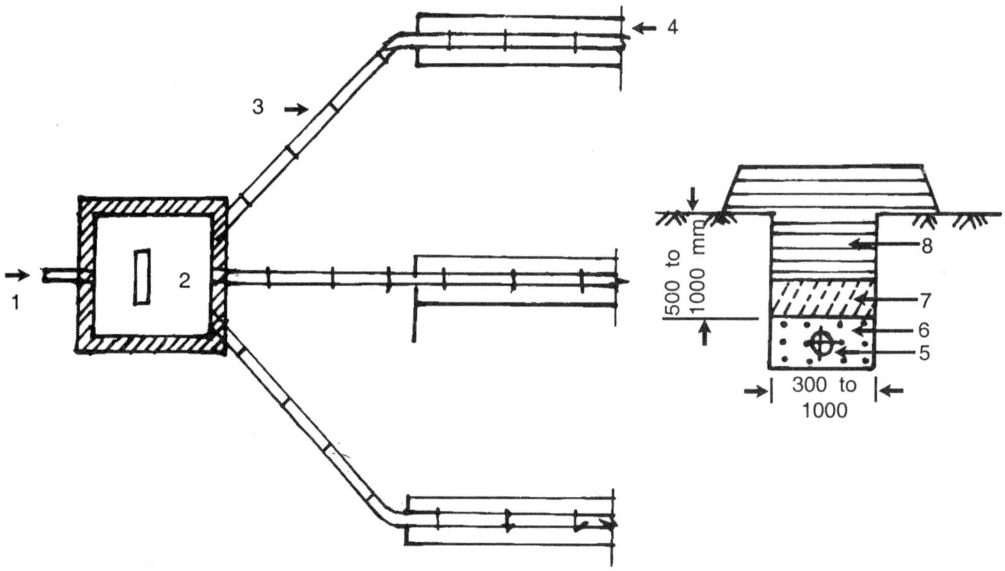

Fig. 31.4 Details of dispersion trench: 1. Effluent from septic tank, 2. Distribution chamber (1 x 1 m) with baffle at centre to distribute the flow, 3. 75 to 100 mm stoneware pipes laid with tight joints, 4. Unglazed stoneware pipes laid with open joints and large end downstream, 5. Stoneware pipe, 6. 15 to 25 cm depth of gravel or crushed stones, 7. 15 cm graded coarse aggregate topped with sand, 8. Earth filled to 15 cm above the ground level and turfed.

31.7 SOAK PITS

Soak pits are used for infiltration of rainwater or wastewater into the ground after introducing the necessary screening devices. They are also used for final disposal of the effluent from septic tanks. They are suitable only for sandy soils and do not work in clayey soils. There are mainly three types of soak pits:

1. Hollow soak pits
2. Filled soak pits
3. Combination of the hollow and filled soak pits

A hollow soak pit consists of a hole built in the ground surrounded by open joint brickwork [Fig. 31.5(a)]. In a filled soak pit, the hole is filled with a graded filter made of broken stones of various sizes [Fig. 31.5(b)]. A commonly used soak pit is a combination of these two.

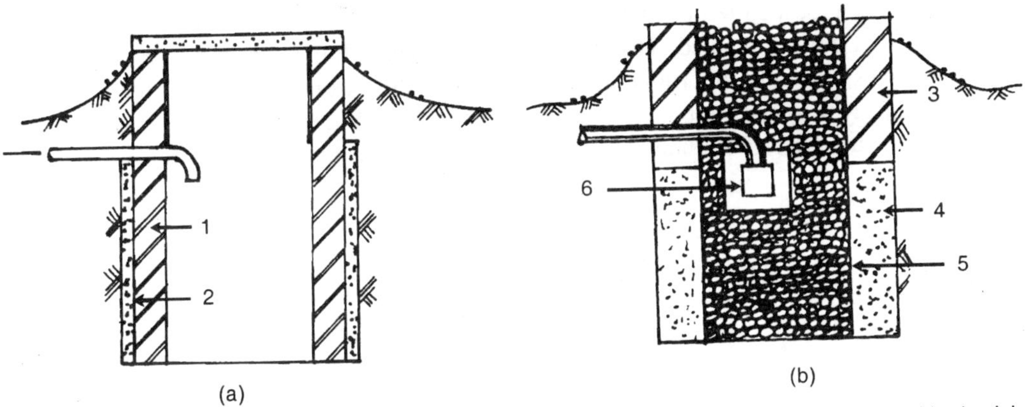

Fig. 31.5 Soak pits: (a) Hollow soak pit 1. Masonry (in brick, stone or concrete block) with dry joints, 2. Minimum 7.5 cm outercase coarse aggregate, (b) Filled soak pit 3. Masonry with good mortar joints over, 4. Outer casing of coarse sand 30 cm thick, 5. Stone or brick aggregate filling, 6. Brick chamber with dry joints.

31.8 OTHER METHODS OF DISPOSAL OF EFFLUENTS

Dispersion trenches and soak pits are not the best methods of disposal of effluents. In many cases, with long time use, thin impermeable linings are formed around the stones and these prevent efficient drainage. This leads to stagnation and bad smell. So it has to be re-laid. Aerobic digestion of effluents can remove odour considerably. In public sanitary system, many devices such as trickling filters, which are studied in Environmental Science, are used for aerobic digestion. However, for local use, where public sewers are not available, two other modern methods that are commonly used to treat effluents for aerobic digestion are oxidation ponds and horizontal planted filters (also called reeded bed filters). These are briefly described further. (As we are dealing with only sanitary systems for residences, we will not deal with public sanitary systems such as trickling filters in this book.)

Oxidation ponds. Oxidation ponds are *shallow ponds* exposed to the sun where the effects of the atmospheric oxygen and the sun's rays are assumed to purify the water. However, these ponds require large spaces for efficient action and are not fit for small units.

Reeded bed filters (planted filters). These filters have come into use more recently and are used successfully to treat and recycle sewage tank effluents as well as grey water

(water discharged from devices other than water closets). The more common type is the horizontal planted filters as shown in Fig. 31.6. It consists of a filter bed on which plants are grown. It should not be deeper than 30 to 60 cm, the depths to which the plant's roots can grow. In deeper depths, the water will tend to flow farther below the dense bed of roots and will not get treated. The bottom of the filter is sealed with concrete and given a slope of about *one per cent* so that the waste water flows down without percolating into the ground. The top is levelled and blinded with a layer of sand laid over the pebbles on which the plants grow. The sand layer also does not allow the foul odour to rise up. A large number of plants such as *Canna Indica* can be used for the planted filters, but the Australian reed plant, *Australis,* is considered as the best suited plant for this purpose as its roots form horizontal rhizomes that form a good root zone filter bed. They are now available in India. The organic matter in the water that flows through the roots undergo aerobic digestion by the bacteria or micro-organisms that grow in the roots of these reeds. They in turn get the oxygen from the air of the atmosphere transferred to them through the hollow stem and the atmosphere.

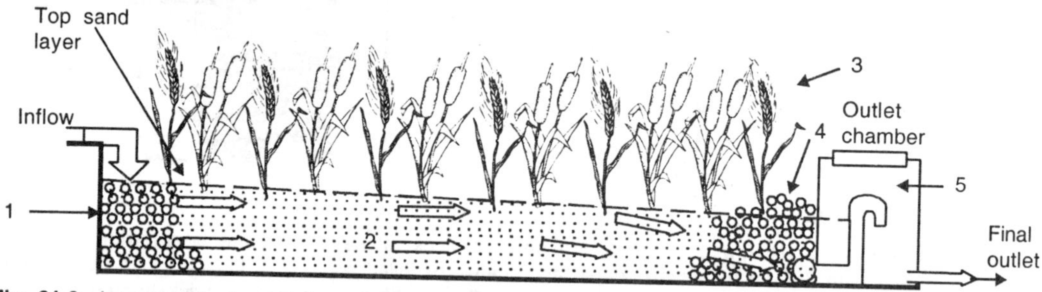

Fig. 31.6 Layout of horizontal planted filter: 1. Distribution trench filled with large stones or gravel, 2. Main filter filled with coarse gravel with sand blinding on top, 3. Reed plants, 4. Collection trench filled with large stones or gravel, 5. Polishing tank.

31.9 LAYOUT OF A MODERN RECYCLING PLANT

The layout of a horizontal planted filter plant to recycle the grey water or sewage effluent will be as shown in Fig. 31.7.

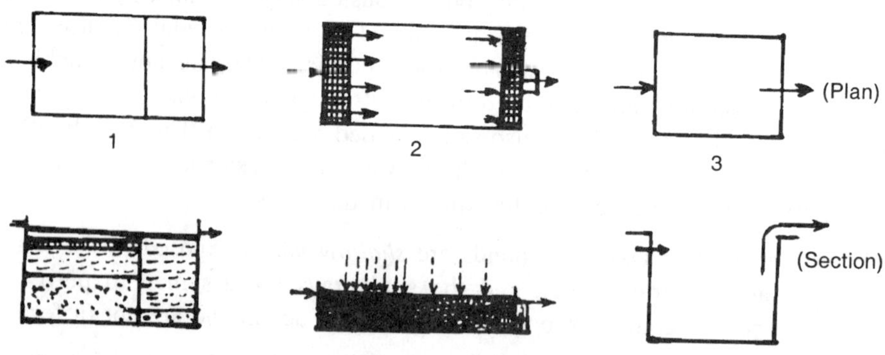

Fig. 31.7 Schematic plant layout for wastewater treatment: 1. Septic tank, 2. Reed bed, 3. Collection or polishing tank.

In general, a modern recycling plant consists of the following (Fig. 31.7):

Collection tank. For direct wastewater treatment, the collection tank is also provided with an anaerobic digestion chamber. This chamber has a filtering device or baffles to remove the suspended organic matter. The treated sewage water can be directly led into this tank. Usually, a small pump is provided to continuously pump the water into the reed bed so that a continuous flow can be maintained.

Reed bed. This bed is provided with pipelines to ensure the water to be treated to pass through the roots of the reeds. Some anaerobic action may also take place at the bottom of the tank.

Collection or polishing tank. The treated water collects in the polishing tank for reuse, irrigation or ground recharge. Additional activated carbon or other filters can be supplemented depending on the recycling used.

The area to be provided for the reed bed depends on the organic content and other impurities in the effluents. An area equal to about 12 to 15 m^2 of filter area per cubic metre of water (used per day) which is to be treated without separation of sewage is an average requirement. For treatment of greywater, only the area required will be very much less. (The minimum requirement of reed bed is about one square metre per person for which the septic tank is designed). The discharge from polishing tank can be used for reuse, for fish culture, irrigation or for recharging the groundwater.

31.10 IMHOFF TANKS AND ANAEROBIC BAFFLED REACTORS

Anaerobic are minute organisms that can live without free oxygen. The septic tank is the most common single decentralized sewage treatment plant. It is an extremely efficient device for anaerobic digestion. It is based on the following principles:

1. Sedimentation and floatation
2. Fermentation of the bottom sludge

The two improved versions that use the principle of septic tank are as follows:

- Imhoff or Emscher tanks
- Anaerobic baffled reactor

Only the principles of their action are described here. Details of design of equipments in the chapter can be found in books on Sanitary Engineering.

Imhoff tank. An imhoff tank works on the same principle as an ordinary septic tank. It has special funnel-shaped baffles inside the tank (Fig. 31.8). They prevent the sludge particles to get mixed with the effluent and also prevent the turbulence in the flow, thus enabling better sedimentation. Usually, the capacity (volume) provided for an imhoff tank is slightly higher than that for a septic tank (7 to 8 times the daily water use) and one or more separate tanks can be provided to handle the flow.

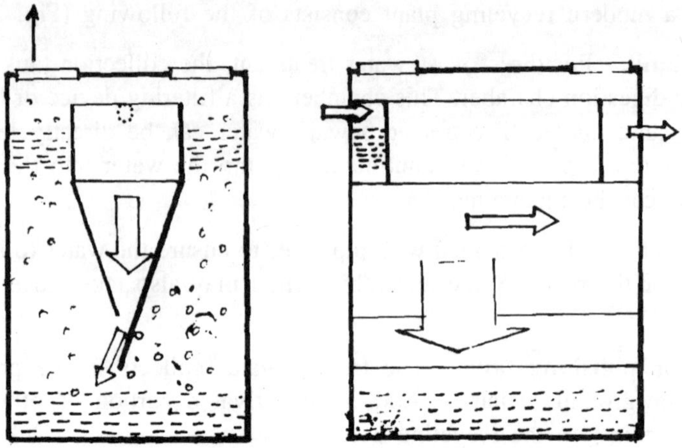

(a) (b)

Fig. 31.8 Principle of Imhoff tank: (a) Cross section, (b) Longitudinal section.

Baffled septic tanks. A baffled septic tank is used for a large system. It incorporates a large initial settling tank followed by at least four and up to ten narrow chambers fitted with baffled walls (Fig. 31.9). This forces the wastewater to flow downwards in each chamber, thus trapping all the sludge for aerobic digestion. Usually, a total capacity of 12 times or more of the daily total water use (in m^3) is provided for these devices. The velocity of upflow in the chambers should not be more than 2 m/hr. Details of the design of baffled septic tanks can be obtained from books on Environmental Engineering.

(See Appendix G for treatments with aerobic bacteria.)

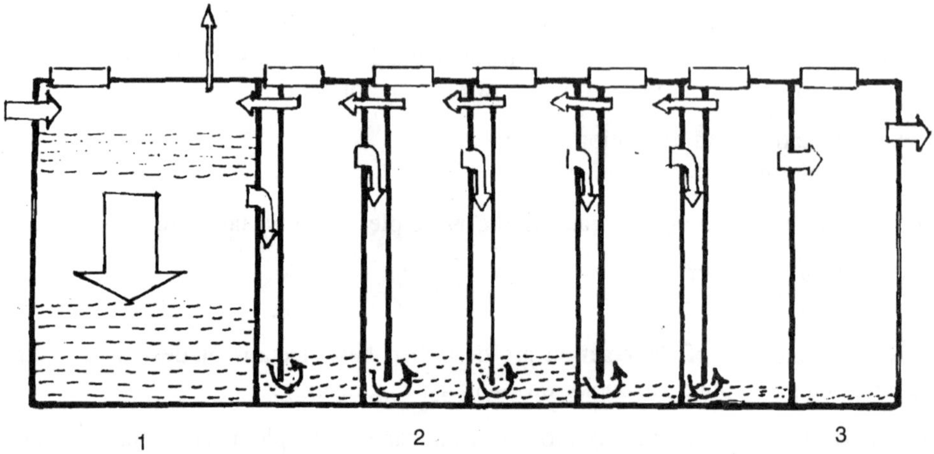

Fig. 31.9 Principle of *Anaerobic* Baffled Reactor: 1. Sedimentation, 2. Anaerobic digestion, 3. Sedimentation.

31.11 WATER SUPPLY AND SEWAGE TREATMENT IN BUILDINGS ON SMALL GROUND AREA WITH NO MUNICIPAL CONNECTIONS

Assuming there is electric supply and we can proceed as follows:

(i) *For drinking water:* Buy in bottles if ground water is good, we can use *Aqua guard equipment* to make water safe.

(ii) *For water for general use:* Use ground water if it is suitable. If it is salty, use *desalination equipment* working on reverse osmosis (See Appendix F). Wells with salt water should be so constructed that the top ground water flows into it.

(iii) *For sewage treatment:* Use small size aerobic sewage treatment plants (now available). They work on oxygen from air (Septic tanks work on anaerobic bacteria) (see Appendix G).

SUMMARY

A very brief account of drainage below the ground level has been given in this chapter. It is an important subject as any defects in the system can lead to pollution of the area and bad odour around the building. As the work is always out of our sight, we should take extreme care to do this work properly. After laying out the sewage pipeline system, it should be tested by *water test* as given in Chapter 37. Detailed specifications for works below the groundwater for drainage are available in the PWD specifications. It should be recommended that septic tanks should be located at least 20 m to 30 m away (depending on the soil) from any drinking water supply such as wells.

REVIEW QUESTIONS

1. What are manholes and inspection chambers? How are they planned for an underground drainage system?

2. Sketch the recommended layout of the foul water disposal system for a residential building. (Assume a typical layout of the building and sanitary appliances.)

3. (a) Explain the principles of design of a septic tank. How many chambers would you provide for a septic tank to serve 5 persons and another to serve 25 persons?
(b) Sketch the cross section of a small septic tank for an individual house in a village to cater for 5 persons.

4. (a) What slopes would you recommend for design of a house sewer?
(b) Sketch how a house sewer system is connected to a public sewer system.

5. Sketch a septic sewage treatment tank for a residence meant for 20 persons. What are aerobic sewage treatment plants and where they are used?

6. (a) What are soak pits? Give the sketch of a suitable soak pit for disposing the water from a septic tank.
(b) What are dispersion trenches? Briefly state how you will lay them.

7. Write short notes on the following:
(a) Imhoff tanks
(b) Anaerobic baffled reactors
(c) Horizontal planted filters
(d) Aerobic treatment of sewage

Chapter 32

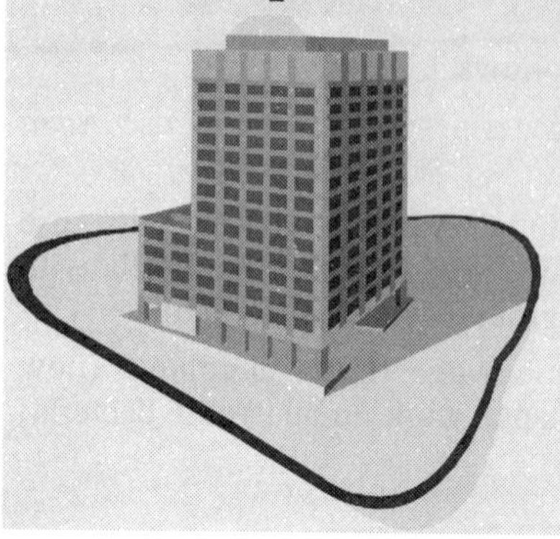

Electricity Supply in Buildings

32.1 INTRODUCTION

In India, electricity is generated in generating stations as alternating current at 50 cycles per second. It is transmitted to the high voltage national grids by 132 kV high voltage lines to reduce the transmission losses. This electricity is stepped down at electric substations to 11 kV, 3 phase, 50 cycles, which is considered economical for local distribution for industries and to the transformers located in various parts of a town. This supply voltage is further reduced by these local transformers and supplied to nearby buildings as single-phase or three-phase supply (50 cycles per second at 220 volts between the phase and the neutral). Thus, in India, electricity is supplied to buildings by the state. The supply is 230 volts between a phase and a neutral and it is 400 (= $\sqrt{3} \times 230$) volts between the phases. [Electrical equipments for use in India should be designed for (220 – 240) volts]. In a single-phase supply (used for low loads), we get one live phase wire and a neutral. In a three-phase supply, we get three live phases and a neutral. The supply agency is responsible for the cables up to and including the electric metre. A system of fuse of 30 or 100 amp capacity is installed between the supply inlet and the electric meterboard by the supply agency to isolate the main supply from the building. From the fuses, the wires are led to the metre and from there, to the consumer unit, from where, the electricity is distributed to the various parts of the building. The owner of the building is responsible for the wiring from the metre to the distribution board and the other parts of the building. It is advisable to put an indicator lamp to the line for each phase, as currents are cut off frequently in India and we can know from the indicator light, when the supplies or phases are cut off. In this chapter, we will briefly deal with the electricity distribution system usually adopted in ordinary residential buildings.

Electricity is distributed in large cities and towns through underground cables, in small towns and villages by overhead wires. With underground supply it is advisable to lay the cables from the compound wall to the building through a PVC or AC pipe to make repair or replacement work easy. If the supply to building is by overhead wires we must use insulated cables and provide a loop at the entry of the building to prevent flow of rainwater to the switchboard.

32.2 SINGLE- AND THREE-PHASE SUPPLY

Electricity is brought to a building by underground cables. When the electricity load is small, and the total current drawn is less than 30 amp, as when the electricity is used only for lighting, we need only a single phase and a neutral. However, when the load is heavy, as when we use many equipments such as air conditioners, cooking ranges, etc. it is necessary to draw current from three live line phases and a neutral. In this way, we distribute the load drawn amongst the three-phases. The electric current metre for the three-phase supply is different from that used for the single-phase supply. The metre and the main supply are located at a convenient point where the meter can be conveniently read by the agency. The ratings of all equipments used in residences must be 220 to 230 volts. When we distribute current from a three-phase supply, we should always avoid proximity of cables of two phases as there is a danger of high voltage leaking into the system.

We should be very careful in cases where electricity is brought inside a building from the supply line by an overhead insulated cable instead of an underground cable. It is to be first run down the walls of the building by properly insulated cable and then turned up at the entry point of the building to prevent rainwater running along the cable into the electric board. This is an important detail to be followed for all lines (electric, telephonic, TV cable, etc.) entering the buildings.

32.3 CONSUMER CONTROL UNIT

From the electric metre, the supply goes to *the consumer control unit* consisting of the ELCB (earth leakage circuit breaker), main isolation switch and distribution points. This unit provides a compact and effective means of controlling and distributing electricity to different parts of the building. Apart from the isolation switch, it contains live phases, neutral and the earth bars as well as individual circuit breakers or old fashioned fuses which are put on the live phases. *It is important to note that the fuses should always be placed on the live phase and not on the neutral.* The modern practice is to use miniature circuit breaker (MCB) instead of the old fuses. The miniature circuit breakers give overload protection only. In addition, earth leakage circuit breaker (ELCB) must be also provided to protect equipments as well as for children's safety. With an ELCB, as soon as there is even a small leakage of current through the earth (or if a child puts its hand in the socket), the power supply is immediately cut off. The ELCB is introduced before the main switch.

32.4 MINIATURE CIRCUIT BREAKER (MCB), EARTH LEAKAGE CIRCUIT BREAKER (ELCB) AND RESIDUAL CURRENT CIRCUIT BREAKER (RCCB)

In older times, fuses were incorporated in the electrical circuits so that when a fault occurs, heavy current will flow through the circuit and the fuse will blow out. Nowadays, instead of fuses, MCB (miniature circuit breakers) are used. The MCB will switch off if there is excess flow of current in the circuit.

Leakage of current to any metal work of an equipment can occur if the wires are displaced or insulation is frayed. It can also occur if the switches or other electric points on a wall get damped or moist from rains or other means. These can produce shock to the user. This can be avoided *by earthing the equipment* and using an earth leakage circuit breaker (ELCB).

With an RCCB in the system, even a very small leakage will operate the circuit breaker and switch off the current in a period of a microsecond. As the neutral wire will not serve this purpose, a separate set of conductors called *earthing* is provided in all buildings. Earthing is carried out as described in Section 32.8. *It should be made mandatory that all house wirings have ELCB or RCCB incorporated in the system* before the main switch.

The MCB, ELCB and RCCB are explained in detail in Section 32.6.

32.5 DISTRIBUTION CIRCUITS

For distribution of electricity from the consumer control unit to other points in an ordinary building, the supply is divided into the following three separate types of circuits:

1. Lighting circuit of low capacity – 6 amp
2. Power circuit through ring circuit
3. Fixed appliance (or individual) circuits (Each appliance is separately wired to the consumer control unit.)

Generally, red wires are used for the live wire. The neutral is black or blue and earth wires are green or striped green and yellow. Brief descriptions of each of these circuits are given below. A phase changer may also be incorporated in the circuit (see section 32.7.4).

32.5.1 Lighting Circuit

Separate circuit is usually provided for the general lighting of the building. One of the following two systems is commonly used for this purpose:

- Loop-in system
- Joint-box system

Loop-in system. In a loop-in system, each loop is rated as 6 amp. A live phase cable runs from a 6 amp fuse (or miniature circuit breaker) from the control board to the live terminals of the first *lighting point or ceiling rose.* From this lighting point, the cable loops out to the next rose and so on till all the lighting points are linked together. Similarly, the neutral and the earth lines are also run. Another length of the same cable runs from each rose to the corresponding switch controlling the light. This is shown in Fig. 32.1. In important places (such as living rooms), more than one lighting circuit (from different phases) should be used so that in the event of a phase failure, some lighting will be in working condition, without the use of a phase danger.

Joint-box system. In a joint-box system for a lighting circuit, the wires run to a series of joint boxes instead of direct to the roses, with seperate joint box for each light switch. The joint box is circular in shape and has four terminals—live L, neutral N, earth E and switch return wire to the switch (SW).

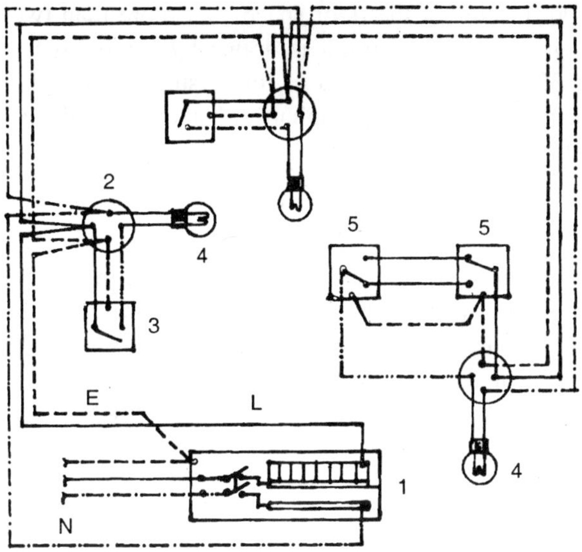

Fig. 32.1 Lighting circuit wiring diagram of Loop in method (N = Neutral, L = Live, E = Earth): 1. Consumer control unit, 2. Ceiling rose, 3. One-way switch, 4. Lamp, 5. Two-way switch.

32.5.2 Power Circuit or Ring Circuit

For small appliances such as refrigerators, mixers, etc., we use a ring circuit. The ring circuit consists of a 30 amp fuse or MCB, protected looped live line as well as *neutral and earth lines*. These are looped from *one socket* to the other as shown in Fig. 32.2. Usually, a separate outlet

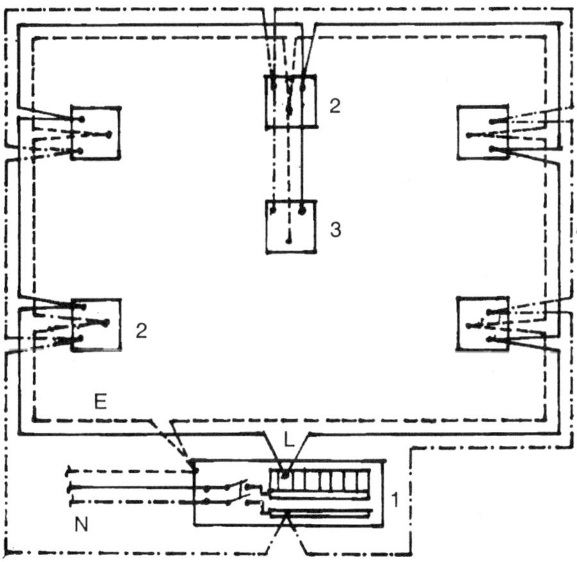

Fig. 32.2 Power circuit wiring diagram (ring main wiring): 1. Consumer control unit, 2. 15 amp socket outlets in ring main, 3. 15 amp spur socket outlet.

is provided for every 100 m^2 area of the floor space. Not more than two outlets are taken for a single spur as shown in Fig. 32.2. Current is drawn by a three-pin plug from the socket. We should note that as 30 amp fuse or MCB of this system is of high rating, it is advisable to give high protection to each individual appliance. This can be done by supplying current through a 13 amp fused plug or fused spur box (i.e. through a fuse of lower rating).

32.5.3 Fixed Appliance Circuits

Appliances such as air-conditioners, electric cookers, etc. consume heavy current for its working. For these appliances, we have to provide *separate circuits from the control unit to the appliance.* These circuits are called *fixed appliance circuits.* Usually, separate switch-off devices are also provided near the equipments to isolate them from the circuit for safety and repairs. Live, neutral and earth lines are connected to the outlet.

32.6 MATERIALS AND DEVICES USED IN WIRING OF BUILDINGS

32.6.1 Cables

Cables consist of conductors surrounded by insulation. The conductors should be preferably of copper. Aluminium conductors are sometimes used to reduce cost but they are inferior to copper cables as uninsulated portions such as connections to switches tend to oxidize and become brittle thus giving trouble in the long run. If possible, aluminium cables should not be used for wiring. Copper-covered aluminium cables are also available and they are considered better than plain aluminium cables as they do not get oxidized so easily. But they are not as good as copper cables. Formerly, single-wire cables were used. Nowadays stranded wires are commonly used for better performance.

Even though rubber was once considered as the standard insulating material, nowadays plastics (especially PVC covered cables) are very much popular. In moist or wet situations, PVC insulation fare better than rubber.

Sizes. At present, cables are manufactured in metric sizes and are referred by the number and diameter of the conductor. Thus, cable designated as 1/1.13 cable contains one conductor of diameter 1.13 mm giving an area of 1.0 sq mm. Table 32.1 gives the recommended sizes of wires for different uses.

Table 32.1 Sizes of wires for different uses

S.No.	*Circuit*	*Fuse*	*Cable size* (area)	*Cable current rating* (amp)
1.	Lighting	5 amp	1.0 to 1.5 mm^2	12 to 15
2.	Immersion heater	15–20	2.5 mm^2	21
3.	Ring circuit and spurs	30	2.5 mm^2	21
4.	Radial circuit	20	2.5 mm^2	21
5.	Radial circuit	30	4 mm^2	27
6.	Cooker	45	6 to 10 mm^2	35 to 48

Notes: (a) The cable current rating indicated above is for rewirable fuses. Rating should be uprated by one third, if cartridge fuse or mini circuitbreakers are used.

(b) Nowadays it is more common to use multistorey cables instead of single-strand wire (see also section 32.9.1).

32.6.2 Fuses

A fuse protects appliances and cables from damage by electrical faults and is similar to a safety valve in a boiler system. Nowadays miniature circuit breakers (MCB) are used instead of fuses in higher class buildings. The current rating of fuses should be lower than that of the cable it has to protect. There are mainly the following two types of fuses:

(b) Rewirable or semi-enclosed fuses
(c) Cartridge fuses

Rewirable fuses. Rewirable fuses are cheap. It takes *twice the current rating of a rewirable fuse to blow it.* Thus, a 5 A rewirable fuse requires a current of 10 A to blow it. As in the same holder different capacity fuse wire is used, one should take care that the same capacity fuse wire is used during the rewiring. This can be done either by using colour-coded fuse wire or making special markings on the fuse holder with the same colour code as given below:

5 A	White
15 A	Blue
20 A	Yellow
30 A	Red
45 A	Green

Cartridge fuses. Cartridge fuses of different ratings (different ratings are of different sizes) are available (Fuse of different ratings are of different sizes.). It is not possible to fit a 20 A cartridge fuse in a 5 A fuse holder. *A 20 A fuse blows at 1.5 times its current rating.* Thus, a 30 A fuse will blow out at 45 A. It is not generally possible by simple inspection to tell when a cartridge fuse has blown.

32.6.3 Miniature Circuit Breaker (MCB)

Miniature circuit breakers are single pole switch which automatically switches off when excess current flows through its operating coil and thus flows in the circuit (overloading or short circuit). It has many advantages over fuses but initial costs are high and hence, they are not generally used for low-cost constructions. It is a modern alternative of fuses described above. *An MCB trips at 1.25 times its rated current capacity.* Thus, a 30 A (amp) MCB trips at 37.5 A (compared to 45 A in case of a cartridge and 60 A in case of a rewirable fuse). So an MCB is more sensitive than a fuse. It is important to note that MCB should not be confused with earth leakage circuit breaker (ELCB). *An MCB is only an overload device.*

32.6.4 Earth Leakage Circuit Breaker (ELCB)

If an electric appliance with a fault is connected to the line and if its metal parts are not earthed, the metal may become live without the fuse blowing. Such a situation can produce an electric shock to the persons touching it. However, if the metal body is earthed, then current will flow through the earth wire. For such a condition, it is common to use a tripping device called the *earth leakage circuit breaker* (ELCB). It is like a main switch which automatically trips even when only a small current flows through the earth wire and the current supply will be cut off completely. This switch is connected to the neutral line of the electric supply in the main distribution board and before the main switch. *Hence, when it trips, the supply to the building will be completely cut off.*

There are two principal types of ELCBs used in home installations—the current operated and the voltage operated. Current-operated ELCBs work on the principle of a core balanced transformer and are more expensive that the voltage-operated ELCBs. It has two ratings—firstly as a main switch (60 or 100 A) and secondly as its operating tripping current rating capacity usually being 30 mA. The voltage-operated device has many disadvantages.

32.6.5 Residual Current Circuit Breaker (RCCB)

This is a much more positive protection device that can be provided instead of ELCB to avoid electric shocks and also current leakage in wiring of buildings. An RCCB is connected to all the three phases and the neutral. It is more expensive than ELCB and provides good protection against even minor defects in the wiring of a building. However, the wiring should be perfect, otherwise even with a small defect, it will trip until it is made good.

32.6.6 Plug Fuses

As indicated earlier in many instances, it is advantageous to provide a low-rated fuse at the plug outlet and hence, modern plugs or sockets are also provided with plug fuses. These are interchangeable fuses and are used to connect expensive electric equipments such as televisions, refrigerators, air conditioners, etc.

Note: When installing junction boxes, switchboards, etc. to walls, it will be better, if possible, to fix them in the inside walls of the rooms rather than the outside walls as the outer walls may get wet during very heavy rains. Alternatively, all such exterior walls, which have electrical switchboards, should be protected by a continuous *chajja* or painted with good exterior grade waterproof paint to protect it from penetration of rainwater.

32.7 ELECTRICAL APPLIANCES

We know W(watts) = A (amps) × V(volt). Thus with 220 volts in a 100 watt bulb the flow is 0.45 amps. In the switchboard main circuit fuses and cutouts are usually designed for a maximum flow of 10 to 15 amps only.

32.7.1 Load Ratings of Usual Household Appliances

Table 32.2 shows load ratings of common household appliances.

The loads of usual household appliances are as follows:

Table 32.2 Load ratings of common household appliances

Appliance	*Load* (in watts)
Electric kettle	1000 – 3000
Fans	25 – 120
Floor polisher	300 – 450
Food mixer	100 – 450
Frying pan	1000 – 1250
Hair drier	350 – 600
Infrared grill	1500 – 3000
Iron (automatic)	750 – 1000
Iron (standard)	450 – 500
Radio	40 – 150
Refrigerators	150 – 300
Sewing machines	75 – 90
Spin drier	100 – 320
Television receiver	200 – 400
Toaster	400 – 600
Vacuum cleaner	220 – 300
Washing machine (no heating)	300 – 500
Washing machine (heating)	3000 – 4000
Water boiler (storage)	1000 – 3000
Water boiler (instantaneous)	3000 – 6000
Immersion water heater	1000 – 3000
Air conditioners	800 – 1500
1/4 hp motor	300 – 400
1/2 hp motor	450 – 600
over 1/2 hp	950 – 1100

W (Watt) = A (amps) × V (voltage)

32.7.2 Electrical Switches

Single pole (SP) switches. A single pole switch is an ordinary switch which connects the two ends of a phase in the 'on' position and disconnects them in he 'off' position. This type of switch is generally put for lighting and also at the normal plug output. There are various types of single pole switches available in the market with different architectural designs but doing the same function. It is very important that the switch is connected to the phase.

Double pole (DP) switches. A double pole switch connects or disconnects both *the line and the neutral from a circuit*. Such type of switch is usually used for heavy currents and as the main switch for the electric installations or large electric appliances such as electric cookers, household pumps, etc. to completely isolate them, if necessary, from the circuit.

32.7.3 Fans

Ceiling fans and extraction fans are the two types of fans generally fixed in buildings. Table fans and pedestal fans can be moved from place to place.

Ceiling fans. Ceiling fans are specified by the diameter of the blades of the fan. The common sizes of fans used in India are 900 mm (36") and 1200 mm (48"). Fans of sizes 1500 mm and 1800 mm are also available. Old types of fan regulators are of the resistance type but now the modern regulators, which are more efficient, are available in the market. The lower speeds of the fans to which they can be regulated are important in their selection. Special care regarding this factor should be taken into account during selection of fans for offices, bedrooms, etc. (see section 35.9.1). Ceiling fan clamps for RC roofs are buried in the slab concrete above the slab reinforcement along with the casting of the roof. For hanging the fans from the wooden beams, we use flat iron, fixed on the two sides of the beam with one or two mild steel bolts passing through the beam to hold both flat irons together. Clamps for steel joists are fabricated from flat iron to fit in rigidly to the bottom flanges of the beam.

In a flat RC ceiling, the hooks extend to about 90 to 100 mm below the ceiling and it is shaped to accommodate a rubber shackle of diameter 50 mm on which the fan is hung. The hanging rods are usually of diameter 16 mm.

Note: For installing a new hook on to an old RC floor (or roof), a 150 mm × 180 mm chase is made exposing the main reinforcements. A steel rod of diameter 16 mm bent into a U-shaped clamp, with hanging hooks on either ends, is hung from the main reinforcements by the hooks. The chased portion is then plastered over with cement plaster after painting the old surface with cement ground or preferably a bonding agent as described in Section 39.7.

Extraction fans. These are used to expel moist or stale air from rooms such as kitchen, bathroom, stores, etc. Sizes go by blade diameters and speeds. These quantities determine the air extraction. Most popular diameters are 150 mm (6"), 175 mm (7") and 200 mm (8"). The speed range is 6000 – 10,000 r.p.m. Usually, about 12 air changes per hour in the kitchen, 20 air changes in bathrooms are aimed at. An extraction fan is generally fitted in a round hole in the wall. Its direction of rotation should be so as to extract air from the room. Their positioning is important to get the maximum efficiency. It should be so arranged that air should be sucked from the rooms and not by drawing outside air from a nearby window.

32.7.4 Phase Changers and Inverters

In India, in a three-phase supply to a building, it is common that one or two of the phases may temporarily go out of action. A phase-changing device (which is a simple 3 way rotatary switch) connected to each phase installed in the circuit near the consumer control unit can help the supply in one circuit given in one phase changed to another. When installing phase-changing devices in a building with a large number of heavy equipments such as air conditioners, it is advisable not to connect all of them to the phase changer as there is a chance of overloading. They should be permanently connected to different phases. Alternatively, an automatic phase changer with *overload protection device should be used.* Generally, only the lights, fans and small equipments (such as refrigerators) are connected to the phase changer.

Battery operated inverters, that restore electric supply (from charged batteries) when the regular supply is cut off for limited periods, are also used nowadays in many buildings. In most cases, they are connected only to limited outlets or light and fan points in the building. *Devices for Uninterrupted Power Supply in case of the main supply failure for very sensitive systems like computers are called UPS.*

32.8 CONSIDERATIONS FOR HOME LIGHTING

32.8.1 Guidelines for Home Lighting

In this section, we will discuss the guidelines for lighting the various rooms in a home.

Living and dining rooms. We combine light fittings which provide good strong light (for reading, sewing, etc.) with decorative light fittings such as floor standing or table lamps. As one needs fairly good lighting over the dining table, we use single or multiple pendent (which should be able to take at least a 100 W lamp) directly over the table. Two or three well-placed wall brackets or table lamps should light up the rest of the room. These need be only of 25, 40 or 60 W. We may also think for providing a 100 W floor lamp that can be moved to any part of the living room for general lighting.

Bedrooms. The lights should be so positioned that we will not be looking at the glaring light when lying in bed. This is particularly important as we may have to switch the light on suddenly in the middle of the night in the room where more than one person may be sleeping and the switching on should not disturb others. Bulbs need not be stronger than 100 W. One strong light and another weak one will be ideal. Lamps fitted for reading should be used to conceal the lighting. For adequate light over the dressing table, two lamps mounted on either side or one lamp above the mirror will be the best.

Kitchen. It should be remembered that kitchen is normally a hot place and hence, cool lighting is the best. Tubelight (fluorescent lamp) will be ideal for general lighting. It should be so placed that it is easily accessible for being cleaned regularly. A 60 W bulb fitted directly over the cooking place will also be helpful.

Bathroom. Here the light should be of uniform level throughout. A simple ceiling mounted fitting is not suitable. The bathroom mirror needs special lighting and for this purpose, we can use a matching wall bracket fitting or a small tubelight over the mirror.

Notes:

(a) Clear lamps should be avoided for interior lighting as they create glare and harsh shadow. Bulbs with inside coating or clear bulbs inside glass fittings which give good diffused light can be used. Clear lamps are to be used for lighting up the outside of buildings.

(b) People need more light to see as they grow older. Persons over forty need three times more light than a child of ten and those over sixty need fifteen times more light than children to read with the same degree of comfort.

(c) The major three factors to be considered in lighting are:

1. The nature of work
2. The age of the person
3. The distance between the light source and the person using it.

32.8.2 Provision of Minimum Number of Outlets in Each Room

The minimum number of outlets to be provided in each room of a building will depend on the lifestyle of the owners. This should be planned well in advance. The following can be taken as the average for a middle class family:

1. Kitchen	4 to 5 (water pump, mixer, oven, etc.)
2. Living room	3 to 4
3. Dining room	2 to 3
4. Bedrooms	2 to 3
5. Bathroom	2 to 3 (heater, for mirror, etc.)
6. Study	2 to 3
7. Staircase	1 to 2 (one- to two-way switch)
8. Stores	1
9. Laundry	2

Switches for bathroom lights and heater should preferably be provided outside the bathroom at its entrance to avoid standing on wet areas to switch them on. The mounting heights of these outlets depend on their use. In kitchen, many of the outlets are fixed just above the kitchen platform. In order that the switches are inaccessible to children, the recommended minimum height is 1 to 1.5 m above the floor level. This also avoids the need to bend to switch on the system.

32.9 WIRING OF BUILDINGS

32.9.1 Types of Wiring

The following three types of wiring are used:

1. Sheathed wiring by surface fixing
2. Conduit installation on walls and ceilings
3. Concealed conduit wiring.

Nowadays concealed PVC conduit wiring is the fashion in most buildings. Conduit wiring can also be made on the surface. It must be remembered that PVC has a high coefficient of expansion. For concealing in concrete, the conduits are laid before concreting and in masonry, a chase is made before the final plastering. In each case, care should be taken to see that there will be no chance for water to enter into these pipes from any place, after they are finally laid. When laying in concrete, it is a good practice to surround the conduit in chicken mesh reinforcement.

32.9.2 Use of Flexible Cords

Flexible cords are used in many electrical appliances such as electric iron, pendent lights, etc. Some important and useful properties of flexible chords are discussed further.

Rating of strands. Flexible cords are made of small strands and not a single wire. The usual sizes of available flexible cords are as given in Table 32.3.

Table 32.3 Rating of strands

Cross-sectional area (mm^2)	*Number and diameter* (mm) *of strands*	*Current rating* (ampere)
0.5	16/0.20	3
0.75	24/0.20	6
1.0	32/0.20	10
1.25	40/0.20	13
1.5	30/0.25	15
2.5	50/0.25	20
4.0	56/0.30	25

From Table 32.3, it can be observed that for lighting and light-duty appliances the sizes to be used are 0.5 and 0.75 mm^2. Typical sizes for various electric ratings can be in terms of cross sectional areas as shown in Table 32.4.

Table 32.4 Sizes of cords of various electrical ratings
(*Note*: The area can be used as from Table 32.3)

Electrical rating (kW)	*Cross-sectional area* (mm^2)
1	0.75
2	1.0
3	1.5

Weight carrying capacity. The maximum weight that may be supported by *twin flexible cords* is as shown in Table 32.5.

Table 32.5 Weight carrying capacity of flexible cords

Area (mm^2)	*Maximum weight* (kg)
0.5	2
0.75	3
1.0	5
1.5	5
2.5	5
4.0	5

When used for suspended lights, it is always preferable to employ additional steel straining wires, suitably positioned, which will take the weight away from the cords. Further, when a non-metallic outlet box of thermoplastic material (like PVC) is used for the suspension of a light fitting, care is necessary to ensure that the box temperature does not exceed 60°C. The mass suspended from the box must not exceed 3.2 kg.

Colour-code identification for flexible cords. Appliances with 3-core flexibles are usually identified by:

Live core	Brown, Red
Neutral core	Blue
Earth	Green/Yellow

Precautions necessary in light fittings and pendents. Heat-resisting cords are necessary for most connections between the ceiling rose and lamp holder where tungsten filament lamps are to be used due to the abnormally high temperatures generated by these. Light fittings and shades, especially if flush-mounted and totally enclosed, require heat-resisting insulation suitable for the temperatures likely to be encountered. Here heat-resisting sleeves are also used. They should be fitted over the individual cores of the flexible cables in such a way that the normal insulation of the cores is not relied upon to prevent a short circuit between the conductors or an earth fault. Similar methods should be employed for accessories and appliances which are subject to such heating conditions.

Temperature ratings for flexible-cord materials. The maximum operating temperatures of insulation or sheath of flexible cords are usually rated as given in Table 32.6.

Table 32.6 Temperature ratings of cables (IEE ratings)

Type of cord	*Normal service* °C	*Contained within fittings and not subject to stress* °C
General purpose rubber compound	60	75
General purpose PVC	70	75
Heat-resisting PVC	85	100
EP or butyl rubber	85	100
Silicone rubber	150	200

Where the insulation and sheath are of different materials, appropriate temperature limits must be observed for the both. If the insulation is not good, it may become hard and brittle so that it lacks flexibility. If allowed to run hot, a twin-twisted cord can thus easily fail and cause a 'short' at the bifurcation point just above the holder, often resulting in a rupture of the flexible wire and collapse of the fitting and lamp.

Flexible-cord sheaths suitable for contact with oil or petrol. Where a risk of fire is present or if a flexible sheath which will not support combustion is required, cords with heat-resisting, oil-resisting and flame retardant (HOFR) sheaths must be used.

32.10 METHODS OF EARTHING OF ELECTRICITY SUPPLY IN BUILDINGS

Earthing of electrical installations is considered in two ways—system earthing and equipment earthing. *System earthing* is the earthing associated with current carrying conductors while *equipment earthing* is the system used for safety of equipments and prevention of shocks. In buildings, generally, earthing is used for the second purpose. There are two systems of earthing, namely pipe electrode type and plate electrode type. These are briefly described here (also refer to Indian National Building Code, Part VIII).

Pipe electrodes. This is installed by digging an auger hole 30 cm in diameter and 3.75 metres depth. Into this pit is inserted a 38 mm diameter GI pipe with holes at 7.5 cm from the centres (staggered on two perpendicular diameters) and 2.5 m from its lower end. The pipe is inserted into the hole and the 2.5 m depth is packed with alternate layers of charcoal (coke) and common salt, each of 50 cm height, with the bottom layer commencing with the charcoal. The rest of 1.25 m depth is filled with earth and the earth wire connection is made on a hole at top of the pipe with GI nuts and washers.

Plate earthing. In plate earthing, a convenient pit is dug to a depth of 2.5 m. Earth connections are made to a plate of 60 cm × 60 cm and 6.30 mm thick of galvanized iron and 3.15 mm thick, if made of copper. This plate is lowered and installed in a bed with 15 cm thick charcoal (or coke) all around, under and above it. Earth connections are made to this plate. The top soil over these earthings may be planted with flower bed which can be watered regularly to keep the soil surrounding the earthing moist.

32.11 TESTING OF ELECTRICAL INSTALLATION IN BUILDINGS

It is good to test the electrical wiring of a building after the wiring is completed. Electrical installation of buildings should satisfy at least the four tests which are discussed further.

Test for insulation resistance. This test is made with *all links in place, all lamps in position and all switches on*. The insulation resistance is measured *between the earth* and the whole system of conductor or any section thereof. The insulation resistance of the installation in megaohm should not be less than 50 divided by the number of points in the circuit except that it need not exceed 1 megaohm for the whole installation.

Test for resistance between the conductors. When making this test, all the switches must be on but the lamps and other loads should be removed. This test is made between all the conductors connected to one pole (or phase conductor of the supply) and all the conductors connected to the other pole (or neutral phases of the supply). Leakage at cable ends is one of the common causes of low reading so before this test, care should be taken to see that the switches, etc. are clean. The result of the test shall not be less than the one specified as in the insulation resistance test with the load on as stated above.

Earth continuity test. The total resistance measured from the earth electrode to any point in the earth continuity conductor in the complete installation should not exceed one ohm.

Checking polarity of single pole (SP) switches. The following checks are needed for the polarity of single pole switches:

(a) In a two wire installation, all non-linked single pole switches should be checked to see whether they are all connected to the phase conductor. (This will ensure total cut off of the current when the switch is in the off position.) It is an important check.

(b) If the installation is three phase (four-wire installation), then test should be made to verify that every non-linked single pole switch is fitted to a conductor connected to one of the phases and that any two of the phases *are not exposed* within the specified distance of two metre. (This ensures that there is no chance for a phase-to-phase connection.)

Note It is interesting to examine how earthing of equipments will save them. We know the relation $V = IR$. If there is any leakage of current with earthing it can flow to earth. R will be small, so that I in amps will be very large. This will blowout the 10 to 15 amps fuse at the switchboard or the automatic trip will switch off the circuit.

SUMMARY

Electrification of buildings is allowed to be carried out only by an electrician with government license. However, all building engineers should have a fair knowledge of the important considerations involved in electricity supply in buildings so that the work can be supervised properly.

REVIEW QUESTIONS

1. Draw a lighting circuit diagram of a single-phase supply for a low-cost house.

2. Draw the lighting and power circuit diagram of a three-phase supply for a residence using electricity for lighting and also for small appliances such as refrigerators. How will you wire the fixed equipment such as air conditioner?

3. Write short notes on the following:
(a) Single pole and double pole switches
(b) MCB and ELCB
(c) Considerations in using flexible chords for electrical appliances
(d) Earthing of electrical circuits

4. Give a short account of the testing of the electrical installation of a building.

5. Write a short account of electric wiring of a residence using electricity for lighting as well as for operating small equipments such as mixers, water pumps and air conditioners.

6. What are Inverter and UPS in the electric supply for a building.

Chapter 33

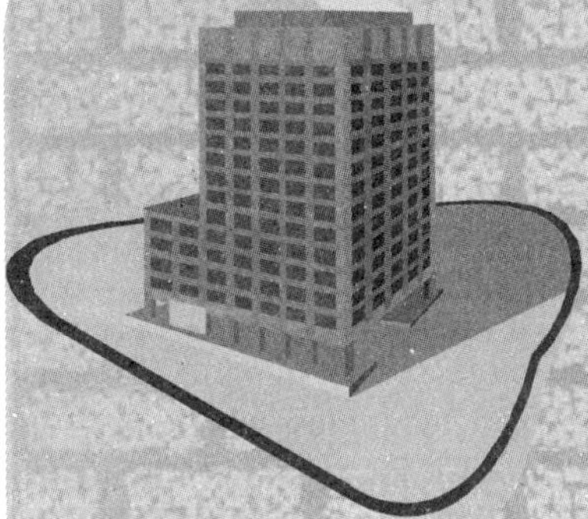

Common Equipment Used in Construction of Ordinary Buildings

33.1 INTRODUCTION

A number of machines such as those used for earth excavation, earth moving, lifting, concrete mixing, placing, etc. are used in civil engineering construction. The study of the selection of the *type and numbers* of various types of equipment needed for a large project is made under a specialized subject called Construction Equipment. Special textbooks are also available on this subject. In this chapter, we will briefly study the nature of the following pieces of equipments which are commonly used in ordinary building construction practices:

1. Concrete mixers
2. Concrete pumps
3. Concrete vibrators
4. Bar bending machine
5. Bulldozers for site clearance
6. Hoists for lifting materials at building sites
7. Pumps for lifting liquids

Temporary works such as scaffolding have been dealt with in Chapter 11.

33.2 CONCRETE MIXERS

Concrete mixers are used for mixing cement, concrete and also cement mortar. Many types of concrete mixers are available in the market. These can be broadly classified as *continuous mixers* and *batch mixers*.

33.2.1 Continuous Mixers

Continuous mixers produce concrete continuously and are used for large works such as dams and ready mix plants. In these machines, the materials are fed continuously (without stoppage) at one end and the mixed concrete comes out at the other end. In building construction, for mixing concrete at site, we generally use batch mixers.

33.2.2 Batch Mixers

Batch mixers are the mixers used to mix concrete in batches. There are two types of batch mixers, namely *the pan type* and *the rotating drum type*.

Pan mixers. Pan mixer consists of a horizontal pan in which the blades revolve at a high speed and the pan may also be rotated about the vertical axis or remain stationary. Pan mixers are considered more efficient for stiff mixes than the rotating drum mixers. Pan mixers are used only in special situations such as to produce low slump, high strength concrete as needed for the production of prestressed concrete sleepers. The mixing time required for these concrete pan mixers is much less than the drum type batch mixers.

Rotating drum mixers. Rotating drum mixers are the most commonly used concrete mixers at building construction sites. They have bladders fixed at angles which help the mixing. The cement, sand, aggregate and water (called the *charge*) are separately loaded in the mixer. As the mixer rotates, the materials are lifted up and given a drop. There are blades attached to the drum which facilitate mixing of the materials. There are three types of drum mixers depending on the method of discharge of the mixed concrete:

1. Tilting mixers usually used for small batches [Fig. 33.1(a)]
2. Non-tilting mixers for large batches [Fig. 33.1(b)]
3. Reverse drum mixers which discharge on their own [Fig. 33.1(c)]

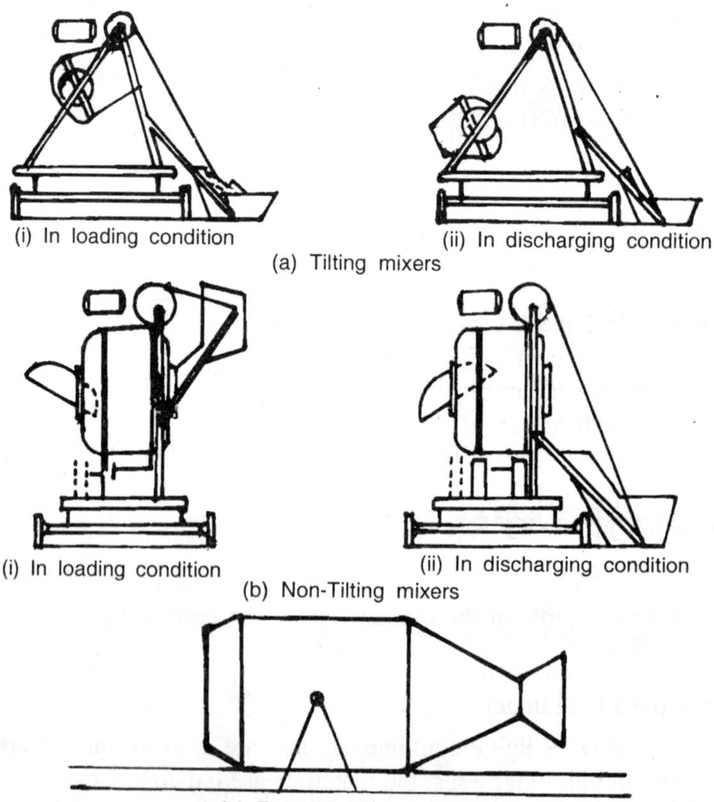

(i) In loading condition (ii) In discharging condition
(a) Tilting mixers

(i) In loading condition (ii) In discharging condition
(b) Non-Tilting mixers

(c) Reverse drum mixer

Fig. 33.1 Concrete mixers.

In the last two types of mixers, the materials are ideally loaded by a skip and in the small tilting mixers, they are directly loaded by labourers. In the tilting mixer, the concrete after mixing is discharged by tilting the mixer usually on the side opposite to the charging side. On the other hand, the non-tilting mixer (also known as rotary mixer) cannot be tilted. Here the mixed concrete is discharged to the side opposite to the loading side by inserting a special inclined chute into the mixer. The non-tilting and reverse drum mixers are designed for large capacity mixes.

It has been observed that in an ordinary non-tilting mixer, there is some segregation of coarse aggregate and so it is the general practice to do a little bit of manual remixing of the concrete on the discharge platform after machine mixing to offset this segregation.

33.2.3 Designating the Concrete Mixers

According to IS 1791–1985, a concrete mixer is designated by a number which shows its nominal (finished) *batch capacity in litres*. The letter T is added for tilting mixers and NT for non-tilting mixers. The tilting mixers are available in lower capacities. Thus, an 85 T mixer has a capacity of 85 litres of mixed concrete and is of the tilting type. Thus, there are 85 T, 100 T, 140 T and 200 T tilting mixers and 200 NT, 280 NT, 375 NT, 500 NT and 1000 NT, non-tilting mixers are available in the market.

The reverse drum mixers are designated by the symbol R. These are the same as the non-tilting mixers except that discharge can be affected without a chute by reversing the rotation so that there will be less segregation. The capacities available are 200 R, 280 R, 375 R, 500 R and 1000 R. A one bag (35 litres) cement mixer for 1 : 2 : 4 concrete needs roughly [(2/3)(35 × 7) ≈ 163] a 200 litre capacity mixer.

Rotating drum concrete mixers are designed to run at a speed of 15 to 20 rpm. For proper mixing in these mixers, 25 to 30 revolutions should be enough to give a well mixed concrete. A mixing time of 1½ to 2½ minutes depending on the size of the mixer in the field gives a good mix (Large batch mixers like those used in ready mix plants can mix concrete in 15 to 30 seconds.)

The mixer should have a capacity of at least 10% in excess of the mixed batch. For example, for a one bag cement (35 litre) 1 : 2 : 4 concrete, the yield is (2/3)(35 + 70 + 140) = 170 litre. Hence, the mixer capacity should be at least 187 litre. We need a mixer of capacity 200 litre for a full one bag mix of concrete.

33.3 CONCRETE PUMPS

Nowadays pumpable concrete is delivered at site by RMC (ready-mixed concrete) trucks and it is placed on the works by concrete pumps. The concrete that can be pumped through special hosepipes is called *pumpable concrete*. Such concrete is specially designed so that it is neither too stiff nor has a high water-cement ratio for segregation to take place. The concrete from the pump travels like a plug under pressure and the *extra fines, chemicals* and water, specially added to make the concrete pumpable, act as lubricants along the walls of the pipe. Such concrete can be pumped up to about 80 m vertically and about 200 m horizontally. The pump is usually supplied by the firm that supplies the RMC. It is mounted on a trailer or a truck.

There are many types of pump, of which the most popular is the direct acting concrete pump shown in Fig. 33.2. Squeeze type of pumps and hydraulic piston pumps are the other types of pumps used for pumping concrete.

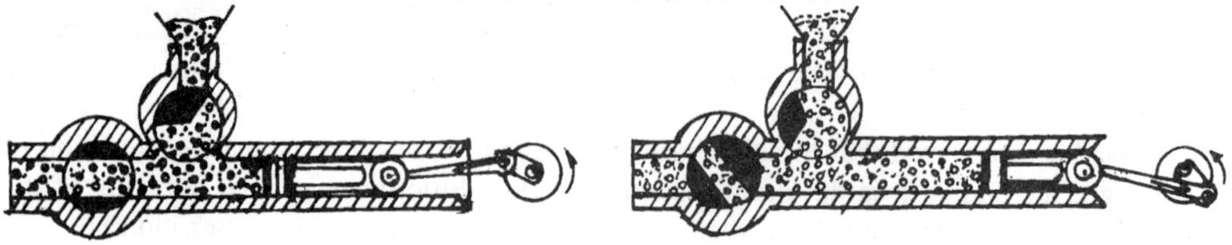

(a) Outlet value open (b) Outlet value closed

Fig. 33.2 Direct acting concrete pump.

33.4 CONCRETE VIBRATORS

In olden days, concrete was placed with high slump and rodded in place. Modern concrete with only moderate slumps has to be placed by vibrators. The most common types of vibrators used in building industry are as follows:

- Immersion vibrators
- Form vibrators
- Tamping board vibrators (surface vibrators)
- Table vibrators

We will briefly examine the use of these vibrators.

33.4.1 Immersion Vibrators

These concrete vibrators are also referred to as *poker* or *needle vibrators*. These are extensively used in building industry. An immersion vibrator consists of a cylindrical casing made of steel and is vibrated by means of a flexible drive with an outer tube run from an operating machine which can be run by electricity, petrol, diesel or compressed air [Fig. 33.3(a)]. The frequency of vibrations of an immersion vibrator is very important as higher frequency gives better compaction. It varies from 3500 to 12,000 cycles per minute. Frequencies below 3500 cycles per minute are inefficient for concrete compaction. The diameter of the vibrating part varies from 20 to 75 mm and its length varies from 25 to 90 cm. The following points are important for efficient use of immersion vibrators:

1. When inserting the vibrator into the freshly laid concrete, we should hold it upright and drop it slowly to an angle. This method will vibrate a larger area.
2. Pull the vibrator out gradually so that the hole is filled up. It is also necessary to restrict the spacing of the introduction of the vibrator to less than 45 cm (18 inch) apart.
3. We should not vibrate the formwork with an internal vibrator as it will damage the formwork. We should place it only up to 75 mm (3 inch) away from the formwork.

4. When compacting concrete on top of another layer already placed, we must push the vibrator to a small distance (say about 75 mm) into the previously placed fresh lower layer to join the two concretes together.
5. The depth to which the vibrator can be immersed is limited to the position of the flexible drive. The vibrator head can be let down into the concrete to the point of the flexible drive.
6. Do not leave the vibrator running when it is not in concrete. Otherwise it can get heated up and the bearings can break down.
7. Put the whole permissible length of the vibrator in the concrete when compacting fresh concrete. This step is necessary to prevent it from heating up. Immersion in the concrete cools it down.
8. Avoid sharp bend of the flexible drive, especially at the vibrator head. Otherwise great strain is placed on the outer tube and internal flexible drive.

The amount of concrete that can be vibrated by one vibrator per hour varies from 4 to 30 m^3 depending on the vibrator and the method of placing concrete.

33.4.2 Form Vibrators (External Vibrators)

Form vibrators are also called *external* or *clamp-on vibrators*. They are clamped onto the formwork and are needed in sections with very heavy reinforcement or for making products such as concrete sleepers, walls, columns, etc [see Fig. 33.3(b)]. They are usually operated by electricity or compressed air. The following points should be noted in their use:

1. They should be firmly clamped to the formwork.
2. They should not be spaced more than 1 m apart. However, the effect of vibration depends on its power. If there are not enough vibrators at the site, those available can be moved from place to place as the concreting proceeds.
3. The concrete to be compacted should be fed little by little and often. Placing in shallow layers gives a more uniform concrete and prevents air being trapped in the concrete.
4. When using for walls or columns, it is advisable to compact the last (top) 60 cm (2 ft) by internal vibrators. This is because, form vibrators, unless the concrete is weighed from above, tend to form a gap between the concrete and the formwork.

33.4.3 Surface Vibrators

These are also known as *tamp board vibrators* and are used for compacting roof slabs, road pavements, etc. There are many types of surface vibrators. The common type of surface vibrator used for slabs in buildings is a screed with a small vibrator (working by eccentrically driven weights) mounted on a single board as shown in Fig. 33.3(c).

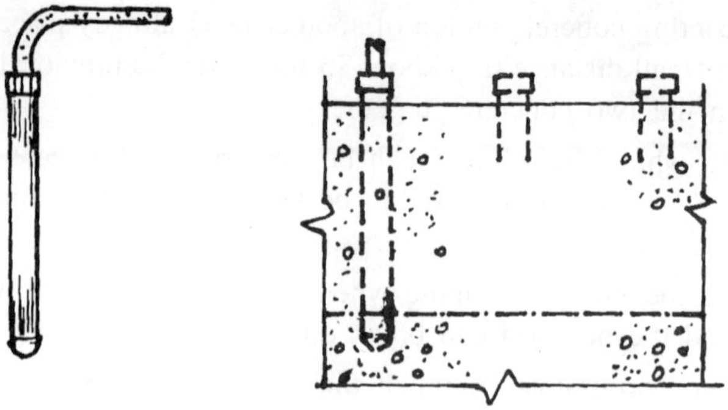

(a) Immersion vibrator (showing the method of vibration to be used)

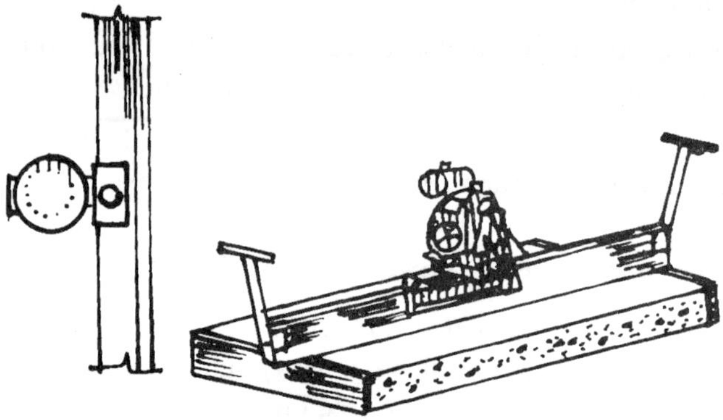

(b) Form vibrator (c) Tamping board (surface) vibrator

Fig. 33.3 Concrete vibrators.

33.4.4 Table Vibrators

Table vibrators are used in factories to make concrete products. The mould is placed on the table which vibrates. We use such vibrators in concrete testing laboratories to compact mortar cubes, concrete, etc. for testing.

33.5 BAR BENDING MACHINES

Steel reinforcement bars are bent to the desired shape by the following methods:

1. Manually using a claw
2. By a bar bending machine

33.5.1 Manual Bending

In manual bending, we use a *claw* which is a lever with a claw arrangement to hold on the bar to be bent. The bar is bent around a rod, called the *mandrel,* after placing it between the mandrel and another rod, called *stop,* as shown in Fig. 33.4. Different set of diameters of mandrel and stop are used for different diameters of bars to be bent. These arrangements are mounted in a platform for easy operation. In practice, a sample bar is first bent and adjustments are made to suit the drawing, and then only other bars are bent (see also Chapter 12).

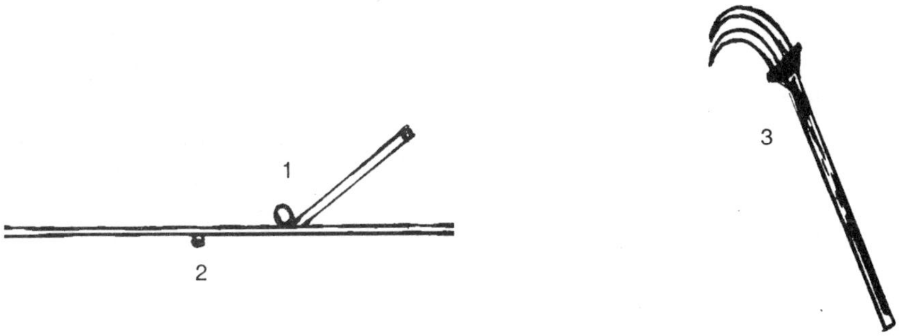

Fig. 33.4 Manual bending of reinforcement bars: 1. Mandrel, 2. Stop, 3. Claw.

33.5.2 Bar Bending Machines

Bar bending machines are used for large projects. They work on the same principle as the manual bending and different sizes of sleeves are placed on the mandrel to suit bending of different sizes of rods. The power to bend is obtained from a machine.

33.6 DOZERS

These machines consist of a rubber wheel or track mounted power unit with a blade at the front end. The blade is controlled by a hydraulic arm. The blade can be tilted at different angles for digging or pushing the earth. For digging, the blade is tilted down and for pushing the earth, it is kept horizontal. The engine has usually eleven forward gears and seven reverse gears. They can dig up to 40 cm at a time.

There are *three types of dozers—angle dozers, bulldozers* and *tree dozers*. In angle dozers, the blades are attached at an *angle other than 90° (usually 30°) to the longitudinal axis of the tractor*. In bulldozers this angle is usually 90°. Angle dozers push the earth to the sides while bulldozers push it to the front. A tree dozer has slightly curved blades, having concavity in the direction of the motion. They are used for uprooting trees, shrubs, etc. The blades can also be moved up or down. Bulldozers can be used for the following purposes:

1. Shallow excavations
2. Clearance of shrubs, trees, etc.
3. Clearance of trees as a pusher
4. As a tilling tractor with rear attachments such as rollers, scarifiers, etc.
5. As a towing tractor to move stuck-up vehicles, etc.
6. As a pusher for scraper machines

33.7 HOISTS

33.7.1 Material Hoists and Passenger Hoists

Hoists are used in building sites for the vertical transportation of materials as in construction of multistorey buildings. Such hoists are called *material hoists*. Hoists are also used at times for transportation of passengers, in which case the hoists should have special safety features. Such hoists are called *passenger hoists*. A material hoist consists of a platform (about 1.5 m wide and 1.2 m deep with a back guard) lifted by means of the hoist rope operated by a power unit. A passenger hoist consists of a cantilever type hoist cage. The latter equipment mounted on trucks is commonly used in large cities for changing street light bulbs, repairing electric posts, etc.

33.7.2 Pulleys and Chain Hoists

These are *systems of chains and pulleys* used to lift heavy objects. They can lift loads up to 50 ton. They are used for lifting precast beams and girders from the ground to the required heights. Unlike cranes, they can work in restricted locations. Such hoists are commonly used in structural engineering laboratories to handle heavy specimens for testing.

33.8 PUMPS

Different types of pumps are used for different purposes and situations. They can be classified in many ways such as water pumps, mud pumps, etc. In general, we will classify pumps into the following four groups:

1. Positive displacement pumps such as reciprocating pumps
2. Rotodynamic pumps such as centrifugal pumps
3. Jet pumps
4. Airlift pumps

As the subject of pumps is vast, we will only deal with the general aspects concerning these pumps.

33.8.1 Different Types of Pumps

Positive displacement pumps. In these pumps, as in piston pumps, the liquid is sucked at the inlet, trapped and then pushed out at the outlet. They can be of reciprocating type, diaphragm type or rotary type.

Rotodynamic pumps. In a rotodynamic pump, such as a centrifugal pump, the pressure head is developed by centrifugal action of a rotating impeller in a casing. Rotodynamic pumps can be of axial flow type, peripheral flow type and other types as per the demand. Electrically operated submersible pumps also work on the same principle.

Jet pumps. Jet pumps are used to raise water from tube wells of moderate depths where the suctions do not work. In a jet pump, a part of the discharged water from delivery side is bypassed into the suction side and leads through *a nozzle on the suction side* (Fig. 33.5). Due to the high velocity in a nozzle, a vacuum is created in the jet unit placed deep in the well, and the high suction can lift the water to large heights.

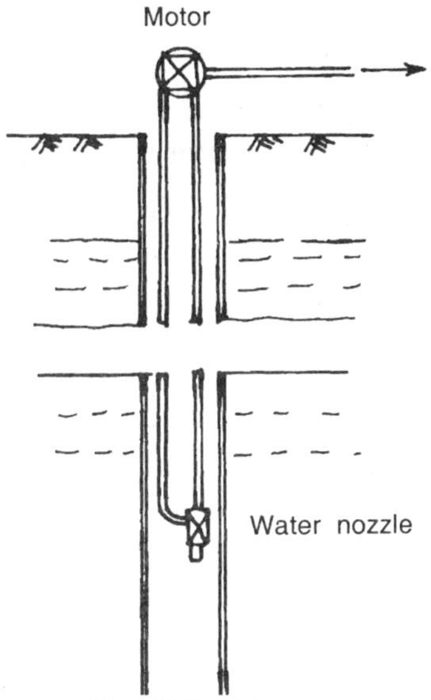

Fig. 33.5 Jet pump.

Airlift pumps. Airlift pumps are used to lift water from larger depths. Air from a *compressor* is led from the ground level to a pipe deep down the bore hole and is allowed to rise up (Fig. 33.6). In the process, a mixture of air and water rises up the pipe, thus lifting the water. The depth of water in the casing should be enough to make the air jet rise through

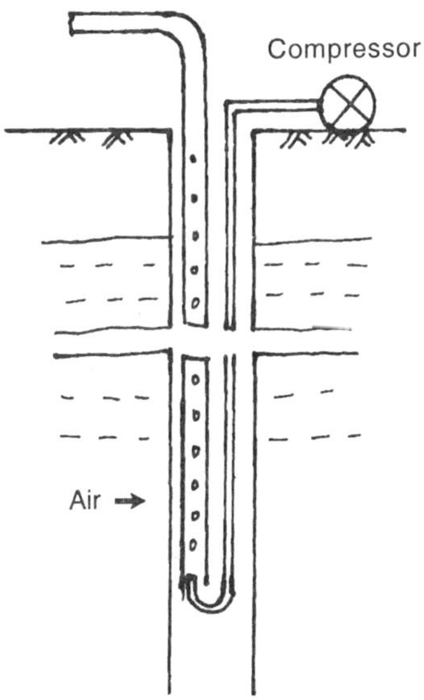

Fig. 33.6 Airlift pump.

the lift pipe and not outside the lift pipe. These pumps do not work in shallow depths of water as then only air will rise up.

33.8.2 Commonly Used Pumps

Some commonly used types of pumps are discussed further.

Piston pumps. These pumps are used for raising clear water from sumps to overhead tanks. The output is low, but it can lift the water to large heights. They are very reliable and require very little maintenance. They can be run by electricity or other means of power.

Mono-block centrifugal pumps. Centrifugal pumps in earlier days were made in two units—the driving motor and the pump unit—as separate and coupled by a flexible flange. Nowadays the mono-block pumps, where the pump and the driving motor are on the same shaft (which eliminates the alignment problem), are mostly used. The pumping rates of these pumps are much higher than those of the piston pumps. When pumping clear water, the number of vanes in the impeller can be increased and clearance can be decreased to give very high efficiency. These pumps are extensively used in buildings and also for agriculture for pumping clear water.

Mud pumps. There are many situations where pumps have to deal with liquid containing solids, sludge, sewage, etc. In pumps used for these purposes, there should be sufficient passage space (clearance between the body and the vanes) for the travel of solids and sludge. This can be accomplished in a centrifugal pump by proper design of the impeller and the number of vanes. Thus, for example, while in a clear water pump, there can be up to six vanes in the impeller, in a mud pump, it may be only 2 or 3.

Jet and airlift pumps. Ordinary piston pumps and centrifugal pumps can lift water only from depths up to 6.5 m (about 22 ft). When water has to be pumped from deeper levels (but not very deep), we use a *water jet pump*. An *air compressor jet pump* can pump water from larger depths than a water jet pump. We have already seen how these pumps work in section 33.8.1. The horsepower of the motor to be used will increase with the depth of water level in the bore well.

Submersible pumps. Before the development of submersible motors, it was difficult to lift water from deep bore wells. Nowadays bore wells of 60 m (200 ft) and above are quite common. After 1970, electric submersible motors started to be manufactured in India, and now submersible pumps are easily available. In a submersible pump, the motor and pump unit can be lowered into the well to be submerged in water, thus enabling it to pump water from very large depths. Pumps of standard sizes—100, 150 and 300 mm diameters—for bore wells are now available in India. They are available as multistage pumps capable of lifting water from large depths.

Hand pumps. These are lift pumps and are commonly used for supply of water in building constructions where the groundwater level is high up to 6 m below the ground level. It works on the simple principle of suction. The pump unit is mounted on the bore hole and a lever manually operates the pump.

33.9 OTHER MACHINES USED IN BUILDING WORKS

There is a tendency to use more and more mechanical devices in building construction works. A mechanical or power float for floating cement/concrete surfaces is shown in Fig. 33.7. This equipment can considerably reduce time and labour for the operation.

Fig. 33.7 Mechanical or power float for floating concrete surface.

SUMMARY

More and more mechanical devices are being used in building construction to increase the productivity and for faster completion of the work. Thus, for example, concreting a roof with RMC and concrete pump considerably reduces the time and labour required. In this chapter, we have dealt with some of the common equipments. We must be aware of the availability of these machines and make use of them in construction practices.

REVIEW QUESTIONS

1. (a) What are concrete batch mixers and how would you designate them?
   (b) Describe briefly the types of concrete batch mixers used for mixing of concrete at construction sites.
   (c) What is the minimum capacity needed for a one bag concrete batch mixer to mix 1 : 3 : 6 concrete?

2. (a) What is pumpable concrete? What are its special features?
   (b) Describe briefly a pump commonly used for pumping concrete.

3. (a) Describe briefly the commonly used concrete vibrators and indicate how an immersion vibrator can be used to compact concrete.

4. Write short notes on the following:
   (a) Dozers
   (b) Concrete mixers
   (c) Concrete vibrators
   (d) Passenger hoists
   (e) Pulleys and chains

5. Describe briefly the working principle of the following types of pumps:
   (a) Jet pump
   (b) Airlift pump
   (c) Submersible pump
   (d) Hand pump

Chapter 34

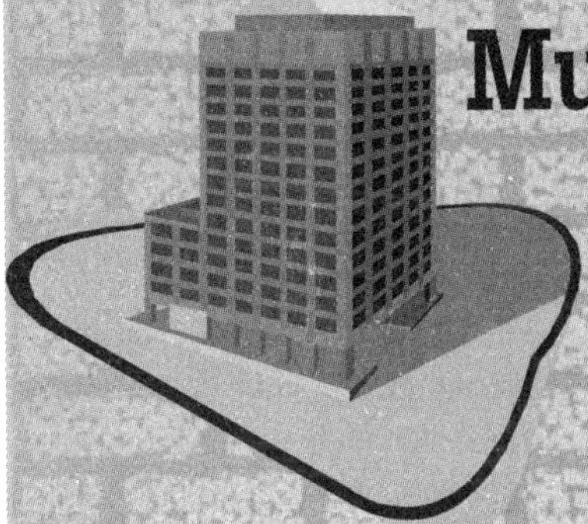

Municipal Requirements in Planning of Buildings

34.1 INTRODUCTION

When planning buildings in cities, towns and municipalities in India, we have to obey certain rules and regulations regarding the minimum size of plot with respect to the width of road in front, spaces to be left around the building for ventilation, etc. In this chapter, we will briefly examine the nature of these requirements that are laid down. As these requirements vary from place to place in the same city and from city to towns, only the basic principles are discussed in this chapter. For the rules applicable to each situation, the regulations published by the concerned authority (panchayat, municipality or corporation) should be consulted.

34.2 CLASSIFICATION OF BUILDINGS

According to the National Building Code of India (1970), buildings are classified into 9 groups *according to its occupancy* as follows:

(i) Group A: Residential buildings
(ii) Group B: Educational buildings
(iii) Group C: Institutional buildings
(iv) Group D: Assembly buildings
(v) Group E: Business buildings
(vi) Group F: Mercantile buildings (whole and retail)
(vii) Group G: Industrial buildings (with fire hazards)
(viii) Group H: Storage buildings
(ix) Group I: Hazardous buildings

These names are self-explanatory. There are many subdivisions in each of these group. For example, Group A (residential) buildings can be further subdivided into private family residences, dormitories, hostels, etc. However, we will not go into these details as they are to be studied under Architectural Planning Requirements. The architect has to plan buildings according to these regulations. These requirements can be obtained from the office of the concerned metropolitan authority or municipality.

34.3 SOME DEFINITIONS

There are some definitions with which we must be familiar.

34.3.1 Building Line

Building line is the line corresponding to the plinth of a building which adjoins the street or extension of the street. According to regulations, there is a minimum distance we have to keep between this line and the adjoining street line.

34.3.2 Building Height

Generally speaking, the height of a building is its height above the ground level. However, in some municipal regulations, it is measured with reference to the middle of the adjacent road level. In cases of buildings abutting a street, heights are measured differently for a flat roof construction from that of a sloped roof construction.

For a flat roof construction adjoining a street, the *height* is defined as the vertical distance measured from the average level of the centre line of the adjoining street to the highest point of the roof of the building.

For a sloped roof *without gables* adjoining a street, the height is taken as the vertical distance measured from the average level of the centre line of adjoining street to the point where the external surface of the outer wall intersects the finished surface of the sloping roof. For a gabled roof, the reference point for height on the building is the mid-point between the eave level and the ridge.

Notes

(a) According to the official criteria of the Council of Tall Buildings and Urban Habitat, "The height of a building is measured from the sidewalk level of the main entrance to the architectural top of the building, including the pent house and tower. Towers include spires and pinnacles. Television and radio antennas and flag poles are not included. Presently, the tallest building in the world is the 508 m high Taipei in Taiwan. Petronas Tower in Malaysia is 452 m high.

(b) The definition of height is important as many corporations classify a building as high and low rise depending on its height. The taxes to be paid for construction permits are different for low and high rise buildings. There are case histories where disputes have arisen with sloped roof buildings regarding their heights.

(c) The so called 63½° rule refers to a diagonal drawn at 63½° to a horizontal line drawn with reference to the rear boundary of the building as shown in Fig. 34.1. No part of the building is to project beyond this diagonal.

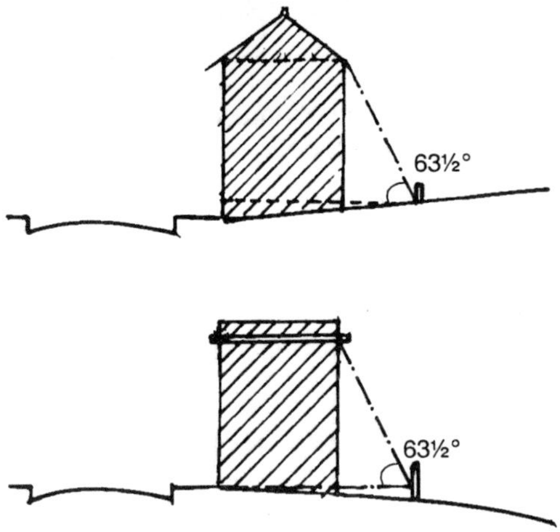

Fig. 34.1 63½° rule at rear space for heights of buildings.

34.3.3 Multistorey Buildings (MSBs)

Buildings *more than* four floors (ground plus three or stilts plus four) or more than 15 m in height are called multistorey buildings. These are allowed to be built only in a plot area not less than 1500 sq m (6.725 grounds in Tamil Nadu).

34.3.4 Carpet Area

This is the *usable area of a building*. In a residential house, it will exclude verandah, bathrooms, staircases, etc. The carpet area of an office building can be 60 to 75% of the plinth area and in a residence, it can be as low as 50 to 65% (65% being the target to be achieved) of the plinth area.

34.3.5 Plinth Area

This is the built up covered area of a building measured at the floor level by taking the external dimensions of the building *excluding the plinth offset*. It also includes areas of porches (other than cantilevered) balconies, etc.

34.3.6 Floor Area

It is the plinth area minus the area occupied by walls, door, openings, etc.

34.3.7 Floor Area Ratio (FAR) or Floor Space Index (FSI)

It is commonly known as the *floor space index* (FSI) and is given by

$$\text{FSI of a flat} = \frac{\text{Total covered area (plinth area) of all floors}}{\text{Total plot area of the building}}$$

An FSI of 1.5 is nowadays allowed for flats in most cities (MSBs require more).

34.3.8 Front Setback (FSB), Rear Setback (RSB) and Side Setback (SSB)

These are the setbacks specified by the competent authority from the boundaries of the building plot.

The minimum front setback (FSB) to be kept when planning a building will be specified with reference to the area where the building is situated and also the width of a road in front. In many old congested commercial areas of cities, it is usually less than that specified for new developing residential areas. The minimum RSB and SSB are usually specified with reference to the height of the building and width of the road in front.

34.4 EXAMPLE OF BUILDING REGULATIONS

As we have already seen that the building regulations vary from state to state and in the same state, from towns to cities and in the same city, from place to place. For example, the regulations at the heart of an old commercial area will be different from the regulations for the new residential area in the same city. As an example, the present regulations for residential buildings in the municipal and township areas of Chennai issued by Madras Metropolitan Development Authority are as shown in Table 34.1.

Table 34.1 Requirements in the city areas for residences G + 1 floors (Maximum for accommodating three dwellings)

S.No.	*Requirement*	*Centre of city*	*Other areas*
1.	Extent of plot (minimum)	80 m^2	150 m^2
2.	Plot frontage (minimum)	4.5 m	9.0 m
3.	FSI (maximum)	1.75	1.5
4.	Plot coverage (maximum)	65%	60%
5.	Maximum height	1½ times the width of the abutting road or 15 m whichever is less.	
6.	Minimum SSB	Nil to 1.5 m depending on the frontage	1.5 m either side or 3 m on one side
7.	Front setback (FSB)	1.5 m	1.5 m to 4.5 m depending on width of the road
8.	Rear setback (RSB)	Nil	1.5 m for 10 m road width to 4.5 m for 30 m road width

The Madras Metropolitan Development Authority has also issued requirements separately for different types of buildings for different areas as given below:

1. Residential (G + 1) and maximum of 3 dwelling units
2. Commercial (G + 1), not exceeding 300 m^2 floor area
3. Institutional, not exceeding 4 floors or 15 m height
4. Industrial, not exceeding 4 floors or 15 m height
5. Theatres/hotels/lodges/religious buildings/Kalyana Mandapams/hospitals, not exceeding 4 floors or 15 m in height

6. Residential layouts
7. Special buildings and group developments

These publications can be obtained from the authority and should be consulted when planning new buildings at a given place.

34.5 OTHER REGULATIONS

The National Building Code (SP 7–2005) and other agencies have laid down many other minimum requirements such as minimum number of rooms, height of rooms, etc. for different types of buildings. These requirements are many, but most normal building practices that we usually follow will satisfy them.

The rules regarding boundary walls and distance from electric supply lines which we have to comply with in cities are briefly discussed below.

Boundary walls. Normally the height of solid boundary wall should be limited to 1.5 m (5 ft). It can also be 2.4 m (8 ft) provided the top 0.9 m (3 ft) is of open type of wall construction approved by the authority. In case of corner plots for a distance of 10 m, the solid wall height should only be up to 0.75 m and the balance should be open type construction.

Distance from the electric lines. The minimum safe distances for various supply voltages specified by the Indian Electricity Supply rules are as shown in Table 34.2.

Table 34.2 Minimum safe distance of building from electricity supply line

S.No.	*Electricity supply voltage*	*Vertical distance*	*Horizontal distance*
1.	Low voltage service lines	2.5 m	1.2 m
2.	Voltage up to and including 11 kV	3.7 m	1.2 m
3.	Above 11 kV and up to 33 kV	3.7 m	2.0 m
4.	Above 33 kV	(3.7 + 0.3) m for every 33 kV or part thereof	(2.0 + 0.3) m for every 33 kV or part thereof

There are many other requirements and these are to be studied in architectural planning of buildings.

SUMMARY

There are many regulations laid down by the competent authorities with regard to the location of a building at a given site. Building plans will be "approved for construction" only if these requirements are met. In addition, before starting actual construction of buildings, we should also go to the site and physically check these requirements from the approved construction drawings so that there will be no difficulty to get completion report from the authority after the building has been completed. There are many case histories where this omission of checking at the site before construction has led to difficulties after completion of the building.

REVIEW QUESTIONS

1. Give the classification of buildings according to the National Building Code of India.
2. Define the following terms:
   (a) FSI
   (b) FSB
   (c) RSB
   (d) SSB
   (e) Plinth area
   (f) Floor area
   (g) Multistorey building (MSB)
3. Show by sketches how you will define the height of the following types of buildings:
   (a) Flat roof construction adjoining a street
   (b) Sloped roof construction adjoining a street
   (c) A gabled roof construction away from the street on a sloping ground

Chapter 35

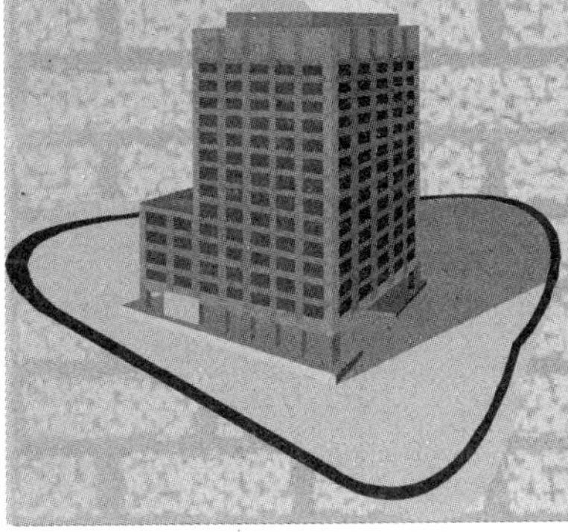

Design of Buildings for Comfort in Hot Climates

35.1 INTRODUCTION

In this chapter, we will briefly examine the following:

1. Factors that affect comfort in hot climates
2. Ventilation requirements of buildings
3. Factors to be considered for building design in hot climates
4. Use of fans and air conditioners for comfort
5. Heat insulation of roofs

35.2 COMFORT: TROPICAL SUMMER INDEX

Our normal body temperature is 37°C (98.6°F) and comfort is to be defined with respect to this value. There are principally three climatic factors that affect human comfort. They are:

1. Air temperature around us
2. Humidity of the air around us
3. Velocity (movement) of the air around us

The first attempt to combine the above factors to form a single factor was the derivation of the term *equatorial comfort index* (ECI) which can be defined as the base temperature of the still saturated air (100% humidity) which will give the same thermal sensation as the air under study. Several combinations of variables can give the same ECI or the same feeling of comfort. Large amount of work has been done along these lines for places such as Singapore.

The results of a large number of observations made in Singapore are reported by British Research Station Overseas Building Notes No. 66. It showed that an ECI value of 25.5°C (78°F) as the temperature acceptable as comfortable to most of the local people in Singapore.

We can convert the given temperature, humidity and air velocity conditions to a single ECI value and compare it with the above ideal value.

35.2.1 Equatorial Comfort Index Diagram

As mentioned earlier, the first attempt to define comfort was the evolution of a nomogram which combines the three factors—temperature, humidity and air movement—to form a single index of warmth defined as *equatorial comfort index.* The original nomogram is shown in Fig. 35.1. It consists of two scales and a set of curves lying between them. The right hand scale is the wet bulb temperature and the left hand scale is the dry bulb temperature. (As the wet bulb temperature is more important, it is represented in larger scale.) The grid of curves in between gives the air velocity.

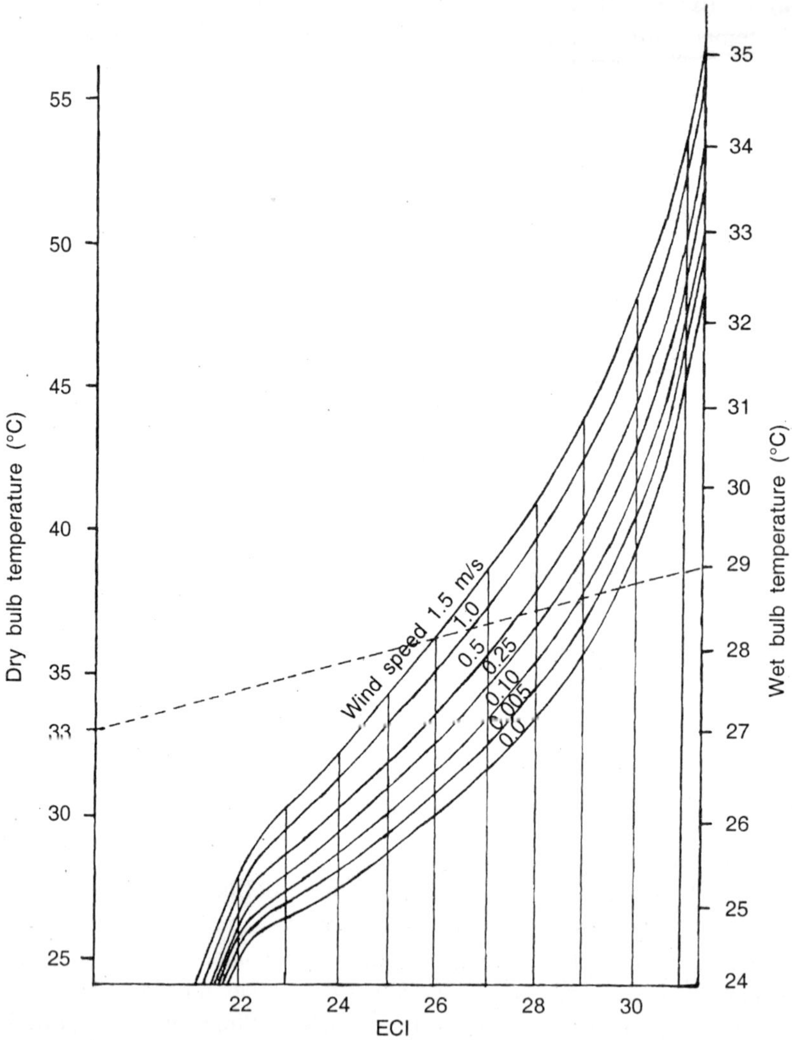

Fig. 35.1 Equatorial comfort index nomogram.

To use this nomogram to find the ECI value for a given condition, we join the points of wet bulb and dry bulb temperatures and find the point where the line meets the given air velocity at the place and read off the equatorial comfort index in degrees from the *X*-axis. For example, with wet bulb reading of 29°C, a dry bulb reading of 33°C and an air speed of about 0.4 m/s, the ECI value is 28°C. In other words, to get an ECI of 28°C for the above temperatures, we need a wind speed of 0.4 m/s which can be provided by a fan. Similarly, to get an ECI of 25.5°C for these temperatures, we need a wind speed of more than 1.5 m/s which is uncomfortable.

35.2.2 Tropical Summer Index

A similar index known as *tropical summer index* (TSI) was developed by the Central Building Research Institute (CBRI), Roorkee, and it is generally accepted for Indian condition. It is defined as the temperature of the *still* air in which an individual feels comfortable at 50% *relative humidity* (not at 100% humidity as in ECI). The TSI is calculated from the following formula:

$$\text{TSI} = 0.308T_w + 0.74T_g - 2.06\sqrt{V + 0.845}$$

where T_w = Wet bulb temperature
T_g = Glob temperature (from a thermometer with the bulb covered with black copper foil)
V = Velocity of air in m/sec

This comfort temperature varies with individuals. From tests on 24 males carried out by CBRI, Roorkee, the values given in Table 35.1 were recommended for Indian conditions.

Table 35.1 CBRI comfort temperatures

Season	*TSI*	*Humidity*	*Temperature*	*Wind*
Summer	26.5 to 29.5	50%	27 to 30°C*	0.5 to 2 m/s
Winter	18 to 21	50%	18 to 22°C	Nil

**Note*. CBRI value of 27° to 30°C (at 50% humidity) is higher than the value of 25.5°C (at 100% humidity) found from the data in Singapore because of the difference in humidity.

The following approximate values can be taken as a practical guide:

1. For temperatures below 29°C and humidities less than 70%, we can be comfortable without a fan.
2. At temperatures 33°C and above, we will be more comfortable under a fan. We need wind speeds as low as 0.1 m/s for 33°C with 40% humidity and that of 3 m/s for 33°C with 60% humidity. Thus, with increase in humidity, we need higher wind speeds for comfort. (Hence, the need of a fan with speed regulator.)

3. Wind speed above 3.5 m/s is not acceptable in practice so that we have to lower the temperature by air conditioning for comfort. For example, at a temperature above 34°C (93.2°F) with 50% humidity, we will not find it to be comfortable even under a fan.

35.3 CLIMATIC ZONES OF INDIA (IS 3792–1978)

India is divided into the following four climatic zones:

1. Hot and arid zones (altitude more than 300 m)
2. Hot and humid zones (altitude less than 300 m)
3. Warm and humid zones (altitude less than 100 m)
4. Cold and very cold zones (high altitudes)

Here we will deal only with the first three groups as the object of house design in the fourth region is to exclude ventilation rather than improve comfort by ventilation. These are indicated in Table 35.2.

Table 35.2 Characteristics of various regions

Region	*Summer condition*	*Winter condition*	*Daily variation*	*Examples*
Hot and arid (altitude > 300 m)	(i) Hot and dry (ii) Low humidity (iii) Low rainfall	(i) Warm days and cold nights (ii) No rain	Large (10 to 20°C)	Allahabad, New Delhi, Sambalpur, Varanasi
Hot and humid (altitude < 300 m)	(i) Hot and wet (ii) High humidity (iii) Seasonal rainfall	(i) Warm to hot (ii) Moderate humidity (iii) No rain	Small (5 to 8°C)	Calcutta, Chennai, Madurai
Warm and humid (altitude < 100 m)	(i) Warm and wet (ii) Very high humidity (iii) Seasonal rainfall	(i) Warm and dry (ii) High humidity (iii) No rain	Small (5 to 8°C)	Cochin, Vizag, Puri, Other coastal areas

Note: We can reduce the three conditions mentioned in Table 35.2 to two conditions, namely 'hot and humid' (coastal regions of India) and 'hot and arid' (North and central India) as given in Section 35.7.

35.4 VENTILATION REQUIREMENTS

It has been shown that except in very crowded rooms, people are able to live healthily in ordinary rooms even when the quantity of fresh air introduced into a room is only very little. Openings in walls are introduced in tropical countries not only to supply air for breathing but also for the sake of day lighting and *for liberal exchange of air in the building by natural atmospheric forces.* Such air movements reduce concentration of bacteria, fungus odours and other airborne contaminants which is necessary for health reasons. In addition, in summer, openings help the cool breeze from outside to enter the building and induce a sensation of freshness in the air as well as remove the heat from within the building. The recommended number of *air changes in a room per hour for health reasons* is 3 to 6 in houses and offices, 6 to 9 in theatres and 12 to 15 in restaurants and public garages.

The two principal agencies that promote ventilation by natural means are as follows:

1. Temperature difference (stack effect as in chimneys)
2. Wind action

In a residential building, we will have to depend on the wind action rather than the stack option so that the problem becomes how to best use cooling breezes. Generally, there is an air movement over 1 m/s at all places on the earth most of the time. But its direction changes with seasons and time of the day. The problem is how to orient the room and its openings so that we get the maximum effect *at the required periods*, say nights in summer.

35.5 WIND ROSE DIAGRAM

For any given place, if we can plot the direction and the speed of the wind for each month throughout the year as shown in Fig. 35.2, we get the wind rose diagram for that month. This can be plotted generally for a given region, but for a particular site of the building, it will also depend on the topographical location of nearby buildings etc. A diagram plotted for the given site will give us an indication of how to best use the cooling breeze in the hot months of the year. It is not necessary to face the building directly against the prevailing winds as it has been observed that the effect is not reduced considerably if the orientation of the building deviates even up to 30°. At a deviation of 45° we get at least 50% of the effect.

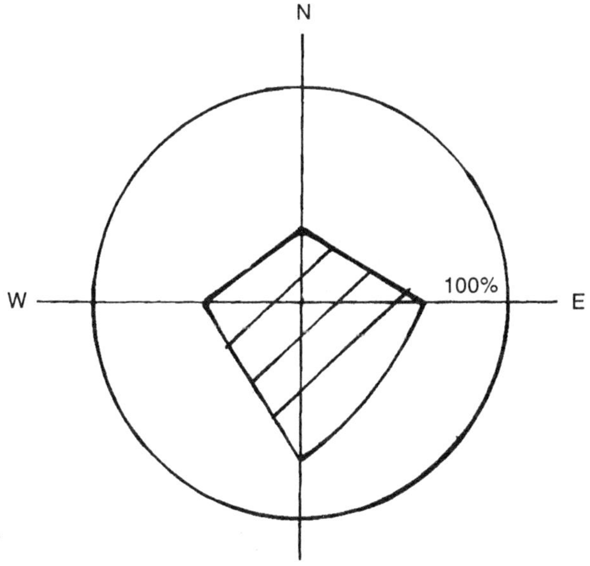

Fig. 35.2 Wind rose diagram for summer (May to July) showing percentage of time (wind speed being more than, say, 3 kph).

35.6 FACTORS TO BE CONSIDERED FOR COMFORT CONDITIONING OF BUILDINGS IN HOT REGIONS

Our objective in design of buildings in hot regions is to keep out as much as possible the warmth of outside from entering the building and also to remove as much as possible the warmth from inside the building and accomplish this task at a reasonable cost. The following seven factors are some of the important factors to be considered for comfort conditioning of a building:

1. Effect of air movement (ventilation)
2. Effect of solar radiation
3. Effect of thermal insulation
4. Effect of mass of construction
5. Effect of shading
6. Effect of orientation of building
7. Effect of types of floors

Let us examine how each of these factors influences the designing of buildings.

35.6.1 Effect of Air Movement (Ventilation)

Proper ventilation can cool human body in hot weather, and it can also reduce the temperature difference between outside and inside the rooms. However, radiant heat from the walls of rooms and the roof is not much removed from the building or impeded by moving air. So these requirements should be considered separately.

35.6.2 Effect of Solar Radiation

When the sun shines on a glass window of a room, the short-wave solar radiation passes through the glass and heats up the walls, floors, etc. of the room. These heated surfaces emit *long-wave heat radiation* which does not pass through the glass. Hence, the room gets heated up with time with no loss of heat. This is known as the *greenhouse effect*. Thus, when solar radiation falls on closed windows, it heats the room in daytime. The heat gained through glazed windows is more than that gained through walls. Hence, windows in east- and west-facing walls should be shaded by sunshades or window curtains. As roofs cannot be shaded, the heat-absorption characteristics of the roof are the most important factors to be considered for heat transfer from solar radiation.

(We now know that global warming is due to greenhouse effect. The carbon dioxide and, to a lesser extent, methane produced in the atmosphere enveloping the earth do not allow the heat from the earth to be radiated back into space. Hence, the world is becoming warmer and warmer with time.)

35.6.3 Effect of Thermal Insulation

Thermal insulation of roofs is desirable in hot countries. Roof insulation reduces the heating of roofs as well as heat flow from the roofs downwards during daytime. However, it also reduces the heat flow to outside during nights. Even whitewashing the flat roof by lime wash can reduce the heating of roofs during daytime by a few degrees.

35.6.4 Effect of Mass in Construction

With light weight construction with low mass (such as timber frames and tile roofing), there is not much difference between outside and inside temperatures. In contrast, heavy weight construction such as brick and concrete is heated up slowly during daytime. It also gives up heat only very slowly during night-time.

35.6.5 Effect of Shading

Shading can be done by the following four methods:

1. External shading by construction
2. External shading by vegetation
3. Internal shading (by curtains)
4. Shading by using special glass (translucent or double) in windows

Effective external shading from the sun's rays greatly reduces the heat gain from solar radiation through walls and glass windows. It can also be done by construction of shading devices such as sunshades. Trees, creepers and constructions and curtains can also be used as shading devices. These are also effective but they reduce ventilation.

35.6.6 Orientation

Orientation of rooms such as bedrooms, living rooms, etc. with respect to the rays of the sun is very important. For example, the position of bedrooms in the west side tends to heat up these rooms in the evenings so that they remain very warm during the nights.

35.6.7 Floor Type

This has only a minor influence. However, cement floors as ground floors seem to have a cooling effect as they are in contact with the earth below.

35.7 DESIGN OF BUILDINGS IN HOT REGIONS

In Table 35.1, we have given characteristics of three hot regions. In general, we can divide the hot regions in India into two categories as follows:

1. Hot and humid climate as in coastal regions of India
2. Hot and arid climate as in north and central regions of India

The principles to be used for design of buildings in the above regions are given in Table 35.3.

Table 35.3 Factors to be considered for producing comfort in buildings in hot climates

Factors	*Hot and humid (coastal regions)*	*Hot and arid (central regions)*
1. General principle	Daily and seasonal changes of temperature are not very high. Hence ventilation and heat insulation to reduce direct heating are the main objectives. Take full advantage of the cooling breeze by cross-ventilated rooms.	Seasonal and daily changes of temperature are high. Hence, exclude heat by day by thermal insulation and closing of windows. Take advantage of falling temperature of outside air during nights by opening all windows and good ventilation from open windows.
2. Ventilation	Plan cross ventilation and large windows in the direction of breeze. Keep windows open all the time.	Provide ample windows and control heat by closing windows during day and opening all windows by night. If needed, increase moist air flow by fan and moist air. (sprinkled water in tatties).
3. Solar radiation and insulation	Solar radiation through windows to be shaded and roofs to be insulated.	Solar radiation through windows to be shaded and roofs to be insulated.
4. Mass of construction	Mass of construction is unimportant. If roofs are not insulated and walls not shaded, use lighter construction.	Use heavy weight construction for rooms used during day and light construction for rooms such as bedrooms used during nights.
5. Shading	Use shading devices to cut off direct heating of roofs and rooms. Natural vegetation or sunshades can be used for shading.	Use shading devices to cut off direct heating of roofs and rooms. Preferably, use natural vegetation for shading as concrete sunshades get heated up.
6. Orientation	Orient rooms such as bedrooms, etc. so that they are not directly heated by western sun.	Orient rooms such that bedrooms, etc. are not directly heated by western sun.
7. Type of floor	Types of floors have no influence.	Concrete floors (not covered with carpets) seem to produce a cooling effect, especially on the ground floor.

Of these factors mentioned in Table 35.3, we will examine ventilation and thermal insulation in more detail.

35.8 NATURAL VENTILATION OF BUILDINGS

The major factors to be considered for producing ventilation in rooms are as follows:

1. Windows should be placed on the side on which the wind blows (windward side). Sufficient openings should be provided on the leeward side also for the wind to blow through the rooms.
2. The maximum wind effect is when the opening is normal to the wind. With the opening at 45° to the direction of the wind, the effect will be reduced to 50 per cent.
3. Even under most favourable condition, the velocity of air in the rooms will only be 50 to 40% of the outside velocity. Generally, it is better to assume it as only around 30%.
4. We may calculate the probable air change by using the formula $Q = K(AV)$

where Q = Quantity of air entering per hour
K = Coefficient of value 0.6 to 0.8 for an opening normal to direction of wind and for equal ratios of openings on opposite sides. For directions of openings other than normal to the windows, it will be reduced. This is a reduction factor.
V = Velocity of wind in m/hr

Example: For a 1 sq m winward opening at 45° to wind with a leeward opening of equal size and with wind velocity 1.5 m/s outside the house, find the air changes in a room of size 4 × 4 m in plan and ceiling height of 3 m.

Wind velocity, $V = 1.5 \times 60 \times 60 = 5400$ m/hr

(Assuming a coefficient of 0.5 for inclination of 45° and value of $K = 0.6$, we get

$$Q = 0.5 \times 0.6 \times 1 \times 5400 = 162 \text{ m}^3$$

Air changes for a 4 × 4 × 3 m room

$$= \frac{162}{48} \approx 3, \text{ i.e. 3 air changes per hour.}$$

This air exchange is sufficient for health conditions but the velocity of air in the room may not be enough for comfort. It will depend on the temperature and humidity. Generally, we prescribe that the total percentage of window openings for a room should be 20 to 30% of the floor area and 30 to 50% of that area should be oriented against the wind during summer. As winds blow from different directions, it is best if windows can be placed on all possible walls.

5. Windows with inlet and outlet placed diagonally opposite ventilate the room better than these placed just opposite to each other.
6. Locate windows of suitable type (casement or adjustable louvers or horizontal pivoted type) *with low sills* in windward and leeward walls for cooling summer winds to ensure reasonable distribution of air indoors.
7. Locate partitions and openings *inside the building* in such a way as to ensure good distribution throughout the spaces inside the building.

35.9 COMFORT CONDITIONING BY MECHANICAL METHODS

We found in Section 35.2 that for certain combinations of temperature and humidity, we need a specified air movement for comfort. In conditions that exceed specified high temperatures, the only method to produce comfort is by cooling of air or by refrigeration which we call air conditioning. Thus, depending on the temperature and humidity, the following mechanical methods can be employed for producing comfort:

1. Ceiling fans
2. Exhaust system
3. Plenum system (ventilation by forcing in of air)
4. Combined extraction and plenum system
5. Air conditioning

As designs of all these systems, except the first one, involve consideration of a large number of factors and are out of the scope of this book, we will briefly deal only with the ceiling fan and air conditioning system as means of producing comfort conditions.

35.9.1 Ceiling Fans

Ceiling fans induce air motion around a person to produce comfort. A fan is found to perform best at a workplace when its height above the flooring is as follows:

$$F_h = (3H + W)/4$$

where F_h = height of fan

H = height of room

W = height of the workplace

Thus, with H = 3 m and W = 0.9 m, we get F_h = 2.5 m or 0.5 m below the ceiling (in 3 m height of rooms) as the best position for the fan. In any case, we should leave 0.3 m between fan and ceiling for good performance. Fans are available with different sweeps (48" and 36", etc.) to suit different sizes of rooms (see section 32.7.3). Their speeds are adjustable by regulators. In old types of fans, speeds are regulated by means of suitable resistances. The newer regulators regulate the speeds of fans by capacitance resulting in less electricity consumption.

35.9.2 Air Conditioning of Buildings

When atmospheric conditions are such that comfort cannot be achieved by use of fans, we have to resort to humidifiers or air conditioners to bring comfort. In modern times, use of air conditioners with or without humidity control is the most common method for comfort conditioning of buildings. As it is a vast subject, we will examine it only in a very elementary way.

There are many types of air conditioning systems for buildings such as the room air conditioners, central air conditioners, etc., which are to be designed to suit the type of the building and requirements. For example, in a five-star hotel and in an air-conditioned theatre or bank, all the spaces in the building are to be air conditioned by the central units. In this section, we will briefly examine only the popular individual room air conditioners used for rooms in offices and residences. They are to be installed on the outside wall of the room to be cooled. According to IS, these room air conditioners are to be made of the nominal cooling capacity expressed in kcal/hr (kilocalories per hour). The standard sizes are 2250, 3000, 3750, 4500 and 6000 kcal/hr. Another unit to express the capacity of room air conditioners is tons. One ton represents the heat required to melt one ton of ice (by giving up its latent heat) in 24 hours. The cooling capacity of one ton capacity will correspond to approximately 3000 kcal/hr.

As a thumb rule, an air conditioner of cooling capacity one ton is quite suitable for a room of area of 10 m^2 and usual ceiling height of 3 m with two to four occupants. Normally, air conditioners have controls only for cooling and the speed of the fan. No humidity controls, which are usually available in the large central air conditioning systems, were generally incorporated till recently in room air conditioners. However, nowadays, this facility is also available in same makes. Window units of different makes fitted with many special facilities are very much used in residences and offices.

35.10 HEAT INSULATION OF ROOFS

After ventilation, the second method to bring down temperature in residences in hot countries like India is the heat insulation of roofs. Unless we insulate the roof of a building built in hot climates, the rooms immediately under the roof gets heated up all through the daytime during summer and remains warm during nights also. The amount of heating depends on the type of roofing material and its exposure. For example, a thatched roof does not get heated up as much as a concrete roof.

The following are the most commonly used methods for heat insulation of concrete roofs:

1. *Applying heat insulating material above the roof.* In this case, the insulating material must be protected by vapour barrier, and the roof must have waterproofing systems also.
2. *Applying heat insulating material under the roof or providing ceilings.* If the heat insulating material is applied directly under the roof, we should also apply a vapour barrier on the hot side for its good performance.
3. *Whitewashing the top of roof.* Each year before onset of summer, the top roof can be whitewashed with fat lime or painted white.
4. *Using reflecting material.* Materials such as aluminium foil or reflecting paint, etc. can be incorporated on top of the roof.
5. *Shading of roof.* Spraying water, growing vegetation (creeper) or other means of shading over the roof can also be used for heat insulation of concrete roofs.
6. *Increasing ceiling height.* Heights less than 2.9 m is not recommended for buildings in hot regions. The decrease in temperature due to heating is about 0.3°C for every 30 cm increase in height. Generally, height of ceilings is kept 3 m (10 ft).

The four thermal properties of a material to be considered are discussed below.

Emissivity or emissive power. It is the time rate of radiant heat *emitted* by unit area of the material with unit difference in temperatures between the surface in question and the surroundings.

Thermal conductivity. It is quantity of heat flowing under steady state condition in unit time in unit area of sheet of unit thickness of the material with unit difference in temperature.

Thermal capacity or specific heat. It is the quantity of heat required to produce unit change of temperature in unit mass of a material.

Reflectivity. If radiation is incident upon the surface of a substance, a part of it will enter the substance and another part will be reflected back. Reflectivity is the capacity to reflect the heat back.

Note. The difference between emissions of heat by radiation from a heated surface and the reflection of heat from a surface being heated by sun's rays should be clearly understood.

The heat reduction felt below the roof due to placing reflective surfaces (such as paint whitewash, aluminium foil) on the top of the roof depends on the *reflectivity and emissivity* of the material. The reflectivity values for some of the materials are as follows:

Asphalt (absorbs most of the heat):	9%
Aluminium sheet and paint:	45 to 60%
Cement concrete:	30 to 40%
White marble:	56%
White paint:	71%
Whitewash:	70 to 90% (Note that this is high.)
AC sheets:	50% (after weathering–15 to 30%)

The principles of insulating roof from solar heat can be understood from the following example:

A polished aluminium surface may reflect as much as 60% of the incident solar radiation. Of the 40% heat absorbed, a *small amount is re-radiated* and lost by convection, etc., but the major part is conducted to the underside of the sheet. The lower side gets heated up, which then radiates its heat to walls, floor, etc. of the rooms below. This radiation to the walls will be smaller if the underside of the sheet is also polished.

White paint has high reflectivity but unlike aluminium, it has, in addition, a *high emissivity* for low temperature radiation. Hence, it re-radiates or emits a large part of the heat it absorbs back to the atmosphere. Thus, when exposed to solar heat, the bottom surface of an aluminium sheet painted white on the upper surface will be a few degrees cooler. Hence, an aluminium sheet painted on the top and polished on the under surface will be more effective for insulation against the solar heat.

In thermal insulation of roof, we try to reduce the heat transmitted to roof by conduction by insulating materials. The insulating property of the material that is to be laid on top of the roof depends on its *thermal conductivity*. Thermocole, fibreglass, mineral wool, etc. give good insulation, whereas cellular concrete, light-weight bricks, vermiculate concrete, etc. have lesser insulating value. The insulation work is to be carried out as described in Section 27.5.

35.11 GREEN BUILDINGS

The newly coined term *green buildings* generally refers to the concept of using renewable resources and materials for construction as well as the management of the building. The features of a green building are as follows:

1. Efficient use of water
2. Energy efficient and ecofriendly equipment
3. Use of renewable energy such as solar or wind energy
4. Use of recycled or recyclable materials
5. Effective use of landscape
6. Effective control and building management system
7. Indoor air quality for human safety and comfort

The benefits of green buildings are as follows:

1. Has less impact on environment
2. Enhances occupants' comfort thus suitable for health and safety
3. Gives long term economy because of less operating costs
4. Improves the output or productivity of occupants

Many of the urban development projects nowadays try to incorporate these concepts into their designs. Near cities where flyash is in plenty, some of the Government agencies such as CPWD insist that the infill material for framed buildings should be flyash bricks instead of clay bricks.

SUMMARY

Proper ventilation of buildings and heat insulation of roofs greatly add to its comfort conditioning by natural means. As air conditioning is expensive and cannot be resorted to except under special circumstances. We should give proper consideration to room ventilation, roof insulation, etc. in construction of buildings in hot climates.

REVIEW QUESTIONS

1. What factors affect comfort of the people living in the tropics?
2. What are the major climatic zones in India and what are the characteristics of each of these regions?
3. Why do we need ventilation in our houses? How can you express these requirements and how do you provide them?

4. What are the factors to be considered for comfort requirement of a residential building by natural means?

5. What mechanical means would you adopt when you cannot attain comfort conditions by natural means?

6. (a) When do you resort to (i) provision of fans and (ii) air conditioning in a residence?
(b) Why is it necessary to provide a speed regulator for fans provided for comfort?
(c) What is meant by a one-ton air conditioner and what will be its capacity in kcal/hr?
(d) What are green buildings?

7. Explain the importance and methods to reduce solar heating in (i) the lower storeys and (ii) the top storey of a residential flat, five storey high.

Chapter 36

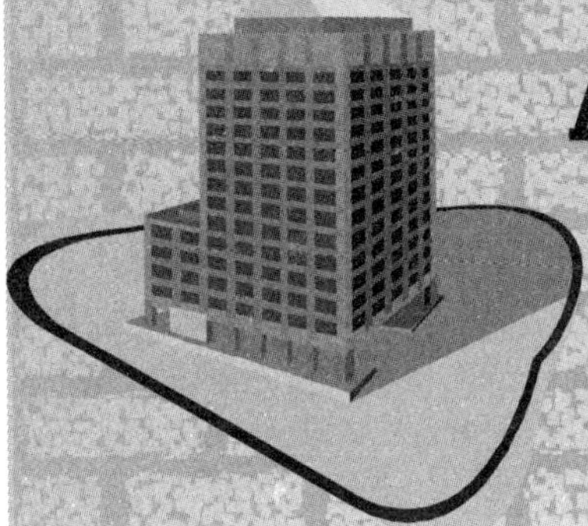

Acoustics of Buildings

36.1 INTRODUCTION

Architectural acoustics is the science of design of buildings such as concert halls and lecture theatres for optimum conditions for producing and listening to music, speech and other sounds produced inside the building. It also deals with the insulation of the room from external noises. In modern days, it has to also deal with the design of means for reinforcing the sound inside the rooms. It is a specialized subject and in this chapter, we will deal only with the basics of the following topics:

1. Basic theory
2. Measurement of sound
3. Reverberation in rooms and sound absorption
4. Echos
5. Room isolation
6. Acoustic design of halls

For detailed design of these aspects in buildings, a specialist or specialized literature on the subject should be consulted.

36.2 BASIC THEORY

Sound in air travels as a series of compression and rarefaction due to air particles put in motion by a vibrating source. The important terms which should be clearly understood are discussed further.

36.2.1 Frequency (or Pitch) of Sound

The frequency or pitch of sound is the number of cycles of this compression and rarefaction.

We express the frequency in cycles per second called hertz (Hz). A young person can hear sounds from 20 to 20,000 Hz but the upper limit diminishes with age, more severely in men than women. In music, we commonly divide sound into eight octave frequency bands identified by their central frequencies of 63, 125, 250, 500, 1000, 2000, 4000 and 8000 Hz. Sound level meters using electric filters can be used to measure the energy at octave frequency bands or 'overall level'. The average velocity of sound in air can be taken as 343 m/s. It travels at higher speeds in liquids (1450 m/s in water) and solids (4000 m/s in concrete).

36.2.2 Intensity of Sound

Intensity of sound (I) is defined as the sound power or the energy crossing per unit time through unit area taken perpendicular to the direction of propagation. Sound power and electrical energy both are expressed in watts for a point at a distance d from the source.

$$\text{Intensity} = \frac{\text{Power}}{4\pi d^2} = \frac{W}{4\pi d^2} \text{ (in watt/cm}^2\text{)}$$

However, 10 watt of electrical energy through an electric bulb will give a very dim light, whereas 10 watt of acoustical energy through a loudspeaker will be very loud.

36.2.3 Loudness of Sound

The nineteenth century German scientists Ernst Weber and Gustav Fechner found that all human sensations such as sound and sight are proportional to logarithm of the intensity of the source. Accordingly, the intensity of a given sound is compared to the minimum sound intensity audible to average human ear I_0 which is taken as 10^{-16} watt/cm^2. Then, the loudness of a given sound of intensity I, denoted by L, can be written as

$$L = \log \frac{I}{I_0} \text{ (in bels)}$$

(The unit *bel* is named after Alexander Graham Bell.) As the unit *bel* is rather large, we adopt one-tenth of that unit called *decibel* (dB).

Hence
$$L = 10 \log_{10} \frac{I}{I_0} \text{ (in dB)}$$

The human hearing has a range of 0 to 130 dB, which is a very astonishing range. Expressing 130 dB in the above form,

$$130 = 10 \log \frac{I}{I_0}$$

So
$$\log \frac{I}{I_0} = 13 \quad \text{or} \quad \frac{I}{I_0} = 10^{13}$$

To give a physical equivalent of the range of human hearing, it is equivalent to the range of a dimension of a human hair to a 30 storey building. Hence, we have to use the logarithmic scale to handle this wide range of numbers in acoustics.

36.2.4 Loudness and Energy

Suppose the loudness of a sound produced by an instrument is 60 dB. Then, the energy of that sound in watt/cm^2 will be calculated as follows:

$$L = 10 \log \frac{I}{I_0} \text{dB}$$

or

$$60 = 10 \log \frac{I}{10^{-16}} \quad \left(\text{as } I_0 = 10^{-6} \text{ watt}\right)$$

or

$$I = 10^6 \times 10^{-16} = 10^{-10} \text{ watt/cm}^2$$

36.2.5 Enhancement of Noise

Let us assume that a sound of loudness 60 dB and intensity (or energy) I_1 is produced by an instrument. Also let another sound of intensity I_2 is produced by another instrument such that $I_2 = 76\ I_1$. Now the loudness oh the second sound L_2 will be calculated as follows:

Given $$I_2 = 76I_1$$

So $$L_2 = 10 \log \frac{76I_1}{10^{-16}}$$

(Since loudness of 60 dB gives $I_1 = 10^{-10}$ watts/cm^2 (as described in section 36.2.4 above),

$$L_2 = 10 \log \frac{76 \times 10^{-10}}{10^{-16}} = 10 \log 76 \times 10^6$$

Hence $$L_2 = 10(7.88) = 79 \text{ dB}$$

This is not much louder than the 60 dB sound of the first instrument. Thus, as we have experienced in practice, a large number of instruments of the same type playing the same melody may not produce a loud music. Hence, we use a large number of different instruments to produce a special blend in an orchestral music.

Acceptable noise levels. The noise level of a jet taking off will be around 90 to 100 dB and that of a traffic on a main road will be 60 to 80 dB. The acceptable levels of noise usually recommended for different situations are given in Table 36.1.

Table 36.1 Acceptable noise levels

S.No.	*Location*	*Comfortable noise level* (dB)
1.	Radio and TV studies	25 to 30
2.	Music room	30 to 35
3.	Hospitals	35 to 50
4.	Conference room	35 to 40
5.	Classrooms	40 to 45
6.	Restaurants and dining rooms	50 to 55

36.2.6 Quality or Timbre of Sound

The quality of a sound is its characteristic which enables us to distinguish between two notes of the same pitch (frequency) and loudness produced by two different sources (instruments or voices.) This is because different sources produce sounds of frequency (say n) and also additional overtones of frequency in multiples of n.

36.3 PHYSICAL MEASUREMENT OF SOUND

Sound level meters as shown in Fig. 36.1 are now available to measure the loudness of sound in decibels. It consists of a microphone which transforms the sound pressure variation in the air into electrical energy and shows it directly in dB There are three scales in a sound level meter—A, B and C. The A scale measures sound by considerably neglecting low frequency sounds, B scale do it by moderately neglecting low frequency and C scale by considering the overall decibel level.

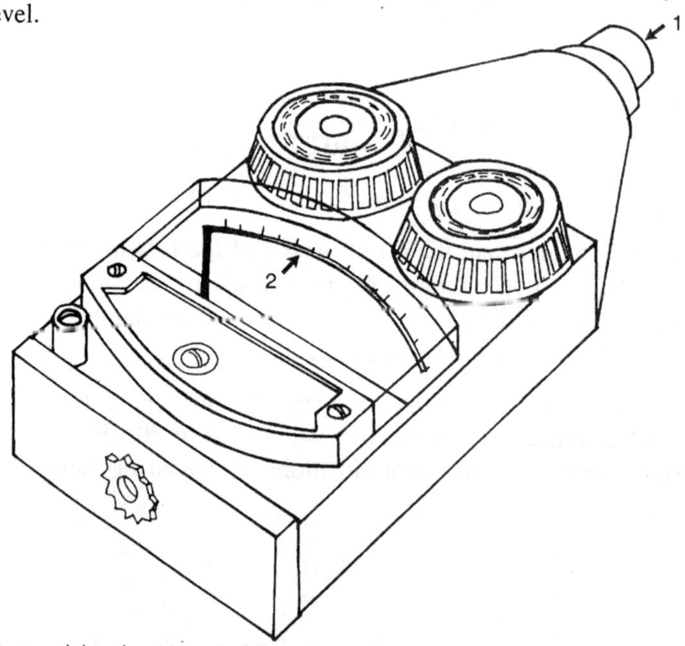

Fig. 36.1 Handheld sound level meter: 1. Microphone to convert sound energy into electrical energy, 2. Measurement in decibels.

36.4 REVERBERATION AND ECHOS

In room acoustics, there are many items to be considered. The *two main defects* to be attended to are reverberation and echos.

36.4.1 Reverberation Time

It was Wallace Clement Sabine, Professor of Physics, Harvard University, who, in 1895, started the work on criteria for good listening conditions in a room. He was asked by the university to improve the very bad listening conditions of a lecture hall of the university attached to its Fogg Art Museum. The sound in the hall was found to persist for about 5½ seconds after its stoppage of speech due to multiple reflection from the surfaces, so that speech was almost unintelligible. Prof. Sabine conducted a number of experiments using an organ pipe with initial sound intensity of 60 dB at 512 Hz and measured the time taken for its decay and called it the *reverberation time* of the sound. (He used his trained ear to measure the sound intensity as then there were no instruments available to measure the sound intensity.) He defined the *reverberation time as the period in seconds required for the sound energy to decay or diminish by 60 dB after the source of the sound is stopped.* Sabine obtained empirically the following Sabine's formula:

$$T = 0.16\frac{V}{A} = \frac{0.16V}{\Sigma\alpha_1 a_1}$$

where T = Reverberation time in seconds
V = Volume of room in m^3
A = Total absorbing power of surfaces in the room
$= \Sigma\alpha_1 a_1$
α_1 = Absorption coefficient of each surface per m^2
a_1 = Area of each absorbing surface in m^2

The coefficient of absorption of a surface is defined as the ratio of the sound energy absorbed to the incident sound energy on the material. Higher the coefficient of absorption, the better is the material as an acoustic material. Even though its value depends on the frequency of the sound energy, usually an average value is given for different materials. Thus, carpet will have a value of 0.4 and cement plaster as low as 0.02.

Figure 36.2 gives the IS 2526(1963) recommended values of reverberation time for acoustic quality in various sounds in a room.

The recommended reverberation time varies with the type of sound to be heard and the volume of the hall. The average values for the same volume increase according to whether it is a hall for speech, motion picture theatre, school auditorium, or a music hall. An average value can be taken as 1 to 1.5 second. The calculation of reverberation time is illustrated in example 36.1 below.

Example 36.1 An auditorium 35 m × 25 m and 9 m high has concrete floor (of absorption coefficient α = 0.03/m^2), suspended ceiling (α = 0.05), cement plastered walls (α = 0.02) and 1000 seats (α = 0.16). Find the reverberation time neglecting the effects of doors, windows, etc.

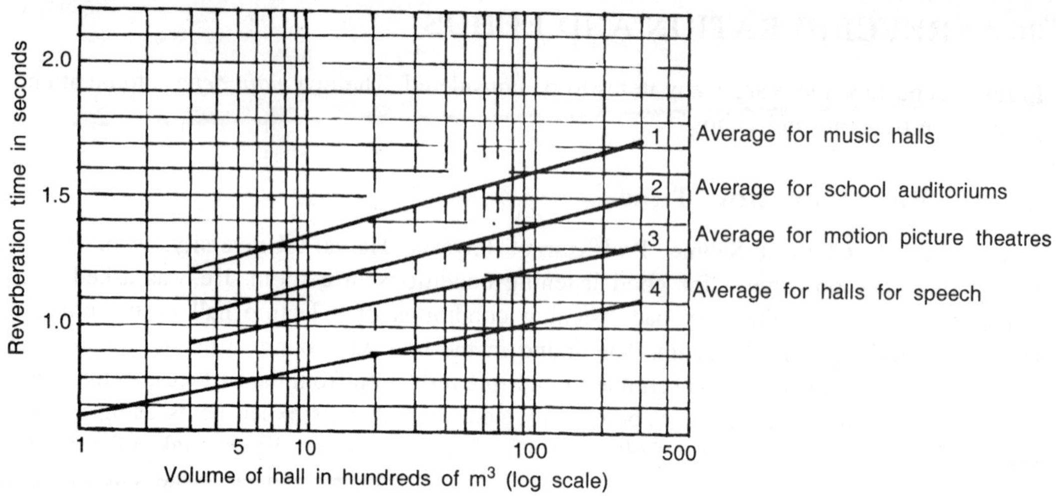

Fig. 36.2 Optimum reverberation times (by Sabine's formula) for 500 cycles for different uses as a function of volume of the room.

Solution: Volume, $V = 35 \times 25 \times 9 = 7875 \text{ m}^3$

Now $A = (\alpha_1 a_1 + \alpha_2 a_2 + \alpha_3 a_3 + \alpha_4 a_4) \text{ m}^2$ sabines

$\therefore$ $A = (0.03 \times 35 \times 25) + (0.05 \times 35 \times 25) + 0.02(120 \times 9) + 0.16 \times 1000$

$= 251.6 \text{ m}^2$ sabines

Hence $$T = \frac{0.16 \times 7875}{251.6} = 5 \text{ seconds}$$

This is much more than 1.0, the reverberation time allowed for speech as shown in Fig. 36.2. Hence, the walls, floor, ceiling have to be treated with better acoustic materials.

36.4.2 Echos

An echo is produced when the sound wave reflected from any surface reaches the ear after the sound wave from the original source has been already heard. This produces a repetition of the same sound. Echos produce unpleasant hearing. This can be avoided by making the reflecting surface absorbent or by reducing the time between the hearings of direct and reflected sounds as given in Section 36.7.

36.5 SOUND ISOLATION

Sound isolation is different from *sound absorption*. Whereas sound absorption is prevention of *reflection,* sound isolation is prevention of *transmission* of sound. Solid brick walls, cavity walls, etc. reduce sound transmission. Similarly, floors of a room immediately above the bedroom or living room should have an impact isolation. For this purpose, we place an isolation layer in upper floors (refer details in concrete floors).

36.6 COMMON ACOUSTICAL DEFECTS AND REMEDIES OF CONFERENCE HALLS

IS 2526 (1963) gives a list of defects, causes and remedies for auditoriums and conference halls. Table 36.2 gives a modified summary of these recommendations.

Table 36.2 Acoustical defects, causes and remedies of auditoriums and conference halls

S.No.	*Defect*	*Cause*	*Remedy*
1.	Excessive reverberation	Size and insufficient absorption	Add absoption.
2.	Echos	Unsuitable shape and reflecting surfaces	Modify shape and add absorption.
3	Sound foci	Concave interior surfaces	Alter shape or add absorbants.
4.	Dead spots	Irregular distribution	Introduce diffusers at uitable points.
5.	Insufficient volume of sound	Large volume of rooms	Introduce loudspeakers.
6.	Bad quality of sound	Selective absorption of sound of certain frequencies or uncontrolled resonance	Use combination of absorption over required frequency range.
7.	High background noise	Bad selection of site, bad isolation and noise from airconditioners, etc.	Provide good sound isolation.

36.7 USE OF RAY DIAGRAM AND ECHO

In spite of many limitations, a ray diagram is often used to determine approximately the listening condition of rooms. If reflected sound reaches the listener *later than* 1/17 (0.06) *second* (travel distance equal to 340/17 = 20 m), there will be echos. Table 36.3 may be used as a guide to good listening with respect to the difference between the direct and reflected sound waves.

Table 36.3 Path difference and listening condition

Sound path difference (m)	*Listening condition*
Less than 8.5	Excellent for speech and music
8.5 to 12.0	Good for speech, fair or music
12.0 to 15.0	Marginally good
15.0 to 20.5	Not good
> 20.5	Echo in case of strong reflected sound

The method of analysis is shown in Fig. 36.3. In locating points, the angle of incidence should be equal to the angle of reflection.

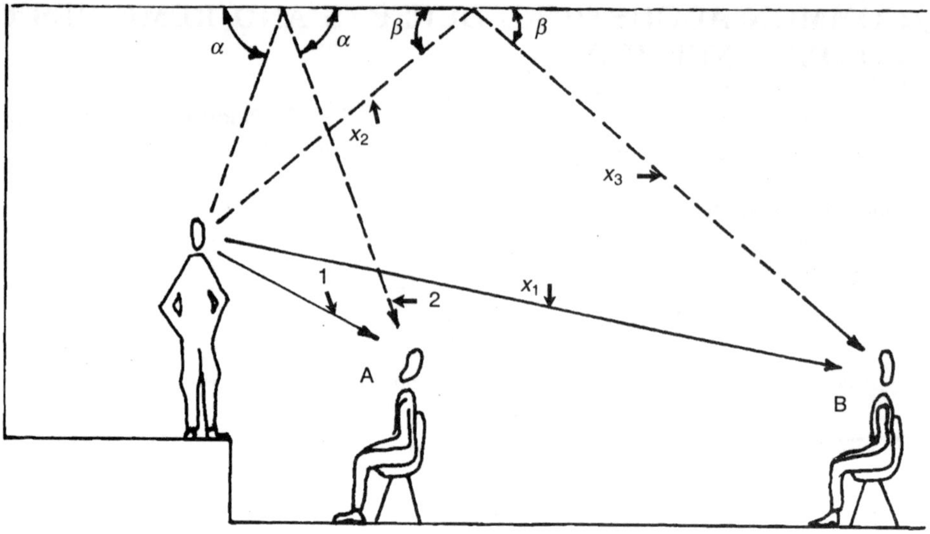

Fig. 36.3 Use of ray diagrams for design of auditoriums: 1. Direct sound path, 2. Reflected sound path. [For excellent hearing of speech and music, the difference between the distances travelled by the reflected and direct paths should be less than 8.5 m (28 ft), i.e. $(x_2 + x_3) - x_1 < 8.5$ m.]

36.8 DESIGN OF AUDITORIUMS

The step-by-step procedure for design of an auditorium can be as follows:

1. *Site selection.* Select a quiet site away from the noise.
2. *Type of building.* Use corridors, closets to buffer the outside noise and produce isolation.
3. *Volume and shape.* Select the dimensions to suit the reverberation time required. The average ceiling height is given by $H = 6T$ m($20T$ ft), where T is the mid-frequency reverberation time in seconds. Fix the shape based on the required seating capacity and to minimize the distance of audience to the performing area. Volumes in m^3 per person of 1.2 to 3 for lecture halls, 4.2 to 5.6 for music halls and 3.5 to 4.5 for cinema halls are generally provided. The ratio of the height to width to length usually varies from 1 : 1½ : 3 to 1 : 2 : 4.
4. *Furnishings.* Eighty per cent of the *ceiling area* should be sound reflecting. The *side walls* should be sound diffusing, the *rear walls* should be sound absorbing and the *floor* should be carpeted except in front of the stage. The *seats* should be absorptive to help provide stable reverberation time with variation of number of audience.
5. *Use of ray diagram analysis.* Use ray diagram analysis for properly shaping the ceiling and walls as shown in Fig. 36.3.
6. *Stage enclosure.* Shape the stage enclosure so as to improve the diffusion of sound.
7. *Sound reinforcing system.* Provide sound reinforcing system such as loudspeakers in a proper manner.
8. *Balcony consideration.* Use balconies as means to increase the seating capacity and also to reduce the distance of the audience from the stage.

36.9 REQUIREMENTS OF AN AUDITORIUM

In general, the following are the factors to be considered in the acoustic design of a theatre or lecture hall:

1. Site selection and planning
2. Determination of capacity and dimensions
3. Shape, depending on the use and capacity
4. Reverberation time
5. Treatment of the interior surfaces
6. Seating arrangement, provision of balconies, etc.
7. Design of acoustical system of sound reinforcement by loudspeakers, etc.
8. Beautification by architectural effects

We have covered most of these factors in Section 36.8.

36.10 ACOUSTICAL MATERIALS

We have seen that on the basis of acoustical properties, materials can be divided into the following groups:

1. Sound reflecting materials
2. Sound absorbing materials
3. Sound insulation materials and construction

We will discuss these further.

Sound reflecting materials. These materials are generally placed in the ceiling, back of the stage and side walls of lecture halls. Materials such as wood, special plaster, concrete are sound reflecting materials. The sound reflecting property of materials is expressed by their sound reflection coefficients.

Sound absorbing materials. Absorption is expressed by the *absorption coefficient.* The materials with loose structure such as carpets, glass, wool mats and perforated hardboards (where the sound energy is lost in the holes) are called *typical acoustic material.*

Sound isolation materials and construction. Dense materials such as bricks do not allow sound to pass through. There are many types of constructions such as solid brick walls, cavity walls, double wall partitions with sound absorbing materials in between. The sound isolation property of a material is expressed by the reduction of noise in dB.

36.11 RECOMMENDATIONS FOR DIFFERENT TYPES OF BUILDINGS FOR GOOD ACOUSTICS

The recommendations applicable to different types of buildings for good acoustics are discussed below.

36.11.1 For Open Air Theatres

An open air theatre should be located in a quiet place. A properly designed shell behind the area, where the sound is produced, is to be provided for two purposes, namely: (i) to reflect the sound to the audience (ii) for the stage performers to hear each other.

The site should be chosen so that there are no reflective surfaces around the site that can produce echos. The auditorium floor can have a level surface in the front for seats and in addition, a sloped surface with slope not less than +12° for good listening and viewing by the audience.

36.11.2 Cinema Theatres

A fan-shaped floor plan with diverging side walls gives good results. A ceiling with a slight *upward slope* towards the rear is recommended. The ideal width recommended is 2 times the height and ideal length recommended is 3 times the height. The reverberation time should be checked. The volume is recommended as 3.5 to 4.5 m^3 per seat.. Echos should be prevented by using sound absorbing in rear walls. Ray diagrams (Figure 36.3) should be used to check the listening conditions of the theatre.

36.11.3 Lecture Halls

A smaller volume 1.2 to 3.0 m^3 per person is enough for lecture rooms. The audience should be seated near the lecturer and seats should be arranged elevating upwards towards the rear. The side walls and ceiling may reflect the sound but the rear walls should be of the absorbing type.

SUMMARY

Room acoustics is important in design and construction of halls for lecture, music, cinema, etc. In this chapter, we have briefly dealt with only the basic concepts of architectural acoustics.

REVIEW QUESTIONS

1. (a) What is meant by the following and in what units, these are measured?
   (i) Intensity of sound
   (ii) Loudness of sound

   (b) What instrument would you use to measure sound level? How does it work?
2. What is the concept of dB in acoustics? What is its allowable value recommended in
   (a) Music rooms
   (b) Public rooms
   (c) Factories

3. (a) What is meant by reverberation time? What is its importance in architectural acoustics?

(b) A room is 12 m × 8 m in plan and the ceiling height is 4 m. The ceiling is covered with acoustic tiles with an absorption coefficient of 0.30. The short walls are covered with thick curtains which fully absorb the sound. The floor and side walls together have an absorption power of 5 m^2 sabines. Determine the reverberation time of the room.

4. (a) List the various factors to be taken into account in design of a conference hall.

(b) List the various defects, causes and remedies in designs of auditoriums and lecture halls.

5. Write short notes on the following indicating in what situations these are used and what type of materials or construction are used for these:

(a) Sound absorption

(b) Sound reflection

(c) Sound isolation

Chapter 37

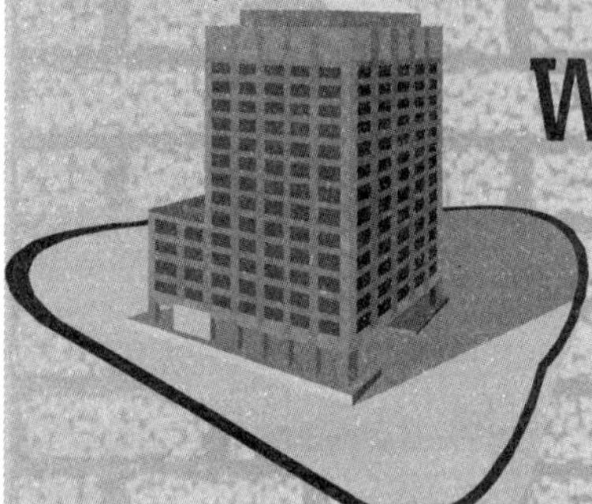

Welding and Structural Steelworks

37.1 INTRODUCTION

In construction of a building, we come across a number of items of steelworks. As welding is important in steel construction, we will first briefly deal with welding. We will also consider steel trusses and steel window grills that are commonly used in buildings.

37.2 WELDING AND INSPECTION OF WELDING OF STEELWORKS

There are three principal forms of welding of steel:

(a) Electric arc welding
(b) Oxyacetylene gas welding
(c) Flash butt welding

The first method uses an electric arc while the second uses an oxyacetylene flame for welding, and the third is carried out only in factories by special machines using electricity. For structural work, electric arc welding is nowadays invariably used. Oxyacetylene flame is used for cutting steel (gas cutting) and also for repair or other work involving small thicknesses of metal, which may burn off with electric arc.

37.3 ELECTRIC ARC WELDING

There are many forms of arc welding but the most commonly used type is the Manual Metal Arc (MMA) welding. In this method, the metal is deposited manually by the operator who controls the direction of the weld and the build up of the weld metal. *The flux surrounding the core wire when melted by the heat of the arc gives a gaseous protection for the molten metal from atmospheric contamination and controls the weld metal reaction.* The temperature of the electric arc is of the order of 10,000°C.

Welding electrodes. The electrodes used for manual arc welding must comply with IS 814 (1991), the specification for "covered electrodes for manual metal arc welding of carbon and carbon-manganese steel". These should be stored properly in dry places. These electrodes consist of a core wire covered on its outside with a flux. The end that is to be connected to the electrode holder has usually no flux coating.

The size of the electrode to be used depends on the type of weld (fillet or butt), as to whether the work is to be done in one or more passes and also the position of welding as to whether it is flat, horizontal, vertical or overhead. The electrodes shall be of proper sizes to ensure thorough fusion and penetration, but, in any case, generally it is to be not more than 4 mm (5/32 inch) in diameter.

Welding Transformer. The welding transformer should give the necessary current and voltage specified for the job.

Types of weld. The two types of welds most commonly used are discussed below.

1. *Butt welds.* They are called single V, double V, single U, double U, single J, single bevel, double bevel depending on shape of the cut of joining surfaces (see Fig. 37.1).

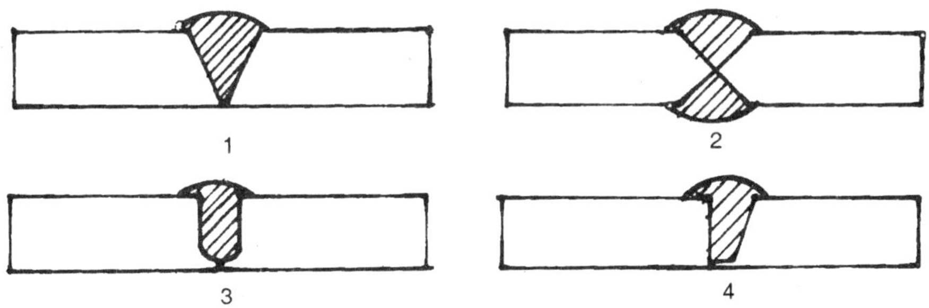

Fig. 37.1 Butt welds: 1. Single V, 2. Double V, 3. Single U, 4. Single J.

2. *Fillet welds.* According to the welding position, fillet welds can be flat, horizontal, vertical or overhead (see Fig. 37.2).

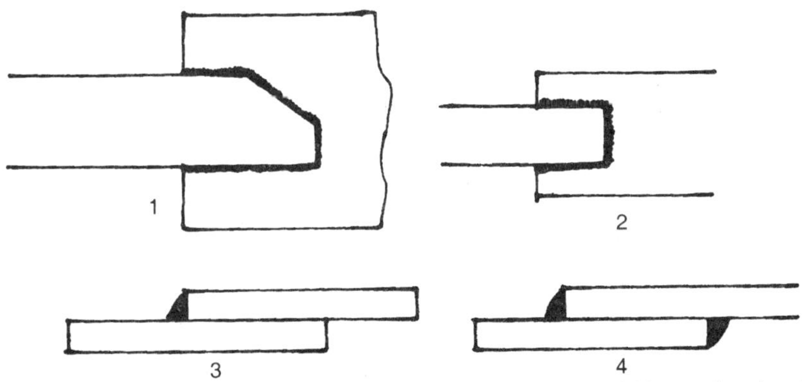

Fig. 37.2 Fillet welding: 1. Sides, diagonal and end weld, 2. Return weld (sides and end weld), 3. Single lap weld, 4. Double lap weld. (The overlaps as a rule should not be less than five times the thickness of the thinner part.)

37.3.1 Specifications of Butt Weld

The specifications of butt welds are briefly discussed below.

(a) Size of a butt weld is its effective throat thickness.

(b) Single V, single U, single J and single bevel butt welds are called *incomplete penetration butt welds* (also called unsealed single butt welds), where the weld metal is not deposited through the full thickness of the joint. Welds like double V welds are full penetration butt welds (the effective throat thickness of incomplete penetration butt weld is taken as 5/8th the thickness of the thinner part joined.)

(c) When structural members of unequal thicknesses are joined by butt welding and the difference in thickness is more than 25 per cent of the thinner part or 3 mm, it should be specially treated by tapering the thicker part or building the weld attached to the thinner part at the junction to 25 per cent greater than the thinner part or equal to the thicker part.

Reinforcement in butt welds. Generally, a surface convexity of 1.0 mm (but not to exceed 1.3 mm) is provided to ensure full strength of the weld. If a flush surface is required, this reinforcement may be removed.

37.3.2 Specification for Fillet Welds

Fillet welds are of triangular cross sections. The standard weld is at 45 degrees (30 to 60 degrees are also used for special fillet welds).

The fillet weld can be of three types:

(a) Side fillet weld

(b) End fillet weld

(c) Diagonal fillet weld

The size of a fillet weld is measured as the minimum '*leg length*' (1.414 times the effective throat for a 45° fillet). The *leg length* of a fillet weld is the distance from the root to the toe of the weld measured along the fusion face. The minimum size of a single-run fillet weld is to be as follows:

(a) Thickness of thicker part of welded members up to 9.5 mm = 3 mm

(b) Thickness of thicker part of welded members up to 19.0 mm = 5 mm

(c) Thickness of thicker part of welded members up to 32 to 50 mm = 8 mm

The *effective lap length* of a weld measured along the weld is taken as the actual length minus twice the size of the weld, as the specified size and throat thickness may not exist at the ends. This should not be less than four times the size of the weld.

When the welds are 'returned', it should be carried around the corners not less than twice the size of the weld. This is specially applicable to side and top fillet welds in tension.

37.4 INSPECTION OF WELDING OF STRUCTURAL STEELWORKS

The primary concerns of inspection of welds are to see that

(a) the welds produced are fully in accordance with the sizes and shapes shown in the drawings,
(b) there are no defects in the welding process and
(c) there are no defects in the finished welds.

Firstly, the steel that is welded should be weldable steel. The presence of increasing carbon can cause cracking of the weld. Hence, weldable steels are those whose carbon content is sufficiently low as not to cause the cracking. Cast iron, which contains a large amount of carbon, for example, cannot be joined by ordinary welding.

Of the two types of welds used to join up members, namely the butt weld and the fillet weld, butt welding is more difficult. In butt welding, the two surfaces to be welded need much more preparation before welding than in fillet welding. The first step in inspection is to see whether the preparation of the surfaces is satisfactory.

In manual butt welding, when welding has to be done on both sides of the joint, the first side is completed first. For welding from the second side, it is first 'back gauged' at the root to get full penetration.

37.5 PROCESS OF WELDING

When a weld is made, some of the parent material melts and combines to form the deposited metal. On cooling, the first to cool is the material adjacent to the parent material and the last to cool is the throat area. The quality of the weld depends on the way the weld material is placed. A skilled welder will be able to make good welds, and an inspector should be able to check each weld for defects and ensure that the welding is satisfactory.

37.6 DEFECTS IN WELDING

In this section, we will briefly discuss various common defects found in welding.

Undercut. This occurs when a grove is melted into the parent material adjacent to the arc formed by the arc action and is not subsequently included in the weld metal.

Porosity. *Porosity* is the presence of cavities in the weld metal caused by the gas entrapment, usually spherical in shape. It is caused by the moisture, scales, oil and other contaminations on the plates, or from damp electrodes. Also, if the arc length is too long, porosity will almost certainly occur when using basic electrodes. Since it takes some time to establish necessary stable condition after the start of welding work porosity is often associated with 'start porosity'. It is usually overcome by striking the arc and travelling for 15 mm and then going back over the initial arc to melt out the start porosity.

Incomplete penetration. In fillet welds, this occurs invariably when larger diameter electrodes than necessary are used to fuse the root. In butt welding, incorrect 'back gauging' can result in lack of penetration.

Lack of fusion at edges. This usually occurs in fillet welds and is due to the failure to fuse together adjacent surfaces of the weld and parent materials.

Slag inclusions. The slag is derived from the flux used with the welding process and if it is entrapped in the weld, it is harmful. This can be prevented by correct plate preparation, correct diameter of electrodes and good technique of welding.

Hot crack. The typical hot crack in a fillet weld is a longitudinal crack characterized by a blue appearance along the crack due to surface oxidation at high temperatures. They are due to bad material composition, weld strain and bad bead shape.

Gas pores. Gases are formed inside the weld.

Surface imperfections. The main surface imperfections that can happen are as follows:
(a) Edge of the plate melted off
(b) Overlap

37.7 METHOD OF INSPECTION

The inspector should be in full attendance supervising the work in case of important works. The following is generally used as a checklist:

(a) Whether the steel is properly cleaned and is free from corrosion, water, oil, scale, dirt, paint, etc.

(b) Whether proper methods were employed when setting up work to ensure tight fit without displacement of competent parts.

(c) When the work is completed, it should be examined for defects and irregularities described in Section 37.6. For this purpose, it may be necessary to hammer the weld or clean it with a wire brush as required.

In all important cases of welding work, as in truss work, it is better to keep a record of the work, the inspection, erection etc. so that during and after the work, systematic checks can be made on the quality and completeness of the work that has been carried out. Such evaluation will ensure a satisfactory quality of the construction work. In very important works, it is possible to make a radiographic examination of the weld joints in steel up to 50 mm in size.

37.8 TACK WELDING

In many cases, as in welded structural work, the structure is divided into parts for easiness of transport or erection and these parts are shop welded, transported to the site and erected at the site first by temporary tack welding. They are finally site welded carefully to form the actual structure. Thus, a long truss may be divided into three parts—the two sides and the central portion—or into halves. These are then aligned and erected at the site first by tack welding and after all the trusses are erected and fully aligned, the joints are field welded to form good and perfect joints, strong enough to take the imposed weights. Under these circumstances, utmost care should be taken to see that records are kept to ensure that these field welds have been actually done and the weldwork is satisfactory in all respects. Cases have occurred where the field staff had forgot to fieldweld the joints after tack welding and upon loading, the structural collapse took place. This should never happen at a site supervised by competent staff.

37.9 GAS WELDING

In gas welding, the flames can be adjusted to three types of flames—reducing, oxidizing and natural. With more acetylene, we get a reducing flame, with more oxygen, an oxidizing flame and in between, we get a neutral flame. Neutral flame is generally used in welding. The temperature varies from 3150 to 3230°C only. Compared to electric arc welding of 10,000°C, it is only one-third the temperature. Hence, it can be used for welding thin plates which otherwise will be burnt off by electric arc.

37.10 GAS CUTTING

Oxyacetylene flame can also be used for gas cutting. This process makes use of the property of oxygen's affinity to ferrous metals at ignition temperatures. The surface to be cut in first heated by an oxyacetylene flame. After heating, a jet of high purity oxygen is directed to the heated surface when combustion takes place with generation of heat. This principle is used in cutting of steel with oxygen. Since only the metal in direct contact with the jet is acted on, a very accurate line of cut can be made by gas cutting.

37.11 TUBULAR ROOFS AND COLUMNS

Details of a roof truss made from angles and gusset plates are shown in Fig. 22.5. In steel tubular truss work made of steel tubes, the recommendations to be followed are discussed below.

Nature of structural steel tubes. The tubes shall conform to IS 1161 – 1998, "Steel tubes for structural purposes – Specifications", and shall be one of the following types:

- Hot finished welded (HFW) type
- Hot finished seamless (HFS) type
- Electric resistance welded (ERW) type

Minimum wall thickness of tube. Structural tubes shall have the minimum wall thickness as indicated below depending upon the exposure:

- Construction not exposed to weather: 3.2 mm
- Construction exposed to weather: 4.0 mm
- Structures not readily accessible for maintenance: 5.0 mm

Caps and base for ends in columns. The ends of all tubes for columns, transmitting loads through the ends, shall be true and square to the axis of the tube and shall be provided with a cap or base accurately fitted to the end of the tube and screwed, welded or shrunk on. The cap or base plate shall be true and square to the axis of the column.

Sealing of tubes. When the end of a tube is not automatically scaled by virtue of its connection by welding to another member, the end shall be properly and completely sealed. Before sealing, the inside of the tube shall be dry and free from loose scale.

Flattened ends. In tubular construction, the ends of tubes are allowed to be flattened or otherwise formed to provide for welded, riveted or bolted connections. These ends are connected to gusset plates. However, the methods adopted for such flattening should not injure the material. The change of sections shall also be gradual.

These tubular trusses are light and are very popular for moderate spans.

37.12 ERECTION OF STEEL TRUSSES

Fabrication and erection of steel trusses are important items of work in building construction. For erection, trusses shall be lifted only at nodes. The trusses below 10 m in span are usually slinged at the apex. However, as this will develop compression stresses in the bottom tie member, it is better if trusses are lifted by slinging at two mid-points on rafters, which are temporarily braced by a wooden member of a suitable section. After the trusses are placed in position, the purlins and wind bracings must be fixed as soon as possible.

The end of the truss which faces the prevailing winds shall be fixed by holding down bolts to the bed plate as shown in Fig. 22.5, and the other end is kept free to move with the bolts in an oblong slot. For this purpose, in case of trusses of spans up to 10 m, the free end of the truss shall be laid on lead sheet or steel plate as per design, and the holes for holding down bolts shall be made in the form of oblong slots, so as to permit the free movement of the truss end. For larger spans, the truss shall be provided with suitable bearing as per the design.

37.13 FIXING OF MILD STEEL BARS AND GRILLS IN WOODEN FRAMES OF WINDOWS AND VENTILATORS

Grills are provided for safety on windows. Addition of grill doors may also be provided as external doors. They are usually made from mild steel bars, squares, flats, etc. according to the required design. In simple design round (12 mm diameter) or square bars can be used. Steel flats of approximately 4 mm can be used to make attractive designs for windows. Floral designs with thinner sections are also very popular.

Bars fixed through window frames give us an economical construction. For fixing these steel bars in wooden window frames (instead of separate grills), through holes can be drilled in one side or both sides of the frame. The bars are then passed into the frame and they shall be of the correct length as to end flush with outside of the frame. If the windows come side by side, the steel rods should be continuously pass through all the frames so that it is difficult to bend. Attractive designs can be made by varying the spacing of these bars.

In case of grills fabricated as per the design, they are fixed to the wooden window frame using round headed bolts and nuts running through the frames in new work before windows are fixed in the wall. If the grills are to be installed in frames that are already attached to walls, they are fixed with countersunk wood screws with heads painted to conceal its position.

37.14 FIXING ROLLING SHUTTERS

The rolling shutters consist of mild steel laths, 1.21 mm thick (18 G) and 80 mm wide or as specified. The laths shall be machine rolled from a continuous strip into an easy curve free from crimps and sharp bends and with an effective bridge depth of 16 mm. These shall be interlocked together throughout their entire length and jointed at the ends in such a way as to maintain alignment and protect the slats against abrasion in the guides. All joints shall be completely air- and weather-tight (see also Section 23.14).

The shutter is supported by means of 'spring barrel' which in turn is supported by cast iron or steel brackets. The shutter slats shall coil round the spring barrel. A galvanized steel sheet hood (not lighter than 18 G reinforced as required for rigidity) shall be provided at the fixing level. The brackets on either side also form the end closures for the hood. The spring shall be preferably of coiled type and shall be manufactured from high tensile spring steel wire or strip of adequate strength to balance the shutters in all positions.

The guide channels shall be a mild steel deep channel section rolled, pressed or built up (fabricated) construction. The thickness of the sheet used shall not be less than 3 mm. The minimum depth for guide channels shall be as follows:

Clear width of shutter	*Depth of guide channel*
Under 3.5 m	60 mm
3.5 m and above	75 mm

The gap between the two legs of the guide channel shall be sufficient to allow the free movement of the shutter and at the same time, close enough to prevent the rattling of the shutter due to wind. Each guide channel shall be provided with a minimum of three fixing cleats or supports for attachment to walls or columns by means of bolts and screws. The spacing of cleats shall not exceed 750 mm. Alternately, the guide channels may be provided with suitable dowels, hooks, or pins for embedding in the walls.

The installation shall be mounted plumb, square and true on the vertical surface of lintels and/or masonry on each side of the opening. When completed, the door shall completely fill the opening for which it was designed and shall not obstruct the opening when in the open position. The shutters should operate easily and smoothly under all conditions.

37.15 FIXING COLLAPSIBLE STEEL GATES

This is one of the works commonly carried out in construction of buildings. Collapsible gates are described in Chapter 23. They are fixed under a lintel or outside the opening or inside the opening. When it is fixed outside or inside, it can be folded clear of the opening, if necessary. The height of the gate should be 150 mm more than that of the opening. The gates are made to roll on bottom or top runners (see also Section 23.13).

SUMMARY

The amount of steelwork to be carried out in construction of a building will depend on the type of the building. Some of these have been briefly discussed in this chapter.

REVIEW QUESTIONS

1. What is the purpose of the covering that is provided on the electrodes used for electric arc welding?
2. (a) Where would you use (i) electric arc welding and (ii) gas welding?
   (b) How can we cut steel plates? Explain the principle used in this process.
3. (a) What are the types of electric arc weldings used?
   (b) How would you form the junction of the steel members of a steel truss (i) made of angle and (ii) made of steel tubes?
4. Explain the following:
   (a) Different ways of fixing steel grills to window frames
   (b) How to fix steel bars to window frames for protection in a low-cost housing scheme
5. Give an account of the common defects to look for in inspection of a welding work?
6. Write a short note on rolling shutters and its fixing for a garage.

Chapter 38

Joining Pipes

38.1 INTRODUCTION

All joints of pipes carrying fresh water as well as wastewater and foul water above the ground should be gastight. The pipes laid below the ground should also be watertight. For this purpose, the standard methods of connecting pipes should be adopted. In this chapter, we will briefly review the common methods of joining different types of pipes and testing them for leakage. It is to be noted that pipes with spigot and sockets used for liquid flow are laid with their sockets upstream and the laying work is carried out from downstream to the upstream.

38.2 JOINING AC PIPES

AC pipes come with spigot (small end) and socket (enlarged end). We use the 'hemp yarn and Portland cement joint' for joining these pipes. For this purpose, the spigot is properly fitted into the socket with uniform space around. One-third depth of this annular space between the socket and the spigot is filled with spun yarn soaked in bitumen or a mixture containing 2 parts cement and 1 part sand and properly pressed in with caulking tool. The remaining two-thirds of the joint is filled with stiff cement mortar 1 : 2 (1 part cement and 2 parts coarse sand). It should be pressed home with a caulking tool and finished smooth at an angle of 45° outwards. The joint should also be cured for 7 days.

38.3 JOINING GLAZED STONEWARE PIPES

These pipes are made with spigot and socket. A gasket of unwoven yarn soaked in thick cement slurry is first placed around the spigot and the spigot is slipped well into the socket of the pipe already laid [See Fig. 38.1(a)]. The above gasket is then rammed home so that it fills about one-fourth the depth of the socket. The remaining space is filled with a stiff

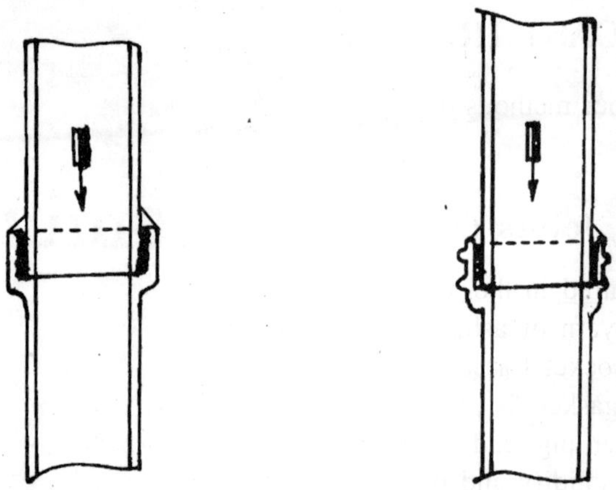

(a) Glazed stoneware-to-stoneware pipes with hemp yarn and cement joint

(b) Iron-to-iron pipes with molten lead well caulked joint

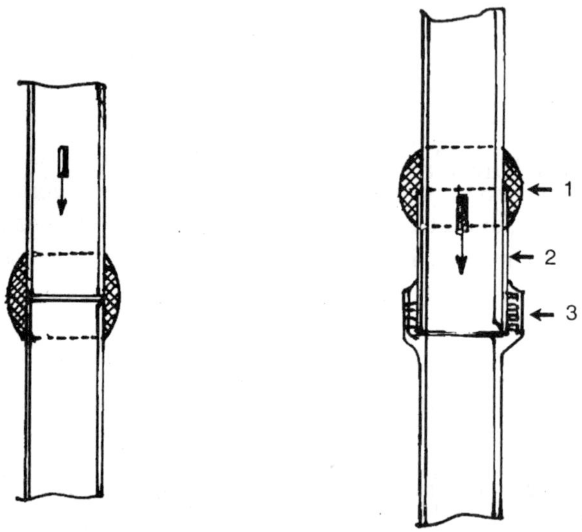

(c) Lead-to-lead pipes with wiped solder joint

(d) Lead pipe (on top) to stoneware pipe: 1. Wiped solder joint, 2. Brass timble, 3. Hemp yarn and cement joint

Fig. 38.1 Joining of pipes.

cement paste containing 1 part of cement and 1 part of fine sand. The joint is finished at an angle of 45° outwards. These pipes are generally used for sewers and placed on the ground. As the joint is rigid, undue settlement should not take place. Hence, in such places, it may have to be bedded completely in concrete as described below.

38.4 JOINING CAST IRON PIPES

One of the following four methods depending on the importance can be used for joining cast iron pipes:

38.4.1 Lead Run Joints (Molten Lead Joints)

The spigot shall be centred in the adjoining socket by tightly caulking in sufficient turns of tarred gasket or hemp yarn or asbestos rope or other suitable material to leave unfilled the required depth of the socket for lead. When gasket or hemp yarn has been caulked tightly home, a jointing ring (gasket) is placed round the barrel and against the faces of the socket [see Fig. 38.1(b)]. Molten pig lead shall then be poured in to fill the remainder of the socket. The lead is allowed to solidify and then be solidly caulked with suitable tools and hammers of not less than 3 kg weight, right round the joint to make up for the shrinkage of the molten metal on cooling. It should be preferably finished 3 mm behind the socket face. Lead for caulking shall conform to the standards. It is essential that the pipes be perfectly dry before lead run joints are made, otherwise blow holes may occur in the lead and because of water turning to steam, injury may result to the pipe mechanic. This method is, therefore, unsuitable for use in wet trenches. Caulking of joints is carried out only after convenient lengths of pipes are paid in position and leaded.

38.4.2 Lead Wool Yarn or Fibre Joints

These joints are suitable for wet conditions where lead run joints are not possible. Special attention is necessary in caulking. The socket shall be caulked with tarred gasket or hemp yarn as described above and the lead thread inserted into the socket and tightly caulked home with suitable tools and hammers of weight not less than 2 kg, until the joint is filled. Lead wool used for caulking shall conform to the standards.

38.4.3 Cement Joints

The following procedure is recommended for cement joints for cast iron pipes:

(a) The joint is first yarned with hemp yarn (or other suitable packing material) dipped in the cement slurry. The yarn is first inserted to a slight depth and well pressed in the same manner as for lead jointing.

(b) Stiff cement mortar 1 : 2 should be rammed into the joint by caulking tools.

(c) The filling to be completed and caulked again finishing at 45° sloping up.

(d) Joints should be kept wet for 24 hours after making

(e) When using cement joints, the use of lead joint at regular intervals is also recommended.

38.4.4 Flanged Joints

If the pipes are manufactured with flanges, they are joined by bolts with a rubber or compressed fibreboard 1.5 to 3 mm or other suitable insertion in between the flanges. For important jobs, use large pipes as in sewage works and also when conveying hot liquids.

38.5 JOINING OF GI PIPES

Galvanized (GI) pipes are joined by means of separate sockets popularly called *couplings*. The pipes can be cut at any place and the couplings have already threads. (There are many other fittings such as elbows, tees union, etc. for joining two pipes where neither of which can be rotated.) The pipes are threaded and cleaned of all foreign matter. The inside of socket and the turned end of pipe are oiled and rubbed with white lead. A few turns of spun yarn is wrapped round the screwed end of the pipe. The end is screwed in the socket or tee, etc. with a pipe wrench.

38.6 JOINING PVC PIPES

Unplasticized PVC (UPVC) pipes are very much used for various types of pipeworks in buildings. They come under various pressure ratings. Complete range of fittings (coupler, elbow, tee, union, end plug, etc.) are also available. The UPVC pipes also come with threaded ends. The commonly used joints are as follows:

1. Solvent welded joints
2. Rubber ring joints
3. Screw or threaded joints
4. Flanged joints

Of these, the solvent welded joints are the most popular. For this purpose, the pipe-end and the socket are smeared with the specified solvent and fitted together to form a leakproof joint. A coupling joint can also be made by using solvent or by rubben rings as shown in Fig. 38.2.

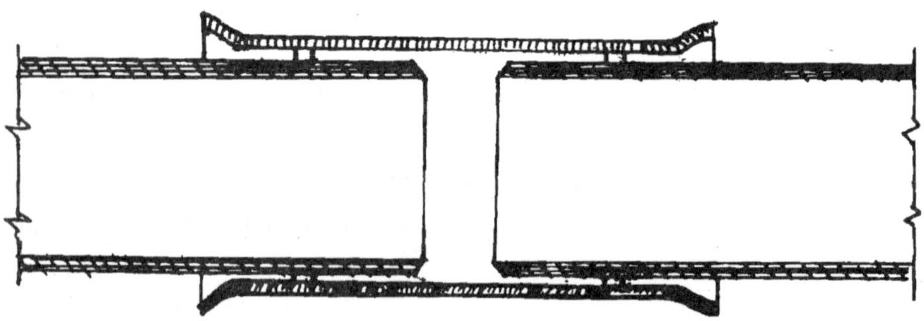

Fig. 38.2 PVC repair coupler for PVC pipes with rubber rings or cemented by special cement.

38.7 JOINING LEAD PIPES (WIPED JOINTS)

Use of lead for water supply is nowadays prohibited due to the dangers of lead poisoning. Lead pipes are joined to lead pipes by wiped solder joint [Fig. 38.1(c)]. The solders consist generally of two parts of lead to one part of tin. (Wiped joint is a solder joint.)

38.8 JOINING CONCRETE PIPES

In building work, concrete pipes used for drainage work can be joined as we join stoneware pipes described in Section 37.3. For joining large concrete pipes used for water supply, we mainly use other joints depending on the type of concrete pipes manufactured.

38.9 JOINING DISSIMILAR PIPES

In many places, PVC pipes have to be connected to brass taps, showers, etc. Some of these connections are shown in Fig. 38.3.

Joining lead and iron pipes. Where a lead pipe (waste pipe, ventilating pipe or trap) is connected with an iron pipe, we use a wiped joint. For this joint, a thimble made of copper or brass is used. The thimble is connected with the pipes by means of a joint made with molten lead, properly caulked, only a sufficient quantity of lead being melted at a time to finish each joint at one pouring.

Joining stoneware with lead pipes. Where any stoneware or semi-vitrified ware trap or pipe is connected with a lead soil pipe (waste pipe or trap communicating with a sewer), there shall be inserted a socket of copper, cast brass or other suitable alloy within such stoneware. At the stoneware (or semi-vitrified ware trap or pipe), the joining is carried out by means of a joint made with mortar consisting of one part of Portland cement and one part of sharp sand and at the lead soil pipe end by means of wiped metallic joint described above [see Fig. 38.1(d)].

Joining stoneware with iron pipes. Where any cast iron soil pipe is connected with a stoneware (or semi-vitrified pipe), the beaded spigot end of such iron pipe shall be inserted into a socket of the stoneware by a joint made with mortar consisting of one part of Portland cement and one part of clean sharp sand. However, on the other hand, where any water closet pan or earthenware trap is connected to a cast iron soil pipe, the joint between the stoneware spigot and the cast iron socket shall always be of a flexible (non-rigid) nature. Such joints shall be made preferably with a mixture of bitumen and chopped asbestos fibre (not dust) or rubber.

Joining PVC pipes to other brass appliances. In many places, PVC pipes will have to be connected to brass taps, showers, etc. Some of these connections are shown in Fig. 38.3.

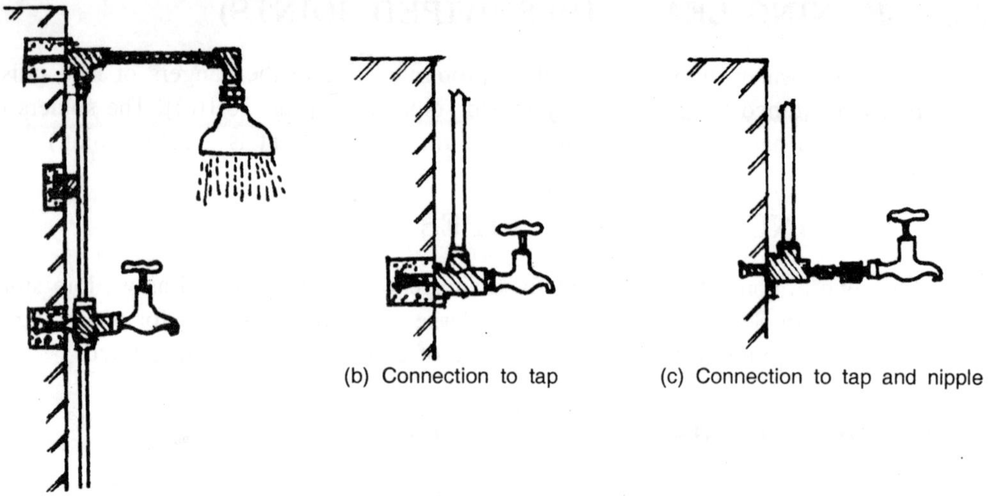

(a) Connection to shower rose and tap

(b) Connection to tap

(c) Connection to tap and nipple

Fig. 38.3 PVC pipe specials.

38.10 CONCRETE SUPPORTS OR PROTECTION FOR PIPES

The laying of pipes depends on ground condition are described below.

38.10.1 Types of Supports Used for Glazed Stoneware Pipes

It may be necessary to support or surround pipes used as drains by means of concrete in certain circumstances and the following methods are adopted depending on the condition of the ground.

Bedding. Bedding shall be rectangular in section and shall extend laterally at least 15 cm beyond and on both sides of the projection of the barrel of the pipe [see Fig. 38.4(a)]. The thickness of the concrete below the barrel of the pipe shall be not less than 10 cm for the pipes under 150 mm in diameter. Where bedding is used alone, the concrete shall be brought at least up to the invert level of the pipe to form a cradle and to avoid line contact between the pipe and the bed.

Haunching. Concrete haunching shall consist of the following [Fig. 38.4(b)]:

(a) A concrete bed as described for bedding
(b) The full width of the bed carried up to the level of the horizontal diameter of the pipe
(c) Splays from this level carried up on both sides of the pipe, from the full width of the bed to meet the pipe barrel tangentially

Surround or encasing. The surround or encasing shall be similar to haunching up to the horizontal diameter of the pipe and the top portion over this shall be finished in a semicircular form to give a uniform encasing for the top half of the pipe [see Fig. 38.4(c)].

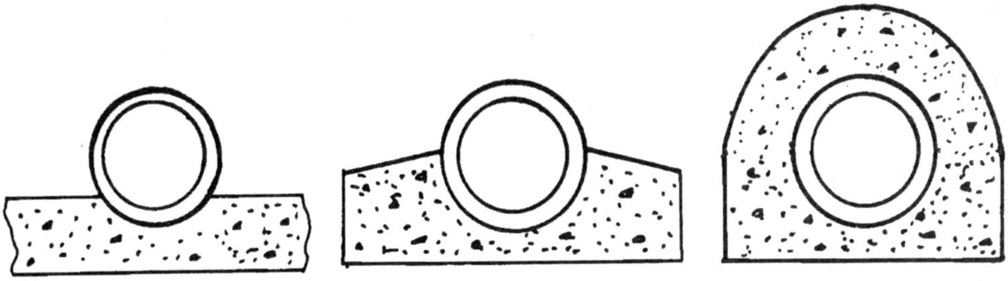

Fig. 38.4 Types of ground support for stoneware pipes.

38.10.2 Conditions for Adoption of Methods of Laying of Glazed Stoneware Pipes

The minimum support or protection for glazed stoneware pipes shall be as follows:

(a) For depth of barrel as laid below the ground level in very shallow and where pipes are unavoidably exposed above the ground surface, the pipes shall be completely encased or surrounded with concrete.

(b) Where the pipes are laid on a soft soil with the maximum water table level lying at the invert level of the pipe, the pipe sewer shall be bedded on concrete.

(c) Where the pipes have to be laid in a soft soil with the maximum water table level rising above the invert level of the pipe but below the top of the barrel, the pipe sewers shall be haunched.

(d) Where the maximum water table level is likely to rise above the top of the barrel, the pipe sewers shall be completely encased or surrounded with concrete.

(e) Where the sewers are to be laid adjacent to growing trees, the pipe sewers shall be encased or surrounded with concrete to avoid damage to the pipes likely to be caused by the roots of the trees.

38.10.3 Supporting Other Types of Pipes

Cement concrete pipes. Chairs made of cement concrete may be used for supporting cement concrete pipes. Chairs shall be provided at suitable intervals which shall not exceed the length of the pipe.

Piers for cast iron pipes. Where supporting piers are specified for cast iron pipes, they shall not be less than 30 cm in length (parallel to the axis of the pipe) and at least equal in section to that described for haunching in section 30.10.1. Piers shall be built just behind the pipe sockets, intermediate piers being provided where necessary. In normal ground, no support or protection to cast iron pipes need be provided. Where concrete haunching surrounds or piers are required, these shall be in accordance with the details given for stoneware pipes.

Support of pipes in bad ground. The recommendations described above apply only to reasonably firm and stable ground conditions. In certain subsoils, rise and fall of the subsoil water level may be the cause of considerable earth movements. This condition along with the other conditions of unstable ground call for additional supports to the pipes in the form of piles, trestles or other suitable means.

38.11 INSPECTION AND TESTING

Inspection should be carried out during installation to ensure correct joining and fixing of pipes according to the drawing and specifications. In addition, after the work is completed and before the appliances are connected, the pipes should be tested section by section for leakage by means of the following tests:

- Water test
- Air test
- Smoke test
- Water seal test

These tests are described below.

38.11.1 Water Test

Water tests are important especially for concealed water supply and drainage pipes in toilets and bathrooms. The pipes are filled with water under pressure and the leakage measured. For a water supply pipe, we should give higher pressures than for a drainpipe. The arrangement for testing a drainpipe is shown in Fig. 38.5(a). Water tests are made in open trenches before filling in so that we can observe leakage, if any.

For testing sewer pipes, 1.5 m head of water is made to stand for two hours and any leakage that has taken place is found by tipping up the pipe. The following leakages are permitted according to the British practice:

(i) 0.05 litre per metre of pipe for 100 m pipe

(ii) 0.08 litre per metre of pipe for 150 mm pipe.

Note. In testing water supply pipes, we can give higher pressures and leakage of joints can be checked easily by visual inspection.

38.11.2 Air Test

In this test the air pressure can be applied after sealing of ends or by inserting a rubber tube through the water seal of a trap. There should be no drop in pressure as indicated by a pressure gauge. For testing a sewer pipe section between two manholes, the ends of the pipes are closed by expanding plugs and the air is pumped in by hand bellows (HB). A manometer is connected by a rubber tube as shown in Fig. 38.5(b). The usual specification for sewers is that the fall in water level as measured by the manometer of 100 mm water gauge should not be more than 25 mm in 5 minutes or with 50 mm water gauge, it should not be more than 12 mm in 5 minutes.

For testing traps in a stack, the air test can be used as shown in Fig. 38.5(c). A 38 mm manometer pressure should be maintained for not less than 3 minutes.

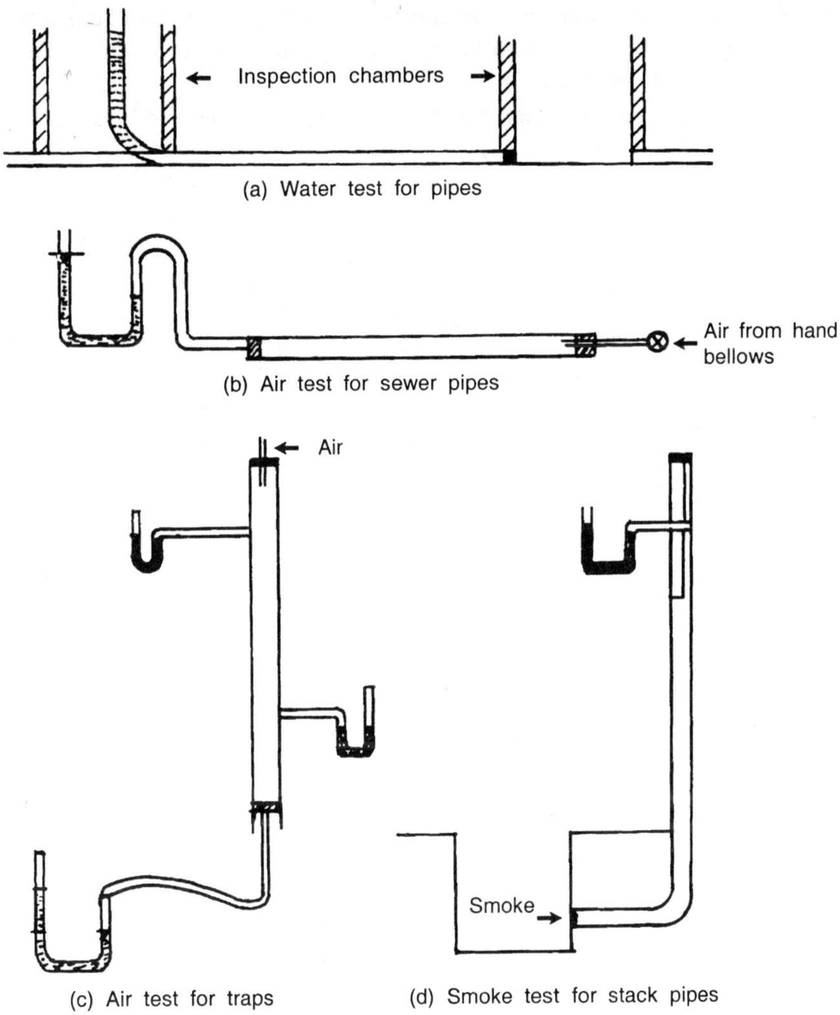

Fig. 38.5 Tests for leakage in drainpipes.

38.11.3 Smoke Test

The *position of the faults* indicated by the water and air tests can be identified by a smoke test. In the smoke test, smoke is pumped into the system by a smoke machine and then sealed with plugs [Fig. 38.5(d)]. Care should be taken to ensure that the system is filled with smoke before sealing with plugs. The escape of the smoke indicates the position of the leakage.

38.11.4 Water Seal Test

This test is intended to check whether the water seals in WC and other traps are intact under adverse circumstances. For this purpose, the following tests are usually suggested. Each test should be carried out at least three times.

1. In small installations, all the appliances and in large installations, sufficiently large numbers that are likely to be operated are discharged together. There should be no emptying of the seals.
2. For testing seals of traps fitted to stationary appliances, the latter should be filled to overflow level and allowed to discharge in the normal way. There should be no loss of water seal.
3. For testing a water closet seal of the flushing type, we may proceed as follows. Six pieces of newspaper about 200 mm × 200 mm (8" × 6") are laid flat one by one over water surface of the seal in the pan. The pan is subjected to normal flush followed *immediately* by pouring 10 litres (2 gallons) of water from a bucket. After this test, the seal should again be functional.

SUMMARY

It is very important that pipes are properly joined according to specifications. Pipeworks, especially the concealed one, should be tested before they are covered.

REVIEW QUESTIONS

1. Describe briefly the following types of pipe joints stating where they are used:
   (a) Rings of hemp yarn and Portland cement mortar joints
   (b) Lead run joints
   (c) Solvent welded joints
   (d) Wiped solder joints

2. (a) Explain how you would test a pipeline for watertightness and gastightness.
   (b) Describe how traps are tested.

3. What are the methods used for laying stoneware pipes under different ground conditions?

4. Write short notes on the following:
   (a) Smoke test
   (b) Wiped solder joint
   (c) Joining PVC pipes

Chapter 39

Miscellaneous Works

39.1 INTRODUCTION

In this chapter, we will briefly deal with the following items which are of general interest in building construction:

1. Rainwater harvesting
2. Repair of cracks in brick masonry in buildings
3. Project management for small building works
4. Boundary walls, fences and gates
5. Garage and gates
6. Minor repair of RC works due to corrosion of steel
7. Gunite and shotcrete works
8. Lightening protection of buildings
9. Fixing of cupboards and shelves

39.2 RAINWATER HARVESTING

As water requirement of the rising population in India cannot be met only by the conventional systems, the need for water harvesting is becoming more and more acute with time. Conservation of water that we get from rains for human consumption and other uses has become a necessity both in our villages and towns. We carry out water harvesting in the following ways:

1. Roof top rainwater harvesting for consumption as well as for recharge of groundwater
2. Rainwater harvesting of water collected on the ground around the buildings for recharge of groundwater level
3. Collection of surface water that may run off by check dams and other devices for storage in surface ponds, dams, etc.

Devices to be used for rainwater harvesting will depend on the local circumstances. For example, rainwater harvesting in cities will be aimed at storing water for human consumption or groundwater improvement and reuse by bore wells. In villages, in addition to the above purposes, it will also be used for cattle and irrigation purposes.

39.2.1 Rooftop Harvesting

For reuse, the rooftop water is collected from the roof and directly lead into the storage tank through pipes. A bypass system may also be provided so that the water from the initial rains during the rainy season (which may contain a lot of dust and leaves) can be used for groundwater improvement and only clean water is led into storage tanks for domestic use.

Roof water may also be used for groundwater improvement. For this purpose, the rainwater can be led to an old well or an existing hole (such as an old bore well) so that it will recharge the groundwater. We should remember that any seepage deep into the ground will take time and hence, water should be stored at least for a few hours in the well, tubewell or a pit before it can slowly percolate into the ground. The rate of percolation will depend on the bottom strata in the storage well. The well, pit or bore should end in permeable strata for proper recharge of the groundwater. A synthetic fiber filter should be provided on top (which can be cleaned) to avoid logging.

39.2.2 Harvesting of Rainwater That Falls on the Ground

The basic principle in harvesting this water is to store the rainwater in a place instead of allowing it to run off and provide time and means for it to percolate slowly into the ground. If the soil at the site is impervious, it may be necessary to have bore holes leading to a permeable layer below. Check dams along the usual flow of the surface water can also be built to store this water.

In many cities and towns, water harvesting is now compulsory for group of flats and for residences with bore water installations and the details of devices that the municipal authority wants can be obtained from local agencies.

39.3 REPAIR OF CRACKS IN BRICKWORK IN BUILDINGS

Cracks occur in walls due to differential movements caused by temperature, settlement of foundation, shrinkage, etc. From the nature of the cracks, we can infer the seat of movement. A line drawn normal to the cracked line will intersect the seat of settlement as shown in Fig. 39.1. Any leakage of drains and consequent softening of soil or other types of settlement or movements can be the cause of these cracks.

The first step before repair of cracks is to find the cause of the cracks and try to remedy the cause. For example, if the cause is due to foundation movement, it is worthwhile to proceed as follows:

(a) If there is any leakage of drains, gulleys, manholes, it can soften the foundation soil and cause differential settlement. Such leakages should stopped.

(b) In clayey soils, the foundation can be strengthened (for example, by plastering the brickwork below the ground level which might not have been plastered). In more

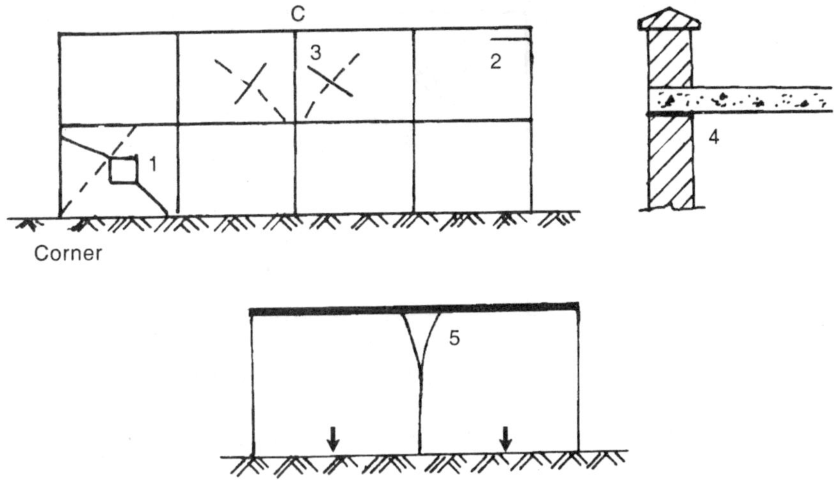

Fig. 39.1 Cracks in masonry due to various causes: 1. Crack due to settlement of corner, 2. Crack under RC slab due to lifting up of corners in two way slabs, 3. Cracks due to settlement of column C, 4. Cracks below slab as a result of the movement due to temperature effects, 5. Crack with larger width at top due to heaving of centre as occurs in black cotton soils.

serious cases, the foundation soil can be improved by lime injection or some other means. Otherwise, the foundation itself can be strengthened by underpinning.

The cracks in brick walls can be repaired by the following methods:

1. If the cracks are less than 1.5 mm in width, they are generally filled up with putty (made at site or commercially available) and painted.
2. Cracks which are wider than 1.5 mm but are not very serious are repaired by raking them out to about 25 mm and refilling with 1 cement, 2 lime, 9 sand mortar or preferably with other patented crack-filling compounds available in the market. Separate types of elastic crack-filling compounds—one suitable for small cracks in the form of a paint and another suitable for large cracks (in the form of a semi-solid paste)—are available for filling the cracks. These are superior to Portland cement mixes.
3. Still wider cracks have to be stitched by using reinforced mortar or concrete stitching blocks. *Thin mortar blocks* can then be stuck on both sides of the brickwork or *thick concrete blocks* can be inserted in the brickwork at every 5th or 6th course as shown in Fig. 39.2.

Thin mortar blocks (1½ to 2 brick length, 100 mm wide and 25 mm thick) are made with 1 : 3 mortar with one 6 mm reinforced embedded at the centre. These blocks after curing are inserted in holes made at every 5th course of the brickwork (0.5 m spacing) on both sides. The masonry needs to be chased only to 25 to 30 mm into the brickwork for its insertion. Alternately, reinforcement rods are inserted instead of these blocks, but it should be made sure that they are covered with a thick coat of mortar and are not in contact with the bricks.

When repairing with thick concrete blocks, the wall has to be cut to its full depth and the RC blocks have to be inserted at regular intervals. The blocks are of 1 : 2 : 4 concrete

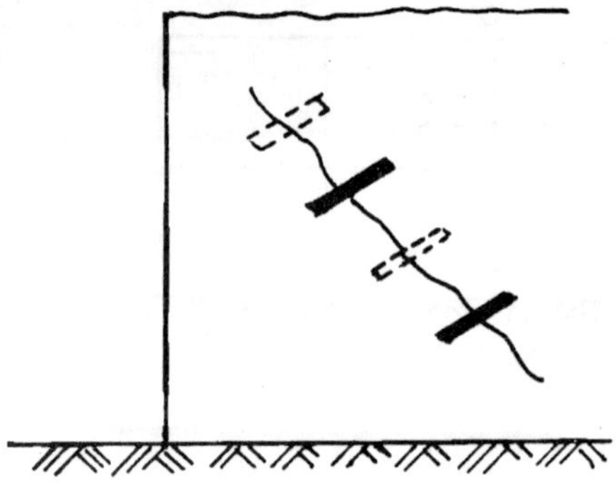

Fig. 39.2 Repair of cracks in masonry by stitching with steel rods.

made one or two brick thick and 1½ to 2 bricks long with width equal to the brickwork. Two 6 mm mild steel reinforcements for each block are placed at the centre. Repair consists in sticking the crack with these blocks at every 5th or 6th course.

If the cracks are due to minor causes, they will stabilize after a period of time.

39.4 PROJECT MANAGEMENT FOR CONSTRUCTION OF SMALL PROJECTS

Successful execution of a project requires the following three phases:

1. Project planning before execution of the project
2. Project scheduling before execution of the project
3. Project management during execution of the project

Each of these phases are important in large projects, but we will consider only the third phase. There are many devices used for project management such as bar chart, milestone chart and critical path method (CPM) which will be studied in the later years. Small projects can be handled by the simple bar charts. Each of the above three phases can be controlled by separate bar charts. For project management, we divide the project into a number of its component activities and indicate the proposed time of start and completion of each activity by a bar on the diagram. For example, project planning and scheduling can broadly consist of items such as:

1. Preparation of project report
2. Preparation of design and plan
3. Getting sanctions
4. Inviting tender
5. Awarding contract
6. Site clearance
7. Construction of buildings
8. Construction of external works, etc.

We may again split the activities of construction or execution of the project of a building into different heads as shown in Chart 39.1 and also put a time for starting and the time for completion of the work (see also Appendix D).

The time is plotted on the *x*-axis and the various items of work is listed on the *y*-axis. The expected start and the expected finish of the work can be indicated in the bar chart. The progress of the work can then be also monitored by colouring the chart as the work progresses. There are many improvements that can be made in this bar chart which will give us more and more information about the progress of the project.

Chart 39.1 Bar chart for construction operations

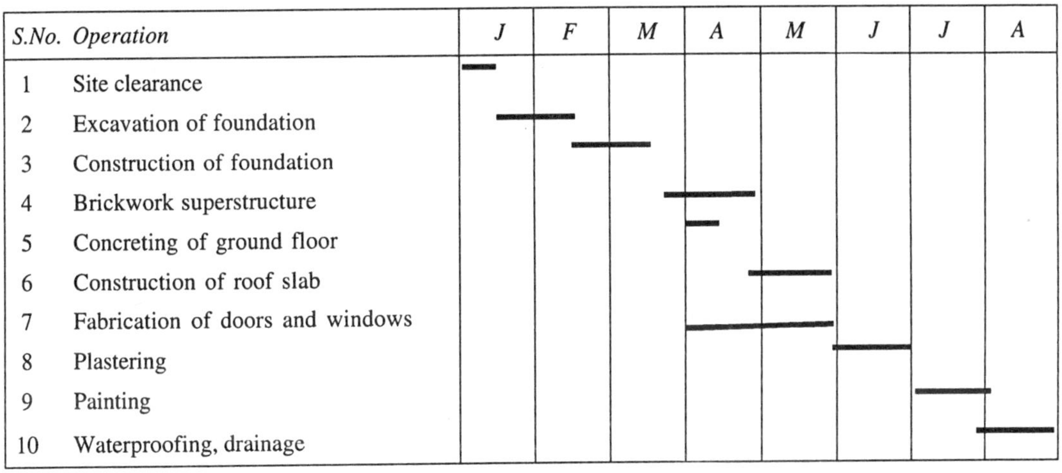

39.5 CONSTRUCTION OF BOUNDARY WALLS AND FENCES

We usually protect the house property by boundary walls or barbed wire fences as described further.

39.5.1 Boundary Walls

As mentioned in Section 34.5, the height of boundary walls in cities should be limited to 1.5 m (5 ft). The boundary walls are generally built inside the line of property and belong to the person who builds it. In order that the foundation should be inside the owner's property, a foundation as shown in Fig. 4.4 in Chapter 4 is usually adopted. However, this type of foundation is unsuitable in clayey soils as it produces unequal settlement due to eccentricity of the load and hence, will tilt sidewards. A compound wall built on underreamed piles and plinth beam is the best solution for a stable wall in clayey foundation soils. If owners of the adjacent properties can agree to build the wall along the boundary line, a more suitable symmetrical and conventional foundation can be built for the wall. The top of these walls should have a coping course. In places of heavy rains, as in Kerala, this course should be broader than the thickness of walls so that rainwater does not drip down these walls (see also Fig. 6.1).

Care should also be taken not to build a long masonry wall over a long continuous grade beam over brick pillars or over underreamed piles not continued to the top of the wall. Due to the difference in thermal expansion of concrete and masonry (brickwork), the top wall will have loose bond and separate from the concrete grade beam. With a horizontal wind load during a storm, the wall can easily topple. Hence, using brick pillars we must use *discontinuous grade beams* such as precast sections on brick pillars to allow for expansion. Bored piles should be continued as columns to the top of the wall to make the wall act one unit connected to the foundation.

39.5.2 Barbed Wire Fences

In many situations, as in the case of enclosing large areas of land, we may only provide a barbed wire fence as shown in Fig. 39.3 fixed on concrete posts or wooden ballies. The

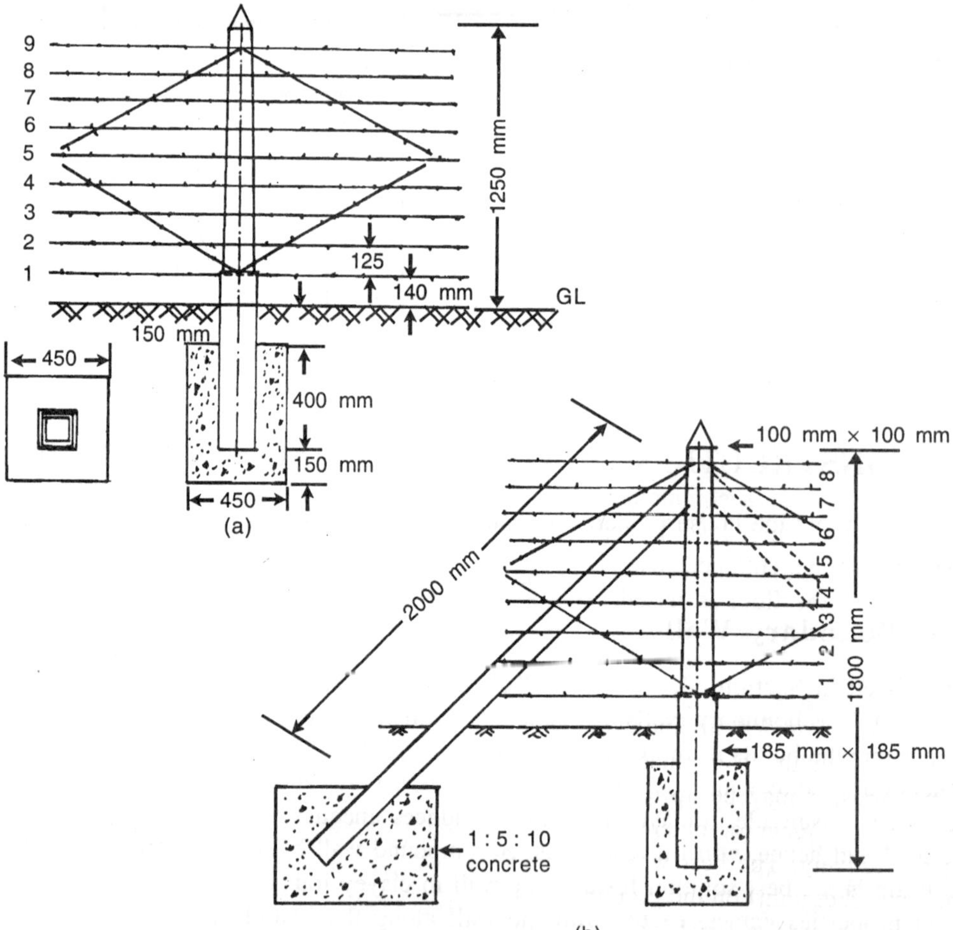

Fig. 39.3 Barbed wire with reinforced concrete posts 185 mm × 185 mm at bottom tapered to 100 mm × 100 mm at top and 1.8 m in length: (a) Intermediate posts (b) End and other anchoring posts in long stretches with struts (horizontal rows 5 to 9 numbers as desired).

spacing of RC posts is usually 3 metres. Every 15th post along straight lengths and corner posts are to be strutted on both sides and the posts near the gates also are strutted on one side only. The first line of barbed wire is fixed at about 15 cm above the ground and it is tied to the post by means of GI binding wires. In concrete posts, holes may be provided for the tie wires, but they weaken the posts. The total height of the post above the ground should be at least 1.2 metres. The total length of the upright posts is usually 1.8 m and that of the inclined struts is 2 m. These can be reinforced with 4 nos. 6 or 8 mm mild steel bars with 6 mm ties at 200 mm spacing. As these posts are fully exposed, it is important to provide at least 25 mm cover to the steel. (For temporary fencing of constructions as in a construction site, we may use wooden ballies tar coated at the buried end instead of concrete posts. The spacing of these wooden ballies can be kept at 2.5 metres.) The CPWD specification as shown in Fig. 39.3 specifies nine parallel rows of barbed wires with two diagonal wires woven through the parallel rows of wires. This weaving can be accomplished by fixing the odd rows of wires first, then the diagonal cross wires and lastly, the even rows of wires. The horizontal rows can be reduced to seven but not less than five. Instead of diagonal wires, we can provide vertical ties at 500 mm spacing woven as described above. Barbed wires are available in two types, the lighter and the heavier types.

39.6 CONSTRUCTION OF GARAGES AND GATES

39.6.1 Garages

The usual clear size of private garages should be 5.4 m × 2.5 m (about 18 × 8 ft) in plan with a clear height of entrance 2.1 m (7 ft) to 2.9 m (9 ft) below the rolling shutter if provided. (A small car such as Alto is about 3495 mm in length, 1495 mm in width and 1460 mm in height while a larger car Toyota Innova is 4555 mm × 1770 mm × 1754 mm. A Tata Safari's dimensions are 4810 mm × 1810 mm × 1924 mm.) The roof height of the garage should be so as to also accommodate the rolling shutter arrangement, if such an arrangement is used. The garage floor level is kept about 35 mm above the general GL with an approach ramp near the garage entrance. The garage floor is given a small slope 1/100 towards the end. The last 1 m of the floor is raised by 60 mm to act as a wheel arrester.

39.6.2 Gates

Main gates are provided in line with garages. A wicket gate may also be provided. The minimum clear width of entrance gates for entry of cars in residences is generally 3 m. In public buildings, it may be up to 3.6 m. Generally, two gate leaves are provided. They are supported on the inner or outer corners of the pillars to permit 180° movement inside or outside respectively. The best way to fix the gates is to have one support on the pillar and the other (a socket joint) on the ground fixed to a block on concrete supported on firm ground. In the pillar, the gate is fixed by a steel rod 375 mm long embedded in a 1 : 2 : 4 concrete slab fixed inside the pillar. The pillar is usually a square 2 or 2½ brick pillar. The gate can be fabricated from steel as in window grills. The bottom 300 mm (1 ft) height of the gate is covered with a plate or close grill to prevent the entry of dogs.

39.7 REPAIRS OF RC WORKS DUE TO CORROSION OF STEEL

There are many types of repairs and the most common repair is the repair of spalling due to corrosion of steel which is discussed further.

We come across a large number of cases where the steel reinforcements corrode due to various causes such as lack of cover, leakage of roof, etc. As steel rust occupies very much larger volume (more than 10 times) than steel, the surrounding concrete cracks and tends to get separated from steel and finally it falls down. This is particularly important in the ceiling plaster under leaking roofs. If we notice any disturbance in this plaster, it should be repaired as soon as possible. There are many cases in which the ceiling plaster has fallen down suddenly. The repair for such cases is carried out as follows:

1. Hammer test the area for loose areas and remove all loose particles of mortar, concrete, rust, etc. from the spalled surface and clean it with a wire brush.
2. Apply two coats of polymer modified active corrosion inhibitor to the remaining steel. (If additional steel is required, tie the new steel to the old steel by binding wires.) Two coats of chromate paint can also be a substitute.
3. Apply one coat of polymer bonding agent (latex based) over the entire area of steel and concrete to be treated.
4. Apply polymer modified 1 : 3 cement mortar over the reinforcement and provide necessary cover. Addition of superplasticizer to reduce water and the incorporation of fibres will help to produce strong mortar. If the area is large, guniting with additive for quick setting can be used for better efficiency of the work.
5. Cure the concrete.
6. Apply a coat of waterproofing slurry coating for protection against environment, if necessary.

39.8 GUNITE AND SHOTCRETE WORK FOR REPAIR OF CONCRETE

39.8.1 Gunite and Shotcrete

Gunite is one of the popular methods used to repair damaged concrete members. *Gunite* is the process of depositing mortar with low water-cement ratio at a high pressure through a nozzle. It was introduced in 1900. Recently the process has been improved so that only the *concrete with small-sized coarse* aggregate (instead of mortar) generally referred to as *shotcrete* can be deposited by this process. In both cases, the damaged part of the member (such as the underside of a roof slab) is cleaned before the operation and gunite or shotcrete is applied.

Gunite is usually used for applying mortar of lesser thickness, and shotcrete is used for placing concrete of larger thickness. The materials used for gunite are cement, sand and water and those used for shotcrete are cement, sand, small aggregate and water. Concrete chemicals (accelerators) are also added to help quick setting of gunite.

There are two different methods of application of this process, namely the 'wet mix' and the 'dry mix'. In the *wet mix method*, the solid materials and the required water are mixed together and the mixture is placed by pneumatic pressure through a nozzle. As the nozzle velocity is 90 to 120 m per second, the resulting compaction of material on the member will be very good. In the *dry mix method*, the mixed solid materials and water pass under pressure through different pipes and are mixed in the nozzle at the final exit point. The dry process is more effective than the wet process as the water-cement ratio required will be less. Hence, nowadays, the dry mix method is more popular. As the work requires special plant and machinery, it is carried out by special agencies only.

39.8.2 Bonding New Concrete or Plaster to Old Surfaces

The old practice of bonding new concrete or plaster to old concrete or plaster surface was to chip the surface at close intervals to ensure bonding as in plastering of ceiling in under surface of slabs. This is not only laborious but can also damage old structure. Instead, we can use the latex-based concrete bonding chemicals by making a slurry of cement with these chemicals and applying it to the surface to be repaired to act as a bonding surface.

Minor leakages in small water tanks and sumps can also be repaired by coating the walls with a slurry made of waterproofing agents, cement and latex-bonding chemicals.

39.9 LIGHTNING PROTECTION OF BUILDINGS

The buildings situated in exposed situations, where lightning can occur or their heights are large compared to the surrounding places, should be provided with lightning-protection systems. The lightning-protection system consists of an *unbroken* chain of conductors from the roof of a building to the ground which will provide an easy path for the large quantity of electricity released by the lightning to discharge to the earth in the shortest possible time. It is very important that the down conductors should follow the most direct path possible between the top air terminal and the bottom earth termination without sharp bends, kinks, etc. Joints should be avoided as far as possible and the conductors should be of pure copper. The area of influence of a lightning conductor is assumed to be a cone with the top most point of the conductor as the apex and a radius related to the height of the apex. This radius has been reported to be up to 3.5 times the height but a value of radius equal to the height of the conductor can be taken as a very safe value.

39.10 CUPBOARDS AND USE OF FERROCEMENT

The following heights taken from floor level are usual for a medium type of building:

1. Ceiling height 3 m (10 ft)
2. Bottom of lintel, top of doors and windows 2.1 m (7 ft)
3. Window sill level: 700 mm –750 mm (2'4" to 2'6")
4. Top of platforms, shelves, tables, etc. 750 mm (2'6")
5. Top of sitting chairs: 450 mm (1'6")

Built up deep cupboards are usually formed by 100 mm Z pattern brick partition walls between rooms. Shallow cupboards upto 1.2 m wide as in drawing rooms can be formed in load-bearing walls under lintels. Instead of 100 mm wall at the back. We can use 25 mm ferrocement walls constructed as follows.

During the construction of the lintels above the cupboard we leave 6 mm rods at 300 mm spacing downwards for the full length of the cupboard and anchor it to the bottom slab or floor with dowel bars. In addition, one layer of 100 mm × 100 mm 10 G weld mesh and a layer of chicken mesh 10 mm × 10 m 20 G are tied to the 6 mm steel.

We apply 1 : 2 cement mortar by holding a 450 m × 1000 m teak plank behind the mesh. The wall is built up in heights of 1 m each day. No plastering is needed as the surfaces can be finished smooth.

In case we need a cupboard with its lower portion to serve one side and the upper portions for the opposite side, we use a precast slab at the middle with 6 mm rods on the opposite sides and finish the wall as indicated above.

SUMMARY

This chapter deals with some of the miscellaneous items of work connected with building construction. There are many such items and we have dealt with only a very limited number of them in this book.

REVIEW QUESTIONS

1. Write short notes on the following:
   (a) Rainwater harvesting
   (b) Bar charts for project management
   (c) Construction of a barbed wire fence for a building site before construction
2. What are gunite and shotcrete? Explain their uses in the repair of buildings.
3. Give a short account of the following:
   (a) Lightning protection of buildings
   (b) Precautions to be taken in brick boundary wall construction over concrete grade beams
   (c) Repairing minor leaks in a water tank in a residential house
4. Detail the procedure for repairing a ceiling in which a portion of roof plaster has fallen down due to corrosion of steel.

Chapter 40

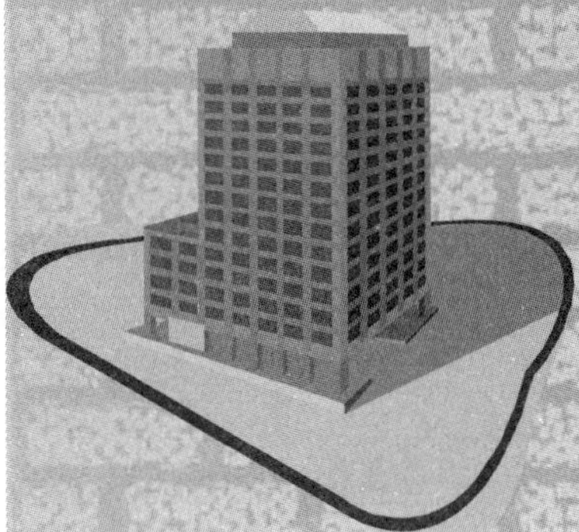

Maintenance of Buildings

40.1 INTRODUCTION

Maintenance and repair of buildings are important and some of them have been already dealt with in the previous chapters. Painting of buildings and polishing of woodwork are given in Chapter 26. We have also dealt with some items of repair such as repair of cracks in walls in Chapter 39. In this chapter, we will very briefly consider some of the other major items of maintenance that we should carry on to keep the building in good shape. Good maintenance enhances not only the beauty and performance but also the useful life of the building.

40.2 CLEANING OF PAINTWORK

40.2.1 Washing Paintwork

Washable paints can be cleaned by washing with clean water. Soda, soft soap and other alkaline substances are injurious to most paints. If at all they are used to remove difficult stains, they should be highly diluted and rinsed off the paint with clean water as soon as the dirt is removed.

40.2.2 Repainting

It is advisable to repaint surfaces before the old paint disintegrates and while a compact and continuous film still exists. The life of paint depends on its quality. For example, today's exterior paints have been developed so that frequent repainting of external surfaces of high rise buildings can be avoided. They are costly but they remain intact and retain their brightness for a long time. We should also choose the correct paint suitable for the exposure condition. Thus, it is better to paint the underside of a wet area such as bathroom and walls where dampness can penetrate with paints which can breathe (such as cement paint).

40.2.3 Paints on Steelwork

Careful observation should be made on painted grills, rolling shutters, etc. They should be regularly cleaned with clean water. If any parts get rusted, it should be removed and the surface repainted. Steel windows, if used in buildings, require special attention.

40.3 MAINTENANCE OF FLOORINGS

There are many types of floors and each has its own methods of maintenance. These are discussed further.

40.3.1 Marble Floors

Daily maintenance is to be made by mopping with mild detergent in water. Marble can easily get stained and scarred. The lighter the colour the more easily the stains can be detected. If there is a spill, it should be wiped out immediately. If these are stubborn stains, proceed as follows:

1. If the stain is from grease, make a paste of chalk dust or whiting with acetone. This mixture is applied to the stain and allowed to stand overnight. Sponge off the mixture and buff the treated area.
2. If the stain is from an organic source such as tea, fruit juice, carbonated beverages mix, use the chalk powder with hydrogen peroxide instead of acetone. Rust stains can also be removed by this paste. If the above process dulls the surface, sprinkle marble polishing powder (tin oxide) and rub it down with thick cloth or an electric buffer. Stubborn dirt can be removed by dry borax and damp cloth followed by rinsing with warm water.

40.3.2 Granite and Other Stone Floors

Mopping with mild detergent and water at regular intervals will keep the floor in good shape.

40.3.3 Terrazzo Floors

For a period of at least three months after the terrazzo floor has been laid, it should be swabbed daily using clean water and a clean rough swab (floor cloth) which should be rinsed frequently in water to avoid dirt being carried back on to the floor. The floor is then allowed to dry. If the floor is dirty, water and a mild soap may be used to clean it. The soap should be completely removed by mopping as otherwise it will be deposited on the floor making it slippery and dull looking. After this initial cleaning, ordinary swabbing will keep the floor shining.

It is seldom possible to remove oil and grease, if spilled on the floor, and if they are allowed to penetrate below the surface. However applying a paste made by a powder such as hydrated lime and marble dust, or whiting with bensol or clear gasoline over the stain, and then washing after 12 hours can be used as described in section 40.3.1.

Acid polishing of terrazzo floors (using oxalic acid) is usually carried out after machine polishing during the first laying of the floor. This can be repeated, if needed. For this purpose, oxalic acid is dusted over the surface at the rate of 35 g per sq m of the surface sprinkled with water and rubbed hard with a pad of woolen rag. The following day the floor is wiped with a moist rag and washed clean with water. Acids such as dilute hydrochloric acid should not be used for cleaning mosaic floors.

40.3.4 PVC Floors

PVC flooring subject to normal usage may be kept clean by mopping with soap solution using a clean damp cloth. The traces of soap should be removed by mopping. Water should not be poured over the PVC flooring for cleaning as it may seep through the joints and cause the adhesive to fail. The floor may be periodically polished to keep up its appearance. It should not be overwaxed and if it develops this condition, it should be cleared with white spirit or paraffin.

40.3.5 Linoleum Floors

Mopping with white kerosene oil can pick up all the dirt from linoleum floors. After removing all the dirt, the floor can be wiped with water and mopped well to get a polished surface. As in the case of PVC floors, excess water and alkaline soaps should not be used for cleaning linoleum floors.

40.3.6 Block-wood Floors

All wood floors should be kept clean and the blocks should not become loose. The floor is cleaned and kept bright by polishing with beeswax or ready-made wax polish.

40.3.7 Ceramic (Glazed and Vitrified) Tile Floors

Ceramic tiles are easy to maintain. Soap and water can be used to clean them. Special care should be taken to clean the joints (unless jointless tiles are used for the flooring). Any stains can be removed by any of the methods used for other floors or even by the household methods of using a cut lemon.

40.4 MAINTENANCE OF DOORS AND WINDOWS

Both the woodwork and glazing have to be regularly maintained.

40.4.1 Care of Woodwork

Most woods, except teakwood, get deteriorated with time unless they are regularly painted and kept free from water and direct exposure to sun. Door frames adjacent to wet areas such as bathroom get affected by rot. Both dry and wet rots grow with moist condition. They can be prevented by proper ventilation and painting. Similarly, doors and windows which are directly exposed to the sun split up due to heating. Hence, these should be shaded and regularly painted.

Usually panelled wood shutters or plastic shutters are provided as bathroom doors. The underside of these wood shutters including the bottom edge is usually protected by U-shaped aluminium sheets and painted with waterproof paint so that water does not affect the bottom edge. This precaution is specially applicable to flush doors for bathrooms. Stains and dirt on painted doors can be removed as in painted walls.

40.4.2 Care of Glazing

While putty glazing is used for ordinary wood, glazing with beads is usually carried in teakwood. Both of these do not get damaged quickly but they should be inspected at regular intervals. Putty glazing can get cracked with time.

40.4.3 Cleaning Glass in Doors and Windows

Glass fitted to doors and windows should be cleaned regularly with clean tepid water. Glass cleaning liquids are available in the market to remove dirt and stains. Less expensive version can be made by mixing 2 tablespoons household ammonia or white vinegar with 1.25 litres of water. This solution is put into a spray bottle and applied with a sponge or rag. Wring them out thoroughly to avoid drips. Start from top and work downwards. A cotton swab or toothbrush can be used to take dirt from corners. Dry the window with crumbled newspaper or paper towels or chamois. Wipe one side horizontally and other vertically so that if streaks are left we know which side it is and can remove them. The streaks are to be removed by a soft, dry cloth. *Ordinary printed newspaper (with printer's ink) dipped in water is considered as a good material for cleaning of window and automobile glass.*

40.5 MAINTENANCE OF SANITARY APPLIANCES

Discolouration of fittings and growth of fungus and algae are the main problems. Sanitary appliances (ceramic wares) need frequent cleaning to maintain them in good condition and also to preserve their appearance. A solution of chloride of lime (bleaching powder) in hot water will remove surface stains from a ceramic ware. To restore the luster of porcelain and glazed surfaces, a cloth moistened with hot water and a little paraffin will be found very effective. Abrasive powders or acid solutions are harmful to these surfaces. There are also a number of ready-to-use chlorinated cleaning materials available in the market for cleaning sanitary appliances. However, they are usually very costly. If the sewage is to be treated by a septic tank, we should not use chemicals and detergents that can kill the bacteria in the septic tank.

40.6 MAINTENANCE OF WATER SUPPLY TAPS AND FIXTURES

Taps and other fixtures should be repaired as soon as they become defective. Taps, chrome fittings, tubs, basins, wall tiles, etc. can be easily cleaned with liquid ammonia in water (one tablespoon of ammonia to one litre, if water). Materials for polishing metals are also available in the market. The sumps and overhead water tanks should be cleaned regularly (see Section 29.2). The overhead tank should be kept well covered with no holes. Malarial mosquitoes breed in fresh water. If the tank is exposed to sunlight algae—which is harmful to human skin—can grow in unchlorinated water.

40.7 MAINTENANCE OF DRAINAGE SYSTEM

In order to avoid nuisance and unhealthy conditions, soil and waste-disposal systems must be kept clean as well as in good working order. The main items to be looked after regarding the maintenance of drainage system are discussed further.

40.7.1 Fittings inside the Building

Water taps, waste pipes (pipes from washbasin or sink to the floor trap), traps in fittings, floor traps, gratings above floor trap, etc. should be *examined periodically* by passing water through them and examining their performance. Otherwise, they are liable to get jammed especially when the water supply is not too good. Gratings in bathrooms require frequent inspection as they are liable to be clogged with hair, oil, etc.

If drains are not working properly, drain cleaning compounds, which are available in the market, should be used to remove the block. Sometimes, physical methods will be required to clean them.

40.7.2 Waste and Soil Pipes

These pipes and their fixity to walls should be examined during painting of external walls. If fixings are damaged, they should be repaired. (Many of the pipes in old buildings are AC pipes which are brittle but cheap. Nowadays, PVC pipes are more often used in Class I and Class II buildings.) Wire balloons on the top should be checked and replaced, if necessary.

40.7.3 Gulleys, Manholes, Manhole Covers and Drainpipes

These should be examined regularly and before the onset of the monsoon. The outlet to the sewer should be checked so that backflow from the street sewer to the house drain does not happen in rainy season. Cockroaches that breed in dark sewers and manholes should be destroyed by chemical sprays. There should be no stagnant water in the drains to breed mosquitoes.

40.7.4 Roof Drainage System

The roof drainage system should be inspected before the rainy season. Gutters, especially (horizontally placed ones) must be cleaned and leakages repaired before the onset of the rains. Any leak from roof drainage system should be identified and repaired during the first pre-monsoon rains.

40.8 MAINTENANCE OF SEPTIC TANKS

For efficient working of a septic tank, grease and slow decomposing matter, etc. should not be put down the drains (a grease trap in the line is desirable). Avoid drain-cleaning chemicals which can kill the bacteria in the septic tanks.

The digested sludge should be removed when the depth of the sludge and scum exceed half the depth of the tank. A portion of the sludge should be left in the tank to act as seed to the fresh sewage. Adding 1/4 kg of brewer's yeast to the tank will hasten the septic tank action if such action is dull. Warning signs of a defective system include foul odours from the drains or the tank and also the growth of lush vegetation over the tank. If water backs up in the drain, it indicates clogging by sludge and scum.

40.9 MAINTENANCE OF WATER SUPPLY SYSTEM

It is very important that the water supply system should be maintained in good and sanitary condition. If the supply is from a municipal water supply line to a sump, from the sump to a overhead water tank and from the water tank to the various distribution points, then periodic draining, cleaning, drying and whitewashing of the sumps at least once in 4 months should be carried out in all public buildings (January, May and September may be chosen for this purpose). The supply pipes can be disinfected as described in Section 29.2.

40.10 MAKING BUILDING LEAKPROOF

Repair of corrosion of RC roof due to corrosion of steel is explained in Section 39.7. Leakage of rainwater has been discussed in detail in Chapter 28 and should be carefully attended to before the rainy seasons. Leakage in wet areas has been discussed in Chapter 27.

40.11 MAINTENANCE OF ELECTRICAL SYSTEM

As already pointed out in Chapter 32, it is advisable to have a ELCB trip switch in the electrical system of all buildings. This will show up any earth leakage. Electrical switches, especially those in the kitchen are liable to be clogged up by grits laid in switches by small red ants. These need cleaning. Anti-ant chemicals for placing in switch boxes are also available in the market. Bearings, condensers, regulators, etc. of fans have to be attended to if their performance is poor. If the connecting plugs of equipments especially of like high amperage like heaters, airconditioners, etc. are loose, they are liable to get heated up and burnt. They should always be plugged in properly (tightly) or should be connected through fused outlets. It is advisable to have phase-changing devices in the electric supply system so that if one phase goes out, the connections can be made to another available phase. Much savings in electricity can be made if we use modern fluorescent lamps for lighting and capacitance type of regulators for fans.

40.12 MAINTENANCE OF TALL TREES NEAR BUILDINGS

Tall trees that grow near foundations of walls of a building can cause a lot of trouble by sending their roots under the building to take away the moisture in the soil. This can produce cracking of walls and flooring in the ground floors. Trees that send their roots laterally cause more trouble than those that send their roots downwards only. If large trees are allowed to grow near the foundation, then trenches along the walls by the side of these trees should be made and the roots should be cut and painted with chemicals (or coal tar) to retard their further growth towards the building. Framed buildings with deep foundations are less affected by tall trees than buildings with load-bearing walls.

SUMMARY

Regular maintenance of buildings is necessary to keep them good in appearance and performance. This is especially necessary in public buildings which are used by many people and are to be maintained by a responsible agency. It is desirable to have a schedule of items to be attended to so that those responsible for the work can attend to them regularly without being told to do the necessary items of work.

REVIEW QUESTIONS

1. Describe briefly how a marble floor is to be maintained in a good condition.
2. Indicate how glass windows are to be cleaned.
3. What precautions would you take while painting the underside of the floor slabs of wet areas such as bathrooms?
4. Indicate the method you would recommend for maintenance of water closets, sinks, washbasins, etc. of a hospital building.
5. Describe briefly the following:
   (a) Maintenance of septic tanks
   (b) Maintenance of water supply system in a public building
   (c) Repair of wall cracks in buildings

Appendix A

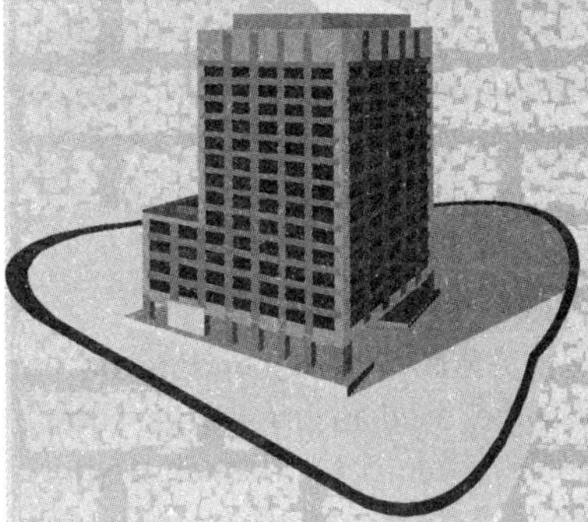

Design of Brick Masonry Walls

A.1 INTRODUCTION

Brick walls are subjected to compression, and IS 1905 – 1987, "Code of Practice for Structural use of Unreinforced Masonry Walls", and SP 20 deal with their design. In this chapter, we will examine the fundamental aspects of design. Details of design can be obtained from the above references.

A.2 EFFECTIVE HEIGHT, EFFECTIVE LENGTHS OF WALLS AND SLENDERNESS RATIO

Designs of all compression members such as columns and walls are based on the *slenderness ratio* of the member, which is defined as the effective length divided by the thickness of the walls. Unlike in a column (which is defined as a member with its breadth less than four times its thickness), for calculating effective length of a wall, we have to consider not only its dimension along the height but also the dimension along its length between supports such as cross walls.

Effective height = 0.75 H to 0.8 H, for walls restrained at both the ends
= 1.5 H, for walls restrained only at one end such as in compound walls and parapets

Effective Length = L, for walls restrained at both the ends
= 1.5 L, for walls restrained at one end
= 2 L, for walls with no restraints

where H = The height of the wall between the restraints
and L = The length of the wall between the cross walls, pilasters and other constraints

The larger of the above values with respect to H and L is taken for calculation of slenderness ratio (for details refer to IS 1905).

$$\text{Slenderness ratio} = \frac{\text{Larger of the values of effective length and effective height of the wall}}{\text{Thickness of the wall}}$$

Brick walls with slenderness ratio up to 6.0 are considered as *short walls* which fail under compression. Walls with slenderness ratio greater than 24 are unstable and are prohibited. To find the reduced load that can be carried on walls with slenderness ratio greater than 6.0, we use a reduction factor explained in Section A.4.

A.3 FAILURE LOAD ON SHORT WALLS

The crushing strength of common bricks vary from 5 N/mm^2 to 35 N/mm^2 depending on the different regions in India. It is recommended that the mortar to be used in brickwork should have a strength very nearly equal to the strength of brick (refer Table 5.1). If such a short-brick pillar is loaded, the failure load of the pillar will be much less than that of the individual brick. This is because when we load the pillar, each brick with its mortar bed behaves as a beam on elastic foundation and the failure is in bending by tension corresponding to its modulus of rupture. Hence, the *maximum allowable strength* p_{max} *of a short column* is as follows:

$$p_{max} = \frac{1}{10} \times \text{compressive strength of the bricks}$$

A.4 REDUCTION FACTOR FOR SAFE STRESS IN BRICK WALLS

The reduction factors given in Table A.1 are recommended in IS 1905, 1987 and SP20 for central loading (e = 0). It will be further reduced for eccentricity of load as shown in Table A.1.

Table A.1 Stress reduction factors for slenderness ratio and eccentricity for brick walls (S_F)

Slenderness ratio	*Eccentricity ÷ Thickness of the wall*				
	0	0.04	0.10	0.20	0.50
6.0	1.000	1.000	1.000	0.990	0.970
8.0	0.920	0.920	0.920	0.910	0.850
10.0	0.840	0.835	0.830	0.810	0.730
12.0	0.760	0.750	0.740	0.706	0.600
16.0	0.580	0.565	0.545	0.500	0.350
21.0	0.470	0.448	0.420	0.354	0.170
24.0	0.440	0.415	0.480	0.310	0.110

A.5 AN EXAMPLE OF APPROXIMATE DESIGN OF BRICK WALLS

Problem. A brick wall of a two-storey building with each floor 3 m high has a thickness of 200 mm. Let the distance between cross walls in the room be 3 m. Assume that the load from each floor is 10 kN per metre length of wall. What should be the minimum strength of bricks that can be used for this building? Assume that there is no eccentricity of load.

Solution:

1. *Find total load on the base.*
   We have
   Total weight of two-storey high wall @ 20 kN/m^3 for brickwork for 0.2 m thick wall = 0.2 × 20(2 × 3) = 24 kN/m (DL)
   Live load from the floors = 2 × 10 = 20 kN/m (for two floors)
   Total load = 44 kN/m
2. *Find effective height, length and slenderness ratio.*
   Effective height = 0.75 × H = 0.75 × 3 = 2.25 m
   Effective length = L = 3 m (larger value = 3 m)

   Larger slenderness ratio $= \dfrac{3}{0.2} = 15$
3. *Find stress factor* S_F for eccentricity, e = 0.
   From Table A.1 for slenderness ratio = 15, S_F = 0.625 (by interpolation)
4. *Find necessary strength of brickwork at the base of brickwork* (σ).

   We have $\dfrac{\text{Load}}{\text{Area}} = \text{Stress} = S_F \times \text{strength of masonry}$

   or $$\frac{44 \times 1000}{200 \times 1000} = S_F\sigma \ \text{N/mm}^2$$

   or $$\sigma = \frac{44}{200 \times 0.625} = 0.352 = 0.4 \ \text{N/mm}^2 \text{ (approx)}$$
5. *Find brick strength for the brickwork of strength 0.4 N/mm^2.*
   We have strength of brick = 10 × strength of the brickwork
   = 10 × 0.4 = 4 N/mm^2

Note. Ordinary well burnt country bricks with a minimum strength of 5 N/mm^2 will be suitable for this building. Further details of design can be obtained from IS 1905.

SUMMARY

Design of masonry walls is important when we build buildings more than one storey in height. The slenderness of the walls and the required strength of the masonry units are important items to be checked in these constructions.

Appendix B

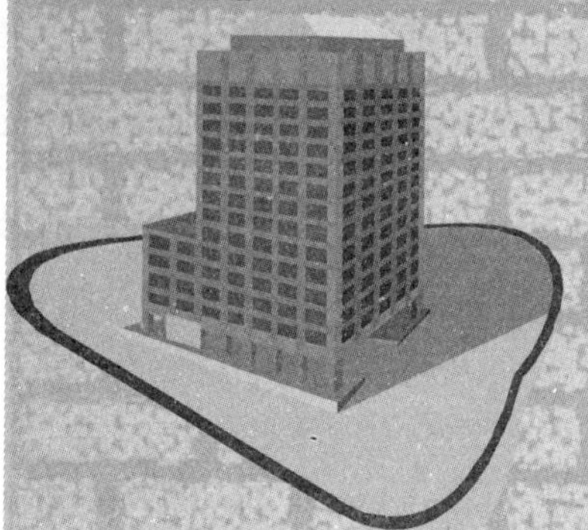

Earthquake-Resistant Design of Masonry Buildings

B.1 INTRODUCTION

Nowadays there is much talk about earthquake-resistant design of buildings and we should know the fundamental of the subject. For disaster-resistant design of buildings (against earthquake or cyclones), we divide the structures into three categories as follows:

1. *Fully resistant structures* which will behave elastically under the worst loads, for example hospital, electrical stations, etc. which should survive the event.
2. *Partially resistant structures* which may undergo damages that can be subsequently repaired, for example other important buildings.
3. *Survival structures* which may get very much damaged during the disaster but there should be features which will exclude any loss of life, for example temporary structures and low-cost buildings.

Again, depending on design and construction of buildings in India, we divide buildings into another three categories as follows:

1. *Engineered buildings.* These buildings are fully designed by engineers and architects like tall multi-storey buildings which have to be designed by engineers.
2. *Non-engineered buildings.* These buildings are fully conceived and constructed by artisans such as carpenters and masons. (A great part of residential houses built in India are built on the basis of empirical knowledge.)
3. *Semi-engineered buildings.* These buildings are not designed fully, but unlike the non-engineered buildings, the codal provisions of IS 4326 (COP of Earthquake-Resistant Design of Buildings) and IS 13828 (Improving Earthquake-Resistant Design of Low Strength Masonry Buildings) are incorporated in their construction. (The necessary features of these buildings are fully explained in a pamphlet "Guidelines for Improving Earthquake Resistance of Housing" published by Building

Materials and Technology Promotion Council, New Delhi.) IS 13828, 1993, "Indian Standards for Improving Earthquake Resistance of Low Strength Masonry Buildings", also explains the features to be incorporated in such buildings.

B.2 DESIGN AND CONSTRUCTION OF EARTHQUAKE RESISTANT BUILDINGS

Multi-storey buildings have to be built as engineered buildings (framed structures). We should strictly adhere to codes in their planning (provision of shear walls and other features with frame construction), in their analysis and design (by providing ductile detailing) as well as in their construction. However, most of the one- or three-storey buildings can be built as semi-engineered buildings to provide seismic resistance. As unreinforced or unconfined masonry will crack under large earthquake forces, earthquake resistance can be provided by confining masonry by RC bands at the plinth, lintel and roof level according to the codes already given in Section A.1. The provision of lintel bands for the various zones is given in Table 8.1 in Chapter 8. The amount of reinforcement will depend on the probable intensity of earthquake or the zone in which the building is built.

B.3 NEED FOR DUCTILE DETAILING

We must have a clear idea about the need of ductile detailing of structures for earthquake resistance. We must remember that it is extremely difficult to predict the magnitude of the earthquake forces. In case of cyclonic wind forces, we can estimate the full force of the wind and we assume these full forces as static forces for analysis. However, in estimation of the seismic forces, what the codes give is an estimate of the probable forces, and the actual earthquake forces are dynamic and can be much larger. The objectives of ductile detailing are as follows:

(a) The structure should resist moderate intensity of earthquake without *any damage.*
(b) The structure should resist exceptionally large intensity of earthquake which can occur without *collapse.*

We have to incorporate ductile behaviour of the structure so that it can take a large amount of overloading beyond the design load and does not collapse, thus ensuring no loss of lives during a large earthquake. This is carried by the following rules laid down for ductile detailing by the code IS 13920.

B.4 RECOMMENDED METHODS OF DESIGN

From the above discussion, we can arrive at the following conclusions:

1. In regions of low earthquake intensity zone 2, normal design according the general codes IS 456–2000 for RCC and IS 1905 for masonry will give enough built in safety.
2. If we have to get a absolutely safe building in other zones, we should adopt a framed fully engineered building according to the codes IS 1893 (1984) and IS 13920 (1993).

3. Otherwise we can have semi-engineered masonry buildings having not more than 4 storeys in zones 3 and 4 and not more than 3 storeys in zone 5. In such buildings, we must use the prescribed guidelines published in references given in Section B.1 which are given in Section B.5.

B.5 BRIEF REVIEW OF RECOMMENDATIONS FOR SEMI-ENGINEERED BUILDINGS

The main recommendations in keeping a building intact are as follows:

1. Provision of good foundations
2. Provision of plinth beams to tie together the foundation
3. Provision of lintel band
4. Provision of roof/floor band
5. Vertical bar at corners
6. Reinforcing around door and window openings

The details of the above recommendations can be obtained from the IS codes and references given in Section B.1.

SUMMARY

Tall and other buildings which should be fully resistant to earthquake forces should be designed and built as fully engineered buildings. However, the ordinary masonry low-rise buildings, which are the majority of houses built in India, can be built as semi-engineered buildings to be detailed according to the recommendations of IS codes and "Guidelines for Improving Earthquake Resistance of Buildings" published by the Building Materials and Technology Promotion Council, Government of India, referred in Section B.1.

Appendix C

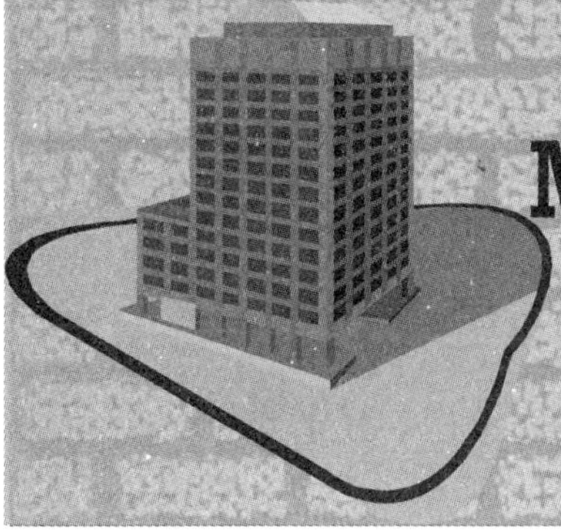

Estimating Costs and Material Requirements

C.1 INTRODUCTION

It is always worthwhile to estimate the cost of the proposed construction so that it does not exceed the available finances. We have already seen in Chapter 1 how a building can be classified as class I, II or III. If the approximate cost to build a building of a particular class in a given locality is known the total cost of the building, can be estimated on an area basis. In this chapter, we will examine the various components of the cost of the building so that we can control the cost during the execution stage itself. Charges for supervision and the supply of labour and materials which is paid for by the owner is usually taken as 7 to 10% of overall cost of the building. In lump sum and itemwise contracts, because of the new tax laws, contractors expect a profit of 15 to 20% of the cost of the work.

C.2 BREAK UP OF COST OF CONSTRUCTION

Table C.1 gives the approximate *average percentage* cost of civil works for a residential building. It also indicates the possible savings that can be made. The actual values will depend on many factors like foundation condition and architectural features.

Table C.1 Percentage cost of civil works

S.No.	*Works*	*Cost* (% of the total cost)	*Possible savings* (% of the total cost)
1.	Foundation and plinth	10 to 15	1 to 3
2.	Superstructure (wall + roof, including plastering)	30 to 40	4 to 13
3.	Painting	5 to 20	1 to 8
4.	Doors and windows	12 to 20	1 to 8
5.	Flooring	10 to 20	2 to 7
6.	Waterproofing of roof and wet areas	8 to 12	1 to 2

In addition, the cost of services is usually estimated *as percentage of civil works* and is taken as follows:

1. Water supply and sanitation — 10 to 12.5 per cent
2. Electrical works — 8 to 12.5 per cent
3. External works — 2 to 5 per cent

Total = 20 to 30.0 per cent (max)

C.3 RATIO OF COST OF MATERIALS TO LABOUR

It has been found that the total cost can be divided into the following percentages for an ordinary building built in India in the traditional way:

1. Material cost – 60 per cent of the total cost
2. Labour cost – 30 per cent of the total cost
3. Other costs – 10 per cent of the total cost

The break up of the cost of materials can be approximated as given in Tables C.2 and C.3.

Table C.2 Cost of principal materials

S.No.	*Item*	*Approximate cost as % cost of the civil work*
1.	Bricks	15 to 18
2.	Cement	12 to 18
3.	Steel	5 to 15
4.	Timber	10 to 15
5.	Sand	7 to 10
6.	Aggregate	8 to 10
7.	Paint	5 to 10

Table C.3 Cost of labour

S.No.	*Worker*	*Approximate cost as % of civil works*
1.	Mason	12
2.	Unskilled labour	12
3.	Carpenter	6

C.4 CONSUMPTION OF MATERIALS FOR BUILDINGS

The data given in Table C.4 can be taken as very approximate quantities of materials required for construction of an ordinary building.

Table C.4 Very approximate quantities of materials required for 1 m^2 plinth area

S.No.	*Materials*	*Quantity per* m^2 *of plinth area*
1.	Bricks (9" ordinary)	250–300 nos.
2.	Sand	0.5 to 0.7 m^3
3.	40 mm aggregates	0.2 m^3
4.	20 mm aggregates	0.2 m^3
5.	Steel for buildings with load bearing walls	10 to 20 kg
6.	Steel for framed flats and hostels	30 to 50 kg
7.	Steel for framed offices	40 to 70 kg
8.	Cement for load bearing walls	3 bags
9.	Cement for framed buildings	4 bags
10.	Lime for brick jelly	40 litres

C.5 ANALYSIS OF ITEMS OF WORKS

The quantities of important *items of works* for buildings can be worked out to control cost as shown in Table C.5 which is for a middle class building on load bearing walls. The values shown are not exact but only indicative.

Table C.5 Items of works for residential buildings on load-bearing walls
(Quantitatives are only indicative. Actuals will depend on the design.)

S.No.	*Materials*	*Approximate quantitity per* m^2 *of the plinth area*	
		G.F. only	*G.F. + F.F.*
1.	Earthwok	0.52 m^3	0.30 m^3
2.	Mass concrete in foundation and under floor	0.22 m^3	0.15 m^3
3.	RCC in foundation	0.11 m^3	0.06 m^3
4.	Brickwork in foundation (varies)	0.16	0.09 m^3
5.	Brickwork in superstructure	0.40 m^3	0.40 m^3
6.	RCC in superstructure:		
	Plinth/lintel/beam	0.05 m^3	0.05 m^3
	Slab and stairs	0.01 m^3	0.18 m^3
	Sunshade/shelves	0.18 m^3	0.01 m^3
7.	Doors and windows	0.45 m^2	0.45 m^2
8.	Plaster	2.60 m^2	2.60 m^2
9.	Painting of walls and ceiling	3.5 m^2	3.5 m^2 (Section C.9)
10.	Waterproofing	0.98 m^2	0.60 m^2
11.	Flooring	0.75 m^2	0.75 m^2

C.6 CEMENT REQUIREMENT IN CEMENT WORKS

Cement is an important material and its consumption must be controlled. Some of the data used for calculation of cement requirements in *concrete, mortar and plaster* are given in Tables C.6, C.7 and C.8.

Table C.6 Cement consumption for concrete

Mix	*W/C ratio*	*Strength* (N/mm^2)		*Cement per* (m^3) *of concrete* (kg)
		Without control	*With control*	
1 : 1½ : 3	0.42	20	30	405
1 : 2 : 4	0.55	15	25	310
1 : 3 : 6	0.75	10	15	215
1 : 4 : 8		Used only for base concrete, etc.		160
1 : 5 : 10		Used only for base concrete, etc.		130

Note. 1 cubic metre = 35.315 cubic feet; Approximate yield = 2/3 × Volume of mix. Hence, volume of cement required per m^3 of concrete is taken as $\frac{1.52}{(\text{Volume of mix})}$ (approx).
(Volume of 1 : 2 : 4 mix is taken as 7)—see also Section C.10

Table C.7 Requirements for 1 m^3 of brick masonry

Mix of mortar	*Cement* (bags)	*Sand* (m^3)
1 : 3	2.6	0.267
1 : 4	1.9	0.275
1 : 6	1.4	0.299
1 : 8	1.1	0.308

Note (Table C.7). For brick masonry, we need 500 metric bricks (and 520 ordinary bricks) *per cubic metre of brickwork* and 25 to 30% of the volume of brickwork will be taken up by mortar. Hence, we need about 0.25 to 0.30 m of mortar per cm of brickwork. *One bag of cement is considered as 50 kg in weight and 35 litres in volume.*

Table C.8 Requirements for 100 m^2 of 12 mm thick cement plaster (1.2 m^3 plaster)

Mix	*Cement* (bags)	*Sand* (m^3)
1 : 2	20.0	1.4
1 : 3	14.2	1.5
1 : 4	11.4	1.6
1 : 6	8.6	1.8

Note. 1 square metre = 10.764 square feet

C.7 ESTIMATION OF STEEL IN RC WORKS

For preliminary *estimation,* it is usual to assume the steel requirement on a percentage basis (or kg/m^3) of the easily calculable volume of concrete based on the data given in Table C.9. The actual requirement by proper design will be much less. (*Note.* Steel rods usually come in lengths of 11 m. Hence we can check the weight of a consignment by counting the length of rods. For example, the weight of a consignment consisting of 8 mm rods each of length 11 m = 11 × 0.395 = 4.4 kg per nod approximately, as weight of 8 mm rod is 0.395 kg per metre.

Table C.9 Estimation of steel in buildings components (Fe 415 steel)

Item	*Coefficients for estimation of steel*	
	As % of steel	As kg/m^3
Footings	0.25 to 0.65	20 to 50
Cast insitu piles	0.40 to 0.75	30 to 60
Rafts	0.75 to 1.50	60 to 120
Columns	1.00 to 4.0	80 to 320
Beams	0.5 to 1.5	30 to 100
Slabs and lintels	0.3 to 0.6	25 to 50
Sunshades	0.3 to 0.5	25 to 30

Note. 1% steel = 78.5 (say 80 kg) of steel per cubic metre of concrete.

Example: Find the steel required for a slab 15 × 10 m and 120 mm thick.

Volume of concrete = 15 × 10 × 0.12 = 18 m^3

Assume % of steel is 0.4 % = (or 0.4 × 78.5 = 31 kg/m^3)

Wt of steel = 18 × 31 = 565 kg

(*Note.* This works out to 565 ÷ 150 = 3.76 kg/m^2 for slab only. In addition we have foundation beams, columns etc. so that the total quantity will workout to 10 to 20 kg/m^2)

C.7.1 Typical Minimum Consumption

The minimum consumption of steel in kg/m^2 by proper design and detailing for a typical four storey building on good soil using to Fe 415 steel can be a total of 20 kg/m^2 as follows.

Table C.10 Minimum steel for a four-storey flat

Item	*Steel consumption*
Foundation	2.0 kg/m^2
Plinth beam	2.0 "
Columns	4.0 "
Beams	5.0 "
Slabs and stairs	5.5 "
Lintels, sunshade etc.	1.5 "

C.8 ANALYSIS OF RATES

One of the topics studied in "estimating" is the analysis of the rates of various items of works. The data for cost of materials, labour and other items such as scaffolding required per unit of work for each item of work in each region of the country is available. From these data, the cost per unit of work can be estimated. As it is a subject to be studied under estimating, we will not go into a detailed study of the subject in this book.

C.9 ESTIMATION OF PAINT REQUIRED FOR PAINTING AN AVERAGE RESIDENTIAL BUILDING

(a) Rough estimate of areas to be painted in terms of plinth area:
Area of ceiling = Floor area (say plinth area)
Area of internal walls = 3 × plinth area
Area of external walls = 2 × plinth area
Area of windows = 30% plinth area each side
Area of doors = 30% plinth area each side

(b) Approximate coverage of different types of paints per coat (Second and third coats will have 25% more coverage).
1. Fat lime for lime-washing 10 m^2/kg
   (For colour wash add 1 to 2 kg colour per 100 m^2)
2. Dry distemper 10 m^2/kg
3. Cement paint 5 m^2/kg
4. Cement wall primer 8 m^2/L (litre)
5. Oil bound distemper 8 m^2/kg
6. Plastic emulsion paint 10 m^2/L
7. Acrylic exterior paint 12 m^2/L
8. Metal primer 10 m^2/L
9. Synthetic enamel paint 10 m^2/L
10. Copal varnish for wood 10 m^2/L

Example: Find the approximate quantity of paint required for painting 200 m^2 of wall with plastic emulsion paint in two coats.

Paint required for first coat = (200/10) = 20.0 litres
Paint required for second coat = 16.0 litres (10% less)
Total = 36.0 litres (approx)

Note. Generally one painter can be assumed to paint 15–20 m^2 of wall surface per day, depending on the type of paint and finish.

C.10 ESTIMATING MATERIALS FOR ONE CUBIC METRE OF CONCRETE (in c.m.)

Concrete mix	*Coarse aggregate*	*Sand*	*Cement*	
			Cum	*Bags*
$1 : 1^1/_2 : 3$	0.85	0.43	0.283	8.0
1 : 2 : 4	0.89	0.45	0.223	6.4
1 : 3 : 6	0.92	0.46	0.154	4.4
1 : 4 : 8	0.95	0.48	0.119	3.4
1 : 5 : 10	0.91	0.46	0.091	2.6

Appendix D

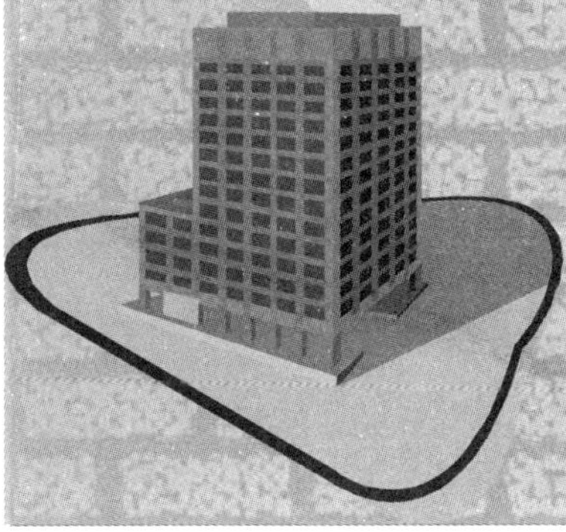

Major Items of Works for Construction of a Single-Storey Residential Building

The following items can be taken as the major items of work to be completed for the construction of a residential single-storey building:

1. Bore well installation for water
2. Site clearance and laying out of building
3. Excavation of foundation and anti-termite treatment
4. Brickwork up to plinth
5. Filling up of trenches
6. Plinth and DPC constructions
7. Filling up of basement and surroundings
8. Brickwork to window sill level and lintel level
9. Construction of lintels
10. Brickwork to roof level and preparation of slab bearings
11. Laying of base concrete for floor slab
12. Construction of roof slab with ducts for electrical wiring
13. Hacking and ceiling for plastering
14. Providing ducts for wiring on walls
15. Installation of door and window frames
16. Wall and ceiling plaster
17. Insulation, waterproofing and drainage of roofs
18. Laying of floor
19. Laying water supply and sewerage pipes
20. Fitting of doors and windows
21. Electrical wiring
22. Painting and varnishing
23. Fitting water supply and sanitary appliances
24. Electricity and water supply connection from municipality
25. External works, fencing, compound wall and gates

Appendix E

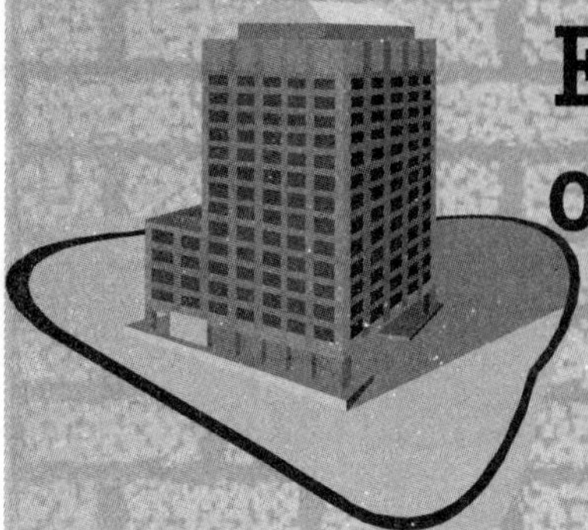

Equivalent Plain Areas of Uneven Surfaces for Payment for Painting of Building Works

CPWD Practice

S.No.	*Description of work*	*How measured*	*Multiplying coefficients*
1	2	3	4
(A)	**Woodwork on doors, windows, etc.**		
1.	Panelled or framed and braced doors, windows, etc.	Measured flat.	1.30 (for each side)
2.	Ledged and battened or ledged, battened and braced doors, windows, etc.	Frame, edges, chocks, cleats etc. shall be deemed to be included in the item.	
3.	Flush doors, etc.	–do–	1.20 (for each side)
4.	Part panelled and part glazed or gauzed doors, window, etc. (excluding painting of wire gauze portion)	–do–	1.00 (for each side)
5.	Fully glazed or gauzed doors, windows, etc. (excluding painting of wire gauze portion)	–do–	0.80 (for each side)
6.	Fully venetioned or louvered doors, windows, etc.	–do–	1.80 (for each side)
7.	Trellis (or Jaffri) work one way or two	Measured flat overall; no deduction shall be made for open spaces; supporting members shall not be measured separately.	2 (for painting all over)
8.	Carved or enriched work	Measured flat	2 (for each side)
9.	Weather boarding	Measured flat (not girthed); supporting framework shall not be measured separately.	1.20 (for each side)

10. Wood shingle roofing	Measured flat (not girthed)	1.10 (for each side)
11. Boarding with cover fillets and match boarding	Measured flat (not girthed)	1.05 (for each side)
12. Tile and slate battening	Measured flat overall; no deduction shall be made for open spaces.	0.80 (for painting all over)
(B) Steelwork on doors, windows, etc.		
13. Plain sheeted steel doors or windows	Measured flat (not girthed) including frame edges etc.	1.10 (for each side)
14. Fully glazed or gauzed steel doors and windows (excluding painting of wire gauze portion)	–do–	0.50 (for each side)
15. Partly panelled and partly glazed or gauzed doors and windows (excluding painting of wire gauze portion)	–do–	0.80 (for each side)
16. Corrugated sheeted steel doors or windows	–do–	1.25 (for each side)
17. Collapsible gates	Measured flat	1.50 (for painting all over)
18. Rolling shuters of interlocked laths	Measured flat (size of opening) all over; jamb guides, bottom rails and locking arrangement etc. shall be included in the item. (Top cover shall be measured separately.)	1.10 (for each side)
(C) General		
19. Expanded metal, hard drawn steel wire fabric of approved quality, grillworks and gratings in guard bars, balustrades, railing partitions and MS bars in windows frames	Measured flat overall; no deduction shall be made for open spaces; supporting members shall not be measured separately.	1 (for paint all over)
20. Open palisade fencing and gates including standards, braces, rails stays etc. in timber or steel	–do– (see S.No. 12)	1 (for paint all over)
21. Corrugated iron sheeting in roofs, side cladding, etc.	–do– (Measured flat not girthed)	1.14 (for each side)
22. AC corrugated sheeting in roofs, side cladding, etc.	–do–	1.20 (for each side)
23. AC semicorrugated sheeting in roofs, side cladding, etc. or Nainital pattern using plain sheets.	–do–	1.10 (for each side)
24. Wire gauze shuters including painting of wire, gauze	–do–	1.00 (for each side)

Appendix F

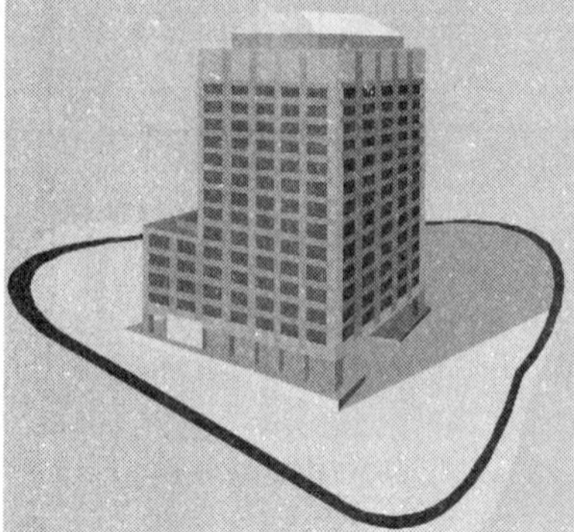

Fresh Water from Brackish Water by Reverse Osmosis

F.1 INTRODUCTION

Chapter 29 of this book deals with the conventional systems of water supply of buildings, where good water is available. Today in many places in the world, we are faced with the problem of getting suitable water we need in large quantities for general purpose from brackish water or even sea water. The limited amount of water we need for drinking can be bought in bottles. Desalination is carried out by the process called *reverse osmosis*. This is briefly described in this Appendix F.

It may be noted that reverse osmosis is different from the many other simple systems like Aquaguard which are used in many houses and offices to make good water supply safe for drinking.

Nowadays in many places in the Middle East countries where fresh water is not available, water supply in buildings for all purposes is derived from sea water. As these countries have cheap natural gas also, supplies for generating the electricity needed to run these machines, reverse osmosis (RO) is not costly.

In places like Singapore, however, where there is no gas or other cheap source of power supply, this process of RO will be costly. They have perfected the method of combining and converting the waste water and the sewage water that are already used, back to good usable water even fit for drinking. This process is called *aerobic treatment of waste water*. This conversion of non-salty waste water to good water by aerobic treatment is described in the next section Appendix G.

F.2 HISTORY OF REVERSE OSMOSIS

Osmosis is a natural physical process. When if concentrated and dilute solutions are separated by a semi-permeable membrane, water passes from the dilute solution side of the membrane to the more concentrated solution side, without any input of external energy. This is simply to produce natural equilibrium and was observed by scientists as early as in 1748.

It was only about 200 years later as late as in 1950 that two American scientists discovered reverse osmosis. They observed that when a concentrated solution of water is separated from a less concentrated solution of water by a semi-permeable membrane and if a high pressure is applied on the concentrated solution side, then pure water will flow from the concentrated solution to the less concentrated side. As this is the reverse of osmosis, this process was named as "reverse osmosis". What is needed is a membrane suitable for the liquid (seawater, brackish water etc.), and a power supply to produce the high pressure.

This discovery also did not lead to commercial use of the process till around 1960, when Srinivasa Sourirajan working under Sydney Leo in the National Research Council Ottawa in Canada discovered a technique for making the suitable membranes. Others also discovered suitable membranes but all of them were patented so that many could not use it. By about 2000, when the patents expired, a number of companies started making suitable membranes and today this RO process is very popular in many parts of the world.

Many cities lying near the sea, particularly in Middle East countries having no rivers very little rainwater, but plentiful supplies of cheap power, this method is used to obtain water supply from sea water.

It is worth mentioning that in Tamil Nadu where there are many villages adjacent to the sea and there is no electric supply, the Central Salt and Marine Chemicals Research Institute (CSMCRI) has developed a bullock driver mechanism to produce high pressure. It is claimed that about 600 litres of drinking water can be produced per hour with this system. In Tamil Nadu also, RO systems using available electricity are being employed in many places near the sea coast for household water supply. A typical unit is shown in Fig. F.1.

Fig. F.1 The two-stage seawater desalination unit.

The RO systems also used in many industries. For instance, very concentrated orange juice can be produced by removing water through reverse osmosis. Or again, water fit for boilers needs to be without salts and it can be produced at any place by this process.

F.3 REVERSE OSMOSIS FOR WATER SUPPLY

The usual standards specified for drinking water and water for general purposes in a household are as given in Table F.1.

Table F.1 Requirements factors for general and drinking purposes

S.No.	*Factor*	*Water for General Use*	*Drinking Water*
1	Total dissolved solids (TDS)	< 2000	< 500
2.	Hardness	< 300	< 100
3.	Iron	< 0.3%	0
4.	pH	6.5 to 8.5	6.5 to 8.5

Notes:

(i) In many places, we need to plan the treatment of salt water only for general purposes since the limited amount of drinking water required for a household may be bought separately in bottled water. Otherwise, we provide a second tap for limited supply of more refined drinking water which has gone for more refinement.

(ii) When we take up water supply work, we must first analyze the available water and compare the results with the requirements shown in Table F.1, and then we should choose a suitable membrane.

(iii) The following type membranes are available

(a) Normal membrane
(b) Membrane for brackish water
(c) Membrane for seawater

In some places (as with seawater), the process may have to be designed for the refinement in two stages.

Only well-known companies that specialize in the installation and maintenance of these projects should be given the work. Since only these companies can reliably take up the work of analyzing available water and designing the membrane and the mechanisms. Small size plants for single buildings to large size plants for the modern high density multiple flat systems are available with such specialized companies.

The membranes can easily retain all the dissolved salts, 99% of particles and most of the bacteria. The membrane also has to be cleaned at intervals of the salts, bacteria etc. that it retains by intermittent reverse flows through the membrane.

In Tamil Nadu, various town panchayats are producing treated drinking water from undrinkable ground water through reverse osmosis. Today, (in 2015), the cost of the plant producing 3000 litres of good water a day is reported as about 6 lakhs. The cost of this water is, therefore, much cheaper than bottled drinking water available commercially for sale.

The RO units should ideally be planned from the start of a construction project and their construction should be entrusted to good firms specialized in such work. Unfortunately, however, water supply and drainage works are most often the last works planned or implemented in a building construction project. Their design and construction receive insufficient attention.

Appendix G

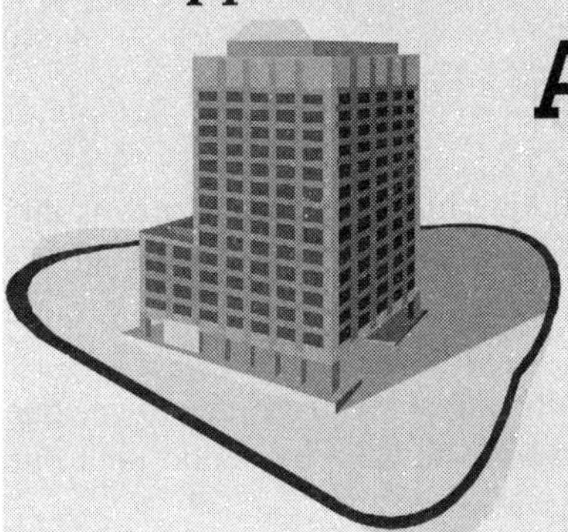

Anaerobic and Aerobic Treatment of Sewage and Waste Water

G.1 INTRODUCTION

In nature a vast member of microbes exist. More than 60% of them are beneficial to humans, while others are toxic and can harm humans. For example fermentation of milk for making curds from milk or fermentation of "dosa mavoo" for making the South Indian dosa is carried out by using microorganisms.

Some of these microbes can live only under conditions where there is no air (oxygen) are called *anaerobic microbes*. Other microbes that can exist only by consuming oxygen are called *aerobic microbes*.

In this appendix, we will examine how humans first discovered the use of anaerobic bacteria for the design of 'septic tanks' and how recently they discovered the use of aerobic bacteria to digest sewage and waste water in very short time and in vessels of small capacity that occupy small space in the building site.

G.2 HISTORY OF THE DISCOVERY OF ANAEROBIC TREATMENT OF SEWAGE IN SEPTIC TANKS

The working of ordinary septic tanks is by anaerobic digestion. It is described in Chapter 31 of this book.

It was as early as 1860 that L.H. Howras of France discovered the septic tank treatment of sewage, which is in use even today. Millions of anaerobic microorganisms break down solid waste into liquid and gas. However, for completion of this treatment, the storage period required can be as large as 36 to 190 hours depending on the temperature of the air.

In all anaerobic digestion processes, two factors play major role in the design of the plants and septic tanks. The first is the space required for storage of the liquid for the considerable time required for completion of the treatment. The second factor is the appropriate management of the hazardous gases produced during the process. However, septic tank system has the advantage that the treated water percolates down the earth and replenishes the level of the ground water table.

G.3 AEROBIC TREATMENT OF SEWAGE

Septic tank treatment of sewage becomes difficult with small size housing plots in towns and cities. Nowadays in towns and cities, the Sewage Board comes in and takes away all the sewage and waste water to a central place. This has the bad effect of lowering of the ground water table at the building site as shown in Section G.6.

Again if one has to build a small cottage in a small plot far away from the town, one has to treat the sewage inside the small plot. It is difficult to treat sewage by septic tank due to lack of space. In such situations also aerobic treatment is ideal.

The discovery of the quick method of treatment of sewage in a smaller space by aerobic bacteria is a great help to people since by this method they can carry out the complete treatment of sewage in a much smaller space and in less time. The treated water could also be used in the building for general use or for gardening so that the ground water level will not be affected much. In Singapore, as described in Section G.7, this treatment work is so perfectly done that the treated water is used even for drinking.

G.4 BRIIEF DESCRPITION OF AEROBIC TREATMENT OF WASTE WATER AND SEWAGE BY AERATION TANK (METHOD 1)

There are many designs for the aerobic treatment. One of them is briefly described step by step as follows: Another method is given in Section G.5.

Step 1. All the waste water from apartments is first collected in the main tank called *equalization tanks.*

Step 2. The water is then pumped into the next tank called *clarification tank* or *sludge remover tank.* Here excess sludge is removed for further treatment.

Step 3. The water is then led into the main aeration tank which is the heart of the process. Here aerobic bacteria and microbes are made to react with the sewage. *Air in which oxygen is present has to be blown to make these organisms multiply, so that fresh organisms need not be supplied.*

Step 4. The treated water is then first pumped through two filters, i.e., pressure sand filter (PSF) and activated carbon filter (ACF).

Step 5. The treated water is then pumped into a collection tank. This water can be used for general purposes except drinking (For drinking water, more purification is required). Thus, in the above process, we have the following tanks through which the water goes.

(i) Equalization tank
(ii) Aeration tank

(iii) Pressure sand filter (PSF)
(iv) Activated carbon filter (ACF)
(v) Collection tank for treated water

G.4.1 Types of Aerobic Treatment Plant

There are two main types of aerobic treatment plants. They are

(i) Continuous flow reactor
(ii) Sequential catch flow reactor

In both cases in a well designed system, the final waste product produced can be as low as 1%. It can be used as manure for plants and trees. The whole system must be designed and constructed by the specialized firms to give good results. (The above data was supplied by Mr. Arvind Kumar and Mr. K. Sakthivel of MORF India Limited, which is a company that specializes in this work.)

G.4.2 Disadvantage of the Aerobic Air Blowing System

As stated above, this system requires air that must be blown into the aeration tank. This is normally possible only by using electricity and where electricity charges are high, the cost also will be high for small units. Hence, for small housing projects in a small piece of land, this system will prove very costly.

The system of aerobic treatment developed by Aqua Engineers of Auroville is suitable for many places. A brief description of this process is given in next section.

G.5 AEROBIC TREATMENT OF WASTE WATER AND SEWAGE BY EFFECTIVE MICROORGANISMS (EM-1) (METHOD 2)

Aqua Engineers of Auroville, Pondicherry have adopted a system of using a combination of these organic microorganisms in a solution that keep them in a healthy ratio to digest sewage. This solution is called *effective micro organisms* (EM-1). Most of the microorganisms in this liquid are used for fermentation of our foods. This aerobic system was developed by a Japanese Agricultural Scientist and has been tried, tested and adopted by Auroville for use in India (Aqua Engineers, P.F. Auroville, should be contacted for its installation).

In this system for treatment of the mixture of domestic sewage and waste water, the mass is taken through three stages of treatment in concrete tanks. In the first tank, the preliminary treatment of separation of solids from the mix takes place. In the second tank, some anaerobic digestion of the mix is undertaken. The mix then flows into the third important tank. It contains the specially made network of plates, made of special material, called *Biogrow Media* as shown in Fig. G.1. On the surface of this Biogrow Media, we pour the aerobic effective microorganisms (EM-1) referred above. This is to be poured daily (on all the days we use the system) so that aerobic action takes place to digest the sewage efficiently. It is claimed that a very efficient aerobic action is caused by EM-1. The advantage of this system is that this

Fig. G.1 Biogrow Media of Auroville.

is very suitable for small units. Also, we will have no electric charges as in the other system where we have to blow air and electric charges will be high.

The treated water goes into a storage tank with a lift pump for the water to be pumped for gardening and other uses of the building.

Notes:

(i) It is rather surprising to note that a substance like yeast (that we use for making bread and other food stuffs) can be stored for a long time and it becomes very active when we mix with water and sugar.

(ii) In many new groups of buildings, far away from towns and cities, water supply and drainage can be provided as follows:

(a) Simple reverse osmosis plant for supplying water for general use like washing, cleaning etc.

(b) Two-stage reverse osmosis plant (a second unit) for supplying drinking water through a limited number of taps in the buildings, or buy good water in cans from the market.

(c) For waste water and sewage treatment, use electrical air blowing aerobic plant.

(iii) For small residential houses in a small plot, we may plan:

(a) Simple reverse osmosis plant for supplying water for general use.

(b) Buy good drinking water in cans or otherwise its refinement is more involved.

(c) Use a small plant for aerobic treatment by EM-1 of Auroville. This treatment does not require use of electricity which is highly expensive.

G.6 EXAMPLE OF GROUND WATER DEPLETION IN TOWNS—HISTORY OF ELLIOTS BEACH IN CHENNAI

The history of the development of the Elliots Beach Township near Adyar on the sea coast of Chennai and its effect on ground water is very interesting.

About thirty years back, there were only a few houses on this beach and one could get fresh ground water at about one meter below ground level. As there were only limited number of house and no municipal sewage connection, people used septic tanks for sewage treatment. The level of the ground water was more or less steady. Below the fresh ground water, it was seawater. Many used well water for their general use. With more and more houses built, the area was put under the city drainage systems which carried away all the sewage and waste to a far away central plant for disposal. The ground water level slowly came down and the sea water crept in. Today the ground water is saltish and is about 6 metres below the ground level. As no ground water from the wells is now usable, all have closed down their wells.

G.7 EXAMPLE OF REUSE OF WASTE AND SEWAGE WATERS

Singapore has no rivers to give fresh water. It has no source of cheap power to convert sea water to fresh water by reverse osmosis as in the countries of the Middle East. It is reported that they have perfected the aerobic treatment of their waste water and sewage to such a level that it can be reused for all purposes ever for drinking.

Appendix H

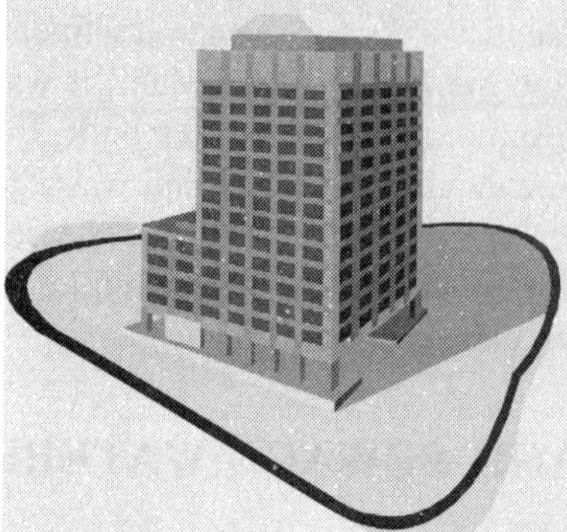

Syllabus for Building Construction

1. INTRODUCTION (5 lectures)

Components of building, major items of construction of a building, classification of buildings into class I, II and III and their brief specifications.

Site preparation, setting out of works, construction of storage sheds, arrangements for supply of water and electricity for construction.

2. CONSTRUCTION OF FOUNDATIONS (5 lectures)

Brief descriptions of shallow and deep foundation used for buildings, load-bearing walls and columns in framed construction.

Earthwork and timbering of excavations, subsoil drainage, dampproofing of walls and ground floor, antitermite treatment. Filling in of trenches.

3. MASONRY CONSTRUCTION (5 lectures)

Brick masonry, bonding and types of bond, block and stone masonry, arches and lintels, specifications for mortars used.

4. PLASTERING AND POINTING (3 lectures)

Internal, external and ceiling plastering, different types of finishes, pointing specifications for plasters.

5. CONCRETE CONSTRUCTION (3 lectures)

Concrete making, placing and curing, temporary works, form work, scaffolding, bending and placement of steel, layout of joints in RC construction.

6. FLOORING (4 lectures)

Different types of floors, Construction of brick cement, concrete, stone, terrazzo, ceramic wood block, parquet, floors, resilient floors.

7. DOORS, WINDOWS AND VENTILATORS (3 lectures)

Timber joints, Types of doors, windows and ventilators, their fixing, Methods of reduction of cost of woodwork in doors and windows.

8. PAINTING AND VARNISHING (3 lectures)

Specifications for painting of walls, woodwork and grills, types of paints used, varnishing and polishing of woodwork.

9. ROOF AND CEILING CONSTRUCTION (4 lectures)

Flat and sloped roofs, tiled and sheeted roofs, waterproofing and weather proofing of flat roofs, ceiling construction.

10. BUILDING SERVICES (4 lectures)

Short description of supply of electricity and water, sanitary works, drainage of waste water and sewage, septic tanks.

11. MISCELLANEOUS BUILDING WORKS (3 lectures)

Rain water harvesting, construction of gates, compound walls, barbed wire fences, garrages, common equipments used in building construction, bar charts and their use.

12. MAINTENANCE OF BUILDINGS (3 lectures)

Usual items of deterioration of buildings, minor repairs of leaks, cracks etc in buildings and other repairs in services.
Total 45 lectures.

References

1. IS 1172, 1993: Code of Basic Requirements for Water Supply, Drainage and Sanitation.
2. IS 1196, 1978: Code of Practice for Laying Bitumen Mastic Flooring.
3. IS 1200, Parts 1–21: Methods of Measurement of Building and Civil Engineering Works.
4. IS 1323, 1982: Code of Practice for Oxyacetylene Welding for Structural Work in Mild Steel.
5. IS 1346, 1991: Code of Practice for Waterproofing of Roofs with Bitumen Felts.
6. IS 1443, 1972: Code of Practice for Laying and Finishing Cement Concrete Flooring Tiles.
7. IS 1477, 1971 (Parts 1 and 2): Code of Practice for Painting of Ferrous Metals in Buildings.
8. IS 1597, 1992: Code of Practice for Stonework, Part 1: Rubble Stone Masonry, Part 2: Ashlar Stone Masonry.
9. IS 1609, 1991: Code of Practice for Laying Dampproofing Treatment Using Bitumen Felts.
10. IS 1641, 1988 and IS 1642, 1989: Code of Practice for Fire Safety of Buildings (Principles for Fire Grading and Classification) and Details of Construction).
11. IS 1661, 1972: Code of Practice for Application of Cement and Concrete Lime Plaster Finishes.
12. IS 1742, 1983: Code of Practice for Building Drainage.
13. IS 1791, 1985: General Requirements for Batch Type Concrete Mixes.
14. IS 1905, 1987: Code of Practice for Structural Use of Unreinforced Masonry.
15. IS 2065, 1983: Code of Practice for Water Supply in Buildings.
16. IS 2114, 1984: Code of Practice for Laying In-Situ Terrazzo Floor Finish.
17. IS 2115, 1980: Code of Practice for Flat Roof Finish Mud Phuska.

18. IS 2250, 1981: Code of Practice for Preparation and Use of Masonry Mortars.
19. IS 2338, Parts 1 and 2: Code of Practice for Finishing of Wood and Wood-Based Materials.
20. IS 2395, Parts 1 and 2, 1944: Code of Practice for Painting Concrete, Masonry and Plastered Surfaces.
21. IS 2402, 1963: Code of Practice for External Rendered Finishes.
22. IS 2440, 1975: Guide for Daylighting of Buildings.
23. IS 2470, Parts 1 and 2: Code of Practice for Installation of Septic Tanks (Design and Secondary Disposal).
24. IS 2502, 1963: Code for Practice for Bending and Fixing of Bars for Concrete Reinforcement.
25. IS 2506, 1992 and IS 2506, 1985: General Requirement of Concrete Vibrators (Concrete Immersion Type and Screed Board Type).
26. IS 2527, 1984: Code for Practice for Fixing Rainwater Gutters and Down Pipes for Roof Drainage.
27. IS 2556, Parts 1–17: Vitreous Sanitary Appliances (Specifications for Various Appliances).
28. IS 2571, 1970: Specifications for Laying In-Situ Concrete Flooring.
29. IS 2750, 1964: Specifications of Steel Scaffoldings.
30. IS 2932, 2003: Enamel, Synthetic Exterior: (a) Undercoating and (b) Finishing Specifications.
31. IS 2933, 1975: Specifications for Enamel Exterior: (a) Undercoating and (b) Finishing.
32. IS 3007, Part 1: Code for Practice for Laying of Asbestos Cement Sheets.
33. IS 3036, 1992: Code of Practice for Laying of Lime Concrete for Waterproofed Finish.
34. IS 3037, 1986: Specification for Bitumen Mastic for Use in Waterproofing of Roofs.
35. IS 3037, 1988: Code of Practice for Dampproofing.
36. IS 3140, 1965: Code of Practice for Painting Asbestos Cement Building Products.
37. IS 3414, 1968: Code of Practice for Design and Installation of Joints in Buildings.
38. IS 3548, 1988: Code of Practice for Glazing in Buildings.
39. IS 3764, 1992: Code of Safety for Excavation Work.
40. IS 4127, 1983: Code of Practice for Laying Glazed Stoneware pipes.
41. IS 4591, 1968: Code of Practice for Installation and Maintenance of Escalators.
42. IS 4634, 1991: Methods of Testing Performance of Batch-Type Concrete Mixers.
43. IS 4926, 2003: Code of Practice for Ready-Mixed Concrete.
44. IS 5491, 1969: Code of Practice for Laying In-Situ Granolithic Concrete Floor Topping.
45. IS 5766, 1970: Code of Practice for Laying Burnt Clay Brick Flooring.

46. IS 6278, 1971: Code of Practice for Whitewashing and Colourwashing.
47. IS 6278, Parts 1–3: Codes of Practice for Anti-termite Treatment of Buildings.
48. IS 6494, 1988: Code of Practice for Waterproofing of Underground Water Reservoirs and Swimming Pools.
49. IS 6598, 1972: Cellular Concrete for Thermal Insulation.
50. IS 7198, 1974: Code of Practice for Dampproofing Using Bitumen Mastic.
51. IS 7634, Parts 1–3: Specification for Materials Laying and Joining Polyethylene(PE) and (unplasticized) UPVC Pipes.
52. IS 9012, 1978: Recommended Practice for Shotcreting.
53. IS 9759, 1981: Guidelines for Dewatering During Construction.
54. IS 9918, 1981: Code of Practice for In-Situ Waterproofing and Dampproofing Treatments with Glass Fibre Tissue Reinforced Bitumen.
55. IS 12054, 1987: Code of Practice for Application of Silicon-Based Water Repellant.
56. IS 12093, 1987: Code of Practice for Laying and Fixing Sloped Roof Covering Using Plain and Corrugated Steel Sheets.
57. IS 13077, 1991: Preparation and Use of Mud Mortar in Masonry – Guide.
58. IS 13182, 1991: Waterproofing and Dampproofing Wet Areas in Buildings – Recommendations.
59. IS 14665, Parts 1–4 (2000): Electric traction lifts.
60. IS 14687, 1999: False Work for Concrete Structures – Guidelines.
61. SP 7, Parts 1–10 (2005): National Building Code of India.
62. SP 20, 1991: Handbook for Masonry Design and Construction.
63. SP 23, 1982: Handbook for Design of Concrete Mixes.
64. SP 35, 1987: Handbook for Water Supply and Drainage.
65. SP 41, 1987: Handbook on Functional Requirements of Buildings (Other than Industrial Buildings).
66. SP 57, 1993: Handbook on Pipes and Fittings for Drinking Water Supply.
67. SP 62, 1997: Handbook for Building Construction Practices (Excluding Electrical Works).
68. Tamil Nadu Building Practice Volumes I and II.
69. CPWD – Specifications 77, Volumes I and II.
70. CPWD – Specifications, Volumes I to VI (Sections 1 to 22).
71. Arora S.P. and Binda S.P., 1988: *Building Construction,* Dhanpat Rai Publications, New Delhi.
72. Punnia B.C., 1993: *Building Construction,* Laxmi Publications, New Delhi.
73. Goyal M.M., 2003: *Handbook of Building Construction,* Amrind Area Consultancy, Faridabad.

Index